IDENTITIES

Sum Identities

$$\sin(x + y) = \sin x \cos y + \cos x \sin y$$
$$\cos(x + y) = \cos x \cos y - \sin x \sin y$$
$$\tan(x + y) = \frac{\tan x + \tan y}{1 - \tan x \tan y}$$

Difference Identities

$$\sin(x - y) = \sin x \cos y - \cos x \sin y$$
$$\cos(x - y) = \cos x \cos y + \sin x \sin y$$
$$\tan(x - y) = \frac{\tan x - \tan y}{1 + \tan x \tan y}$$

Double-Angle Identities

$$\sin 2x = 2 \sin x \cos x$$
$$\cos 2x = \begin{cases} \cos^2 x - \sin^2 x \\ 1 - 2\sin^2 x \\ 2\cos^2 x - 1 \end{cases}$$
$$\tan 2x = \frac{2\tan x}{1 - \tan^2 x} = \frac{2\cot x}{\cot^2 x - 1} = \frac{2}{\cot x - \tan x}$$

Half-Angle Identities

$$\sin \frac{x}{2} = \pm\sqrt{\frac{1 - \cos x}{2}}$$

Sign ($+/-$) is determined by quadrant in which $x/2$ lies

$$\cos \frac{x}{2} = \pm\sqrt{\frac{1 + \cos x}{2}}$$
$$\tan \frac{x}{2} = \frac{1 - \cos x}{\sin x} = \frac{\sin x}{1 + \cos x} = \pm\sqrt{\frac{1 - \cos x}{1 + \cos x}}$$

Identities for Reducing Powers

$$\sin^2 x = \frac{1 - \cos 2x}{2} \qquad \cos^2 x = \frac{1 + \cos 2x}{2}$$
$$\tan^2 x = \frac{1 - \cos 2x}{1 + \cos 2x}$$

Cofunction Identities

(Replace $\pi/2$ with $90°$ if

$$\sin\left(\frac{\pi}{2} - x\right) = \cos x$$
$$\tan\left(\frac{\pi}{2} - x\right) = \cot x \qquad \cot\left(\frac{\pi}{2} - x\right) = \tan x$$
$$\sec\left(\frac{\pi}{2} - x\right) = \csc x \qquad \csc\left(\frac{\pi}{2} - x\right) = \sec x$$

Product–Sum Identities

$$\sin x \cos y = \tfrac{1}{2}[\sin(x + y) + \sin(x - y)]$$
$$\cos x \sin y = \tfrac{1}{2}[\sin(x + y) - \sin(x - y)]$$
$$\sin x \sin y = \tfrac{1}{2}[\cos(x - y) - \cos(x + y)]$$
$$\cos x \cos y = \tfrac{1}{2}[\cos(x + y) + \cos(x - y)]$$

Sum–Product Identities

$$\sin x + \sin y = 2\sin\frac{x + y}{2}\cos\frac{x - y}{2}$$
$$\sin x - \sin y = 2\cos\frac{x + y}{2}\sin\frac{x - y}{2}$$
$$\cos x + \cos y = 2\cos\frac{x + y}{2}\cos\frac{x - y}{2}$$
$$\cos x - \cos y = -2\sin\frac{x + y}{2}\sin\frac{x - y}{2}$$

BASIC TRIGONOMETRIC IDENTITIES

Reciprocal Identities

$$\csc x = \frac{1}{\sin x} \qquad \sec x = \frac{1}{\cos x} \qquad \cot x = \frac{1}{\tan x}$$

Quotient Identities

$$\tan x = \frac{\sin x}{\cos x} \qquad \cot x = \frac{\cos x}{\sin x}$$

Identities for Negatives

$$\sin(-x) = -\sin x \qquad \cos(-x) = \cos x$$
$$\tan(-x) = -\tan x$$

•Pythagorean Identities

$$\sin^2 x + \cos^2 x = 1 \qquad \tan^2 x + 1 = \sec^2 x$$
$$1 + \cot^2 x = \csc^2 x$$

The first person to invent a car that runs on water…

… may be sitting right in your classroom! Every one of your students has the potential to make a difference. And realizing that potential starts right here, in your course.

When students succeed in your course—when they stay on-task and make the breakthrough that turns confusion into confidence—they are empowered to realize the possibilities for greatness that lie within each of them. We know your goal is to create an environment where students reach their full potential and experience the exhilaration of academic success that will last them a lifetime. *WileyPLUS* can help you reach that goal.

Wiley**PLUS** is an online suite of resources—including the complete text—that will help your students:

- come to class better prepared for your lectures
- get immediate feedback and context-sensitive help on assignments and quizzes
- track their progress throughout the course

"I just wanted to say how much this program helped me in studying… I was able to actually see my mistakes and correct them. … I really think that other students should have the chance to use *WileyPLUS*."

Ashlee Krisko, *Oakland University*

www.wiley.com/college/wileyplus

80% of students surveyed said it improved their understanding of the material. *

TO THE INSTRUCTOR

WileyPLUS is built around the activities you perforr

Prepare & Present

Create outstanding class presentations using a wealth of resources, such as PowerPoint™ slides, image galleries, interactive simulations, and more. Plus you can easily upload any materials you have created into your course, and combine them with the resources Wiley provides you with.

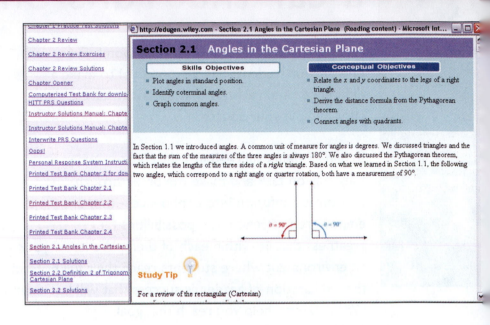

Create Assignments

Automate the assigning and grading of homework or quizzes by using the provided question banks, or by writing your own. Student results will be automatically graded and recorded in your gradebook. *WileyPLUS* also links homework problems to relevant sections of the online text, hints, or solutions— context-sensitive help where students need it most!

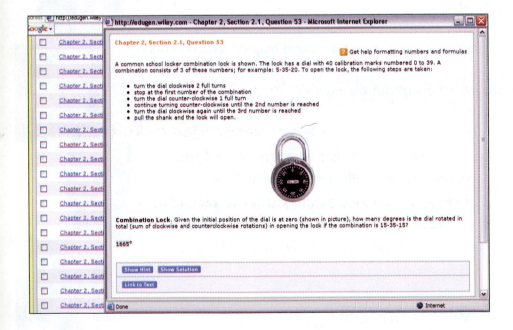

*Based on a spring 2005 survey of 972 student users of *WileyPLUS*

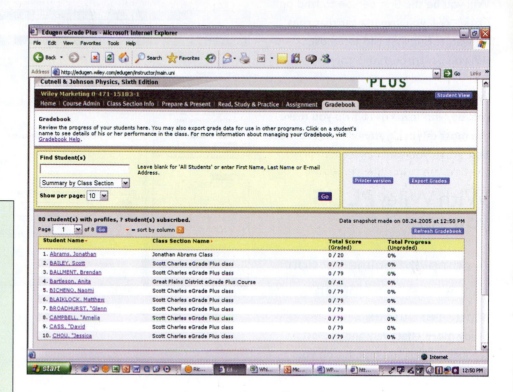

TO THE STUDENT

You have the potential to make a difference!

Will you be the first person to land on Mars? Will you invent a car that runs on water? But, first and foremost, will you get through this course?

WileyPLUS is a powerful online system packed with features to help you make the most of your potential, and get the best grade you can!

With Wiley**PLUS** you get:

A complete online version of your text and other study resources

Study more effectively and get instant feedback when you practice on your own. Resources like self-assessment quizzes, tutorials, and animations bring the subject matter to life, and help you master the material.

Problem-solving help, instant grading, and feedback on your homework and quizzes

You can keep all of your assigned work in one location, making it easy for you to stay on task. Plus, many homework problems contain direct links to the relevant portion of your text to help you deal with problem-solving obstacles at the moment they come up.

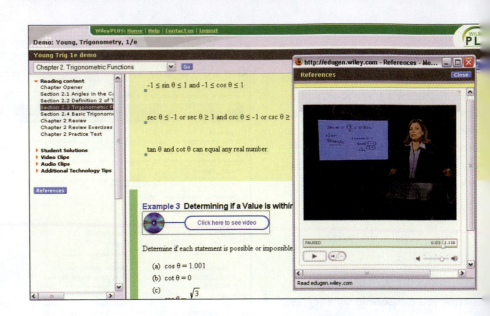

The ability to track your progress and grades throughout the term

A personal gradebook allows you to monitor your results from past assignments at any time. You'll always know exactly where you stand.

If your instructor uses *WileyPLUS*, you will receive a URL for your class. If not, your instructor can get more information about *WileyPLUS* by visiting www.wiley.com/college/wileyplus

---

"It has been a great help, and I believe it has helped me to achieve a better grade."

Michael Morris, *Columbia Basin College*

69% of students surveyed said it helped them get a better grade. *

The Wiley Faculty Network

Where Faculty Connect

The Wiley Faculty Network is a faculty-to-faculty network promoting the effective use of technology to enrich the teaching experience. The Wiley Faculty Network facilitates the exchange of best practices, connects teachers with technology, and helps to enhance instructional efficiency and effectiveness. The network provides technology training and tutorials, including *WileyPLUS* training, online seminars, peer-to-peer exchanges of experiences and ideas, personalized consulting, and sharing of resources.

Connect with a Colleague

Wiley Faculty Network mentors are faculty like you, from educational institutions around the country, who are passionate about enhancing instructional efficiency and effectiveness through best practices. You can engage a faculty mentor in an online conversation at **www.wherefacultyconnect.com**

Participate in a Faculty-Led Online Seminar

The Wiley Faculty Network provides you with virtual seminars led by faculty using the latest teaching technologies. In these seminars, faculty share their knowledge and experiences on discipline-specific teaching and learning issues. All you need to participate in a virtual seminar is high-speed internet access and a phone line. To register for a seminar, go to **www.wherefacultyconnect.com**

Connect with the Wiley Faculty Network

Web: **www.wherefacultyconnect.com**
Phone: 1-866-4FACULTY

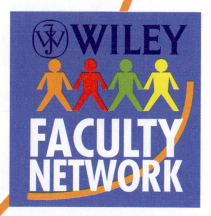

BICENTENNIAL
1807
⊕ WILEY
2007
BICENTENNIAL

THE WILEY BICENTENNIAL—KNOWLEDGE FOR GENERATIONS

*E*ach generation has its unique needs and aspirations. When Charles Wiley first opened his small printing shop in lower Manhattan in 1807, it was a generation of boundless potential searching for an identity. And we were there, helping to define a new American literary tradition. Over half a century later, in the midst of the Second Industrial Revolution, it was a generation focused on building the future. Once again, we were there, supplying the critical scientific, technical, and engineering knowledge that helped frame the world. Throughout the 20th Century, and into the new millennium, nations began to reach out beyond their own borders and a new international community was born. Wiley was there, expanding its operations around the world to enable a global exchange of ideas, opinions, and know-how.

For 200 years, Wiley has been an integral part of each generation's journey, enabling the flow of information and understanding necessary to meet their needs and fulfill their aspirations. Today, bold new technologies are changing the way we live and learn. Wiley will be there, providing you the must-have knowledge you need to imagine new worlds, new possibilities, and new opportunities.

Generations come and go, but you can always count on Wiley to provide you the knowledge you need, when and where you need it!

WILLIAM J. PESCE
PRESIDENT AND CHIEF EXECUTIVE OFFICER

PETER BOOTH WILEY
CHAIRMAN OF THE BOARD

TRIGONOMETRY

CYNTHIA Y. YOUNG
University of Central Florida

John Wiley & Sons, Inc.

> *This book is dedicated to teachers and students*

PUBLISHER	Laurie Rosatone
SENIOR ACQUISITIONS EDITOR	Angela Y. Battle
SENIOR DEVELOPMENT EDITOR	Ellen Ford
PROJECT EDITOR	Jennifer Battista
SENIOR PRODUCTION EDITOR	Sujin Hong
DIRECTOR OF MARKETING	Frank Lyman
MARKETING MANAGER	Amy Sell
SENIOR ILLUSTRATION EDITOR	Sigmund Malinowski
SENIOR PHOTO EDITOR	Felicia Ruocco
MEDIA EDITOR	Stefanie Liebman
DEVELOPMENT ASSISTANT	Justin Bow
MARKETING ASSISTANT	Tara Martinho
DESIGN DIRECTOR	Harry Nolan
COVER DESIGNER	Hope Miller
INTERIOR DESIGNER	Brian Salisbury
COVER PHOTO	Gary Brettnacher/Getty Images, Inc.

This book was set in 10/12 Times by Techbooks and printed and bound by Von Hoffmann. The cover was printed by Von Hoffmann.

This book is printed on acid free paper. ∞

Library of Congress Cataloging-in-Publication Data

Young, Cynthia Y.
  Trigonometry / Cynthia Y. Young.
    p. cm.
  Includes indexes.
  ISBN-13: 978-0-471-75685-9 (acid-free paper)
  ISBN-10: 0-471-75685-7 (acid-free paper)
    1. Trigonometry—Textbooks.   I. Title.
  QA531.Y68 2007
516.24--dc22

                                        2006033465

Printed in the United States of America

10 9 8 7 6 5 4 3 2 1

ABOUT THE AUTHOR

Cynthia Y. Young

EDUCATION

- Ph.D. **Applied Mathematics** University of Washington
- M.S. **Electrical Engineering** University of Washington
- M.S. **Mathematical Sciences** University of Central Florida
- B.A. **Mathematics Education** University of North Carolina

RESEARCH

Associate Professor Department of Mathematics, University of Central Florida
Sabbaticals/Fellowships Naval Research Lab, Boeing, Kennedy Space Center
Principal Investigator Turbulence Effects on LIDAR, Office of Naval Research
Laser Propagation in the Marine Environment, Naval
Research Lab

TEACHING

- University of Central Florida
- Shoreline Community College
- Courses taught: College Algebra, Trigonometry, Calculus, Differential Equations, Applied Boundary Value Problems, Advanced Engineering Mathematics, Special Functions, Optical Wave Propagation. Honors Courses: College Algebra, Calculus

AWARDS

2005–2010	Science Talent Expansion Program	*National Science Foundation*
2003–2004	Research Incentive Program	*University of Central Florida*
2002–2003	CAS Distinguished Researcher Award	*University of Central Florida*
2002–2003	University Professional Service Award	*University of Central Florida*
2001–2002	Teaching Incentive Program	*University of Central Florida*
2001	Young Investigator Award	*Office of Naval Research*
2001	Excellence Undergrad. Teaching Award	*University of Central Florida*
1993–1997	NPSC Four-Year Doctoral Fellowship	*Kennedy Space Center*

PERSONAL INTERESTS

- Her Labrador Retrievers' (Molly, Blue, Ellie, and Wiley) Field Trials
- Tampa Bay Buccaneer Fan
- Golf
- Boating

In Trigonometry it may be difficult to get students to love mathematics, but you can get students to love *succeeding* at mathematics. Students walk into Trigonometry with different experiences and knowledge from Algebra and Geometry courses. Trigonometry is a fun course to teach but I found my students were still saying "I understand you in class but when I get home I'm lost." There is a gap that exists between when we have our students in class and when they are at home doing homework. I wrote this book to help you fill that gap with your students.

I expect students to move beyond what they have learned in the past and as they grow confident in their mathematical ability, progress to a more mature thought process of mathematics. This book starts with the concrete and lures students to the more abstract. This is why, in this book, we start with triangles and angles in degrees which are more familiar to students and then lure them to the more abstract concepts such as unit circle approach and radian measure.

My goal was to write a book that students can read without sacrificing rigor. I believe I have accomplished that goal. I sincerely hope this book meets your expectation and I welcome your feedback.

Goals

How many times have we heard our students complain that they can't read the book? I've tried several texts but could not find a text my students could read. So, I decided to write my own book. I know there is no perfect text, but my overarching goal was to write a book that students can read that has the level of rigor necessary to prepare them for Calculus. The following principles represent how this book is designed to be an effective teaching and learning aid.

Encouraging Approach

Many students in Trigonometry suffer from math anxiety. In order to conquer the math anxiety hurdle the student has to experience success. Mathematics is a difficult subject to teach, even to students who like it. It requires an instructor and a textbook that present information in a clear, concise, inviting, and engaging format. Students typically do not read mathematics textbooks, so I have chosen a design and tone that will encourage students, paying particular attention to different learning styles.

Appropriate Rigor

Trigonometry is a foundation course for Calculus. Although this text includes helpful student pedagogy and focuses on a balanced approach to skills and concepts, it also maintains the rigor appropriate for a true Trigonometry course serving as preparation for Calculus. Examples of this rigor include proving formulas and identities as opposed to simply stating them; giving specific attention to domain restrictions; and offering a variety of challenge problems designed to test students' depth of understanding.

Overcoming Differences

Students have different learning styles such as audio, visual, and kinesthetic. When multiple learning styles are addressed, student learning can be improved. In addition, some students learn by seeing repeated examples of the same type, while others learn better from counter example (what not to do). In this book we use both visual and audio techniques (the text has accompanying audio and visual clips). In addition, color is carefully and consistently used for pedagogical reasons.

Topical Coverage

Trigonometry topics are very standard with one exception: books lead with either the right triangle or unit circle approach. All Trigonometry books discuss the three definitions of trigonometry functions (right triangle trigonometry, Cartesian plane definition, and circular functions) but texts are differentiated by which definition they lead with. I have chosen to lead with right triangle trigonometry because it is concrete. Students are familiar with triangles and angles in degrees. So trigonometric functions of acute angles are first defined as ratios of sides of right triangles. Then, trigonometric functions are defined for any angle (acute or nonacute) in terms of the Cartesian plane and ratios of coordinates and distances. Right triangles are superimposed on the Cartesian plane so students see that both definitions are consistent with one another. Lastly, trigonometric functions are defined as circular functions, the unit circle approach. Again, the unit circle and right triangles are superimposed onto the Cartesian plane so the students see that all three definitions are consistent with each other. The unit circle is introduced in Chapter 3 which is where radian measure is defined. I have found that starting with the concrete right triangle trigonometry and luring students to the more abstract unit circle approach gives students more confidence and a better understanding.

Innovative Pedagogy

I have developed the following features to address specific course issues. I hope that the benefits of these features help fill the gap that exists between class time and homework.

Skills and Conceptual Objectives

Every section opens with a list of both skills and conceptual objectives that students are expected to master in that section. This emphasizes and encourages not just the rote memorization of mathematical processes, but also the understanding of the concepts.

Parallel Words and Math

Most texts will present examples with equations on the left accompanied by brief running marginal annotations on the right. This text reverses that presentation so that an explanation in words is provided on the left and the mathematics is parallel to the right. This format reflects how students read naturally and makes it easier for them to read through examples. It also makes the presentation in the text similar to your presentation in class. In class we say the step and then we do the step. We don't do it and then tell the students what we did.

Color

Color is used for pedagogical reasons in this text. For example, if $f(x) = \sin x$ is red in the text then the graph of that function is also in red. This makes it clearer to the student and easier to follow, particularly when period, amplitude, and phase shifts are discussed.

Applications

In Trigonometry, there are wonderful applications in all areas: business, sports, engineering, physics, optics, astronomy, biology, chemistry, and so forth. For this text we asked faculty to submit their favorite applications, several of which are included. All of these areas are represented in applications in this text. See the Applications Index for categories of topics covered.

Common Mistakes

Some students learn better with lots of examples and repetition, and some learn more effectively with counterexamples. Traditional books will illustrate several different examples and then put a warning box with a totally different problem. For students to make the connection, it is important to show them the *same problem* worked both correctly and incorrectly. This helps students avoid most common mistakes.

Homework Exercises

There are five categories of homework exercises: *Skills, Applications, Catch the Mistake, Challenge,* and *Technology.* The Skills exercises strengthen students' ability to solve basic problems similar to worked examples. Once students have strengthened their skills and have gained confidence in the material they can apply those skills in solving Applications exercises. Then a Catch the Mistake set of exercises helps students who learn from counterexample and also helps students play the role of teacher and identify mistakes, a process which truly assesses understanding. The Challenge exercises, which require a more mature analytical thought process, push students deeper into their understanding. Finally, there are Technology Exercises that make use of graphing calculators to complement analytical procedures.

Technology Using Graphing Calculators

We have placed numerous Technology Tips in the margins throughout the text to illustrate techniques for using graphing calculators on the same problems that are worked in the examples. By keeping Technology Tips in the margins and identifying Technology Exercises, this book can be customized for instructors with varying levels of graphing calculator use, from light to heavy.

Features and Benefits at a Glance

All of the course issues discussed above reflect the gap that exists between course work and homework. The features in this book are designed to help instructors fill that gap.

Feature	Benefit to Student
Chapter opening vignette Chapter overview Organizational flow chart Supplement navigation chart	Preview what will be covered and how the topics are related. This manages students' expectations and improves their understanding. The supplement chart highlights available resources to develop students' study skills.
Skills and Conceptual Objectives	Emphasize the importance of conceptual understanding as well as skills.
Clear, concise, and inviting writing style, tone, and layout	Reduce math anxiety, encourage student success.
Parallel words and math	Increase students' ability to read the examples.
Common Mistake/ Correct Incorrect boxes	Demonstrate common mistakes so that the students strengthen their understanding (and avoid making the same mistakes).
Concept Checks and Your Turn exercises	Engage students during class time.
Catch the Mistake exercises	Encourage students to play the role of teacher and assess their understanding.
OOPS! Tying It All Together	Provide students with a fun and interesting way to apply what they have learned to real-life situations.
Chapter Review Review Exercises Practice Test	Improve student study skills, allow for student self-assessment and practice.

Supplements

Instructor Supplements

INSTRUCTOR'S SOLUTIONS MANUAL (ISBN: 0-470-06873-6)
- Contains worked out solutions to all exercises in the text.

POWERPOINT SLIDES
- For each section of the book a corresponding set of lecture notes and worked out examples are presented as PowerPoint slides. These are available on the book companion website at www.wiley.com/college/young.

TEST BANK (ISBN: 0-470-06874-4)
- Contains a variety of questions and answers for every section of the text.

COMPUTERIZED TEST BANK (ISBN: 0-470-10853-3)
Electronically enhanced version of the Test Bank that
- provides varied question types, most of which are algorithmically-generated questions,
- allows instructors to freely edit, randomize, and create questions, and
- allows instructors to create and print different versions of a quiz or exam.

BOOK COMPANION WEBSITE (WWW.WILEY.COM/COLLEGE/YOUNG)
- Contains all instructor supplements listed above, plus a selection of personal response system questions and a Technology Appendix.

WileyPLUS
- Provides additional resources for instructors, such as assignable homework exercises, tutorials, gradebook, integrated links between the online version of the text and supplements (see description below).

Student Supplements

DIGITAL VIDEO TUTOR (DVD) (ISBN: 0-470-04234-6)
- Streaming video of the author presenting chapter overviews, chapter summaries, and working selected examples step by step that are tied to specific textbook sections. DVD will be fully integrated with the textbook. An icon in the text indicates an example that has a video clip available. The Digital Video Tutor can be optionally packaged with the text.

STUDENT SOLUTIONS MANUAL (ISBN: 0-471-78847-3)
- Includes worked out solutions for all odd problems in the text.

BOOK COMPANION WEBSITE (WWW.WILEY.COM/COLLEGE/YOUNG)
- Provides additional resources for students, including web quizzes, technology tips, and audio clips.

WileyPLUS
- Presents additional resources for students such as additional self-practice exercises, tutorials, audio clips, integrated links between the online version of the text and supplements (see description below).

WileyPLUS

Expect More from Your Classroom Technology

Cynthia Young's *Trigonometry* is supported by *WileyPLUS*—a powerful and highly integrated suite of teaching and learning resources designed to bridge the gap between what happens in the classroom and what happens at home. *WileyPLUS* includes a complete online version of the text, algorithmically generated exercises, all of the text supplements, plus course and homework management tools, in one easy-to-use website.

Organized around the everyday activities you perform in class, *WileyPLUS* helps you:

Prepare and Present: *WileyPLUS* lets you create class presentations quickly and easily using a wealth of Wiley-provided resources, including an online version of the textbook, PowerPoint slides, and more. You can adapt this content to meet the needs of your course.

Create Assignments: *WileyPLUS* enables you to automate the process of assigning and grading homework or quizzes. You can use algorithmically generated end-of-section and end-of-chapter problems from the text, or write your own.

Track Student Progress: An instructor's gradebook allows you to analyze individual and overall class results to determine students' progress and level of understanding.

Promote Strong Problem-Solving Skills: *WileyPLUS* can link homework problems to the relevant section of the online text, providing students with context-sensitive help. *WileyPLUS* also features GO^{TM} (Guided Online) Tutorial problems that promote conceptual understanding of key topics and video walkthroughs of example problems.

Provide numerous practice opportunities: Algorithmically generated problems provide unlimited self-practice opportunities for students, as well as problems for homework and testing.

Support Varied Learning Styles: *WileyPLUS* includes the entire text in digital format, enhanced with varied problem types, audio, and video walkthroughs to support the array of different student learning styles in today's classrooms.

Administer Your Course: You can easily integrate *WileyPLUS* with another course management system, gradebooks, or other resources you are using in your class, enabling you to build your course, your way.

WileyPLUS **includes a wealth of instructor and student resources:**

Digital Video Tutor: Presents chapter overviews, chapter summaries, and step-by-step walkthroughs of selected examples that are tied to specific textbook sections. Icons throughout the text direct the student to the video examples that the author works out "live." (Also available on DVD.)

GO^{TM} Tutorial Problems: Presented in a way that requires students to demonstrate complete understanding, or mastery, of the topics. They teach and assess at the same time, and provide students with hints and offer stepped-out tutorials based on their input attempt at solving the problem.

Computerized Test Bank: Includes questions from the printed test bank, many of which are algorithmically generated.

Student Solutions Manual: Includes worked-out solutions for all odd-numbered problems and study tips.

Instructor's Solutions Manual: Presents worked out solutions to all problems.

PowerPoint Lecture Notes: In each section of the book a corresponding set of lecture notes and worked out examples are presented as PowerPoint slides that are tied to the examples in the text.

View an online demo at www.wiley.com/college/mathwp or contact your local Wiley representative for more details.

The Wiley Faculty Network–Where Faculty Connect

The Wiley Faculty Network is a faculty-to-faculty network promoting the effective use of technology to enrich the teaching experience. The Wiley Faculty Network facilitates the exchange of best practices, connects teachers with technology, and helps to enhance instructional efficiency and effectiveness. The network provides technology training and tutorials, including *WileyPLUS* training, online seminars, peer-to-peer exchanges of experiences and ideas, personalized consulting, and sharing of resources.

Connect with a Colleague

Wiley Faculty Network mentors are faculty like you, from educational institutions around the country, who are passionate about enhancing instructional efficiency and effectiveness through best practices. You can engage a faculty mentor in an online conversation at www.wherefacultyconnect.com.

Participate in a Faculty-Led Online Seminar

The Wiley Faculty Network provides you with virtual seminars led by faculty using the latest teaching technologies. In these seminars, faculty share their knowledge and experiences on discipline-specific teaching and learning issues. All you need to participate in a virtual seminar is high-speed Internet access and a phone line. To register for a seminar, go to www.wherefacultyconnect.com.

Connect with the Wiley Faculty Network

Web: www.wherefacultyconnect.com
Phone: 1-866-4FACULTY

ACKNOWLEDGMENTS

I'd like to thank the entire Wiley team. To my editor and dear friend, Angela Battle, there are no words to express the gratitude in my heart. We share more than a middle name; this book is *ours*. To my developmental editor, Ellen Ford, thanks for helping me find my writing voice and for always finding the positive in every review. You kept me going. To Laurie Rosatone, publisher, thanks for all of the encouragement. To Jennifer Battista, project editor, thanks for all of your hard work coordinating supplements, recruiting reviewers, contributors, and class testers, and creating a community of colleagues. To Harry Nolan and Brian Salisbury for their creative and elegant cover and text design. To Sigmund Malinowski, Felicia Ruocco, and the rest of the photo and illustration team for developing exactly what I was thinking. To the media editor, Stefanie Liebman, videographer Wayne Ferguson, and the advertising team for their support while filming the videos. To the production editor, Sujin Hong, who helped us all meet the deadlines. To Frank Lyman, Chris Ruel, Amy Sell, and the marketing team for your tireless efforts. And I'd especially like to thank Julia Flohr who inspired me to think about writing and all of the other Wiley sales reps who will help us share this book with instructors around the country.

To my husband, Dr. Christopher Laird Parkinson, thanks for making dinner so many nights while I typed. After the first book you knew how much time my writing would consume, and still supported me throughout this project. Thank you. To our Labrador Retrievers (Molly, Blue, Wiley, and Ellie) for snuggling on my feet while I worked on this book, and to Wiley, who forced me to take breaks so I could play with her.

Dad, thanks for always reminding me that I was named after the only two teachers in the family. Mom, thanks for giving me confidence.

Writing this text has been the most humbling experience I've ever had. I have some very strong notions about what the book needs to do for students and instructors. The text development process has allowed me to hone my approach and test my convictions. Finding the balance between what instructors want and what students need has been challenging, but I think I've succeeded. The feedback I've received so far has been extremely gratifying.

I'd like to thank Lori Dunlop-Pyle for writing the Tying It All Together and OOPS! features at the end of every chapter and Pauline Chow for creating all of the Technology Tips throughout the text. A special thanks to Dr. Mark McKibben, who worked with me on developing the Solutions Manual. Mark and Pauline—thanks for taking such ownership. I'd also like to thank all of the reviewers, accuracy checkers, class testers, contributors, and focus group participants. You have all contributed to this book. Your effort has made this text an ally for both students and instructors.

REVIEWERS

Ebrahim Ahmadizadeh, *Northampton Community College*
Khadija Ahmed, *Monroe County Community College*
Gerardo Aladro, *Florida International University*
Jan Archibald, *Ventura College*
Mary Benson, *Pensacola Junior College*
Patricia Bezona, *Valdosta State University*
Michael Butros, *Victor Valley College*
Elsie Campbell, *Angelo State University*
Mathews Chakkanakuzhi, *Palomar College*
Michael Corral, *Schoolcraft College*
Matthew Cropper, *Eastern Kentucky University*
Donna Densmore, *Bossier Parish Community College*
Jacqueline Donofrio, *Monroe Community College*
Karen Ernst, *Hawkeye Community College*
Nancy Eschen, *Florida Community College*
Heather Gamber, *Cy-Fair College*
Antanas Gilvydas, *Richard J. Daley College*
David Goldberg, *Washtenaw Community College*
Carolyn Hamilton, *Utah Valley State College*
Angela Heiden, *St. Clair Community College*
Celeste Hernandez, *Richland College*
Laura Hillerbrand, *Broward Community College*
Jo Beth Horney, *South Plains College*
Sharon Jackson, *Brookhaven College*
Mohamed Jamaloodeen, *Golden West College*
Raja Khoury, *Collin County Community College*
Robert Korte, *Kent State University*
Anne Landry, *Florida Community College, Jackson*
Denise LeGrand, *University of Arkansas, Little Rock*
Gene Majors, *Fullerton College*
Charles Mathers, *Northampton Community College*
Erika Miller, *Villa Julie College*
Linda Myers, *Harrisburg Area Community College*
Sue Neal, *Wichita State University*
Stephen Nicoloff, *Paradise Valley Community College*
Donna Nordstrom, *Pasadena City College*
Karen Pagel, *Dona Ana Branch Community College*
Ron Palcic, *Johnson County Community College*

Mohammed Pasha, *Del Mar College*
Jeanne Pirie, *Erie Community College—North*
William Radulovich, *Florida Community College*
Richard Rupp, *Del Mar College*
Kathleen Shepherd, *Monroe County Community College*
Min Soe, *Rogers State University*
Virginia Starkenburg, *San Diego City College*
Jo Tucker, *Tarrant County Junior College*
Andrea Vorwark, *Maplewoods Community College*
Jim Voss, *Front Range Community College*
Rebecca Walls, *West Texas A&M University*
James Wang, *University of Alabama, Tuscaloosa*
Jianzhong Wang, *Sam Houston State University*
Kathryn Wetzel, *Amarillo College*
Cheryl Whitelaw, *Southern Utah University*
Randy Wills, *Southeastern Louisiana University*
Yan Wu, *Georgia Southern University*

CONTRIBUTORS

O. Pauline Chow, *Harrisburg Area Community College*
Lori Dunlop-Pyle, *University of Central Florida*
Charles Mathers, *Northampton Community College*
Mark McKibben, *Goucher College*
Mary Jane Sterling, *Bradley University*

CLASS TESTERS

Yvonne Aucoin, *Tidewater Community College*
Karen Ernst, *Hawkeye Community College*
William Radulovich, *Florida Community College at Jacksonville*
Virginia Starkenburg, *San Diego City College*
Rob Wylie, *Carl Albert State College*

SUPPLEMENT AUTHORS

O. Pauline Chow, *Harrisburg Area Community College*
Mark McKibben, *Goucher College*
M. Dee Medley, *Augusta State University*
Virginia Starkenburg, *San Diego City College*

Commitment to Accuracy

From the beginning, the editorial team and I have been committed to providing an accurate and error-free text. In this goal, we have benefited from the help of many people. In addition to the reviewer feedback on this issue and normal proofreading, we enlisted the help of several extra sets of eyes during the production stages and during video filming. I wish to thank them for their dedicated reading of various rounds of pages to help me insure the highest level of accuracy.

ACCURACY CHECKERS

Felix Apfaltrer, *Borough of Manhattan Community College*
Jan Archibald, *Ventura College*
Antanas Gilvydis, *Richard J. Daly College*
Angela Heiden, *St. Clair Community College*
Ann Landry, *Florida Community College, Jackson*
Kathryn Lavelle, *Westchester Community College*
Evan Siegel, *New Jersey City University*
Jim Voss, *Front Range Community College*

TABLE OF CONTENTS

A NOTE FROM THE AUTHOR TO THE STUDENT

I wrote this text with careful attention to ways of making your learning experience more successful. If you take full advantage of the unique features and elements of the textbook, I believe your experience in Trigonometry will be fulfilling and enjoyable. Let's walk through some of the special book features that will help you in your study of Trigonometry.

Prerequisites and Review (Appendix)

A review of prerequisite knowledge (college algebra topics) is included in the Appendix to provide a brush-up on knowledge and skills necessary for success in the course. In addition, reviews of prerequisite concepts are provided throughout chapters on an as-needed basis.

Clear, Concise, and Inviting Writing

Special attention has been given to presenting an engaging and clear narrative in a layout that is designed to reduce math anxiety in students.

APPENDIX

Algebraic Prerequisites and Review

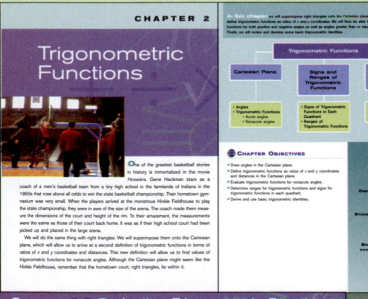

Chapter Introduction Flow Chart and Objectives

A flow chart and list of chapter objectives give you an overview of the chapter to help you see the big picture and the relationships between topics.

Skills/Concepts Objectives

For every section, objectives are divided by skills *and* concepts so that you can learn the difference between solving problems and truly understanding concepts.

Examples

Examples pose a specific problem using concepts already presented and then work through the solution. These serve to enhance your understanding of the subject matter.

Concept Check

The Concept Check provides a periodic stopping point that asks you to stop and think about a related question to extend your conceptual understanding.

Your Turn

Immediately following an example, you are asked to solve a similar problem to reinforce and check your understanding. This feature helps build confidence as you progress in the chapter and is ideal for in-class activity and for preparing to do homework later.

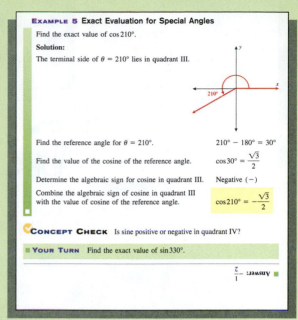

EXAMPLE 5 Exact Evaluation for Special Angles

Find the exact value of $\cos 210°$.

Solution:

The terminal side of $\theta = 210°$ lies in quadrant III.

Find the reference angle for $\theta = 210°$.	$210° - 180° = 30°$
Find the value of the cosine of the reference angle.	$\cos 30° = \dfrac{\sqrt{3}}{2}$
Determine the algebraic sign for cosine in quadrant III.	Negative $(-)$
Combine the algebraic sign of cosine in quadrant III with the value of cosine of the reference angle.	$\cos 210° = -\dfrac{\sqrt{3}}{2}$

CONCEPT CHECK   Is sine positive or negative in quadrant IV?

YOUR TURN   Find the exact value of $\sin 330°$.

■ Answer: $-\dfrac{1}{2}$

COMMON MISTAKE

A common mistake when given the tangent function is assuming that the numerator is the value of sine and the denominator is the value of cosine.

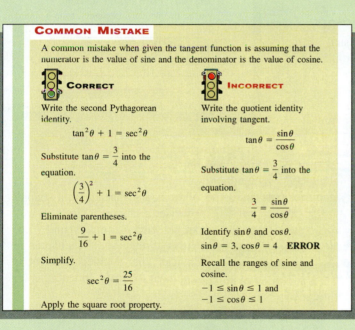

CORRECT

Write the second Pythagorean identity.

$$\tan^2\theta + 1 = \sec^2\theta$$

Substitute $\tan\theta = \dfrac{3}{4}$ into the equation.

$$\left(\dfrac{3}{4}\right)^2 + 1 = \sec^2\theta$$

Eliminate parentheses.

$$\dfrac{9}{16} + 1 = \sec^2\theta$$

Simplify.

$$\sec^2\theta = \dfrac{25}{16}$$

Apply the square root property.

INCORRECT

Write the quotient identity involving tangent.

$$\tan\theta = \dfrac{\sin\theta}{\cos\theta}$$

Substitute $\tan\theta = \dfrac{3}{4}$ into the equation.

$$\dfrac{3}{4} = \dfrac{\sin\theta}{\cos\theta}$$

Identify $\sin\theta$ and $\cos\theta$.

$\sin\theta = 3, \cos\theta = 4$ **ERROR**

Recall the ranges of sine and cosine.

$-1 \le \sin\theta \le 1$ and
$-1 \le \cos\theta \le 1$

Common Mistake/ Correct vs. Incorrect

In addition to standard examples, some problems are worked both correctly and incorrectly to highlight common errors students make. Counterexamples like these are often an effective learning approach for many students.

Parallel Words and Math

This text reverses the common presentation of examples by placing the explanation in words *on the left* and the mathematics in parallel *on the right*. This makes it easier for students to read through examples as the material flows more naturally as is commonly presented in lecture.

WORDS	**MATH**
Write the definition of the cosecant function.	$\csc\theta = \dfrac{r}{y} \quad (y \neq 0)$
The reciprocal identity holds for $y \neq 0$.	$\csc\theta = \dfrac{r}{y} = \dfrac{1}{\frac{y}{r}} \quad y \neq 0$
Substitute the definition of sine, $\sin\theta = \dfrac{y}{r}$.	$\csc\theta = \dfrac{r}{y} = \dfrac{1}{\frac{y}{y}} = \dfrac{1}{\sin\theta}$

STUDY TIP
Reciprocals always have the same algebraic sign.

Study Tips and Caution Notes
These marginal reminders call out important hints or warnings you should be aware of relating to the topic or problem.

CAUTION
If no unit of angle measure is specified, then radian measure is implied.

TECHNOLOGY TIP
When $\sin\theta = -0.6293$, use the $\boxed{\sin^{-1}}$ key and the absolute value to find the reference angle.

```
sin⁻¹(0.6293)
         38.99849667
```

Technology Tips
These marginal notes provide instruction and visual examples using graphing and scientific calculators to solve problems.

Trigonometric Functions

CHAPTER OBJECTIVES
- Draw angles in the Cartesian plane.
- Define trigonometric functions as ratios of x and y coordinates and distances in the Cartesian plane.
- Evaluate trigonometric functions for nonacute angles.
- Determine ranges for trigonometric functions and signs for trigonometric functions in each quadrant.

NAVIGATION
THROUGH
SUPPLEMENTS

DIGITAL VIDEO SERIES #2

Video Icons
Video icons appear on all chapter overviews, chapter reviews, as well as selected examples and Your Turns throughout the chapter to indicate that a video segment is available for that element.

EXAMPLE 3 Finding Measures of Coterminal Angles
Determine the angles of the smallest possible positive measure that are coterminal with the following angles.

a. 830°   **b.** −520°

Solution (a):

Since 830° is positive, subtract 360°.

Subtract 360° again.

The angle with measure 110° is t... that is coterminal with the angle with...

Solution (b):

Since −520° is negative, add 360°.

Add 360° again.

The angle with measure 200° is t...

SOLUTIONS MANUAL
CHAPTER 2

YOUR TURN   Calculate the values for the six trigonometric functions when $\theta = 360°$.

The following table summarizes the trigonometric function values for common quadrantal angles: 0°, 90°, 180°, 270°, and 360°.

θ	$\sin\theta$	$\cos\theta$	$\tan\theta$	$\cot\theta$	$\sec\theta$	$\csc\theta$
0°	0	1	0	undefined	1	undefined
90°	1	0	undefined	0	undefined	1
180°	0	−1	0	undefined	−1	undefined

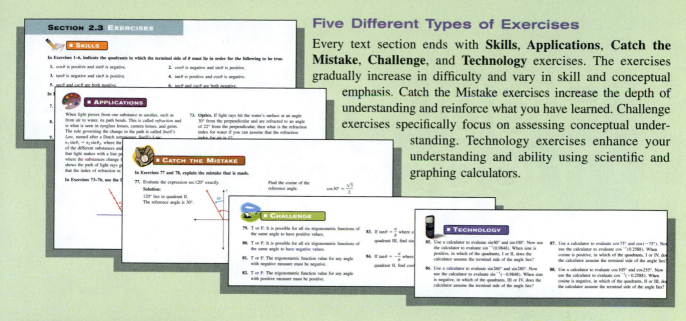

Five Different Types of Exercises

Every text section ends with **Skills**, **Applications**, **Catch the Mistake**, **Challenge**, and **Technology** exercises. The exercises gradually increase in difficulty and vary in skill and conceptual emphasis. Catch the Mistake exercises increase the depth of understanding and reinforce what you have learned. Challenge exercises specifically focus on assessing conceptual understanding. Technology exercises enhance your understanding and ability using scientific and graphing calculators.

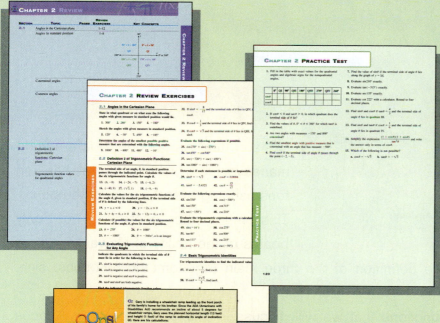

Chapter Review— Summary, Exercises, Practice Test

At the end of every chapter, a review chart organizes the topics in an easy-to-use one-page layout. This feature includes key concepts and formulas, as well as indicating relevant pages and review exercises so that you can quickly summarize a chapter and study smarter. Chapter Review Exercises are provided for extra study and practice. A Practice Test is also included to give you even more practice before moving on.

OOPS! and Tying It All Together

These unique end-of-chapter elements provide a fun and interesting way to apply what you have learned. OOPS! takes a situation in which doing math incorrectly results in a real-world mistake. Tying It All Together presents a question or problem that requires you to use more than one concept to solve the problem.

Right Triangle Trigonometry

Getty Images News and Sport Services

Surfers of Hawaii Can Always Hang Ten on Anything

"**H**ang ten" is a long board-surfing maneuver in which the surfer has ten toes hanging over the front of the board. This mnemonic sentence not only boasts of the ability of Hawaiian surfers but also helps us remember the definition of the three main trigonometric functions: sine, cosine, and tangent.

There are six trigonometric functions that we will discuss in this book, but we will focus primarily on these three. We will define the trigonometric functions in terms of ratios of the three sides of a right triangle: SOHCAHTOA.

SOH **S**ine equals **O**pposite over **H**ypotenuse

CAH **C**osine equals **A**djacent over **H**ypotenuse

TOA **T**angent equals **O**pposite over **A**djacent

The ancient Greeks used right triangle trigonometry, and we still use it today in sports, surveying, navigation, and engineering.

In this chapter we will review angles, degree measure, and special right triangles. We will discuss the properties of similar triangles. We will use the concept of similar right triangles to define the six basic trigonometric functions as ratios of sides of right triangles (right triangle trigonometry).

Right Triangle Trigonometry

Angles

- **Degree Measure**
- **Types of Angles**
  - Right
  - Acute
  - Obtuse

Triangles

- **Right Triangles**
  - Pythagorean theorem
- **Special Right Triangles**
  - 30-60-90
  - 45-45-90
- **Similar Triangles**

Trigonometric Functions

- Sine (sin)
- Cosine (cos)
- Tangent (tan)
- Secant (sec)
- Cosecant (csc)
- Cotangent (cot)

CHAPTER OBJECTIVES

- Understand degree measure.
- Use the Pythagorean theorem to determine the sides of right triangles.
- Understand the difference between congruent and similar triangles.
- Define the six trigonometric functions as ratios of sides of right triangles.
- Learn trigonometric function values for special angles.
- Evaluate trigonometric functions exactly and with calculators.
- Solve right triangles.

NAVIGATION THROUGH SUPPLEMENTS

DIGITAL VIDEO SERIES #1

STUDENT SOLUTIONS MANUAL CHAPTER 1

BOOK COMPANION SITE
www.wiley.com/college/young

Angles

The study of trigonometry relies heavily on the concept of angles. Before we define angles, let us review some basic terminology. A **line** is the straight path connecting two points (A and B) and extending beyond the points in both directions. The portion of the line between the two points (including the points) is called a **line segment**. A **ray** is the portion of the line that starts at one point (A) and extends to infinity (beyond B). A is called the **endpoint of the ray**.

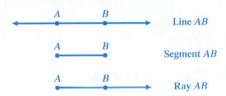

An **angle** is formed when two rays share the same endpoint. The common endpoint is called the **vertex**.

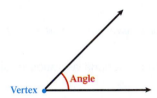

We say that an angle is formed when a ray is rotated around its endpoint. The ray in its original position is called the **initial side**. In the Cartesian plane the initial side is usually the positive *x*-axis. The ray after it is rotated is called the **terminal side**. Rotation in a counterclockwise direction corresponds to a **positive angle**, whereas rotation in a clockwise direction corresponds to a **negative angle**.

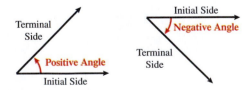

Degree Measure

Lengths, or distances, can be measured in different units: feet, miles, and meters are three common units. In order to compare angles of different sizes we need a standard unit of measure. One way to measure the size of an angle is with **degree measure**. We will discuss **degrees** now and in Chapter 3 discuss another angle measure called *radians*.

DEFINITION **DEGREE MEASURE OF ANGLES**

An angle formed by one complete counterclockwise rotation has **measure 360 degrees**, denoted 360°.

One complete revolution = 360°

WORDS **MATH**

360° represents 1 complete rotation.
$$\frac{360°}{360°} = 1$$

180° represents a $\frac{1}{2}$ rotation.
$$\frac{180°}{360°} = \frac{1}{2}$$

90° represents a $\frac{1}{4}$ rotation.
$$\frac{90°}{360°} = \frac{1}{4}$$

The Greek letter theta, θ, is the most common name for an angle. Other common names of angles are alpha, α, beta, β, and gamma, γ.

WORDS **MATH**

An angle measuring exactly 90° is called a **right angle**.

A right angle is often represented by the adjacent sides of a rectangle, indicating that the two rays are *perpendicular*.

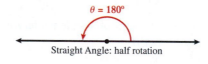

Right Angle: quarter rotation

An angle measuring exactly 180° is called a **straight angle**.

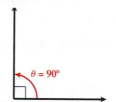

Straight Angle: half rotation

An angle measuring greater than 0° but less than 90° is called an **acute angle**.

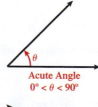

Acute Angle
0° < θ < 90°

An angle measuring greater than 90° but less than 180° is called an **obtuse angle**.

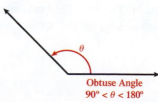

Obtuse Angle
90° < θ < 180°

If the sum of the measures of two positive angles is 90° the angles are called **complementary**.

Complementary Angles
$\alpha + \beta = 90°$

If the sum of the measures of two positive angles is 180° the angles are called **supplementary**.

Supplementary Angles
$\alpha + \beta = 180°$

EXAMPLE 1 Finding Measures of Complementary and Supplementary Angles

Find the measure of each angle.

a. Find the complement of 50°.
b. Find the supplement of 110°.
c. Find the complement of α (in terms of α).
d. Find two supplementary angles where one angle is twice as large as the second angle.

Solution:

a. The sum of complementary angles is 90°.

Solve for θ.

$\theta + 50° = 90°$
$\theta = 40°$

b. The sum of supplementary angles is 180°.

Solve for θ.

$\theta + 110° = 180°$
$\theta = 70°$

c. Let β be the complement of α.
The sum of complementary angles is 90°.

Solve for β.

$\alpha + \beta = 90°$
$\beta = 90° - \alpha$

d. The sum of supplementary angles is 180°.
Let $\beta = 2\alpha$.
Solve for α.

$\alpha + \beta = 180°$
$\alpha + 2\alpha = 180°$
$3\alpha = 180°$
$\alpha = 60°$

Substitute $\alpha = 60°$ into $\beta = 2\alpha$.

$\beta = 120°$

The angles have measures 60° and 120° .

■ **YOUR TURN** Find two supplementary angles where one angle is three times as large as the second angle.

Triangles

Trigonometry is the study of triangles, with emphasis on calculations involving the lengths of sides and the measures of angles. What do you already know about triangles? You know that triangles are three-sided closed plane figures. An important property of triangles is that the sum of the measures of the three angles of any triangle is 180°.

ANGLE SUM OF A TRIANGLE

The sum of the measures of the angles of any triangle is 180°.

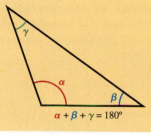

$$\alpha + \beta + \gamma = 180°$$

EXAMPLE 2 Finding an Angle of a Triangle

If two angles of a triangle have measures 32° and 68°, what is the measure of the third angle?

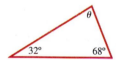

Solution:

The sum of the measures of all three angles is 180°. $32° + 68° + \theta = 180°$

Solve for θ. $\theta = 80°$

■ **YOUR TURN**   If two angles of a triangle have measures 16° and 96°, what is the measure of the third angle?

There are several special triangles discussed in geometry. An **equilateral triangle** has three equal sides and three equal angles (60°-60°-60°). An **isosceles triangle** has two equal sides (legs) and two equal angles opposite those legs. The most important triangle that we will discuss in this course is a *right triangle*. A **right triangle** is a triangle in which one of the angles is a right angle (90°). Since one angle is 90°, the other two angles must be complementary (sum to 90°), so that the sum of all three angles is 180°. The longest side of a right triangle, called the **hypotenuse**, is opposite the right angle. The other two sides are called the **legs** of the right triangle.

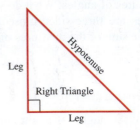

The *Pythagorean theorem* relates the sides of a right triangle. It says that the sum of the squares of the lengths of the two legs is equal to the square of the length of the hypotenuse. It is important to note that length (a synonym to distance) is always taken to be positive.

PYTHAGOREAN THEOREM

In any right triangle, the square of the length of the longest side (hypotenuse) is equal to the sum of the squares of the lengths of the other two sides (legs).

$$a^2 + b^2 = c^2$$

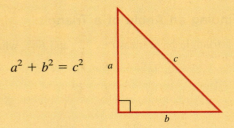

It is important to note that the Pythagorean theorem *only* applies to right triangles. It is also important to note that it does not matter which side is called *a* or *b* as long as the longest squared side is equal to the sum of the smaller squared sides.

EXAMPLE 3 Using the Pythagorean Theorem to Find the Side of a Right Triangle

Suppose you have a 10 foot ladder and want to reach a height of 8 feet to clean out the gutters on your house. How far from the base of the house should the base of the ladder be?

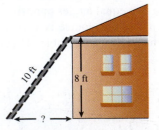

Solution:

Label the unknown side as x.

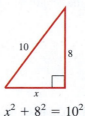

Apply the Pythagorean theorem. $\qquad x^2 + 8^2 = 10^2$

Simplify. $\qquad x^2 + 64 = 100$

Solve for x. $\qquad x^2 = 36$

$\qquad x = \pm 6$

Length must be positive. $\qquad x = 6$

The ladder should be 6 feet from the base of the house along the ground.

> **YOUR TURN** A ramp is being built to accommodate wheelchair access to a home's front porch. The front porch is 3 feet off the ground and the driveway is 36 feet from the front porch. If the ramp starts at the driveway edge and continues to the front porch, how long is the ramp? Round to the nearest foot.

Special Right Triangles

Right triangles such as the 3-4-5, the 5-12-13, and the 8-15-17 are special in that all of these are right triangles with side lengths equal to whole numbers that satisfy the Pythagorean theorem. There are two special right triangles that warrant additional attention: the 30°-60°-90° triangle and the 45°-45°-90° triangle. Although in trigonometry we focus more on the angles than on the side lengths, we are interested in special relationships between the lengths of the sides of these right triangles. We will start with the 45°-45°-90° triangle.

WORDS	MATH
A 45°-45°-90° triangle is an isosceles (two legs are equal) right triangle.	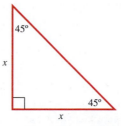

Use the Pythagorean theorem. $\qquad x^2 + x^2 = \text{hypotenuse}^2$

Simplify the left side of the equation. $\qquad 2x^2 = \text{hypotenuse}^2$

Solve for the hypotenuse. $\qquad \text{hypotenuse} = \pm\sqrt{2x^2} = \pm\sqrt{2}\,|x|$

x and the hypotenuse are lengths and must be positive. $\qquad \text{hypotenuse} = \sqrt{2}\,x$

The hypotenuse of a 45°-45°-90°
is $\sqrt{2}$ times the length of either leg.

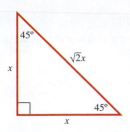

If we let $x = 1$, then the triangle will have legs with length equal to 1 and the hypotenuse will have length $\sqrt{2}$. Notice that these satisfy the Pythagorean theorem: $1^2 + 1^2 = (\sqrt{2})^2$ or $2 = 2$.

EXAMPLE 4 Solving a 45°-45°-90° Triangle

A house has a roof with a 45° pitch (the angle the roof makes with the house). If the house is 60 feet wide, what are the lengths of the sides of the roof that form the attic? Round to the nearest foot.

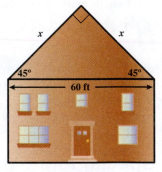

Solution:

Draw the 45°-45°-90° triangle.

Let x represent the length of the unknown legs.

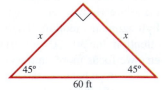

Recall that the hypotenuse of a 45°-45°-90° triangle is $\sqrt{2}$ times the length of either leg.

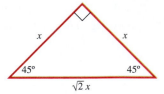

Let the hypotenuse equal
60 feet. $60 = \sqrt{2}x$

Solve for x. $x = 42.42641$

Round to the nearest foot. $x \approx 42$ ft

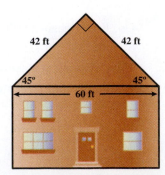

Let us now determine the relationship of the sides of a 30°-60°-90° triangle. We start with an equilateral triangle (equal sides and equal 60° angles).

WORDS

Draw an equilateral triangle with sides $2x$.

MATH

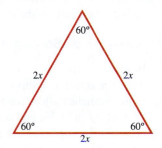

Draw a segment (which bisects the base) representing the height of the triangle, h. There are now two equal 30°-60°-90° triangles.

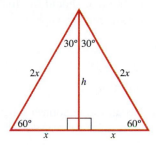

Notice that the hypotenuse is twice the shortest leg, which is opposite the 30° angle.

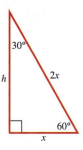

Use the Pythagorean theorem.

Solve for h.

$$h^2 + x^2 = (2x)^2$$
$$h^2 + x^2 = 4x^2$$
$$h^2 = 3x^2$$
$$h = \pm\sqrt{3x^2} = \sqrt{3}|x|$$

h and x are lengths and must be positive.

$$h = \sqrt{3}x$$

The hypotenuse of a 30°-60°-90° triangle is two times the length of the leg opposite the 30° angle.

The leg opposite the 60° angle is $\sqrt{3}$ times the length of the leg opposite the 30° angle.

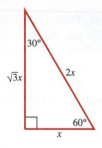

If we let $x = 1$, then the triangle will have legs with lengths 1 and $\sqrt{3}$ and hypotenuse of length 2. These satisfy the Pythagorean theorem: $1^2 + (\sqrt{3})^2 = 2^2$ or $4 = 4$.

EXAMPLE 5 Solving a 30°-60°-90° Triangle

Before a hurricane it is wise to stake down trees for additional support during the storm. If the branches allow for the rope to be tied 15 feet up the tree, and a desired angle between the rope and the ground is 60°, how much rope is needed? How far from the base of the tree should the four stakes be hammered?

Solution:

Recall the relationship between the sides of a 30°-60°-90° triangle.

In this case, the leg opposite the 60° angle is 15 feet.

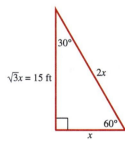

Solve for x.

$$\sqrt{3}x = 15 \text{ ft}$$

$$x = \frac{15}{\sqrt{3}} \approx 8.7 \text{ ft}$$

Find the length of the hypotenuse.

$$\text{hypotenuse} = 2x = 2\left(\frac{15}{\sqrt{3}}\right) \approx 17 \text{ ft}$$

The ropes should be staked approximately 8.7 feet from the base of the tree, and approximately 4(17) = 68 feet of rope will be needed.

■ **YOUR TURN** Rework Example 5 with a tree height (where the ropes are tied) of 20 feet. How far from the base of the tree should the ropes be staked and how much rope will be needed?

SECTION 1.1 SUMMARY

In this section you have practiced working with angles. One unit of measure of angles is degrees. An angle measuring exactly 90° is called a right angle. The sum of the three angles of any triangle is always 180°. Triangles that contain a right angle are called right triangles. The Pythagorean theorem was used to solve right triangles. Two special right triangles are the 30°-60°-90° and 45°-45°-90° triangles.

SECTION 1.1 EXERCISES

■ **SKILLS**

In Exercises 1–8, specify the measure of the angle in degrees.

1. $\frac{1}{2}$ rotation counterclockwise   2. $\frac{1}{4}$ rotation counterclockwise

3. $\frac{1}{3}$ rotation clockwise   4. $\frac{2}{3}$ rotation clockwise

5. $\frac{5}{6}$ rotation counterclockwise   6. $\frac{7}{12}$ rotation counterclockwise

7. $\frac{4}{5}$ rotation clockwise   8. $\frac{5}{9}$ rotation clockwise

In Exercises 9–14, find (a) the complement and (b) the supplement of the given angles.

9. 18°   10. 39°   11. 42°

12. 57°   13. 89°   14. 75°

In Exercises 15–18, find the measure of each angle.

15.

16.

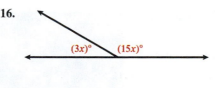

17. Supplementary angles with measures $8x$ degrees and $4x$ degrees

18. Complementary angles with measures $3x + 15$ degrees and $10x + 10$ degrees

In Exercises 19–22, refer to the following triangle.

19. If $\alpha = 117°$ and $\beta = 33°$ find γ.

20. If $\alpha = 110°$ and $\beta = 45°$ find γ.

21. If $\gamma = \beta$ and $\alpha = 4\beta$ find all three angles.

22. If $\gamma = \beta$ and $\alpha = 3\beta$ find all three angles.

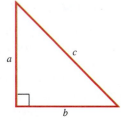

$\alpha + \beta + \gamma = 180°$

In Exercises 23–28, refer to the following right triangle.

23. If $a = 4$ and $b = 3$, find c. **24.** If $a = 3$ and $b = 3$, find c.

25. If $a = 6$ and $c = 10$, find b. **26.** If $b = 7$ and $c = 12$, find a.

27. If $a = 8$ and $b = 5$, find c. **28.** If $a = 6$ and $b = 5$, find c.

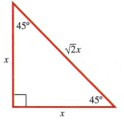

In Exercises 29–32, refer to the following 45°-45°-90° triangle.

29. If each leg has length 10 inches, how long is the hypotenuse?

30. If each leg has length 8 meters, how long is the hypotenuse?

31. If the hypotenuse has length $2\sqrt{2}$ centimeters, how long are the legs?

32. If the hypotenuse has length $\sqrt{10}$ feet, how long are the legs?

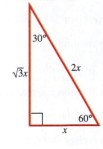

In Exercises 33–38, refer to the following 30°-60°-90° triangle.

33. If the shortest leg has length 5 meters, what are the lengths of the other leg and the hypotenuse?

34. If the shortest leg has length 9 feet, what are the lengths of the other leg and the hypotenuse?

35. If the longer leg has length 12 yards, what are the lengths of the other leg and the hypotenuse?

36. If the longer leg has length n units, what are the lengths of the other leg and the hypotenuse?

37. If the hypotenuse has length 10 inches, what are the lengths of the two legs?

38. If the hypotenuse has length 8 centimeters, what are the lengths of the two legs?

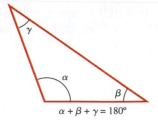

■ APPLICATIONS

39. Clock. What is the measure (in degrees) that the minute hand makes in 20 minutes?

40. Clock. What is the measure (in degrees) that the minute hand makes in 25 minutes?

41. London Eye. The London Eye (similar to a bicycle wheel) makes 1 rotation in approximately 30 minutes. What is the measure of the angle (in degrees) that your cart (spoke) will rotate in 12 minutes?

Index Stock

42. London Eye. The London Eye (similar to a bicycle wheel) makes 1 rotation in approximately 30 minutes. What is the measure of the angle (in degrees) that your cart (spoke) will rotate in 5 minutes?

43. Revolving Restaurant. If a revolving restaurant overlooking a waterfall can rotate 270° in 45 minutes, how long does it take to make a complete revolution?

44. Revolving Restaurant. If a revolving restaurant overlooking a waterfall can rotate 72° in 9 minutes, how long does it take to make a complete revolution?

45. Field Trial. In a Labrador retriever field trial, a dog is judged by the straight line it takes to a fallen bird. If the direct line is through water, it is considered "cheating" if the dog goes around the water and up the shore. If the judge wants to calculate how far a dog will travel along a straight path, she walks the two legs of the right triangle and uses the Pythagorean theorem. How far would this dog travel (run and swim) if it traveled along the hypotenuse?

46. Field Trial. How far would the dog in Exercise 45 run/swim if it traveled along the hypotenuse? The judge walks 25 feet along the shore and then 100 feet out to the bird.

47. Christmas Lights. A couple want to put up Christmas lights along the roofline of their house. If the front of the house is 100 feet wide and the roof has a 45 degree pitch, how many linear feet of Christmas lights should the couple buy? Round to the nearest foot.

48. Christmas Lights. Repeat Exercise 47 for a house that is 60 feet wide. Round to the nearest foot.

49. Tree Stake. A tree needs to be staked down before a storm. If the ropes can be tied on the tree branches 17 feet above the ground and the staked rope should make a 60° angle with the ground, how far from the base of the tree should the ropes be staked?

50. Tree Stake. How many feet of rope are needed to have two stakes supporting the tree in Exercise 49?

51. Tree Stake. A tree needs to be staked down before a storm. If the ropes can be tied on the tree branches 10 feet above the ground and the staked rope should make a 30° angle with the ground, how far from the base of the tree should the ropes be staked?

52. Tree Stake. How many feet of rope are needed to have four stakes supporting the tree in Exercise 51?

53. Party Tent. Skipper and Michelle want to rent a 40 foot by 20 foot tent for their backyard to host a barbeque. The base of the tent is supported 7 feet above the ground by poles and then roped stakes are used for support. The ropes make a 60° angle with the ground. How large a footprint in their yard would they need for this tent (and staked ropes)? In other words, what are the dimensions of the rectangle formed by the stakes on the ground? Round to the nearest foot.

54. Party Tent. Ashley's parents are throwing a graduation party and are renting a 40 foot by 80 foot tent for their backyard. The base of the tent is supported 7 feet above the ground by poles and then roped stakes are used for support. The ropes make a 45° angle with the ground. How large a footprint in their yard will they need for this tent (and staked ropes)? Round to the nearest foot.

LuckyPix

■ CATCH THE MISTAKE

In Exercises 55 and 56, explain the mistake that is made.

55. In a 30°-60°-90° triangle find the length of the side opposite the 60° angle if the side opposite the 30° angle is 10 inches.

Solution:

The length opposite the 60° angle is twice the length opposite the 30° angle. $2(10) = 20$

The side opposite the 60° angle has length 20 inches.

This is incorrect. What mistake was made?

56. In a 45°-45°-90° triangle find the length of the hypotenuse if each leg has length 5 centimeters.

Solution:

Use the Pythagorean theorem. $5^2 + 5^2 = \text{hypotenuse}^2$

Simplify. $50 = \text{hypotenuse}^2$

Solve for the hypotenuse. $\text{hypotenuse} = \pm 5\sqrt{2}$

The hypotenuse has length $\pm 5\sqrt{2}$ centimeters.

This is incorrect. What mistake was made?

■ CHALLENGE

57. T or F: The Pythagorean theorem holds for any equilateral triangle.

58. T or F: The Pythagorean theorem holds for all isosceles triangles.

59. T or F: The two angles opposite the legs of a right triangle are complementary.

60. T or F: In a 30°-60°-90° triangle the length of the side opposite the 60° angle is twice the length of the side opposite the 30° angle.

61. What is the measure (in degrees) of the smaller angle the hour and minute hands make when it is 12:20?

62. What is the measure (in degrees) of the smaller angle the hour and minute hands make when it is 9:10?

In Exercises 63–66, use the following figure.

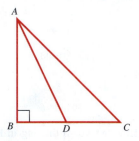

63. If $AB = 3$, $AD = 5$, and $AC = \sqrt{58}$ find DC.

64. If $AB = 4$, $AD = 5$, and $AC = \sqrt{41}$ find DC.

65. If $AC = 30$, $AB = 24$, and $DC = 11$, find AD.

66. If $AB = 60$, $AD = 61$, and $DC = 36$, find AC.

67. Given a square with side length x, draw the two diagonals. The result is 4 special triangles. Describe these triangles. What are the angle measures?

68. Solve for x in the following triangle.

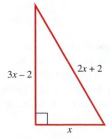

Skills Objectives	Conceptual Objective
▪ Label sides and angles of triangles as equal. ▪ Use similarity to determine the length of a side of a triangle.	▪ Understand the difference between congruent and similar triangles.

Geometry Review

We stated in Section 1.1 that the sum of the measures of three angles of any triangle is 180°. How do we know that is true? You will soon see why that is true, but first we need some relationships from geometry.

Vertical angles are angles opposite one another at the intersection of any two lines. In the diagram below, angles **1** and **3** are vertical angles. Angles **2** and **4** are also vertical angles. Vertical angles have equal measure.

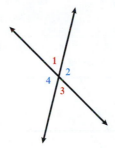

A **transversal** is a line that intersects two lines if the two lines are parallel (noninter-secting) lines, m and n. We say $m\|n$. There are eight resulting angles that have special properties. Angles 3, 4, 5, and 6 are classified as **interior angles**, whereas angles 1, 2, 7, and 8 are classified as **exterior angles**.

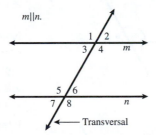

In diagrams, there are two traditional ways to indicate angles having equal measure:

▪   with the same number of arcs *or*

▪   with a single arc and the same number of hash marks.

In this text we will use the same number of arcs.

NAME	PICTURE	RULE
Vertical angles		Vertical angles have equal measure. $\angle 1 = \angle 3$ and $\angle 2 = \angle 4$
Alternate interior angles	$m\|n$ 	Alternate interior angles have equal measure. $\angle 3 = \angle 6$ and $\angle 4 = \angle 5$
Alternate exterior angles	$m\|n$ 	Alternate exterior angles have equal measure. $\angle 1 = \angle 8$ and $\angle 2 = \angle 7$
Corresponding angles	$m\|n$ 	Corresponding angles have equal measure. $\angle 1 = \angle 5$ Note that the following are also corresponding angles: $\angle 2$ and $\angle 6$ $\angle 3$ and $\angle 7$ $\angle 4$ and $\angle 8$
Interior angles on the same side of transversal	$m\|n$ $\angle 3 + \angle 5 = 180°$ $\angle 4 + \angle 6 = 180°$	Interior angles on the same side of the transversal have measures that sum to $180°$ (they are supplementary).

EXAMPLE 1 Finding Angle Measures

Given that $\angle 1 = 110°$ and $m \| n$, find the measure of angle 7.

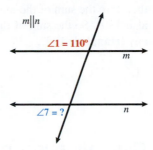

Solution:

Corresponding angles have equal measure, $\angle 1 = \angle 5$.

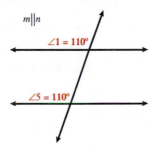

$\angle 5$ and $\angle 7$ are supplementary angles.

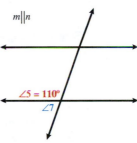

Measures of supplementary angles sum to 180°. $\angle 5 + \angle 7 = 180°$

Substitute $\angle 5 = 110°$. $110° + \angle 7 = 180°$

Solve for $\angle 7$. $\angle 7 = 70°$

We will now use these properties in the above table to illustrate why the measures of angles in a triangle sum to 180°.

Draw two parallel lines (solid) and draw two transverse lines (dashed).

The **corresponding angles 1** and **6** have equal measure.

The **vertical angles 2** and **4** have equal measure.

The **corresponding angles 3** and **5** have equal measure.

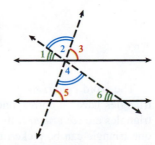

The sum of the measures of angles **1**, **2**, and **3** is 180°, therefore the sum of the measures of angles **4**, **5**, and **6** is also 180°. So the sum of the measures of three angles of any **triangle** is 180°.

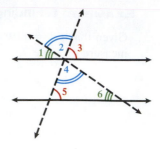

Classification of Triangles

Angles with equal measure are labeled with the same number of arcs. Similarly, sides of equal length are labeled with the same number of hash marks.

NAME	PICTURE	RULE
Equilateral triangle		All sides are equal.
Isosceles triangle		Two sides are equal.
Scalene triangle		No sides are equal.

The word *similar* in mathematics means identical in shape, although not necessarily the same size. It is important to note that two triangles can have the exact same shape (same angles) but have different sizes.

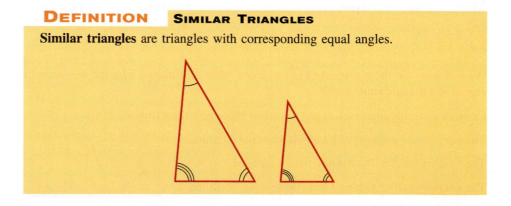

DEFINITION   **SIMILAR TRIANGLES**

Similar triangles are triangles with corresponding equal angles.

The word *congruent* means equal in all corresponding parts; therefore, congruent triangles have all corresponding angles equal and all corresponding sides equal. Two triangles are *congruent* if they have exactly the same shape *and* size. In other words, if one triangle can be picked up and situated on top of another triangle so that the two triangles coincide, they are said to be *congruent*.

DEFINITION   **CONGRUENT TRIANGLES**

Congruent triangles are triangles with corresponding equal angles *and* corresponding equal sides.

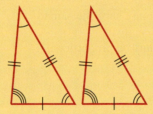

It is important to note that all congruent triangles are also similar triangles, but not all similar triangles are congruent triangles.

Trigonometry (as you will see in Section 1.3) relies on the properties of similar triangles. Since similar triangles have the same shape (congruent corresponding angles), the sides opposite the corresponding angles must be proportional.

Recall in Section 1.1 that the sides of any 30°-60°-90° triangle are given by the relationship that corresponds to sides opposite each of the angles.

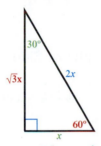

For example, if we let $x = 1$ correspond to a triangle A and $x = 5$ correspond to a triangle B then we would have the following two triangles:

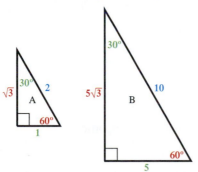

Notice that the sides opposite the same angles are proportional. For example, if we take the ratio of sides of triangle B to sides of triangle A we get the same number.

Sides opposite **30°**: $\dfrac{\text{Triangle B}}{\text{Triangle A}} = \dfrac{5}{1} = 5$

Sides opposite **60°**: $\dfrac{\text{Triangle B}}{\text{Triangle A}} = \dfrac{5\sqrt{3}}{\sqrt{3}} = 5$

Sides opposite **90°**: $\dfrac{\text{Triangle B}}{\text{Triangle A}} = \dfrac{10}{2} = 5$

This proportionality property holds for all similar triangles.

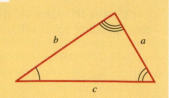

CONDITIONS FOR SIMILAR TRIANGLES

For two triangles to be similar the following must *both* be true:

1. Corresponding angles must have the same measure.
2. Corresponding sides must be proportional (ratios must be equal)

$$\frac{a}{a'} = \frac{b}{b'} = \frac{c}{c'}$$

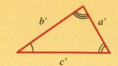

Let us now use properties of similar triangles to determine lengths of sides of similar triangles.

EXAMPLE 2 Finding Lengths of Sides in Similar Triangles

Given that the two triangles are similar, find the length of the unknown sides (b and c).

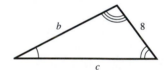

Solution:

Solve for b.

The corresponding sides are proportional.	$\dfrac{8}{2} = \dfrac{b}{5}$
Multiply by 5.	$b = 4(5)$
Solve for b.	$b = 20$

Solve for c.

The corresponding sides are proportional.	$\dfrac{8}{2} = \dfrac{c}{6}$
Multiply by 6.	$c = 4(6)$
Solve for c.	$c = 24$

Check that ratios are equal: $\dfrac{8}{2} = \dfrac{20}{5} = \dfrac{24}{6} = 4$

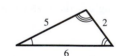

■ **YOUR TURN**   Given that the two triangles are similar, find the length of the unknown sides (*a* and *b*).

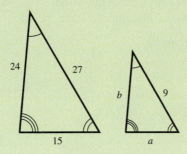

Applications Involving Similar Triangles

The common ratios associated with similar triangles have been very useful when you can measure the sides of a smaller triangle and then calculate the measures of larger triangles if you measure one side. For example, shadows can be used to quickly estimate the height of flagpoles, trees, and any other tall objects. Surveyors use similar triangles to determine distances that are difficult to measure.

EXAMPLE 3 Calculating the Height of a Tree

Billy wants to rent a lift to trim his tall trees. However, he must decide which lift he needs: one that will lift him 25 feet or a more expensive lift that will lift him 50 feet. He would prefer to rent the less expensive lift but is not sure exactly how tall the tree is. His wife, Jeanine, hammered a yard stake in the ground and measured its shadow to be $1\frac{3}{4}$ feet long and measured the tree's shadow to be 19 feet. If the stake was standing 3 feet above the ground, how high is the tree? More important, which lift should Billy rent?

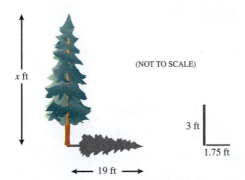

Solution:

Rays of sunlight are straight and parallel to each other. Therefore, the rays make the same angle with the tree that they do with the stake.

(NOT TO SCALE)

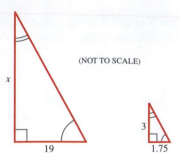

We have two similar triangles.

The ratios of corresponding sides of similar triangles are equal.

$$\frac{x}{3} = \frac{19}{1.75}$$

Solve for x.

$$x = \frac{3(19)}{1.75}$$

Simplify.

$$x \approx 32.57$$

The tree is approximately 33 feet. Billy should rent the more expensive lift to be safe.

YOUR TURN Billy's neighbor decides to do the same thing. He borrows Jeanine's stake and measures the shadows. If the shadow of his tree is 15 feet and the shadow of the stake (3 feet above the ground) is 1.2 feet, how tall is Billy's neighbor's tree?

SECTION 1.2 SUMMARY

In this section you have learned to label angles with equal measure with the same number of arcs and sides of equal length with the same number of hash marks. Similar triangles (the same shape) have corresponding angles with equal measure. Congruent triangles (the same shape and size) are similar triangles that also have equal corresponding side lengths. Similar triangles have the property that the ratios of their corresponding side lengths are equal.

SECTION 1.2 EXERCISES

■ SKILLS

In Exercises 1–10, find the measure of the indicated angles.

1. $C = 80°$, find B. **2.** $C = 80°$, find E. **3.** $C = 80°$, find F.

4. $C = 80°$, find G. **5.** $F = 75°$, find B. **6.** $F = 75°$, find A.

7. $A = (8x)°$ and $D = (9x - 15)°$, find the measures of A and D.

8. $B = (9x + 7)°$ and $F = (11x - 7)°$, find the measures of B and F.

9. $A = (12x + 14)°$ and $G = (9x - 2)°$, find the measures of A and G.

10. $C = (22x + 3)°$ and $E = (30x - 5)°$, find the measures of C and E.

$m \| n$

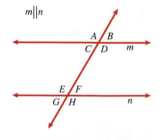

In Exercises 11–16, match the corresponding triangle with the appropriate name.

11. Equilateral triangle

12. Right triangle (nonisosceles)

13. Isosceles triangle (nonright)

14. Acute scalene triangle

15. Obtuse scalene triangle

16. Isosceles right triangle

In Exercises 17–22, calculate the specified lengths given that the two triangles are similar.

17. $a = 4, c = 6, d = 2, f = ?$

18. $a = 12, b = 9, e = 3, d = ?$

19. $d = 5, e = 2.5, b = 7.5, a = ?$

20. $e = 1.4, f = 2.6, c = 3.9, b = ?$

21. $d = 2.5$ m, $f = 1.1$ m, $a = 26.25$ km, $c = ?$

22. $e = 10$ cm, $f = 14$ cm, $c = 35$ m, $b = ?$

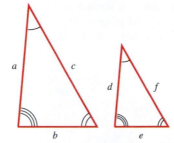

 ■ **APPLICATIONS**

23. Height of a Tree. The shadow of a tree measures $14\frac{1}{4}$ feet. At the same time of day the shadow of a 4 foot pole measures 1.5 feet. How tall is the tree?

24. Height of a Flagpole. The shadow of a flagpole measures 15 feet. At the same time of day the shadow of a 2 foot stake measures $\frac{3}{4}$ feet. How tall is the flagpole?

25. Height of a Lighthouse. The Cape Hatteras Lighthouse in the Outer Banks of North Carolina is the tallest lighthouse in North America. If a 5 foot woman casts a 1.2 foot shadow and the lighthouse casts a 54 foot shadow, approximately how tall is the Cape Hatteras Lighthouse?

26. Height of a Man. If a 6 foot man casts a 1 foot shadow, how long a shadow will his 4 foot son cast?

Although most people know that a list exists of the Seven Wonders of the Ancient World, only a few can name them: the Great Pyramid of Giza, the Hanging Gardens of Babylon, the Statue of Zeus at Olympia, the Temple of Artemis at Ephesus, the Mausoleum of Halicarnassus, the Colossus of Rhodes, and the Lighthouse of Alexandria.

27. Seven Wonders. One of the Seven Wonders of the Ancient World was a lighthouse, on the Island of Pharos in Alexandria, Egypt. It is the first lighthouse in recorded history and was built about 280 BC. Those records tell us that it was the tallest one ever built. It survived for

1500 years until it was completely destroyed by an earthquake in the 14th century. On a sunny day if a 2 meter tall man casts a shadow approximately 5 centimeters (0.05 meters) long, then the lighthouse would have cast approximately a 3 meter shadow. How tall was this fantastic structure?

28. **Seven Wonders.** Only one of the great Seven Wonders of the Ancient World is still standing—the Great Pyramid of Giza. Each of the base sides along the ground measures 230 meters. If a 1 meter child casts a 90 centimeter shadow at the same time the shadow of the pyramid extends 16 meters along the ground (beyond the base), approximately how tall is the Great Pyramid of Giza?

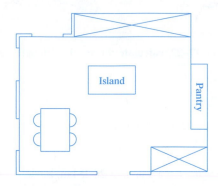

Lonely Planet Images/Getty Images, Inc.

In Exercises 29–30, use the following drawing:

In a home remodeling project your architect gives you plans that have an indicated distance of $2'4''$, which measures $\frac{1}{4}$ inch with a ruler.

29. **Measurement.** How long is the pantry in the kitchen if it measures $\frac{11}{16}$ inch with a ruler?

30. **Measurement.** How wide is the island if it measures $\frac{7}{16}$ inch with a ruler?

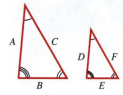

■ **CATCH THE MISTAKE**

In Exercises 31 and 32, explain the mistake that is made.

31. If $A = 8$, $B = 5$, and $D = 3$, find E.

Solution:

Set up a ratio of similar triangles: $\dfrac{A}{E} = \dfrac{B}{D}$

Substitute $A = 8$, $B = 5$, $D = 3$: $\dfrac{8}{E} = \dfrac{5}{3}$

Cross multiply: $5E = 8(3)$

Solve for E: $E = \dfrac{24}{5}$

This is incorrect. What mistake was made?

32. If $A = 8$, $B = 5$, $D = 3$, find F.

Solution:

Set up a ratio of similar triangles: $\dfrac{A}{B} = \dfrac{D}{F}$

Substitute $A = 8$, $B = 5$, $D = 3$: $\dfrac{8}{5} = \dfrac{3}{F}$

Cross multiply: $8F = 5(3)$

Solve for F: $F = \dfrac{15}{8}$

This is incorrect. What mistake was made?

■ **CHALLENGE**

33. T or F: Two similar triangles must have equal corresponding angles.

34. T or F: All congruent triangles are similar, but not all similar triangles are congruent.

35. T or F: Two angles in a triangle cannot have measures 82° and 67°.

36. T or F: Two equilateral triangles are similar but do not have to be congruent.

37. Find x.

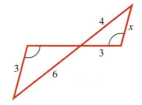

38. Find x and y.

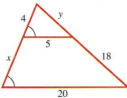

For Exercises 39–41, refer to the following:

The lens law relates three quantities: the distance from the object to the lens, D_0, the distance from the lens to the image, D_i, and the focal length of the lens, f.

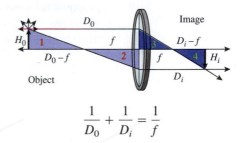

$$\frac{1}{D_0} + \frac{1}{D_i} = \frac{1}{f}$$

39. Explain why triangles 1 and 2 are similar triangles and why triangles 3 and 4 are similar triangles.

40. Set up the similar triangle ratios for:

a. triangles 1 and 2

b. triangles 3 and 4

41. Use the ratios in Exercise 40 to derive the lens law.

SECTION 1.3 Definition 1 of Trigonometric Functions: Right Triangle Ratios

Skills Objectives

- Learn the trigonometric functions as ratios of sides of a right triangle.
- Calculate trigonometric ratios of general angles.
- Relate trigonometric cofunctions through complementary angles.

Conceptual Objectives

- Understand that right triangle ratios are based on similar triangles.
- Understand that the three definitions of trigonometric functions are consistent.

Why Three Definitions?

The word **trigonometry** stems from the Greek words *trigonon*, which means triangle, and *metrein*, which means to measure. It began as a branch of geometry and was utilized extensively by early Greek mathematicians to determine unknown distances. The major *trigonometric functions*, including *sine*, *cosine*, and *tangent*, were first defined as ratios of sides in a right triangle. This is the definition we will use in this section. Since the two angles besides the right angle in a right triangle have to be acute, a second definition was needed to extend the domain of trigonometric functions to nonacute angles using the Cartesian plane (Sections 2.2 and 2.3). In the 18th century, the definitions of trigonometric functions were broadened by being defined as points on a unit circle (Section 3.4). All three approaches give consistent definitions, as you will see.

Trigonometric Functions: Right Triangle Ratios

In Section 1.2 we learned that triangles which have corresponding equal angles have proportional sides and are called similar triangles. The concept of similar triangles, one of the basic insights in trigonometry, allows us to determine the length of a side of one triangle if we know the length of certain sides of a similar triangle.

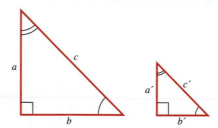

Since the two right triangles above have equal angles, they are similar triangles, and the following ratios hold true.

$$\frac{a}{a'} = \frac{b}{b'} = \frac{c}{c'}$$

Separate the common ratio into three equations.

$$\frac{a}{a'} = \frac{b}{b'} = \frac{c}{c'} \quad \text{or} \quad \frac{a}{a'} = \frac{b}{b'} \qquad \frac{b}{b'} = \frac{c}{c'} \qquad \frac{a}{a'} = \frac{c}{c'}$$

WORDS	MATH
Start with the first ratio.	$\dfrac{a}{a'} = \dfrac{b}{b'}$
Cross multiply.	$ab' = a'b$
Divide both sides by bb'.	$\dfrac{ab'}{bb'} = \dfrac{a'b}{bb'}$
Simplify.	$\dfrac{a}{b} = \dfrac{a'}{b'}$

Similarly, it can be shown that $\dfrac{b}{c} = \dfrac{b'}{c'}$ and $\dfrac{a}{c} = \dfrac{a'}{c'}$.

Notice that even though the sizes of the triangles are different, since the corresponding angles are equal, the ratio of the two legs of the large triangle is equal to the ratio of the legs of the small triangle, $\dfrac{a}{b} = \dfrac{a'}{b'}$. Similarly, the ratios of a leg and the hypotenuse of the large triangle and a leg and the hypotenuse of the small triangle are also equal, $\dfrac{b}{c} = \dfrac{b'}{c'}$ and $\dfrac{a}{c} = \dfrac{a'}{c'}$.

For any right triangle, there are six possible ratios of sides that can be calculated for each angle, θ:

$$\frac{b}{c} \quad \frac{a}{c} \quad \frac{b}{a}$$

$$\frac{c}{b} \quad \frac{c}{a} \quad \frac{a}{b}$$

These ratios are referred to as **trigonometric ratios** or **trigonometric functions,** since they depend on θ, and each is given a name:

FUNCTION NAME	ABBREVIATION
Sine	sin
Cosine	cos
Tangent	tan
Cosecant	csc
Secant	sec
Cotangent	cot

WORDS	MATH
The sine of θ	$\sin\theta$
The cosine of θ	$\cos\theta$
The tangent of θ	$\tan\theta$
The cosecant of θ	$\csc\theta$
The secant of θ	$\sec\theta$
The cotangent of θ	$\cot\theta$

Sine, cosine, tangent, cotangent, secant, and cosecant are names given to specific ratios of lengths of sides of right triangles.

DEFINITION (1)   TRIGONOMETRIC FUNCTIONS

$$\sin\theta = \frac{b}{c} \qquad \cos\theta = \frac{a}{c} \qquad \tan\theta = \frac{b}{a}$$

$$\csc\theta = \frac{c}{b} \qquad \sec\theta = \frac{c}{a} \qquad \cot\theta = \frac{a}{b}$$

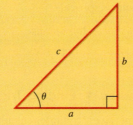

The following terminology will be used throughout this text (refer to the right triangle above):

- Side c is the **hypotenuse**.
- Side b is the side (leg) **opposite** angle θ.
- Side a is the side (leg) **adjacent** to angle θ.

Also notice that since $\sin\theta = \dfrac{b}{c}$ and $\cos\theta = \dfrac{a}{c}$, then $\tan\theta = \dfrac{\sin\theta}{\cos\theta} = \dfrac{\frac{b}{c}}{\frac{a}{c}} = \dfrac{b}{a}$.

The main three trigonometric functions should be learned in terms of the ratios.

$$\sin\theta = \frac{\text{opposite}}{\text{hypotenuse}} \qquad \cos\theta = \frac{\text{adjacent}}{\text{hypotenuse}} \qquad \tan\theta = \frac{\text{opposite}}{\text{adjacent}}$$

The remaining three trigonometric functions can be derived from these using the *reciprocal identities*. Recall that the **reciprocal** of x is $\dfrac{1}{x}$ for $x \neq 0$.

RECIPROCAL IDENTITIES

$$\csc\theta = \frac{1}{\sin\theta} \qquad \sec\theta = \frac{1}{\cos\theta} \qquad \cot\theta = \frac{1}{\tan\theta}$$

$$\theta \neq n\pi \qquad \theta \neq \frac{2n+1}{\pi} \qquad \theta \neq \frac{2n+1}{\pi}$$

It is important to note that the reciprocal identities only hold for values of θ that do not make the denominator equal to zero. Using this terminology, we have an alternate definition that is easier to remember.

DEFINITION (1) **TRIGONOMETRIC FUNCTIONS (ALTERNATE FORM)**

$$\sin\theta = \frac{\text{opposite}}{\text{hypotenuse}} \qquad \text{SOH}$$

$$\cos\theta = \frac{\text{adjacent}}{\text{hypotenuse}} \qquad \text{CAH}$$

$$\tan\theta = \frac{\text{opposite}}{\text{adjacent}} \qquad \text{TOA}$$

and their reciprocals:

$$\csc\theta = \frac{1}{\sin\theta} = \frac{\text{hypotenuse}}{\text{opposite}} \qquad \sec\theta = \frac{1}{\cos\theta} = \frac{\text{hypotenuse}}{\text{adjacent}} \qquad \cot\theta = \frac{1}{\tan\theta} = \frac{\text{adjacent}}{\text{opposite}}$$

EXAMPLE 1 Finding Trigonometric Functions of a General Angle θ.

For the given triangle, calculate $\sin\theta$, $\tan\theta$, and $\csc\theta$.

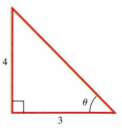

Solution:

STEP 1 Solve for the hypotenuse.

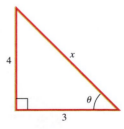

Use the Pythagorean theorem.

$$4^2 + 3^2 = x^2$$
$$16 + 9 = x^2$$
$$x^2 = 25$$
$$x = \pm 5$$

Lengths of sides can only be positive. $x = 5$

STEP 2 **Label the sides of the triangle**
- **with numbers representing lengths**
- **as hypotenuse or opposite/adjacent with respect to θ.**

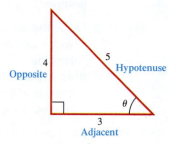

STEP 3 **Set up the trigonometric functions as ratios.**

Sine is opposite over hypotenuse.
$$\sin\theta = \frac{\text{opposite}}{\text{hypotenuse}} = \frac{4}{5}$$

Tangent is opposite over adjacent.
$$\tan\theta = \frac{\text{opposite}}{\text{adjacent}} = \frac{4}{3}$$

Cosecant is the reciprocal of sine.
$$\csc\theta = \frac{1}{\sin\theta} = \frac{1}{\frac{4}{5}} = \frac{5}{4}$$

CONCEPT CHECK   What trigonometric function can be found using cosine and reciprocal properties?

YOUR TURN   For the triangle in Example 1, calculate $\cos\theta$, $\sec\theta$, and $\cot\theta$.

EXAMPLE 2 **Finding Trigonometric Functions of a General Angle θ.**

For the given triangle, calculate $\cos\theta$, $\tan\theta$, and $\sec\theta$.

Solution:

STEP 1 **Solve for the unknown leg.**

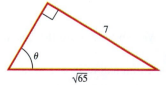

Answer: $\cos\theta = \dfrac{3}{5}$; $\sec\theta = \dfrac{5}{3}$; and $\cot\theta = \dfrac{3}{4}$

Use the Pythagorean theorem.

$$x^2 + 7^2 = (\sqrt{65})^2$$
$$x^2 + 49 = 65$$
$$x^2 = 16$$
$$x = \pm 4$$
$$x = 4$$

Lengths of sides can only be positive.

STEP 2 **Label the sides of the triangle**
- **with numbers representing lengths**
- **as hypotenuse or opposite/adjacent with respect to θ.**

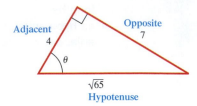

STEP 3 **Set up the trigonometric functions as ratios.**

Cosine is adjacent over hypotenuse. $\quad \cos\theta = \dfrac{\text{adjacent}}{\text{hypotenuse}} = \dfrac{4}{\sqrt{65}}$

Tangent is opposite over adjacent. $\quad \tan\theta = \dfrac{\text{opposite}}{\text{adjacent}} = \dfrac{7}{4}$

Secant is the reciprocal of cosine. $\quad \sec\theta = \dfrac{1}{\cos\theta} = \dfrac{1}{\dfrac{4}{\sqrt{65}}} = \dfrac{\sqrt{65}}{4}$

Note: In mathematics, when we come across expressions that contain a radical in the denominator, like $\dfrac{4}{\sqrt{65}}$, we rationalize the denominator by multiplying both the numerator and denominator by the radical, $\sqrt{65}$.

$$\frac{4}{\sqrt{65}} \cdot \underbrace{\left(\frac{\sqrt{65}}{\sqrt{65}}\right)}_{1} = \frac{4\sqrt{65}}{65}$$

We can now write the cosine function as $\quad \cos\theta = \dfrac{4\sqrt{65}}{65}$. In this text, denominators in expressions will always be rationalized.

✓ **CONCEPT CHECK** What trigonometric function can be found using sine and reciprocal properties?

■ **YOUR TURN** For the triangle in Example 2, calculate $\sin\theta$, $\csc\theta$, and $\cot\theta$.

■ **Answer:** $\sin\theta = \dfrac{7\sqrt{65}}{65}$, $\csc\theta = \dfrac{\sqrt{65}}{7}$, and $\cot\theta = \dfrac{4}{7}$

Cofunctions

Notice the *co* in *co*sine, *co*secant, and *co*tangent functions. These *cofunctions* are based on the relationship of *complementary* angles. Let us look at a right triangle with labeled sides and angles.

Remember that SOHCAHTOA represents ratios of sides *with respect to an angle*. In this triangle we see that

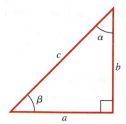

$$\left.\begin{array}{l} \sin \beta = \dfrac{\text{opposite of } \beta}{\text{hypotenuse}} = \dfrac{b}{c} \\[2mm] \cos \alpha = \dfrac{\text{adjacent to } \alpha}{\text{hypotenuse}} = \dfrac{b}{c} \end{array}\right\} \quad \sin\beta = \cos\alpha$$

Recall that the sum of the measures of angles in a triangle is 180°. In a right triangle, one angle is 90°; therefore the two acute angles are complementary angles (the measures sum to 90°). Therefore in the triangle above, β and α are *complementary* angles. In other words, the sine of an angle is the same as the *co*sine of the *co*mplement of that angle. This is true for all *trigonometric cofunctions*.

COFUNCTION THEOREM

A trigonometric function of an angle is always equal to the cofunction of the complement of the angle. If $\alpha + \beta = 90°$, then

$$\sin\beta = \cos\alpha$$

$$\sec\beta = \csc\alpha$$

$$\tan\beta = \cot\alpha$$

COFUNCTION IDENTITIES

$\sin\theta = \cos(90° - \theta)$ $\cos\theta = \sin(90° - \theta)$

$\tan\theta = \cot(90° - \theta)$ $\cot\theta = \tan(90° - \theta)$

$\sec\theta = \csc(90° - \theta)$ $\csc\theta = \sec(90° - \theta)$

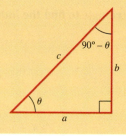

EXAMPLE 3 Writing Trigonometric Functions in Terms of Their Cofunctions

Write each function in terms of its cofunction.

a. $\sin 30°$ **b.** $\tan x$ **c.** $\csc 40°$

Solution (a):

Cosine is the cofunction of sine. $\sin\theta = \cos(90° - \theta)$

Substitute $\theta = 30°$. $\sin 30° = \cos(90° - 30°)$

Simplify. $\sin 30° = \cos 60°$

Solution (b):

Cotangent is the cofunction of tangent. $\tan\theta = \cot(90° - \theta)$

Substitute $\theta = x$. $\tan x = \cot(90° - x)$

Solution (c):

Secant is the cofunction of cosecant. $\csc\theta = \sec(90° - \theta)$

Substitute $\theta = 40°$. $\csc 40° = \sec(90° - 40°)$

Simplify. $\csc 40° = \sec 50°$

■ **YOUR TURN**   Write each function in terms of its cofunction.

    a. $\cos 45°$ **b.** $\csc y$

SECTION 1.3   SUMMARY

In this section we have defined trigonometric functions as ratios of sides of right triangles. This approach is called *right triangle trigonometry*. This is the first of three definitions of trigonometric functions (others will follow in Chapters 2 and 3). We now can find trigonometric functions of an acute angle by taking ratios of the three sides: adjacent, opposite, and hypotenuse. It is important to remember that adjacent and opposite are with respect to one of the acute angles. We learned that trigonometric functions of an angle are equal to the cofunctions of the complement to the angle.

SECTION 1.3 EXERCISES

■ SKILLS

In Exercises 1–6, use the following triangle to find the indicated trigonometric functions.

1. $\sin\theta$ **2.** $\cos\theta$

3. $\csc\theta$ **4.** $\sec\theta$

5. $\tan\theta$ **6.** $\cot\theta$

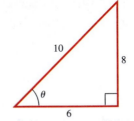

In Exercises 7–12, use the following triangle to find the indicated trigonometric functions. Rationalize any denominators containing radicals that you encounter in the answers.

7. $\cos\theta$ **8.** $\sin\theta$

9. $\sec\theta$ **10.** $\csc\theta$

11. $\tan\theta$ **12.** $\cot\theta$

■ **Answer: a.** $\sin 45°$ **b.** $\sec(90° - y)$

In Exercises 13–18, use the cofunction identities to fill in the blanks.

13. $\sin 60° = \cos$ _____

14. $\sin 45° = \cos$ _____

15. $\cos x = \sin$ _____

16. $\cot A = \tan$ _____

17. $\csc 30° = \sec$ _____

18. $\sec B = \csc$ _____

In Exercises 19–24, write the trigonometric function in terms of its cofunction.

19. $\sin(x + y)$

20. $\sin(60° - x)$

21. $\cos(20° + A)$

22. $\cos(A + B)$

23. $\cot(45° - x)$

24. $\sec(30° - \theta)$

■ **APPLICATIONS**

A man lives in a house that borders a pasture. He decides to go to the grocery store to get some milk. He is trying to decide whether to drive along the roads in his car or take his All Terrain Vehicle (ATV) across the pasture. His car drives faster than the ATV but the distance the ATV would travel is less than the distance he would travel in his car.

25. Shortcut. If $\sin \theta = \frac{3}{5}$ and $\cos \theta = \frac{4}{5}$ and if he drove his car along the streets, it would be 14 miles round trip. How far would he have to go on his ATV round trip?

26. Shortcut. If $\tan \theta = 1$ and if he drove his car along the streets, it would be 200 yards round trip. How far would he have to go on his ATV round trip? Round your answer to the nearest yard.

■ **CATCH THE MISTAKE**

In Exercises 27–30, calculate the indicated trigonometric functions for the given triangle and explain the mistake that is made.

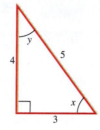

27. Calculate $\sin y$.

Solution:

Formulate sine in terms of trigonometric ratios.

$$\sin y = \frac{\text{opposite}}{\text{hypotenuse}}$$

The opposite side is 4, the hypotenuse side is 5.

$$\sin y = \frac{4}{5}$$

This is incorrect. What mistake was made?

28. Calculate $\tan x$.

Solution:

Formulate tangent in terms of trigonometric ratios.

$$\tan x = \frac{\text{adjacent}}{\text{opposite}}$$

The adjacent side is 3, the opposite side is 4.

$$\tan x = \frac{3}{4}$$

This is incorrect. What mistake was made?

29. Calculate $\sec x$.

Solution:

First, formulate sine in terms of trigonometric ratios.

$$\sin x = \frac{\text{opposite}}{\text{hypotenuse}}$$

The opposite side is 4, the hypotenuse side is 5.

$$\sin x = \frac{4}{5}$$

Write secant as the reciprocal of sine.

$$\sec x = \frac{1}{\sin x}$$

Simplify.

$$\sec x = \frac{1}{\frac{4}{5}} = \frac{5}{4}$$

This is incorrect. What mistake was made?

30. Calculate $\csc y$.

Solution:

First, formulate cosine in terms of trigonometric ratios.

$$\cos y = \frac{\text{adjacent}}{\text{hypotenuse}}$$

The adjacent side is 4, the hypotenuse side is 5.

$$\cos y = \frac{4}{5}$$

Write cosecant as the reciprocal of cosine.

$$\csc y = \frac{1}{\cos y}$$

Simplify.

$$\csc y = \frac{1}{\frac{4}{5}} = \frac{5}{4}$$

This is incorrect. What mistake was made?

■ CHALLENGE

In Exercises 31–38, use the following special triangles (30°-60°-90° and 45°-45°-90°) to help you answer the questions.

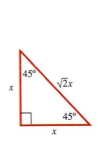
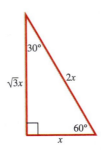

31. T or F: $\sin 45° = \cos 45°$.

32. T or F: $\sin 60° = \cos 30°$.

33. Calculate $\sin 30°$ and $\cos 30°$.

34. Calculate $\sin 60°$ and $\cos 60°$.

35. Calculate $\tan 30°$ and $\tan 60°$.

36. Calculate $\sin 45°$, $\cos 45°$, and $\tan 45°$.

37. Calculate $\sec 45°$ and $\csc 45°$.

38. Calculate $\tan 60°$ and $\cot 30°$.

SECTION 1.4 Evaluating Trigonometric Functions: Exactly and with Calculators

Skills Objectives

■ Evaluate trigonometric functions exactly for special values.
■ Evaluate trigonometric functions with calculators.
■ Represent partial degrees in either decimal degrees (DD) or degree-minute-second (DMS).

Conceptual Objective

■ Understand the difference between *exact* and *approximate* values for trigonometric functions.

In Section 1.3 we defined trigonometric functions as ratios of sides of right triangles. We even calculated what the value of a trigonometric function was for a general angle, but we haven't yet discussed the trigonometric function values for specific angles.

We now turn our attention to evaluating trigonometric functions for known angles. In this section we will distinguish between evaluating a trigonometric function *exactly* or *approximating* the value of a trigonometric function *with a calculator*. Throughout this text instructions will typically specify which is desired.

Trigonometric Function Exact Values for Special Angle Measures: 30°, 45°, and 60°

There are three special acute angles that are very important in trigonometry: 30°, 45°, and 60°. In Section 1.1 we discussed two important triangles: 30°-60°-90° and 45°-45°-90° Recall the relationships between the sides of these two right triangles.

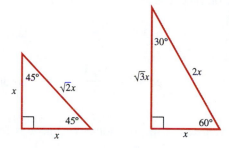

We can combine these relationships with the trigonometric ratios developed in Section 1.3 to evaluate the trigonometric functions for the special values of 30°, 45°, and 60°.

 EXAMPLE 1 Evaluating the Trigonometry Functions Exactly for 30°

Evaluate the six trigonometric functions for an angle that measures 30°.

Solution:

Label the 30°-60°-90° sides with respect to the 30° angle.

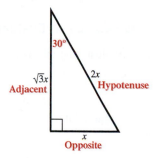

Use the trigonometric ratio definitions of sine, cosine, and tangent.

$$\sin 30° = \frac{\text{opposite}}{\text{hypotenuse}} = \frac{x}{2x} = \frac{1}{2}$$

$$\cos 30° = \frac{\text{adjacent}}{\text{hypotenuse}} = \frac{\sqrt{3}x}{2x} = \frac{\sqrt{3}}{2}$$

$$\tan 30° = \frac{\text{opposite}}{\text{adjacent}} = \frac{x}{\sqrt{3}x} = \frac{1}{\sqrt{3}} = \frac{1}{\sqrt{3}} \cdot \frac{\sqrt{3}}{\sqrt{3}} = \frac{\sqrt{3}}{3}$$

Use the reciprocal identities to obtain values for cosecant, secant, and cotangent.

$$\csc 30° = \frac{1}{\sin 30°} = \frac{1}{\frac{1}{2}} = 2$$

$$\sec 30° = \frac{1}{\cos 30°} = \frac{1}{\frac{\sqrt{3}}{2}} = \frac{2}{\sqrt{3}} = \frac{2}{\sqrt{3}} \cdot \frac{\sqrt{3}}{\sqrt{3}} = \frac{2\sqrt{3}}{3}$$

$$\cot 30° = \frac{1}{\tan 30°} = \frac{1}{\frac{\sqrt{3}}{3}} = \frac{3}{\sqrt{3}} = \frac{3}{\sqrt{3}} \cdot \frac{\sqrt{3}}{\sqrt{3}} = \sqrt{3}$$

The six trigonometric functions evaluated at 30° are:

$$\sin 30° = \frac{1}{2}$$ $$\cos 30° = \frac{\sqrt{3}}{2}$$ $$\tan 30° = \frac{\sqrt{3}}{3}$$

$$\csc 30° = 2$$ $$\sec 30° = \frac{2\sqrt{3}}{3}$$ $$\cot 30° = \sqrt{3}$$

■ **YOUR TURN** Evaluate the six trigonometric functions for an angle that measures 60°.

In comparing our answers in Example 1 and Your Turn, we see that the following cofunction relationships are true:

$$\sin 30° = \cos 60° \quad \sec 30° = \csc 60° \quad \tan 30° = \cot 60°$$
$$\sin 60° = \cos 30° \quad \sec 60° = \csc 30° \quad \tan 60° = \cot 30°$$

which is expected, since 30° and 60° are complementary angles.

EXAMPLE 2 Evaluating the Trigonometry Functions Exactly for 45°

Evaluate the six trigonometric functions for an angle that measures 45°.

Solution:

Label the 45°-45°-90° sides with respect to one of the 45° angles.

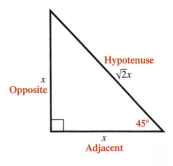

Answer: $\sin 60° = \frac{\sqrt{3}}{2}$ $\cos 60° = \frac{1}{2}$ $\tan 60° = \sqrt{3}$

$\csc 60° = \frac{2\sqrt{3}}{3}$ $\sec 60° = 2$ $\cot 60° = \frac{\sqrt{3}}{3}$

Use the trigonometric ratio definitions of sine, cosine, and tangent.

$$\sin 45° = \frac{\text{opposite}}{\text{hypotenuse}} = \frac{x}{\sqrt{2}x} = \frac{1}{\sqrt{2}} = \frac{1}{\sqrt{2}} \cdot \frac{\sqrt{2}}{\sqrt{2}} = \frac{\sqrt{2}}{2}$$

$$\cos 45° = \frac{\text{adjacent}}{\text{hypotenuse}} = \frac{x}{\sqrt{2}x} = \frac{1}{\sqrt{2}} = \frac{1}{\sqrt{2}} \cdot \frac{\sqrt{2}}{\sqrt{2}} = \frac{\sqrt{2}}{2}$$

$$\tan 45° = \frac{\text{opposite}}{\text{adjacent}} = \frac{x}{x} = 1$$

Use the reciprocal identities to obtain cosecant, secant, and cotangent.

$$\csc 45° = \frac{1}{\sin 45°} = \frac{1}{\frac{\sqrt{2}}{2}} = \frac{2}{\sqrt{2}} = \frac{2}{\sqrt{2}} \cdot \frac{\sqrt{2}}{\sqrt{2}} = \sqrt{2}$$

$$\sec 45° = \frac{1}{\cos 45°} = \frac{1}{\frac{\sqrt{2}}{2}} = \frac{2}{\sqrt{2}} = \frac{2}{\sqrt{2}} \cdot \frac{\sqrt{2}}{\sqrt{2}} = \sqrt{2}$$

$$\cot 45° = \frac{1}{\tan 45°} = \frac{1}{1} = 1$$

The six trigonometric functions evaluated at 45° are:

$$\sin 45° = \frac{\sqrt{2}}{2} \qquad \cos 45° = \frac{\sqrt{2}}{2} \qquad \tan 45° = 1$$

$$\csc 45° = \sqrt{2} \qquad \sec 45° = \sqrt{2} \qquad \cot 45° = 1$$

We see that the following cofunction relationships are true:

$$\sin 45° = \cos 45° \qquad \sec 45° = \csc 45° \qquad \tan 45° = \cot 45°$$

which is expected, since 45° and 45° are complementary angles.

The trigonometric function values for the three special angles are summarized in the following table.

Trigonometric Values for Special Angles

θ	$\sin\theta$	$\cos\theta$	$\tan\theta$	$\cot\theta$	$\sec\theta$	$\csc\theta$
30°	$\frac{1}{2}$	$\frac{\sqrt{3}}{2}$	$\frac{\sqrt{3}}{3}$	$\sqrt{3}$	$\frac{2\sqrt{3}}{3}$	2
45°	$\frac{\sqrt{2}}{2}$	$\frac{\sqrt{2}}{2}$	1	1	$\sqrt{2}$	$\sqrt{2}$
60°	$\frac{\sqrt{3}}{2}$	$\frac{1}{2}$	$\sqrt{3}$	$\frac{\sqrt{3}}{3}$	2	$\frac{2\sqrt{3}}{3}$

It is important to **learn** the special values in red for sine and cosine. All other values in the table can be found through reciprocals or quotients of these two functions. Notice that tangent is the ratio of sine to cosine.

$$\sin\theta = \frac{\text{opposite}}{\text{hypotenuse}} \quad \cos\theta = \frac{\text{adjacent}}{\text{hypotenuse}} \quad \tan\theta = \frac{\sin\theta}{\cos\theta} = \frac{\frac{\text{opposite}}{\text{hypotenuse}}}{\frac{\text{adjacent}}{\text{hypotenuse}}} = \frac{\text{opposite}}{\text{adjacent}}$$

Using Calculators to Evaluate Trigonometric Functions

We will now turn our attention to using calculators to approximate trigonometric functions for acute angles ($0° < \theta < 90°$). Scientific and graphing calculators have buttons for sine (sin), cosine (cos), and tangent (tan). Let us start with what we already know and confirm with our calculators. Before we start, make sure that your calculator is in degree mode (deg). To find sin 30°, most calculators press sin first and then 30, but others press 30 first and then sin. Consult your owner's manual.

EXAMPLE 3 Approximating Values of Trigonometric Functions with a Calculator

Use a calculator to find the values of

a. sin 75°   **b.** tan 67°   **c.** sec 52°

Round your answers to four decimal places.

Solution:

a. 0.965925826   ≈ 0.9659

b. 2.355852366   ≈ 2.3559

c. 0.615661475 $1/x$ (or x^{-1})   1.62469245   ≈ 1.6247

STUDY TIP

In calculating secant, cosecant, and cotangent function values with a calculator, it is important not to round the number until after using the reciprocal, $1/x$.

■ **YOUR TURN** Use a calculator to find the values of

a. cos 22°   **b.** tan 81°   **c.** csc 37°

Round your answers to four decimal places.

When calculating secant, cosecant, and cotangent function values with a calculator, it is important not to round the number until after using the reciprocal, $1/x$ (or x^{-1}).

Representing Partial Degrees: DD or DMS

Since one revolution is equal to 360°, an angle of 1° seems very small. However, there are times when we want to break down a degree even further.

For example, if we are off even one thousandth of a degree in pointing our antenna toward a geostationary satellite, we won't receive a signal. That is because the distance the signal travels (35,000 kilometers) is so much larger than the size of the satellite (5 meters).

■ **Answers: a.** 0.9272   **b.** 6.3137   **c.** 1.6616

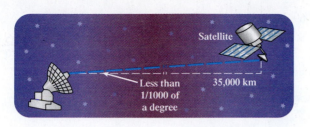

There are two traditional ways of representing part of a degree: *degree-minute-second* (DMS) and *decimal degrees* (DD). Most calculators allow you to enter angles in either DD or DMS format, and some even make the conversion between them. However, we will illustrate a manual conversion technique.

The **degree-minute-second** way of breaking down degrees is similar to how we break down time in hours-minutes-seconds. There are 60 minutes in an hour and 60 seconds in a minute, which results in an hour being broken down into 3,600 seconds. Similarly, we can think of degrees like hours. We can divide $1°$ into 60 equal parts. Each part is called a **minute** and is denoted $1'$. One minute is therefore $\frac{1}{60}$ of a degree.

$$1' = \left(\frac{1}{60}\right)° \quad \text{or} \quad 60' = 1°$$

We can then break down each minute into 60 seconds. Therefore 1 **second**, denoted $1''$, is $\frac{1}{60}$ of a minute or $\frac{1}{3600}$ of a degree.

$$1'' = \left(\frac{1}{60}\right)' = \left(\frac{1}{3600}\right)° \quad \text{or} \quad 60'' = 1'$$

The following represents how DMS expressions are stated.

WORDS	MATH
22 degrees, 5 minutes	$22°\,5'$
49 degrees, 21 minutes, 17 seconds	$49°\,21'\,17''$

We add and subtract fractional values in DMS form similar to how we add and subtract time. For example, if Carol and Donna both run the Boston Marathon and Carol crosses the finish line in 5 hours, 42 minutes, and 10 seconds and Donna crosses the finish line in 6 hours, 55 minutes, and 22 seconds, how much time elapsed between the two women crossing the finish line? The answer is 1 hour, 13 minutes, and 12 seconds. We combine hours with hours, minutes with minutes, and seconds with seconds.

EXAMPLE 4 Adding Degrees in DMS Form

Add $27°\,5'\,17''$ and $35°\,17'\,52''$.

Solution:

Align degrees with degrees, minutes with minutes, and seconds with seconds.

$$\begin{array}{rrr} 27° & 5' & 17'' \\ +\ 35° & 17' & 52'' \\ \hline \end{array}$$

Add degrees, minutes, and seconds, respectively.

$$\begin{array}{rrr} 62° & 22' & \underline{69''} \\ & & 1' + 9'' \end{array}$$

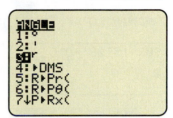

Note that 69 seconds is equal to 1 minute, 9 seconds.

Simplify.

$$62° \quad 23' \quad 9''$$

YOUR TURN Add $35°21'42''$ and $7°5'30''$.

TECHNOLOGY TIP

```
90°-15°28'▶DMS
         74°32'0"
■
```

EXAMPLE 5 Subtracting Degrees in DMS Form

Subtract $15°28'$ from $90°$.

Solution:

Align degrees with degrees and minutes with minutes.

$$\begin{array}{rr} 90° & 0' \\ - \quad 15° & 28' \end{array}$$

Borrow $1°$ from $90°$ and write it as 60 minutes.

$$\begin{array}{rr} 89° & 60' \\ - \quad 15° & 28' \end{array}$$

Subtract degrees and minutes, respectively.

$$74° \quad 32'$$

YOUR TURN Subtract $23°8'$ from $90°$.

An alternative way of representing parts of degrees is with *decimal degrees*. For example, $33.4°$ and $91.725°$ are measures of angles in **decimal degrees**. To convert from DD to DMS, multiply once by 60 to get the minutes and—if necessary—again by 60 to get the seconds. A similar two-stage reverse process is necessary for the opposite conversion of DMS to DD.

TECHNOLOGY TIP

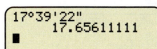

```
17°39'22"
      17.65611111
■
```

EXAMPLE 6 Converting from Degrees-Minutes-Seconds to Decimal Degrees

Convert $17°39'22''$ to decimal degrees. Round to the nearest thousandth.

Solution:

Write the number of minutes in decimal form, where $1' = \left(\dfrac{1}{60}\right)°$.

$$39' = \left(\dfrac{39}{60}\right)° \approx 0.65°$$

Write the number of seconds in decimal form, where $1'' = \left(\dfrac{1}{3600}\right)°$.

$$22'' = \left(\dfrac{22}{3600}\right)° \approx 0.0061°$$

Write the expression as a sum. $17°39'22'' = 17° + 0.65° + 0.0061°$

Add and round to the nearest $17°39'22'' ≈ 17.656°$
thousandth.

■ **YOUR TURN**   Convert $62°8'15''$ to decimal degrees. Round to the nearest
thousandth.

EXAMPLE 7 Converting from Decimal Degrees to
Degrees-Minutes-Seconds

Convert $29.538°$ to degrees, minutes, and seconds. Round to the nearest
second.

Solution:

Write as a sum. $29.538° = 29° + 0.538°$

Multiply decimal part by 60,
since $1° = 60'$. $29.538° = 29° + 0.538°\left(\dfrac{60'}{1°}\right)$

Simplify. $29.538° = 29° + 32.28'$

Write as a sum. $29.538° = 29° + 32' + 0.28'$

Multiply decimal by 60,
since $1' = 60''$. $29.538° = 29° + 32' + 0.28'\left(\dfrac{60''}{1'}\right)$

Simplify. $29.538° = 29°32'16.8''$

Round to the nearest second. $29.538° = 29°32'17''$

■ **YOUR TURN**   Convert $35.426°$ to degrees, minutes, and seconds. Round
to the nearest second.

In this text we will primarily use decimal degrees for angle measure, and we will also
use decimal approximations to trigonometric functions. A common question that arises
is how to round the decimals. There is a difference between specifying that a number be
rounded to a particular *decimal place* and specifying rounding to a certain number of
significant digits. More discussion on that topic will follow in the next section on solving right triangles. For now, we will round angle measures to the nearest minute in DMS
or the nearest hundredth in DD, and we will round trigonometric function values to four
decimal places.

TECHNOLOGY TIP

(a)

```
sin(18°10')
      .3117821943
```

(b)

```
cos(29.524°)⁻¹
      1.149227998
■
```

If the TI calculator is in degree mode and angles are entered without °, then degrees will be used.

```
cos(29.524)⁻¹
      1.149227998
```

EXAMPLE 8 Evaluating Trigonometric Functions with Calculators

Evaluate the following trigonometric functions for the specified angle measurements. Round your answers to four decimal places.

a. $\sin(18°10')$ **b.** $\sec(29.524°)$

Solution (a):

Write $18°10'$ in decimal degrees.

$$18°10' = 18° + \left(\frac{10}{60}\right)°$$

$$\approx 18° + 0.167°$$

$$\approx 18.17°$$

Use a calculator to evaluate the sine function. $\sin(18.167°) = 0.311787722$

Round to four decimal places. ≈ 0.3118

Solution (b):

Write secant as the reciprocal of cosine. $\sec(29.524°) = \dfrac{1}{\cos(29.524°)}$

Use a calculator to evaluate the expression. $\sec(29.524°) = 1.149227998$

Round to four decimal places. $\sec(29.524°) \approx 1.1492$

SECTION 1.4 **SUMMARY**

In this section you have learned the *exact* values of the sine, cosine, and tangent functions for special angle measures: 30°, 45°, and 60°. The values for the other trigonometric functions can be determined through reciprocal properties. Calculators are used to *approximate* trigonometric values of any acute angle. Degrees were broken down into smaller parts using one of two systems: decimal degrees and degrees-minutes-seconds.

SECTION 1.4 EXERCISES

 **■ SKILLS**

In Exercises 1–6, label the trigonometric function with the corresponding value.

a. $\dfrac{1}{2}$ **b.** $\dfrac{\sqrt{3}}{2}$ **c.** $\dfrac{\sqrt{2}}{2}$

1. $\sin 30°$ **2.** $\sin 60°$ **3.** $\cos 30°$ **4.** $\cos 60°$ **5.** $\sin 45°$ **6.** $\cos 45°$

For Exercises 7–9, use the results in Exercises 1–6 and the trigonometric quotient identity $\tan\theta = \dfrac{\sin\theta}{\cos\theta}$ to calculate the following values.

7. $\tan 30°$ **8.** $\tan 45°$ **9.** $\tan 60°$

For Exercises 10–18, use the results in Exercises 1–9 and the reciprocal identities $\csc\theta = \dfrac{1}{\sin\theta}$, $\sec\theta = \dfrac{1}{\cos\theta}$, and $\cot\theta = \dfrac{1}{\tan\theta}$ to calculate the following values.

10. $\csc 30°$ **11.** $\sec 30°$ **12.** $\cot 30°$ **13.** $\csc 60°$ **14.** $\sec 60°$ **15.** $\cot 60°$

16. $\csc 45°$ **17.** $\sec 45°$ **18.** $\cot 45°$

In Exercises 19–30, use a calculator to approximate the trigonometric functions for the indicated values. **Round your answers to four decimal places.**

19. $\sin 37°$ **20.** $\sin 17.8°$ **21.** $\cos 82°$ **22.** $\cos 21.9°$

23. $\tan 54°$ **24.** $\tan 43.2°$ **25.** $\sec 8°$ **26.** $\sec 75°$

27. $\csc 89°$ **28.** $\csc 51°$ **29.** $\cot 55°$ **30.** $\cot 29°$

In Exercises 31–38, perform the indicated operations using the following angles.

$$\angle A = 5°17'29'' \qquad \angle B = 63°28'35'' \qquad \angle C = 16°11'30''$$

31. $\angle A + \angle B$ **32.** $\angle A + \angle C$ **33.** $\angle B + \angle C$ **34.** $\angle B - \angle A$

35. $\angle B - \angle C$ **36.** $\angle C - \angle A$ **37.** $90° - \angle A$ **38.** $90° - \angle B$

In Exercises 39–46, convert from degrees-minutes-seconds to decimal degrees. Round to the nearest hundredth if only minutes are given and to the thousandth if seconds are given.

39. $33°20'$ **40.** $89°45'$ **41.** $59°27'$ **42.** $72°13'$

43. $27°45'15''$ **44.** $36°5'30''$ **45.** $42°28'12''$ **46.** $63°10'9''$

In Exercises 47–54, convert from decimal degrees to degrees-minutes-seconds.

47. $15.75°$ round to the nearest minute   **48.** $15.50°$ round to the nearest minute

49. $22.35°$ round to the nearest minute   **50.** $80.47°$ round to the nearest minute

51. $30.175°$ round to the nearest second   **52.** $25.258°$ round to the nearest second

53. $77.535°$ round to the nearest second   **54.** $5.995°$ round to the nearest second

In Exercises 55–60, evaluate the trigonometric functions for the indicated values. **Round your answers to four decimal places.**

55. $\sin(10°25')$ **56.** $\cos(75°13')$ **57.** $\tan(22°15')$

58. $\sec(68°22')$ **59.** $\csc(28°25'35'')$ **60.** $\sec(50°20'19'')$

 ■ APPLICATIONS

Have you ever noticed that if you put a stick in the water it looks bent? We know the stick didn't bend. Instead, the light rays bent, which made the image appear to bend. Light rays propagating from one medium (like air) to another medium (like water) experience refraction, or "bending," with respect to the surface. Light bends according to Snell's law, which states:

$$n_i \sin(\theta_i) = n_r \sin(\theta_r)$$

- n_i is the refractive index of the medium the light is leaving.
- θ_i is the incident angle between the light ray and the normal (perpendicular) to the interface between mediums.
- n_r is the refractive index of the medium the light is entering.
- θ_r is the refractive angle between the light ray and the normal (perpendicular) to the interface between mediums.

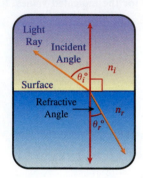

In Exercises 61–64, calculate the index of refraction, n_r, of the indicated refractive medium given the following assumptions:

- The incident medium is air.
- Air has an index of refraction value of $n_i = 1.00$.
- The incidence angle is $\theta_i = 30°$.

61. Diamond, $\theta_r = 12°$

62. Emerald, $\theta_r = 18.5°$

63. Water, $\theta_r = 22°$

64. Plastic, $\theta_r = 20°$

■ CATCH THE MISTAKE

In Exercises 65 and 66, calculate the indicated values and explain the mistake that is made.

65. $\sec 60°$

Solution:

Write secant as the reciprocal of cosine.

$$\sec\theta = \frac{1}{\cos\theta}$$

Substitute $\theta = 60°$.

$$\sec 60° = \frac{1}{\cos 60°}$$

Recall $\cos 60° = \frac{\sqrt{3}}{2}$.

$$\sec 60° = \frac{1}{\frac{\sqrt{3}}{2}}$$

Simplify.

$$\sec 60° = \frac{2}{\sqrt{3}}$$

Rationalize the denominator.

$$\sec 60° = \frac{2\sqrt{3}}{3}$$

This is incorrect. What mistake was made?

66. $\sec(36°25')$

Solution:

Convert $36°25'$ to decimal degrees.

$$\frac{25}{100} = 0.25$$

$$36°25' = 36.25°$$

Use the reciprocal identity, $\sec\theta = \frac{1}{\cos\theta}$.

$$\sec(36.25°) = \frac{1}{\cos(36.25°)}$$

Approximate with a calculator. $\sec(36.25°) \approx 1.2400$

This is incorrect. What mistake was made?

■ CHALLENGE

Thus far in this text we have only discussed trigonometric values of acute angles, $0° < \theta < 90°$. What about when θ is approximately 0° or 90°? We will formally consider these cases in the next chapter, but for now, draw and label a right triangle that has one angle very close to 0°.

In Exercises 67–70, use trigonometric ratios and the assumption that a is much larger than b to approximate the values without using a calculator.

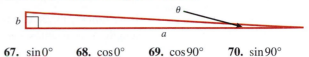

67. $\sin 0°$ **68.** $\cos 0°$ **69.** $\cos 90°$ **70.** $\sin 90°$

■ TECHNOLOGY

In Exercises 71 and 72, perform the indicated operations. Which is the more accurate way of calculating?

71. Calculate $\sec 70°$ the following two ways:

   a. Write down $\cos 70°$ (round to three decimal places), then divide 1 by that number. Write the number to five decimal places.

   b. Using a calculator: 70, cos, 1/x, round to five decimal places.

72. Calculate $\csc 40°$ the following two ways:

   a. Write down $\sin 40°$ (round to three decimal places), then divide 1 by that number. Write the number to five decimal places.

   b. Using a calculator: 40, sin, 1/x, round to five decimal places.

In Exercises 73 and 74, illustrate a calculator procedure for converting between DMS and DD.

73. Convert $3°14'25''$ to decimal degrees. Round to three decimal places.

   $25 \div 3600 = +14 \div 60 = +3 =$

74. Convert $27.683°$ to degrees-minutes-seconds. Round to three decimal places.

   $0.683 \times 60 =$ The result is a decimal in the form $A.B$

   $B \times 60 =$ seconds (round to the nearest second).

SECTION 1.5 Solving Right Triangles

Skills Objectives

■ Solve right triangles.
■ Identify correct significant digits for sides and angles.

Conceptual Objective

■ Understand the importance of significant digits in solving right triangles.

To *solve a triangle* means to find the measure of the three angles and three sides of the triangle. In this section we will only discuss right triangles (therefore we know one angle has a measure of 90°). We know some information (measures of two sides or measures of a side and an acute angle) and we determine the measures of the unknown sides and angles. However, before we start solving right triangles and determining measures, we must first discuss accuracy and significant digits.

Accuracy and Significant Digits

If we quickly measure a room to be 10 feet by 12 feet and want to calculate the diagonal length of the room, we use the Pythagorean theorem.

WORDS	MATH
Apply the Pythagorean theorem.	$10^2 + 12^2 = d^2$
Simplify.	$d^2 = 244$
Use the square root property.	$d = \pm\sqrt{244}$
The length of the diagonal must be positive.	$d = \sqrt{244}$
Approximate the radical with a calculator.	$d \approx 15.62049935$

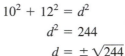
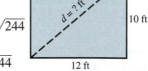

10 ft

12 ft

$d = ?$ ft

STUDY TIP

The least accurate number in your calculation determines the accuracy of your result.

Would you say that the 10 foot by 12 foot room has a diagonal of 15.62049935 feet? No, because the known room measurements were only given with an accuracy of a foot and the diagonal above is calculated to eight decimal places. *Your results are no more accurate than the least accurate number in your calculation.* In this example we round to the nearest foot, and hence we say that the diagonal of the 10 foot by 12 foot room is about 16 feet.

Significant digits are used to determine the precision of a measurement.

DEFINITION SIGNIFICANT DIGITS

The number of significant digits in a number is found by counting all of the digits from left to right starting with the first nonzero digit.

NUMBER	SIGNIFICANT DIGITS
0.04	1
0.276	3
0.2076	4
1.23	3
17	2
17.00	4
17.000	5
6.25	3
8000	?

The reason for the question mark next to 8000 is that we don't know. If 8000 is a result from rounding to the nearest thousand, then it has one significant digit. If 8000 is the result of rounding to the nearest ten, then it has three significant digits, and if there are 8000 people surveyed then 8000 is an exact value and it has four significant digits. In this text we will assume that integers have the greatest number of significant digits. Therefore, 8000 has four significant digits.

Values obtained from trigonometric functions are almost always approximations, and so it is important to understand the accuracy relationship between side lengths and angle measures.

Angle to Nearest	Significant Digits for Side Measure
1°	two
10′ or 0.1°	three
1′ or 0.01°	four
10″ or 0.001°	five

Solving Right Triangles

In solving right triangles, we first determine which of the given measurements has the *least* number of significant digits and round our answers to the same number of significant digits.

EXAMPLE 1 Solving a Right Triangle Given an Angle and a Side

Given the triangle to the right, solve the right triangle—find a, b, and θ.

Solution:

STEP 1 Determine accuracy.

56° is rounded to the nearest degree (corresponds to two significant digits).

15 feet is accurate to two significant digits.

Answers: Round angles to the nearest degree and sides to two significant digits.

STEP 2 Solve for θ.

Two acute angles in a right triangle are complementary.

$$\theta + 56° = 90°$$

Solve for θ.

$$\theta = 34°$$

The answer is already rounded to the nearest degree.

STEP 3 Solve for a.

The cosine of an angle is adjacent over the hypotenuse.

$$\cos 56° = \frac{a}{15}$$

Solve for a.

$$a = 15 \cos 56°$$

Evaluate the right side of the expression using a calculator.

$$a \approx 8.38789$$

Round a to two significant digits of accuracy.

$$a = 8.4 \text{ feet}$$

TECHNOLOGY TIP

Using a TI calculator to find $15\cos(56)$ and $15\sin(56)$, press

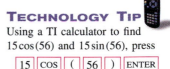

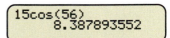

Using a scientific calculator, press

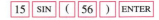

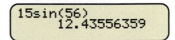

STEP 4 Solve for *b*.

Notice that there are two ways to solve for *b*: trigonometric functions or the Pythagorean theorem. Although it is tempting to use the Pythagorean theorem, it is better to use the given information with trigonometric functions than to use a value that has already been rounded, which could make results less accurate.

The sine of an angle is opposite over the hypotenuse.

$$\sin 56° = \frac{b}{15}$$

Solve for *b*.

$$b = 15\sin 56°$$

Evaluate the right side of the expression using a calculator.

$$b \approx 12.43556$$

Round *b* to two significant digits of accuracy.

$$b = 12 \text{ feet}$$

STEP 5 Check the solution.

Angles are rounded to the nearest degree and sides are rounded to two significant digits of accuracy.

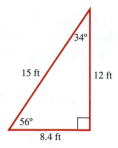

Check trigonometric values of specific angles with trigonometric ratio values.

$$\sin 34° \overset{?}{\equiv} \frac{8.4}{15} \qquad \cos 34° \overset{?}{\equiv} \frac{12}{15} \qquad \tan 34° \overset{?}{\equiv} \frac{8.4}{12}$$

$$0.5529 \approx 0.56 \qquad 0.8290 \approx 0.8 \qquad 0.6745 \approx 0.7$$

■ **YOUR TURN**   Given the triangle below, solve the right triangle—find *a*, *b*, and *θ*.

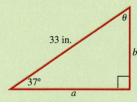

Note: If an angle is given in degrees-minutes-seconds, first convert it to decimal degrees.

In Example 1, an angle and a side are given. In Example 2, two sides are given. Your calculator has three inverse trigonometric buttons—$\sin^{-1}$, $\cos^{-1}$, and $\tan^{-1}$—which are used when you are given the value of the trigonometric function and seek the angle. For example, we know that $\sin 30° = \frac{1}{2} = 0.5$. If we know the trigonometric value, 0.5, we can use the inverse button to find the angle, $\sin^{-1} 0.5$, which yields the result of 30°. We will discuss inverse trigonometric functions in more detail in Chapter 6, but for now we will use the corresponding buttons on calculators to find angles when trigonometric values are known.

It is important to note that $\sin^{-1}\theta \neq \dfrac{1}{\sin\theta}$. We use the -1 exponent to indicate the inverse function. For example, if we want the cosecant function, which is the reciprocal of sine, we use the following procedure:

$$\csc 22° = \dfrac{1}{\sin 22°} \qquad \text{Keystrokes: } \sin(22) = 1/x = \qquad \text{Result: real number}$$

However, if we want the inverse sine function, we use the $\sin^{-1}$ button.

$$\sin^{-1} 0.6 \qquad \text{Keystrokes: } \sin^{-1}(0.6) = \qquad \text{Result: degrees}$$

When using the inverse trigonometric buttons, don't forget that the result is the measure of an angle in degrees.

EXAMPLE 2 Solving a Right Triangle Given Two Sides

Given the triangle to the right, solve the right triangle—find a, α, and β.

Solution:

STEP 1 Determine accuracy.

The given sides are accurate to four significant digits.

Answers: Round angles to the nearest hundredth of a degree, 0.01°.

STEP 2 Solve for α.

The cosine of an angle is adjacent over the hypotenuse.

$$\cos\alpha = \dfrac{19.67 \text{ cm}}{37.21 \text{ cm}}$$

Evaluate the right side with a calculator. $\qquad \cos\alpha = 0.528621338$

Write the angle in terms of the inverse function.

$$\alpha = \cos^{-1} 0.528621338$$

Use a calculator to evaluate the inverse. $\qquad \alpha = 58.08764854°$

Round α to the nearest hundredth of a degree. $\qquad \boxed{\alpha \approx 58.09°}$

TECHNOLOGY TIP

```
19.67/37.21
        .5286213383
cos⁻¹(Ans)
        58.08764854
```

STEP 3 Solve for β.

Two acute angles in a right triangle are complementary. $\qquad \alpha + \beta = 90°$

Substitute $\alpha = 58.09°$. $\qquad 58.09 + \beta = 90°$

Solve for β. $\qquad \boxed{\beta = 31.91°}$

The answer is already rounded to the nearest hundredth of a degree.

STEP 4 Solve for a.

Use the Pythagorean theorem. $\qquad a^2 + b^2 = c^2$

Substitute given values for b and c. $\qquad a^2 + 19.67^2 = 37.21^2$

Solve for a. $\qquad a = 31.5859969$

Round a to four significant digits of accuracy. $\qquad \boxed{a = 31.59 \text{ cm}}$

Step 5 **Check the solution.**

Angles are rounded to the nearest hundredth degree, and sides are rounded to four significant digits of accuracy.

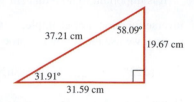

Check trigonometric values of specific angles with trigonometric ratio values.

$$\sin 31.91° \stackrel{?}{=} \frac{19.67}{37.21} \qquad \sin 58.09° \stackrel{?}{=} \frac{31.59}{37.21}$$
$$0.5286 \approx 0.5286 \qquad\qquad 0.8489 \approx 0.84896$$

■ **Your Turn**  Given the triangle shown, solve the right triangle— find a, α, and β.

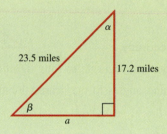

Applications

In many applications of solving right triangles, you are given a side and an acute angle and asked to find one of the other sides. Two common examples involve an observer (or point of reference) located on the horizontal and an object that is either above or below the horizontal. If the object is above the horizontal, then the angle made is called **angle of elevation**, and if the object is below the horizontal, then the angle made is called the **angle of depression**.

For example, if a race car driver is looking straight ahead (horizontal line of sight), then looking up is elevation and looking down is depression.

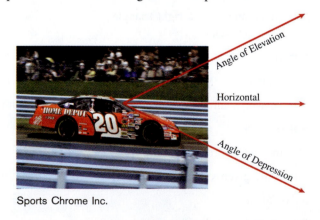

Sports Chrome Inc.

■ **Answer:** $a = 16.0$ miles, $\alpha = 43.0°$, and $\beta = 47.0°$

If the angle is a physical one (like a skateboard ramp) then the appropriate name is **angle of inclination**.

Angle of Inclination

EXAMPLE 3 Angle of Depression (NASCAR)

In this picture, Michael Waltrip (#15) is behind Dale Earnhardt Jr. (#8). If the angle of depression is 18°, and Michael gets so close that he can only see the 3 foot back end of Dale Jr.'s car, how far apart are their bumpers? Assume that the horizontal distance from Michael Waltrip's eyes to the front of his car is 5 feet.

Stephen M. Dowell/Orlando Sentinel Archives

Solution:

Draw the right triangle and label the known quantities.

Identify the tangent ratio.

$$\tan 18° = \frac{3}{x}$$

Solve for x.

$$x = \frac{3}{\tan 18°}$$

Evaluate the right side.

$$x = 9.233 \text{ ft}$$

Round to the nearest foot.

$$x = 9 \text{ ft}$$

Subtract 5 feet from x.

$$9 - 5 = \boxed{4 \text{ ft}}$$

Their bumpers are 4 feet apart.

Suppose NASA wants to talk with astronauts on the International Space Station (ISS), which is traveling at a speed of 17,700 mph, 400 kilometers above the surface of the Earth. If the antennas at the ground station in Houston have a pointing error of even one minute, 1′, they will miss the chance to talk with the astronauts.

EXAMPLE 4 Pointing Error

Assume that the ISS (which is 108 meters long and 73 meters wide) is in a 400 kilometer low Earth orbit. If the communications antennas have a 1 minute pointing error, how many meters off will the communications link be?

Solution:

Draw a right triangle that depicts this scenario.

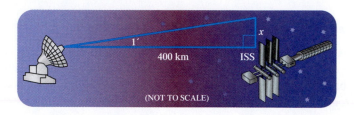

(NOT TO SCALE)

TECHNOLOGY TIP

```
0°1'
         .0166666667
400tan(Ans)
         .1163552867
```

Identify the tangent ratio.

$$\tan 1' = \frac{x}{400}$$

Solve for x.

$$x = (400 \text{ km}) \tan 1'$$

Convert $1'$ to decimal degrees.

$$1' = \left(\frac{1}{60}\right)^{\circ} = 0.016666667^{\circ}$$

Convert the equation for x in terms of decimal degrees.

$$x = (400 \text{ km}) \tan 0.016666667^{\circ}$$

Evaluate the expression on the right.

$$x = 0.116355 \text{ km}$$

400 kilometers is accurate to three significant digits, and $1'$ corresponds to four significant digits, so we express the answer to three significant digits.

The pointing error causes the signal to be off by 116 meters. Since ISS is only 108 meters long, it is expected that the signal will be missed by the astronaut crew.

Direction is often given as the measure of an acute angle with respect to the north–south vertical line. "The plane has a **bearing** N 20° E" means that the plane is bearing 20° to the east of due north.

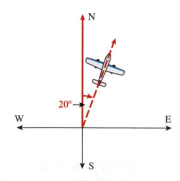

EXAMPLE 5 Bearing (Navigation)

A jet takes off bearing (N 28° E), flies 5 miles, then makes a left (90°) turn and flies 12 miles. If the control tower operator wanted to locate the plane, what bearing would she use?

Solution:

Draw a picture that represents this scenario.

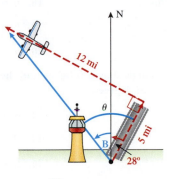

Identify the tangent ratio. $\tan\theta = \dfrac{12}{5}$

Use the inverse tangent function to solve for θ. $\theta = \tan^{-1}\left(\dfrac{12}{5}\right) \approx 67.4°$

Subtract 28° from θ to find the bearing. $B = 67.4° - 28° = 39.4°$

Round to the nearest degree. $\boxed{B = \text{N } 39° \text{ W}}$

SECTION 1.5 SUMMARY

In this section we have solved right triangles. When either a side length and an acute angle measure are given or when two side lengths are given, it is possible to solve the right triangle (find all unknown side and angle measurements). The least accurate number used in your calculations determines the accuracy of the results.

SECTION 1.5 EXERCISES

 ■ **SKILLS**

In Exercises 1–20, use the right triangle diagram below and the information given to find the indicated measure. Write your answers for angle measures in decimal degrees.

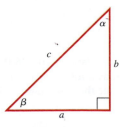

1. $\beta = 35°$, $c = 17$ inches, find a.

2. $\beta = 35°$, $c = 17$ inches, find b.

3. $\alpha = 55°$, $c = 22$ feet, find a.

4. $\alpha = 55°$, $c = 22$ feet, find b.

5. $\alpha = 20.5°$, $b = 14.7$ miles, find a.

6. $\beta = 69.3°$, $a = 0.752$ miles, find b.

7. $\beta = 25°$, $a = 11$ km, find c.

8. $\beta = 75°$, $b = 26$ km, find c.

9. $\alpha = 48.25°$, $a = 15.37$ cm, find c.

10. $\alpha = 29.80°$, $b = 16.79$ cm, find c.

11. $a = 29$ mm, $c = 38$ mm, find α.

12. $a = 89$ mm, $c = 99$ mm, find β.

13. $b = 2.3$ meters, $c = 4.9$ meters, find α.

14. $b = 7.8$ meters, $c = 13$ meters, find β

15. $\alpha = 21°17'$, $b = 210.8$ yards, find a.

16. $\beta = 27°21'$, $a = 117.0$ yards, find b.

17. $\beta = 15°20'$, $a = 10.2$ km, find c.

18. $\beta = 65°30'$, $b = 18.6$ km, find c.

19. $\alpha = 40°28'10''$, $a = 12{,}522$ km, find c.

20. $\alpha = 28°32'50''$, $b = 17{,}986$ km, find c.

In Exercises 21–32, use the right triangle diagram below and the information given to solve the right triangle. Write your answers for angle measures in decimal degrees.

21. $\alpha = 32°$ and $c = 12$ feet

22. $\alpha = 65°$ and $c = 37$ feet

23. $\beta = 72°$ and $c = 9.7$ mm

24. $\beta = 45°$ and $c = 7.8$ mm

25. $\alpha = 54.2°$ and $a = 111$ miles

26. $\beta = 47.2°$ and $a = 9.75$ miles

27. $\alpha = 28°23'$ and $b = 1734$ feet

28. $\alpha = 72°59'$ and $a = 2175$ feet

29. $a = 42.5$ feet and $b = 28.7$ feet

30. $a = 19.8$ feet and $c = 48.7$ feet

31. $a = 35{,}236$ km and $c = 42{,}766$ km

32. $b = 0.1245$ mm and $c = 0.8763$ mm

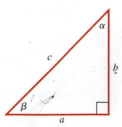

■ **APPLICATIONS**

5 ft

(NOT TO SCALE)

1°

33. **Golf.*** If the flagpole that a golfer aims at on a green measures 5 feet from the ground to the top of the flag, and a golfer measures a 1 degree angle from top to bottom, how far (horizontal distance) is the golfer from the flag?

34. **Golf.*** If the flagpole that a golfer aims at on a green measures 5 feet from the ground to the top of the flag, and a golfer measures a 3 degree angle from top to bottom, how far (horizontal distance) is the golfer from the flag?

Exercises 35 and 36 illustrate a mid-air refueling scenario that military aircraft often use. Assume the elevation angle that the hose makes with the plane being fueled is $\theta = 36°$.

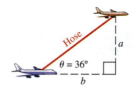

Hose

a

$\theta = 36°$

b

35. **Mid-Air Refueling.** If the hose is 150 feet long, what should be the altitude difference, a, between the two planes?

36. **Mid-Air Refueling.** If the smallest acceptable altitude difference, a, between the two planes is 100 feet, how long should the hose be?

---

*Exercises 33 and 34 are courtesy of Mr. Charles Mathers, Department of Mathematics, Northampton Community College.

Exercises 37–40 use the idea of a glide slope (the angle the flight path makes with the ground).

Precision Approach Path Indicator (PAPI) lights are used as a visual approach slope aid for pilots landing aircraft. Typical glide path for commercial jet airliners is 3°. The space shuttle has an outer glide approach of 18–20°. PAPI lights typically are a row of four lights. All four lights are on, but they have different combinations of red or white. If all four lights are white, then the angle of descent is too high, if all four lights are red, then the angle of descent is too low, and if there are two white and two red, then the approach is perfect.

37. **Glide Path of Commercial Jet Airliner.** If a commercial jetliner is 5000 feet (about 1 mile) ground distance from the runway, what should be the altitude of the plane to achieve 2 red/2 white PAPI lights? (Assume that this corresponds to a 3° glide path.)

38. **Glide Path of Commercial Jet Airliner.** If a commercial jetliner has an altitude of 450 feet when it is 5200 feet from the runway (approximately 1 mile ground distance), what is the glide slope angle? Will the pilot see white lights, red lights, or both?

39. **Glide Path of Space Shuttle Orbiter.** If the pilot of the space shuttle orbiter is at an altitude of 3000 feet when she is 15,500 (approximately 3 miles) from the shuttle landing facility, what is her glide slope angle? Is she too high or too low?

40. **Glide Path of Space Shuttle Orbiter.** If the same pilot in Exercise 39 raises the nose of the gliding shuttle so that she only drops 500 feet by the time she is 7800 feet (approximately 1.5 miles ground distance) from the shuttle landing strip, what is her glide angle then? Is she within the specs (18°−20°) to land the shuttle?

In Exercises 41 and 42, use the illustration below that shows a search and rescue helicopter with a 30° field of view with a search light.

41. **Search and Rescue.** If the search and rescue helicopter is flying at an altitude of 150 feet above sea level, what is

the diameter of the circle that is illuminated on the surface of the water?

42. **Search and Rescue.** If the search and rescue helicopter is flying at an altitude of 500 feet above sea level, what is the diameter of the circle that is illuminated on the surface of the water?

For Exercises 43–46, refer to the following:

Geostationary orbits are useful because they cause a satellite to appear stationary with respect to a fixed point on the rotating Earth. As a result, an antenna (dish TV) can point in a fixed direction and maintain a link with the satellite. The satellite orbits in the direction of the Earth's rotation, at an altitude of approximately 35,000 kilometers.

43. **Dish TV.** If your dish TV antenna has a pointing error of 1″ (second), how long would the satellite have to be in order to maintain a link? Round your answer to the nearest meter.

44. **Dish TV.** If your dish TV antenna has a pointing error of $\frac{1}{2}$″ (half a second), how long would the satellite have to be in order to maintain a link? Round your answer to the nearest meter.

45. **Dish TV.** If the satellite in a geostationary orbit (35,000 km) was only 10 meters long, about how accurately pointed would the dish have to be? Give the answer in degrees to two significant digits.

46. **Dish TV.** If the satellite in a geostationary orbit (35,000 km) was only 30 meters long, about how accurately pointed would the dish have to be? Give the answer in degrees to two significant digits.

47. **Angle of Elevation (Traffic).** A person driving in a sedan is driving too close to the back of an 18 wheeler on an interstate highway. He decides to back off until he can see the entire truck (at the top). If the height of the trailer is 15 feet, and the sedan driver's angle of elevation is roughly 30°, how far is he sitting from the end of the trailer?

48. **Angle of Depression (Opera).** The balcony seats at the opera house have an angle of depression of 55°. If the horizontal (ground) distance to the center of the stage is 50 feet, how far are the patrons in the balcony to the singer at center stage?

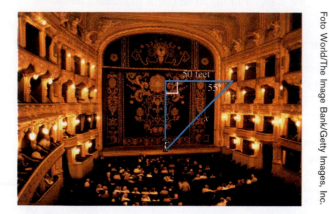

49. **Angle of Inclination (Skiing).** The angle of inclination of a mountain with triple black diamond ski paths is 65°. If a skier at the top of the mountain is at an elevation of 4000 feet, how long is the ski run from the top to the base of the mountain?

50. **Bearing (Navigation).** If a plane takes off bearing N 33° W and flies 6 miles, then makes a right (90°) turn and flies 10 miles, what bearing will the traffic controller use to locate the plane?

For Exercises 51 and 52, refer to the following:

With the advent of new technology, tennis racquets can now be constructed to permit a player to serve at speeds in excess of 120 mph (as demonstrated by Andy Roddick and Pete Sampras, to name just two). One of the most effective serves in tennis is a power serve that is hit at top speed directly at the top left corner of the right service court (or top right corner of the left service court). When attempting this serve, a player will toss the ball rather high into the air, bring the racquet back, and then make contact with the ball at the precise moment when the position of the ball in the air coincides with the top of the netted part of the racquet when the player's arm is fully stretched over his or her head.

51. **Tennis.*** Assume that the player is serving into the right service court, and stands just *two inches* to the right of the center line behind the base line. If, at the moment the racquet strikes the ball, both are *72 inches from the ground* and the serve actually hits the top left corner of the right service court, determine the angle at which the ball meets the ground in the right service court.

52. **Tennis.*** Assume that the player is serving into the right service court, and stands just *two inches* to the right of the center line behind the base line. If the ball hits the top left corner of the right service court at an angle of *44°*, at what height above the ground must the ball be struck?

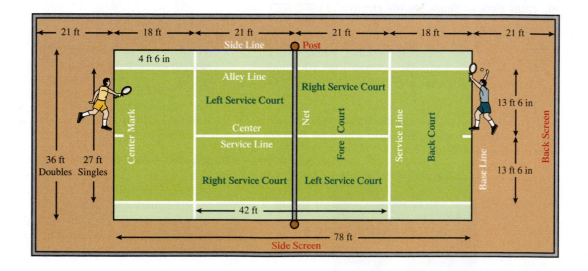

---

*Exercises 51 and 52 are courtesy of Dr. Mark McKibben, Department of Mathematics, Goucher College.

For Exercises 53 and 54, refer to the following:

The structures of molecules is critical to the study of materials science and organic chemistry, and has countless applications to a variety of interesting phenomena, including gemstones, catalysts, and hemoglobin. Trigonometry plays a critical role in determining bonding angles of molecules. For instance, the structure of the $(FeCl_4Br_2)^{-3}$ ion (dibromatetetrachlorideferrate III) is shown in the figure below.

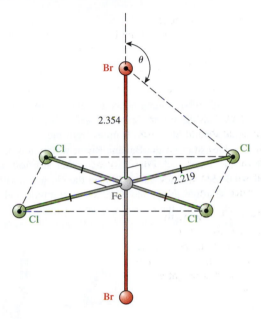

53. Chemistry.* Determine the angle θ (i.e., the angle between the axis containing the *apical atom* bromide (Br) and the segment connecting Br to Cl).

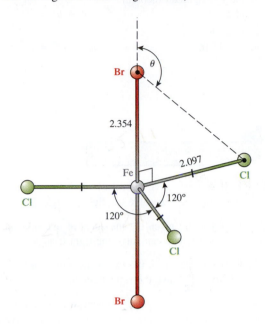

54. Chemistry.* Now, suppose one of the chlorides (Cl) is removed. The resulting structure is triagonal in nature, resulting in the following structure. Does the angle θ change? If so, what is its new value?

 ■ CATCH THE MISTAKE

In Exercises 55 and 56, use the right triangle diagram below and explain the mistake that is made.

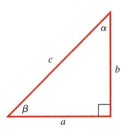

55. If $b = 800$ ft and $a = 10$ ft, find β.

Solution:

Represent tangent as opposite over adjacent.

$$\tan \beta = \frac{b}{a}$$

Substitute $b = 800$ ft and $a = 10$ ft.

$$\tan \beta = \frac{800}{10}$$

Use a calculator to evaluate β.

$$\beta = \tan 80 = 5.67°$$

This is incorrect. What mistake was made?

56. If $\beta = 56°$ and $c = 15$ ft, find b and then find a.

Solution:

Write sine as opposite over hypotenuse.

$$\sin 56° = \frac{b}{15}$$

Solve for b.

$$b = 15 \sin 56°$$

Use a calculator to approximate b.

$$b = 12.4356$$

*Exercises 53 and 54 are courtesy of Dr. Mark McKibben, Department of Mathematics, Goucher College.

Round the answer to two significant digits.

$$b = 12 \text{ ft}$$

Use the Pythagorean theorem to find a.

$$a^2 + b^2 = c^2$$

Substitute $b = 12$ ft and $c = 15$ ft.

$$a^2 + 12^2 = 15^2$$

Solve for a. $\qquad a = 9$

Round the answer to two significant digits. $\qquad a = 9.0$ ft

Compare this with the results from Example 1. Why did we get a different value for a here?

■ **CHALLENGE**

57. T or F: If you are given the measures of two sides of a right triangle, you can solve the right triangle.

58. T or F: If you are given the measures of one side and one acute angle, you can solve the right triangle.

59. T or F: If you are given the two acute angles of a right triangle, you can solve the right triangle.

60. T or F: If you are given the hypotenuse of a right triangle and the angle opposite the hypotenuse, you can solve the right triangle.

61. Use the information in the picture below to determine the height of the mountain.

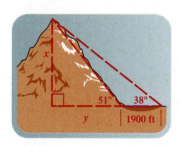

62. Two friends who are engineers at Kennedy Space Center (KSC) watch the shuttle launch. Carolyn is at the Vehicle Assembly Building (VAB) 3 miles from the launch pad and Jackie is across the Banana River, which is 8 miles from the launch pad. They call each other at liftoff, and after 10 seconds they each estimate the elevation with the ground. Carolyn thinks the elevation is approximately 40° and Jackie thinks the elevation is approximately 15°. Approximately how high is the shuttle after 10 seconds?

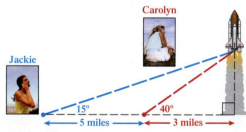

Image Bank/Getty Images, Inc.
Photonica/Getty Images

■ **TECHNOLOGY**

63. Use a calculator to find $\sin^{-1}(\sin 40°)$ by pressing $40 \sin \sin^{-1}$.

64. Use a calculator to find $\cos^{-1}(\cos 17°)$ by pressing $17 \cos \cos^{-1}$.

65. Use a calculator to find $\cos(\cos^{-1}.8)$ by pressing $.8 \cos^{-1} \cos$.

66. Use a calculator to find $\sin(\sin^{-1}.3)$ by pressing $.3 \sin^{-1} \sin$.

67. Based on the result from Exercise 63, what would $\sin^{-1}(\sin\theta)$ be for an acute angle θ?

68. Based on the result from Exercise 64, what would $\cos^{-1}(\cos\theta)$ be for an acute angle θ?

Q: Two computers in an office are connected to the same printer. Each computer is 30 feet away from the printer. Nate wants to connect the two computers to each other and needs to calculate how much cable will be necessary. He knows that 30 feet of cable already join each computer to the printer, and that the measure of the angles α and β is 40 degrees. Calling the distance between the computers x, he uses the following calculations to determine that 36 feet of cable are needed:

$$\tan 40° = \frac{30}{x}$$

$$x \tan 40° = 30$$

$$x = \frac{30}{\tan 40°} \approx 36 \text{ ft}$$

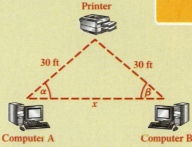

However, the cable is too short. What is wrong with Nate's calculations?

A: Nate used $\tan 40°$ even though the triangle is not a right triangle. Divide the triangle into two right triangles, and use trigonometric functions to find the distance between the computers.

TYING IT ALL TOGETHER

Ted manages a department store that is doing some renovations. The solid lines in the diagram below represent current walls in the store. (*Note:* The drawing is not necessarily to scale.) He would like to place a partition where the dashed line is so that the store can have a separate section for women's jewelry that will be inside the right triangle below. However, he is having trouble taking measurements in that area because racks and shelves are in his way. Fortunately, angles α and γ are along a walkway so he is easily able to measure angle α as 51° and angle γ as 129°. The side of the triangle that is adjacent to angle β is also along a walkway, so he is able to measure it as 22 feet. The area of what is currently the women's department (which includes the triangle) is about 1920 square feet. Ted has about $7,000 in his budget to build the partition. How long will the partition need to be?

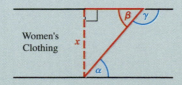

SECTION	TOPIC	PAGES	REVIEW EXERCISES	KEY CONCEPTS
1.1	Angles, degrees, and triangles	4–13	1–18	Terminal Side / Positive Angle / Initial Side — Initial Side / Negative Angle / Terminal Side
	Degree measure	4–6	17 and 18	One complete rotation = 360° — $\theta = 90°$ Right Angle: quarter rotation — Acute Angle $0° < \theta < 90°$
	Complementary and supplementary angles	6	1–4	Complementary Angles $\alpha + \beta = 90°$ — Supplementary Angles $\alpha + \beta = 180°$
	Triangles	7–8	5–8	$\alpha + \beta + \gamma = 180°$
	Right triangles: Pythagorean theorem	8–9	9–12	$a^2 + b^2 = c^2$

SECTION	TOPIC	PAGES	REVIEW EXERCISES	KEY CONCEPTS
	Special triangles	9–13	13–16	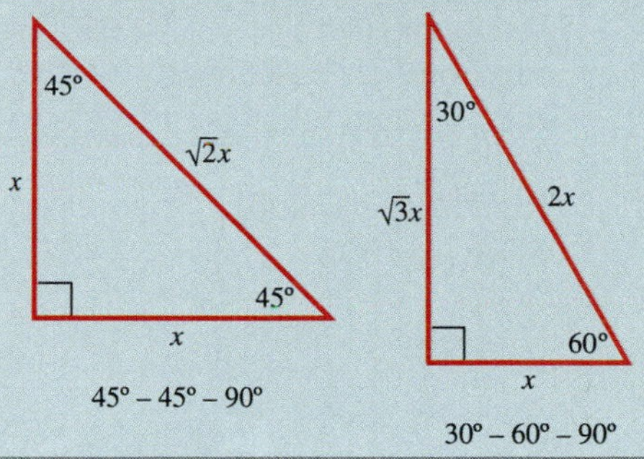
	Applications	10–13	17 and 18	
1.2	Similar triangles	17–24	19–32	
	Angle relationships	17–20	19–24	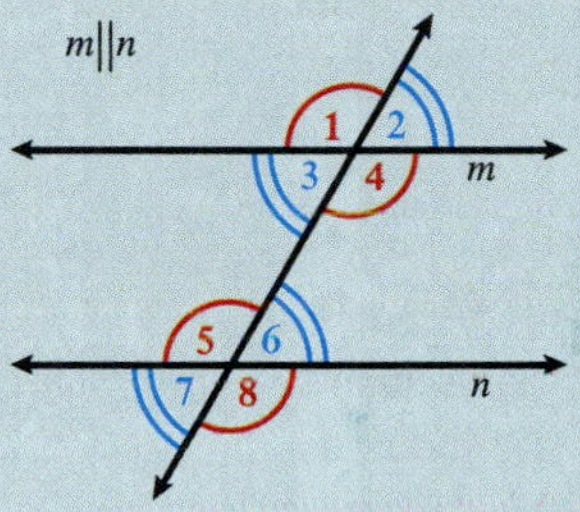 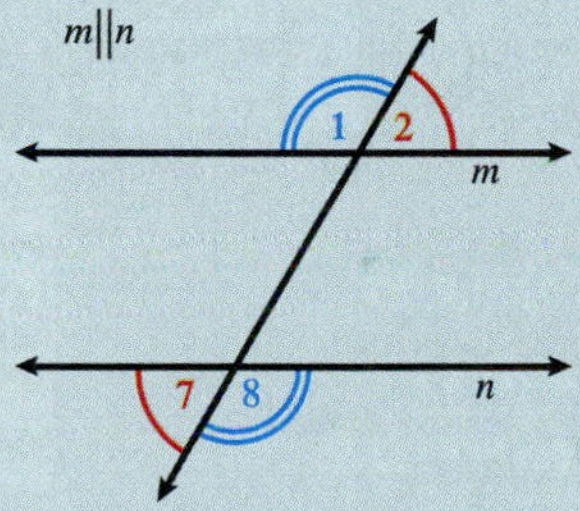
	Similar triangles	20–23	25–28	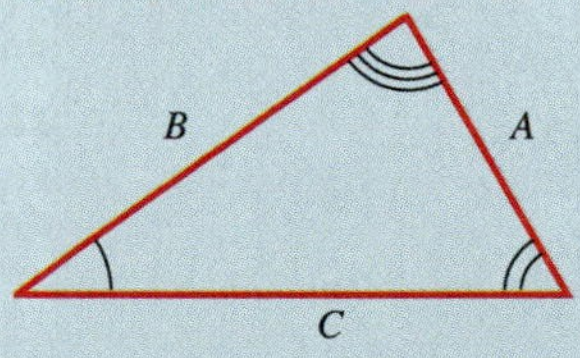 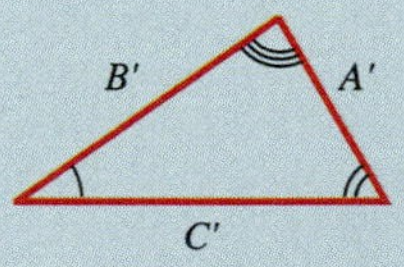
	Applications	23–24	29–32	

$45° - 45° - 90°$

$30° - 60° - 90°$

$m \| n$

$\dfrac{A}{A'} = \dfrac{B}{B'} = \dfrac{C}{C'}$

SECTION	TOPIC	PAGES	REVIEW EXERCISES	KEY CONCEPTS
1.3	Definition 1 of trigonometric functions: right triangle ratios	27–34	33–46	Definition 1 defines trigonometric functions of acute angles as ratios of sides in a right triangle.
	Right triangle ratios: SOHCAHTOA	28–32	33–38	SOH $\sin\theta = \dfrac{\text{opposite}}{\text{hypotenuse}}$ CAH $\cos\theta = \dfrac{\text{adjacent}}{\text{hypotenuse}}$ TOA $\tan\theta = \dfrac{\text{opposite}}{\text{adjacent}}$
	Reciprocal identities	30	35–38	$\cot\theta = \dfrac{1}{\tan\theta}$ $\csc\theta = \dfrac{1}{\sin\theta} \qquad \sec\theta = \dfrac{1}{\cos\theta}$ *c* and *s* go together
	Cofunctions	33–34	39–46	If $\alpha + \beta = 90°$: $\begin{aligned}\sin\alpha &= \cos\beta\\ \sec\alpha &= \csc\beta\\ \tan\alpha &= \cot\beta\end{aligned}$
1.4	Evaluating trigonometric functions: exactly and with calculators	36–44	47–82	
	Exact values for special angles	37–39	47–64	(see table below)

θ	$\sin\theta$	$\cos\theta$
30°	$\dfrac{1}{2}$	$\dfrac{\sqrt{3}}{2}$
45°	$\dfrac{\sqrt{2}}{2}$	$\dfrac{\sqrt{2}}{2}$
60°	$\dfrac{\sqrt{3}}{2}$	$\dfrac{1}{2}$

The other trigonometric functions can be found for these values using

$$\tan\theta = \frac{\sin\theta}{\cos\theta}$$ and reciprocal identities.

SECTION	TOPIC	PAGES	REVIEW EXERCISES	KEY CONCEPTS
	Approximate trigonometric values with a calculator	40	65–72	Make sure the calculator is in degrees mode. Sin, Cos, and Tan buttons can be combined with the reciprocal button 1/x to get Csc, Sec, and Cot.
	Partial degrees: DD or DMS	40–44	73–80	$1' = \left(\dfrac{1}{60}\right)^{\circ}$ or $60' = 1°$ $1'' = \left(\dfrac{1}{60}\right)' = \left(\dfrac{1}{3600}\right)^{\circ}$ or $60'' = 1'$ DMS $\rightarrow$ DD: Divide by multiples of 60 DD $\rightarrow$ DMS: Multiply by multiples of 60
	Applications	40–41	81 and 82	
1.5	Solving right triangles	47–55	83–96	

ANGLE TO NEAREST	SIGNIFICANT DIGITS FOR SIDE MEASURE
1°	two
10' or 0.1°	three
1' or 0.01°	four
10'' or 0.001°	five

	Given a side and an acute angle or given two sides	47–52	83–94	Always use given measurements if possible for accurate results.
	Applications	52–55	95 and 96	

CHAPTER 1 REVIEW EXERCISES

1.1 Angles, Degrees, and Triangles

Find (a) the complement and (b) the supplement of the given angles.

1. 28°   2. 17°   3. 35°   4. 78°

Refer to the following triangle.

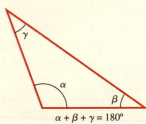

5. If $\alpha = 120°$ and $\beta = 35°$, find γ.

6. If $\alpha = 105°$ and $\beta = 25°$, find γ.

7. If $\gamma = \beta$ and $\alpha = 7\beta$, find all three angles.

8. If $\gamma = \beta$ and $\alpha = 6\beta$, find all three angles.

Refer to the following right triangle.

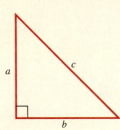

9. If $a = 4$ and $c = 12$, find b.

10. If $b = 9$ and $c = 15$, find a.

11. If $a = 7$ and $b = 4$, find c.

12. If $a = 10$ and $b = 8$, find c.

Refer to the following 45°-45°-90° triangle.

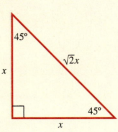

13. If the two legs have length 12 yards, how long is the hypotenuse?

14. If the hypotenuse has length $\sqrt{8}$ feet, how long are the legs?

Refer to the following 30°-60°-90° triangle.

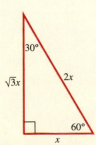

15. If the shorter leg has length 3 feet, what are the lengths of the other leg and the hypotenuse?

16. If the hypotenuse has length 12 kilometers, what are the lengths of the two legs?

Applications

17. **Clock.** What is the measure (in degrees) that the minute hand makes in exactly 25 minutes?

18. **Clock.** What is the measure (in degrees) that the second hand makes in exactly 15 seconds?

1.2 Similar Triangles

Find the measure of the indicated angle.

19. $\angle F = 75°$, find $\angle G$.

20. $\angle F = 75°$, find $\angle D$.

21. $\angle F = 75°$, find $\angle C$.

22. $\angle F = 75°$, find $\angle E$.

23. $\angle F = 75°$, find $\angle B$.

24. $\angle F = 75°$, find $\angle A$.

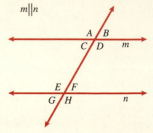

Calculate the specified lengths given that the two triangles are similar.

25. $A = 10$, $C = 8$, $D = 5$, $F = ?$

26. $A = 15$, $B = 12$, $E = 4$, $D = ?$

27. $D = 4.5$ m, $F = 8.2$ m, $A = 81$ km, $C = ?$

28. $E = 8$ cm, $F = 14$ cm, $C = 8$ m, $B = ?$

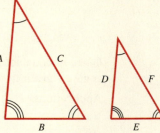

Applications

29. **Height of a Tree.** The shadow of a tree measures 9.6 meters. At the same time of day, the shadow of a 4 meter basketball backboard measures 1.2 meters. How tall is the tree?

30. **Height of a Man.** If an NBA center casts a 1 foot 9 inch shadow, and his 4 foot son casts a 1 foot shadow, how tall is the NBA center?

In a home remodeling project, your architect gives you plans that have an indicated distance of 3 feet measuring 1 inch with a ruler.

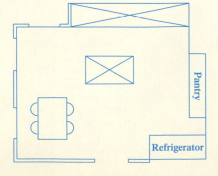

31. Home Renovation. How wide is the built-in refrigerator if it measures $1\frac{1}{3}$ inches with a ruler?

32. Home Renovation. How wide is the pantry if it measures $1\frac{1}{4}$ inches with a ruler?

1.3 Definition 1 of Trigonometric Functions: Right Triangle Ratios

Use the following triangle to find the indicated trigonometric functions. Rationalize any denominators that you encounter in your answers.

33. $\cos\theta$

34. $\sin\theta$

35. $\sec\theta$

36. $\csc\theta$

37. $\tan\theta$

38. $\cot\theta$

Use the cofunction identities to fill in the blanks.

39. $\sin 30° = \cos\underline{\hspace{2cm}}$

40. $\cos A = \sin\underline{\hspace{2cm}}$

41. $\tan 45° = \cot\underline{\hspace{2cm}}$

42. $\csc 60° = \sec\underline{\hspace{2cm}}$

Write the trigonometric function in terms of its cofunction.

43. $\sin(30° - x)$

44. $\cos(55° + A)$

45. $\csc(45° - x)$

46. $\sec(60° - \theta)$

1.4 Evaluating Trigonometric Functions: Exactly and with Calculators

Label the trigonometric function with the corresponding value.

a. $\dfrac{\sqrt{3}}{2}$ **b.** $\dfrac{1}{2}$ **c.** $\dfrac{\sqrt{2}}{2}$

47. $\sin 30°$ **48.** $\cos 30°$ **49.** $\cos 60°$

50. $\sin 60°$ **51.** $\sin 45°$ **52.** $\cos 45°$

Use the results in Exercises 47–52 and the trigonometric quotient identity $\tan\theta = \dfrac{\sin\theta}{\cos\theta}$ to calculate the following values.

53. $\tan 30°$ **54.** $\tan 45°$ **55.** $\tan 60°$

Use the results in Exercises 47–55 and the reciprocal identities $\csc\theta = \dfrac{1}{\sin\theta}$, $\sec\theta = \dfrac{1}{\cos\theta}$, and $\cot\theta = \dfrac{1}{\tan\theta}$ to calculate the following values.

56. $\csc 30°$ **57.** $\csc 45°$ **58.** $\csc 60°$

59. $\sec 30°$ **60.** $\sec 45°$ **61.** $\sec 60°$

62. $\cot 30°$ **63.** $\cot 45°$ **64.** $\cot 60°$

Use a calculator to approximate the trigonometric functions for the indicated values. Round answers to four decimal places.

65. $\sin 42°$ **66.** $\cos 57°$ **67.** $\cos 17.3°$

68. $\tan 25.2°$ **69.** $\cot 33°$ **70.** $\sec 16.8°$

71. $\csc 40.25°$ **72.** $\cot 19.76°$

Convert from degrees-minutes-seconds to decimal degrees. Round to the nearest hundredth if only minutes are given and to the thousandth if seconds are given.

73. $39° 17'$ **74.** $68° 15'$

75. $29° 30' 25''$ **76.** $25° 45' 15''$

Convert from decimal degrees to degrees-minutes-seconds.

77. $42.25°$ round to the nearest minute

78. $60.45°$ round to the nearest minute

79. $30.175°$ round to the nearest second

80. $25.258°$ round to the nearest second

Applications

Light bends according to Snell's law, which states:

$$n_i\sin(\theta_i) = n_r\sin(\theta_r)$$

- n_i is the refractive index of the medium the light is leaving.

- θ_i is the incident angle between the light ray and the normal (perpendicular) to the interface between mediums.

- n_r is the refractive index of the medium the light is entering.

- θ_r is the refractive angle between the light ray and the normal (perpendicular) to the interface between mediums.

Calculate the index of refraction, n_r, of the indicated refractive medium given the following assumptions:

■ The incident medium is air.

   ■ Air has an index of refraction value of $n_i = 1.00$.

   ■ The incidence angle is $\theta_i = 60°$.

81. Optics. Glass, $\theta_r = 35.26°$

82. Optics. Glycerin, $\theta_r = 36.09°$

1.5 Solving Right Triangles

Use the right triangle diagram below and the information given to find the indicated measure. Write your answers for angle measures in decimal degrees.

83. $\beta = 25°$, $c = 15$ in., find a.

84. $\alpha = 50°$, $c = 27$ ft, find a.

85. $\alpha = 33.5°$, $b = 21.9$ miles, find a.

86. $\alpha = 47.45°$, $a = 19.22$ cm, find c.

87. $\beta = 37°45'$, $a = 120.0$ yards, find b.

88. $\beta = 75°10'$, $b = 96.5$ km, find c.

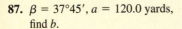
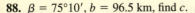

Use the right triangle diagram below and the information given to solve the right triangle. Write your answers for angle measures in decimal degrees.

89. $\alpha = 30°$ and $c = 21$ ft

90. $\beta = 65°$ and $c = 8.5$ mm

91. $\alpha = 48.5°$ and $a = 215$ miles

92. $\alpha = 30°15'$ and $b = 2154$ ft

93. $a = 30.5$ ft and $b = 45.7$ ft

94. $a = 11,798$ km and $c = 32,525$ km

Applications

Illustrate a mid-air refueling scenario that our military aircraft often use. Assume the elevation angle that the hose makes with the plane being fueled is $\theta = 32°$.

95. Mid-Air Refueling. If the hose is 150 feet long, what should the altitude difference, a, be between the two planes?

96. Mid-Air Refueling. If the smallest acceptable altitude difference, a, between the two planes is 100 feet, how long should the hose be?

1. Calculate the measure of three angles in a triangle if the following are true:

   a. The measure of the largest angle is 5 times the measure of the smallest angle.

   b. The larger of the two acute angles is 3 times the measure of the smallest angle.

2. In a right triangle, if the side opposite a 30° angle has a length of 5 cm, what is the length of the other leg and the hypotenuse?

3. A 5 foot girl is standing *in* the Grand Canyon, and she wants to estimate the height (depth) of the canyon. Her shadow is 6 inches. To measure the shadow cast by the top of the canyon, she walks the length of the shadow. She takes 200 steps and estimates that each step is roughly 3 feet. Approximately how tall is the Grand Canyon?

For Exercises 4 and 5, use the following triangle.

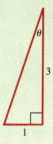

4. Find the exact values for the indicated functions.

   a. $\sin\theta$ b. $\cos\theta$ c. $\tan\theta$

   d. $\sec\theta$ e. $\csc\theta$ f. $\cot\theta$

5. Find the exact values for the indicated functions.

   a. $\sin(90 - \theta)$ b. $\sec(90 - \theta)$ c. $\cot(90 - \theta)$

6. Fill in the values in the table:

θ	$\sin\theta$	$\cos\theta$	$\tan\theta$	$\cot\theta$	$\sec\theta$	$\csc\theta$
30°						
45°						
60°						

7. Use a calculator to approximate $\sec 42.8°$. Round your answer to four decimal places.

8. What is the difference between $\cos\theta = \frac{2}{3}$ and $\cos\theta = 0.66667$?

9. Convert $33°45'20''$ to decimal degrees. Round to the appropriate decimal place.

For Exercises 10 and 11, refer to the following triangle.

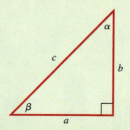

10. If $\alpha = 20°$ and $a = 10$ cm, find b.

11. If $a = 9.2$ km and $c = 23$ km, find β.

12. What is the measure (in degrees) that a second hand makes in 5 seconds?

13. Light going from air to quartz crystal appears to bend according to Snell's law: $n_i \sin(\theta_i) = n_r \sin(\theta_r)$, where air has an index of refraction value of $n_i = 1.00$. If the incidence angle is 25°, $\theta_i = 25°$, and the refraction angle in the quartz is $\theta_r = 16°$, what is the index of refraction of quartz crystal?

14. If the search and rescue helicopter has a field of view of 40° and is flying at an altitude of 150 feet above sea level, what is the diameter of the circle that is illuminated on the surface of the water?

CHAPTER 2

Trigonometric Functions

Courtesy MGM Home Entertainment Distribution Corp.

One of the greatest basketball stories in history is immortalized in the movie *Hoosiers.* Gene Hackman stars as a coach of a men's basketball team from a tiny high school in the farmlands of Indiana in the 1950s that rose above all odds to win the state basketball championship. Their hometown gymnasium was very small. When the players arrived at the monstrous Hinkle Fieldhouse to play the state championship, they were in awe of the size of the arena. The coach made them measure the dimensions of the court and height of the rim. To their amazement, the measurements were the same as those of their court back home. It was as if their high school court had been picked up and placed in the large arena.

We will do the same thing with right triangles. We will superimpose them onto the Cartesian plane, which will allow us to arrive at a second definition of trigonometric functions in terms of ratios of *x* and *y* coordinates and distances. This new definition will allow us to find values of trigonometric functions for nonacute angles. Although the Cartesian plane might seem like the Hinkle Fieldhouse, remember that the hometown court, right triangles, lie within it.

In this chapter we will superimpose right triangles onto the Cartesian plane, which will allow us to define trigonometric functions as ratios of x and y coordinates. We will then be able to find values of trigonometric functions for both positive and negative angles as well as angles greater than or equal to 90° (nonacute angles). Finally, we will review and develop some basic trigonometric identities.

Trigonometric Functions

Cartesian Plane

- Angles
- Trigonometric Functions
  - Acute angles
  - Nonacute angles

Signs and Ranges of Trigonometric Functions

- Signs of Trigonometric Functions in Each Quadrant
- Ranges of Trigonometric Functions

Basic Trigonometric Identities

- Pythagorean
- Reciprocal
- Quotient

CHAPTER OBJECTIVES

- Draw angles in the Cartesian plane.
- Define trigonometric functions as ratios of x and y coordinates and distances in the Cartesian plane.
- Evaluate trigonometric functions for nonacute angles.
- Determine ranges for trigonometric functions and signs for trigonometric functions in each quadrant.
- Derive and use basic trigonometric identities.

NAVIGATION THROUGH SUPPLEMENTS

DIGITAL VIDEO SERIES #2

STUDENT SOLUTIONS MANUAL CHAPTER 2

BOOK COMPANION SITE
www.wiley.com/college/young

Skills Objectives	Conceptual Objectives
■ Plot angles in standard position. ■ Identify coterminal angles. ■ Graph common angles.	■ Relate the x and y coordinates to the legs of a right triangle. ■ Derive the distance formula from the Pythagorean theorem.

In Section 1.1 we introduced angles. A common unit of measure for angles is degrees. We discussed triangles and the fact that the sum of the measures of the three angles is always 180°. We also discussed the Pythagorean theorem, which relates the lengths of the three sides of a *right* triangle. Based on what we learned in Section 1.1, the following two angles, which correspond to a right angle or quarter rotation, both have a measurement of 90°.

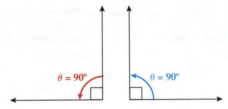

STUDY TIP

For a review of the rectangular (Cartesian) coordinate system see Appendix A.1.

We need a frame of reference. In this section we use the Cartesian plane as our frame of reference by superimposing angles onto the *Cartesian coordinate system*, or *rectangular coordinate system* (Appendix A.1).

Angles in Standard Position

Let us now bridge two things together that you already know. First, recall that an angle is found when a ray (*initial side*) is rotated around its endpoint (*vertex*). The ray after it is rotated is called the *terminal side*. Also recall the Cartesian (rectangular) coordinate system with the *x*-axis, *y*-axis, and origin. Using the positive *x*-axis combined with the origin as a frame of reference, we can graph angles in the Cartesian plane. If the *initial side* of the angle is aligned along the *positive x-axis* and the *vertex* of the angle is positioned at the *origin*, then the angle is said to be in *standard position*.

DEFINITION   STANDARD POSITION

An angle is said to be in **standard position** if its initial side is along the positive *x*-axis and its vertex is at the origin.

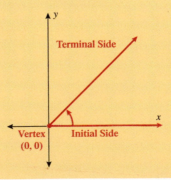

We say that an angle lies in the quadrant in which its terminal side lies. For example, an acute angle $(0° < \theta < 90°)$ lies in **quadrant I**, whereas an obtuse angle $(90° < \theta < 180°)$ lies in **quadrant II**.

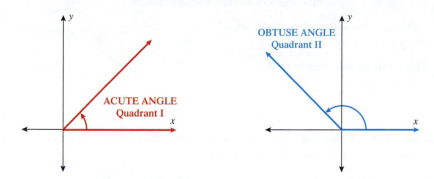

Similarly, angles with measure $180° < \theta < 270°$ lie in **quadrant III**, and angles with measure $270° < \theta < 360°$ lie in **quadrant IV**. Angles in standard position with terminal sides along the x-axis or y-axis (90°, 180°, 270°, or 360°) are called **quadrantal angles**. An abbreviated way to represent an angle, θ, that lies in quadrant I is $\theta \in$ QI. Similarly, if an angle lies in quadrant II we say $\theta \in$ QII, and so forth.

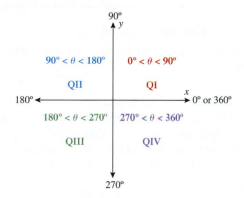

Recall that rotation in a counterclockwise direction corresponds to a *positive angle*, whereas rotation in a clockwise direction corresponds to a *negative angle*.

EXAMPLE 1 Sketching Angles in Standard Position

Sketch the following angles in standard position, and state the quadrant in (or axis on) which the terminal side lies.

a. −90°   **b.** 210°

Solution (a):

The initial side lies on the x-axis.

The negative angle indicates clockwise rotation.

90° is a right angle.

The terminal side lies on the y-axis.

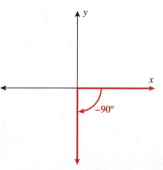

Solution (b):

The initial side lies on the *x*-axis.

The positive angle indicates counterclockwise rotation.

180° represents a straight angle, an additional 30° yields a 210° angle.

The terminal side lies in QIII.

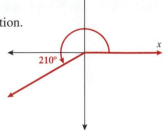

CONCEPT CHECK What quadrant would you expect the terminal side of a 135° angle to lie in if it is in standard position?

YOUR TURN Sketch the following angles in standard position, and state the quadrant in which the terminal side lies.

a. −300°   **b.** 135°

Coterminal Angles

DEFINITION **COTERMINAL ANGLES**

Two angles in standard position with the same terminal side are called **coterminal angles**.

For example, −40° and 320° are coterminal; their terminal rays are identical even though they are formed by rotation in opposite directions. The angles 60° and 420° are also coterminal; angles larger than 360° or less than −360° are generated by continuing the rotation beyond a full rotation. Thus, all coterminal angles have the same initial side (positive *x*-axis) and the same terminal side, just different rotations.

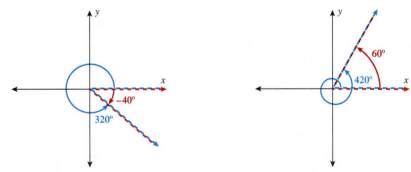

The terminal side lies in QII.   The terminal side lies in QI.

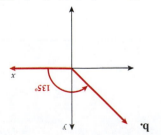

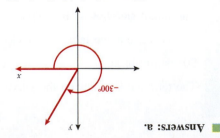

Answers: a.

EXAMPLE 2 Recognizing Coterminal Angles

Determine if the following pairs of angles are coterminal.

a. $\alpha = 120°, \beta = -180°$ **b.** $\alpha = 20°, \beta = 740°$

Solution (a):

The terminal side of α is in QII.

The terminal side of β is along the x-axis.

α and β are   not coterminal angles .

Solution (b):

Since 360° represents one rotation, 720° represents two rotations. Therefore, after rotating 720° the angle is again along the positive x-axis. Rotating an additional 20° achieves a 740° angle. Since $\alpha = 20°$ and $\beta = 740°$ have the same terminal side, they are   coterminal angles .

YOUR TURN   Determine if the following pairs of angles are coterminal.

a. $\alpha = 240°, \beta = -120°$ **b.** $\alpha = 20°, \beta = -380°$

To find measures of coterminal angles, if the given angle is positive, subtract 360° repeatedly until the result is a positive angle less than or equal to 360°. If the given angle is negative, add 360° repeatedly until the result is a positive angle less than or equal to 360°.

EXAMPLE 3 Finding Measures of Coterminal Angles

Determine the angles of the smallest possible positive measure that are coterminal with the following angles.

a. 830°   **b.** −520°

Solution (a):

Since 830° is positive, subtract 360°. $830° - 360° = 470°$

Subtract 360° again. $470° - 360° = 110°$

The angle with measure   110°   is the angle with the smallest positive measure that is coterminal with the angle with measure 830°.

Solution (b):

Since −520° is negative, add 360°. $-520° + 360° = -160°$

Add 360° again. $-160° + 360° = 200°$

The angle with measure   200°   is the angle with the smallest positive measure that is coterminal with the angle with measure −520°.

YOUR TURN   Determine the angles of the smallest possible positive measure that are coterminal with the following angles.

a. 900°   **b.** −430°

Common Angles in Standard Position

The common angles for which we determined exact values for trigonometric functions in Chapter 1 are 30°, 45°, and 60°. Recall the relationships between the sides of 30°-60°-90° and 45°-45°-90° triangles.

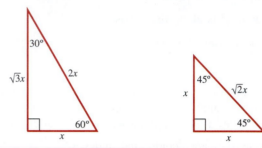

Let us assume the hypotenuse is equal to 1. Then we have the following triangles:

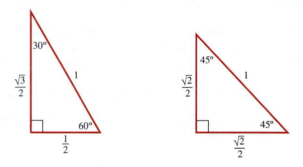

We can position these triangles on the Cartesian plane, so that we have three angles (30°, 45°, and 60°) in standard position. Notice that the x and y coordinates correspond to the side lengths.

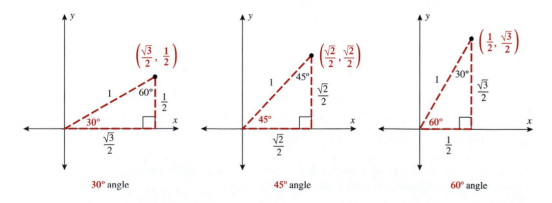

| 30° angle | 45° angle | 60° angle |

If we graph the three angles on the same Cartesian coordinate system, we get the following in the first quadrant:

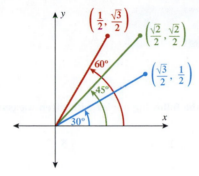

Using symmetry (Appendix A.2) and the angles and coordinates in quadrant 1 (QI), we get the following angles and coordinates in QII, QIII, and QIV. Notice that all of these coordinate pairs satisfy the equation of a unit circle (radius equal to 1): $x^2 + y^2 = 1$.

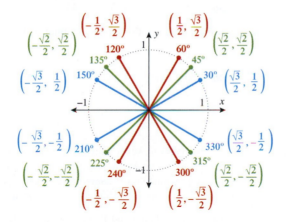

Section 2.1 SUMMARY

In this section we classified an angle as being in standard position if its initial side lies along the positive x-axis and its vertex is located at the origin. Angles in standard position have terminal sides that lie either in one of the four quadrants or along one of the two axes. Coterminal angles are angles that have the same initial side and the same terminal side. The special triangles, 30°-60°-90° and 45°-45°-90°, with hypotenuses having measure 1, were used to develop the coordinates in QI for the special angles 30°, 45°, and 60°. Symmetry was then used to locate similar pairs of coordinates in the other quadrants. All right triangles with one vertex located at the origin and hypotenuse equal to 1 have another vertex that lies on the unit circle.

SECTION 2.1 EXERCISES

■ SKILLS

In Exercises 1–20, state in what quadrant or on what axes the following angles with given measure in standard position would lie.

1. 89°	**2.** 91°	**3.** 145°	**4.** 175°	**5.** 310°
6. 355°	**7.** 270°	**8.** 180°	**9.** −540°	**10.** −450°
11. 210.5°	**12.** 270.5°	**13.** 12.34°	**14.** 100.001°	**15.** 595°
16. 620°	**17.** 525°	**18.** 1085°	**19.** −905°	**20.** −640°

In Exercises 21–32, sketch the angles with given measure in standard position.

21. 135°	**22.** 225°	**23.** −405°	**24.** −450°	**25.** −225°	**26.** −330°
27. 330°	**28.** −150°	**29.** 510°	**30.** −720°	**31.** 840°	**32.** −380°

In Exercises 33–38, match the angles with coterminal angles.

    a. 30°   **b.** −95°   **c.** 185°   **d.** −560°   **e.** 780°   **f.** 75°

33. −535°	**34.** −690°	**35.** 60°	**36.** 265°	**37.** −645°	**38.** 160°

In Exercises 39–46, determine the angles of the smallest possible positive measure that are coterminal with the following angles.

39. 412°	**40.** 379°	**41.** −92°	**42.** −187°
43. −390°	**44.** 945°	**45.** 510°	**46.** 1395°

■ APPLICATIONS

47. Clock. What is the measure of the angle swept out by the second hand if it starts on the 3 and continues for 3 minutes and 20 seconds?

48. Clock. What is the measure of the angle swept out by the hour hand if it starts at 3 P.M. on Wednesday and continues until 5 P.M. on Thursday.

49. Rodeo. Jake (who is left handed) and Blake (who is right handed) decide to have a roping competition. Each one stands on a designated spot while the other releases a steer. Jake goes first, and his lasso makes a −930° angle before the steer is looped by the rope. Blake follows, and his lasso makes a 1230° angle before the steer is looped by the rope. Are these angles coterminal? Was the steer in approximately the same place when roped by both boys? Assume the ropes are approximately the same length.

50. Rodeo. Emma and Eva decide to do the same competition their twin brothers Jake and Blake did (Exercise 49). The girls decide to mark off four fields (similar to QI, QII, QIII, and QIV, with the spot where they stand located at the origin). If Emma's lasso makes a −1000° angle before the steer is looped, what "quadrant" is the steer in when roped? If Eva's lasso makes a 1000° angle before the steer is looped, what "quadrant" is the steer in when roped?

51. Track. Don and Ron both started running around a circular track, starting at the same point. But Don ran counterclockwise and Ron ran clockwise. The paths they ran swept through angles of 900° and −900°, respectively. Did they end up in the same spot when they finished?

52. Track. Dan and Stan both started running around a circular track, starting at the same point. The paths they ran swept through angles of 3640° and 1890°, respectively. Did they end up in the same spot when they finished?

For Exercises 53 and 54, refer to the following:

A common school locker combination lock is shown. The lock has a dial with 40 calibration marks numbered 0 to 39. A combination consists of 3 of these numbers (e.g., 5-35-20). To open the lock, the following steps are taken:

iStockphoto

- Turn the dial clockwise 2 full turns.

- Stop at the first number of the combination.

- Turn the dial counterclockwise one full turn.

- Continue turning counterclockwise until the 2nd number is reached.

- Turn the dial clockwise again until the 3rd number is reached.

- Pull the shank and the lock will open.

53. Combination Lock. Given that the initial position of the dial is at zero (shown in the picture), how many degrees is the dial rotated in total (sum of clockwise and counterclockwise rotations) in opening the lock if the combination is 35-5-20?

54. Combination Lock. Given that the initial position of the dial is at zero (shown in the picture), how many degrees is the dial rotated in total (sum of clockwise and counterclockwise rotations) in opening the lock if the combination is 20-15-5?

■ CATCH THE MISTAKE

In Exercises 55 and 56, explain the mistake that is made.

55. Find the angle with smallest positive measure that is coterminal with the angle with measure $-45°$. Assume that both angles are in standard position.

Solution:

Coterminal angles are complementary angles. $-45° + \alpha = 180°$

Add $45°$ to both sides. $\alpha = 225°$

This is incorrect. What mistake was made?

56. In what quadrant (or axes) does the terminal side of an angle in standard position lie if the angle has measure $-1950°$?

Solution:

Keep adding $360°$.

$$-1950° + 360° = -1590°$$
$$-1590° + 360° = -1230°$$
$$-1230° + 360° = -870°$$
$$-870° + 360° = -510°$$
$$-510° + 360° = -150°$$

$-150°$ lies in QII.

This is incorrect. What mistake was made?

■ CHALLENGE

57. T or F: The terminal sides of two coterminal angles must lie in the same quadrant or on the same axes.

58. T or F: An acute angle in standard position and an obtuse angle in standard position cannot be coterminal.

59. T or F: If the measures of two angles are $n°$ and $(-n)°$, then the angles are coterminal.

60. T or F: The difference in measure between two positive coterminal angles must be a multiple of $360°$.

61. Write an expression that represents all angles with (a) positive measure or (b) negative measure that are coterminal with an angle that has measure $30°$.

62. Find the measure of the angle in standard position with all of the following characteristics:

- Negative measure
- Coterminal with the supplement of an angle with measure $130°$
- Less than one rotation

SECTION 2.2 Definition 2 of Trigonometric Functions: Cartesian Plane

In Chapter 1 we defined trigonometric functions as ratios of side lengths of right triangles. This definition only holds for acute ($0° < \theta < 90°$) angles, since the two angles in a right triangle other than the right angle must be acute. In this chapter, we define trigonometric functions as ratios of x and y coordinates and distances in the Cartesian plane, which is consistent with right triangle trigonometry for acute angles. However, this second approach enables us to formulate trigonometric functions for quadrantal angles (terminal side lies along an axis) and nonacute angles.

Trigonometric Functions: Cartesian Plane

To define the trigonometric functions in the Cartesian plane, let us start with an acute angle, θ, in standard position. Choose any point (x, y) on the terminal side of the angle as long as it is not the vertex (origin).

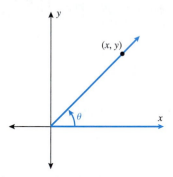

A right triangle can be drawn so that the right angle is made when a perpendicular segment connects the point (x, y) to the x-axis. Notice that the side opposite θ has length y and the other leg of the triangle has length x.

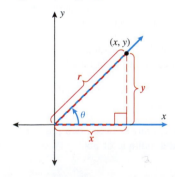

The distance, **r**, from the origin, $(0, 0)$, to the point (x, y) can be found using the distance formula (Appendix A.1):

$$r = \sqrt{(x - 0)^2 + (y - 0)^2}$$
$$r = \sqrt{x^2 + y^2}$$

Since r is a distance, it is always positive. $\quad r > 0$

Using our first definition of trigonometric functions in terms of right triangle ratios (Section 1.3), we say that $\sin\theta = \dfrac{\text{opposite}}{\text{hypotenuse}}$. From this picture we see that sine can also be defined as $\sin\theta = \dfrac{y}{r}$. Similar reasoning holds for all six trigonometric functions and leads us to the second definition of the trigonometric functions in terms of ratios of coordinates and distances in the Cartesian plane.

DEFINITION (2)   TRIGONOMETRIC FUNCTIONS

Let (x, y) be a point other than the origin on the terminal side of an angle θ in standard position. Let r be the distance from the point (x, y) to the origin; then the six trigonometric functions are defined as

$$\sin\theta = \frac{y}{r} \qquad\qquad \cos\theta = \frac{x}{r} \qquad\qquad \tan\theta = \frac{y}{x} \quad (x \neq 0)$$

$$\csc\theta = \frac{r}{y} \quad (y \neq 0) \qquad \sec\theta = \frac{r}{x} \quad (x \neq 0) \qquad \cot\theta = \frac{x}{y} \quad (y \neq 0)$$

where $r = \sqrt{x^2 + y^2}$, or $x^2 + y^2 = r^2$. The distance, r, is positive: $r > 0$.

EXAMPLE 1 Calculating Trigonometric Function Values for Acute Angles

The terminal side of an angle, θ, in standard position passes through the point $(2, 5)$. Calculate the values of the six trigonometric functions for angle θ.

Solution:

STEP **1** Draw the angle and label the point $(2, 5)$.

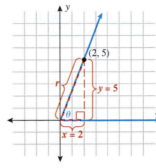

STEP **2** Calculate the distance, **r**. $\qquad r = \sqrt{2^2 + 5^2} = \sqrt{29}$

STEP **3** Formulate the trigonometric functions in terms of **x**, **y**, and **r**.
Let $x = 2, y = 5, r = \sqrt{29}$.

$$\sin\theta = \frac{y}{r} = \frac{5}{\sqrt{29}} \qquad \cos\theta = \frac{x}{r} = \frac{2}{\sqrt{29}} \qquad \tan\theta = \frac{y}{x} = \frac{5}{2}$$

$$\csc\theta = \frac{r}{y} = \frac{\sqrt{29}}{5} \qquad \sec\theta = \frac{r}{x} = \frac{\sqrt{29}}{2} \qquad \cot\theta = \frac{x}{y} = \frac{2}{5}$$

STEP 4 Rationalize radical denominators in the sine and cosine functions.

$$\sin\theta = \frac{5}{\sqrt{29}} \cdot \frac{\sqrt{29}}{\sqrt{29}} = \frac{5\sqrt{29}}{29} \qquad \cos\theta = \frac{2}{\sqrt{29}} \cdot \frac{\sqrt{29}}{\sqrt{29}} = \frac{2\sqrt{29}}{29}$$

STEP 5 Write the values of the six trigonometric functions for θ.

$\sin\theta = \dfrac{5\sqrt{29}}{29}$	$\cos\theta = \dfrac{2\sqrt{29}}{29}$	$\tan\theta = \dfrac{5}{2}$
$\csc\theta = \dfrac{\sqrt{29}}{5}$	$\sec\theta = \dfrac{\sqrt{29}}{2}$	$\cot\theta = \dfrac{2}{5}$

STUDY TIP

There is no need to memorize definitions for secant, cosecant, and cotangent functions, since they can be derived from reciprocals of sine, cosine, and tangent functions.

Note: In Example 1, we could have used the definitions of sine, cosine, and tangent functions along with the reciprocal identities to calculate cosecant, secant, and cotangent functions.

■**YOUR TURN** The terminal side of an angle, θ, in standard position passes through the point (3, 7). Calculate the values of the six trigonometric functions for angle θ.

We can now use this second definition of trigonometric functions to find values for nonacute angles (angles with measure greater than or equal to 90°).

EXAMPLE 2 Calculating Trigonometric Function Values for Nonacute Angles

The terminal side of an angle, θ, in standard position passes through the point $(-4, -7)$. Calculate the values of the six trigonometric functions for angle θ.

Solution:

STEP 1 Draw the angle and label the point $(-4, -7)$.

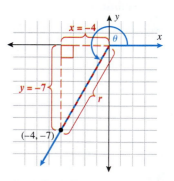

STEP 2 Calculate the distance, r. $\qquad r = \sqrt{(-4)^2 + (-7)^2} = \sqrt{65}$

■ **Answer:** $\sin\theta = \dfrac{7\sqrt{58}}{58}$ $\cos\theta = \dfrac{3\sqrt{58}}{58}$ $\tan\theta = \dfrac{7}{3}$

$\csc\theta = \dfrac{\sqrt{58}}{7}$ $\sec\theta = \dfrac{\sqrt{58}}{3}$ $\cot\theta = \dfrac{3}{7}$

STEP 3 Formulate the trigonometric functions in terms of *x*, *y*, and *r*.

Let $x = -4$, $y = -7$, and $r = \sqrt{65}$.

$$\sin\theta = \frac{y}{r} = \frac{-7}{\sqrt{65}} \qquad \cos\theta = \frac{x}{r} = \frac{-4}{\sqrt{65}} \qquad \tan\theta = \frac{y}{x} = \frac{-7}{-4} = \frac{7}{4}$$

$$\csc\theta = \frac{r}{y} = \frac{\sqrt{65}}{-7} \qquad \sec\theta = \frac{r}{x} = \frac{\sqrt{65}}{-4} \qquad \cot\theta = \frac{x}{y} = \frac{-4}{-7} = \frac{4}{7}$$

STEP 4 Rationalize radical denominators in the sine and cosine functions.

$$\sin\theta = \frac{y}{r} = \frac{-7}{\sqrt{65}} \cdot \frac{\sqrt{65}}{\sqrt{65}} = -\frac{7\sqrt{65}}{65}$$

$$\cos\theta = \frac{x}{r} = \frac{-4}{\sqrt{65}} \cdot \frac{\sqrt{65}}{\sqrt{65}} = -\frac{4\sqrt{65}}{65}$$

STEP 5 Write the values of the six trigonometric functions for *θ*.

$$\sin\theta = -\frac{7\sqrt{65}}{65} \qquad \cos\theta = -\frac{4\sqrt{65}}{65} \qquad \tan\theta = \frac{7}{4}$$

$$\csc\theta = -\frac{\sqrt{65}}{7} \qquad \sec\theta = -\frac{\sqrt{65}}{4} \qquad \cot\theta = \frac{4}{7}$$

Note: In Example 2, we could have used the definitions of the sine, cosine, and tangent functions along with the reciprocal identities to calculate the cosecant, secant, and cotangent functions.

YOUR TURN   The terminal side of an angle, *θ*, in standard position passes through the point $(-3, -5)$. Calculate the values of the six trigonometric functions for angle *θ*.

If we say that the terminal side of an angle is given by an equation of a line that passes through the origin, can we determine the angle? No, we must specify which part of the line represents the terminal ray in order to determine the angle. For example, in the diagram below, if we specify $x > 0$, then we know that the terminal side lies in QI. Alternatively, if we specify $x < 0$, then we know that the terminal side lies in QIII.

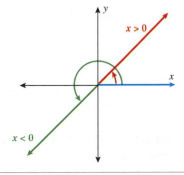

Answer: $\sin\theta = -\dfrac{5\sqrt{34}}{34}$ $\cos\theta = -\dfrac{3\sqrt{34}}{34}$ $\tan\theta = \dfrac{5}{3}$

$\csc\theta = -\dfrac{\sqrt{34}}{5}$ $\sec\theta = -\dfrac{\sqrt{34}}{3}$ $\cot\theta = \dfrac{3}{5}$

Once we know which part of the line represents the terminal side of the angle, does it matter which point on the line we use to formulate the trigonometric function values? No, because corresponding sides of similar triangles are proportional.

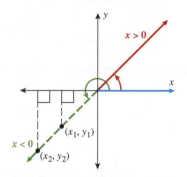

EXAMPLE 3 Calculating Trigonometric Function Values for Nonacute Angles

Calculate the values for the six trigonometric functions of the angle θ, given in standard position, if the terminal side of θ is defined by $y = 3x$, $x \leq 0$.

Solution:

STEP 1 Draw the line and label a point on the terminal side.

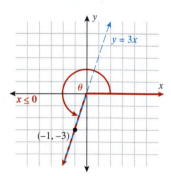

STEP 2 Calculate the distance, r. 　　　$r = \sqrt{(-1)^2 + (-3)^2} = \sqrt{10}$

STEP 3 Formulate the trigonometric functions in terms of x, y, and r.
Let $x = -1$, $y = -3$, and $r = \sqrt{10}$.

$$\sin\theta = \frac{y}{r} = \frac{-3}{\sqrt{10}} \qquad \cos\theta = \frac{x}{r} = \frac{-1}{\sqrt{10}} \qquad \tan\theta = \frac{y}{x} = \frac{-3}{-1} = 3$$

$$\csc\theta = \frac{r}{y} = \frac{\sqrt{10}}{-3} \qquad \sec\theta = \frac{r}{x} = \frac{\sqrt{10}}{-1} \qquad \cot\theta = \frac{x}{y} = \frac{-1}{-3} = \frac{1}{3}$$

STEP 4 Rationalize radical denominators in the sine and cosine functions.

$$\sin\theta = \frac{y}{r} = \frac{-3}{\sqrt{10}} \cdot \frac{\sqrt{10}}{\sqrt{10}} = -\frac{3\sqrt{10}}{10}$$

$$\cos\theta = \frac{x}{r} = \frac{-1}{\sqrt{10}} \cdot \frac{\sqrt{10}}{\sqrt{10}} = -\frac{\sqrt{10}}{10}$$

STEP 5 **Write the values of the six trigonometric functions for θ.**

$$\sin\theta = -\frac{3\sqrt{10}}{10}$$

$$\cos\theta = -\frac{\sqrt{10}}{10}$$

$$\tan\theta = 3$$

$$\csc\theta = -\frac{\sqrt{10}}{3}$$

$$\sec\theta = -\sqrt{10}$$

$$\cot\theta = \frac{1}{3}$$

Note: In Example 3, we could have used the definitions of the sine, cosine, and tangent functions along with the reciprocal identities to calculate the cosecant, secant, and cotangent functions.

■ **YOUR TURN** Calculate the values for the six trigonometric functions of the angle θ, given in standard position, if the terminal side of θ is defined by $y = 2x$, $x \leq 0$.

Thus far in this book we have avoided evaluating expressions such as $\sin 90°$ because $90°$ is not an acute angle. However, with our second definition of trigonometric functions in the Cartesian plane we now are able to formulate the trigonometric functions for quadrantal angles (angles in standard position whose terminal sides coincide with an axis) such as $90°$, $180°$, $270°$, and $360°$. Notice that $90°$ and $270°$ lie along the y-axis and therefore have an x-coordinate value equal to 0. Similarly, $180°$ and $360°$ lie along the x-axis and have a y-coordinate value equal to 0. Some of the trigonometric functions are defined with x or y coordinates in the denominator, and since dividing by 0 is undefined in mathematics, not all trigonometric functions are defined for some quadrantal angles.

EXAMPLE 4 Calculating Trigonometric Function Values for Quadrantal Angles

Calculate the values for the six trigonometric functions when $\theta = 90°$.

Solution:

STEP 1 **Draw the angle and label a point on the terminal side.**
Note: A convenient point on the terminal side is $(0, 1)$.

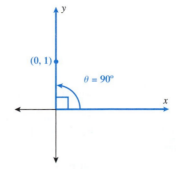

TECHNOLOGY TIP

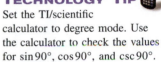

Set the TI/scientific calculator to degree mode. Use the calculator to check the values for $\sin 90°$, $\cos 90°$, and $\csc 90°$. Use the reciprocal identities to enter $\csc 90°$ as $(\sin 90)^{-1}$.

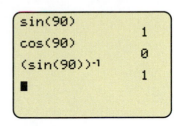

Since $\tan 90°$ is undefined, the TI calculator will display error message as:

```
tan(90)
■
```

```
ERR:DOMAIN
1■Quit
2:Goto
```

For most scientific calculators, "error" or "err" will be displayed if the value is not defined. A TI calculator will display the error message for $\sec 90°$ as:

```
(cos(90))⁻¹
■
```

```
ERR:DIVIDE BY 0
1■Quit
2:Goto
```

It will not display the correct value for $\cot 90°$.

```
(tan(90))⁻¹
```

```
ERR:DOMAIN
1■Quit
2:Goto
```

■ **Answers:** $\sin\theta = -\dfrac{2\sqrt{5}}{5}$ $\cos\theta = -\dfrac{\sqrt{5}}{5}$ $\csc\theta = -\dfrac{\sqrt{5}}{2}$

$\tan\theta = 2$ $\sec\theta = -\sqrt{5}$ $\cot\theta = \dfrac{1}{2}$

STEP 2 **Calculate the distance, *r*.** $\qquad r = \sqrt{(0)^2 + (1)^2} = \sqrt{1} = 1$

STEP 3 **Formulate the trigonometric functions in terms of *x*, *y*, and *r*.**

Let $x = 0, y = 1, r = 1$.

$$\sin\theta = \frac{y}{r} = \frac{1}{1} \qquad \cos\theta = \frac{x}{r} = \frac{0}{1} \qquad \tan\theta = \frac{y}{x} = \frac{1}{0}$$

$$\csc\theta = \frac{r}{y} = \frac{1}{1} \qquad \sec\theta = \frac{r}{x} = \frac{1}{0} \qquad \cot\theta = \frac{x}{y} = \frac{0}{1}$$

STEP 4 **Write the values of the six trigonometric functions for *θ*.**

$\sin\theta = 1$	$\cos\theta = 0$	$\tan\theta$ is undefined
$\csc\theta = 1$	$\sec\theta$ is undefined	$\cot\theta = 0$

Note: In Example 4, we could have used the definitions of sine, cosine, and tangent along with the reciprocal identities to calculate cosecant, secant, and cotangent.

■ **YOUR TURN** Calculate the values for the six trigonometric functions when $\theta = 270°$.

TECHNOLOGY TIP
Use a TI/scientific calculator to check the values for $\sin 180°$, $\cos 180°$, and $\tan 180°$.

```
sin(180)
                    0
cos(180)
                   -1
tan(180)
                    0
```

Use the reciprocal identities to enter $\sec 180°$ as $(\cos 180)^{-1}$.

```
(cos(180))-1
                   -1
```

A TI calculator displays the values of $\csc 180°$ and $\cot 180°$ as:

```
(sin(180))-1
(tan(180))-1
■
```

```
ERR:DIVIDE BY 0
1▮Quit
2:Goto
```

EXAMPLE 5 **Undefined Trigonometric Function Values for Quadrantal Angles**

Calculate the values for the six trigonometric functions when $\theta = 180°$.

Solution:

STEP 1 **Draw the angle and label a point on the terminal side.**

Note: A convenient point on the terminal side is $(-1, 0)$.

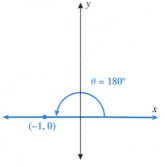

STEP 2 **Calculate the distance, *r*.** $\qquad r = \sqrt{(-1)^2 + (0)^2} = \sqrt{1} = 1$

STEP 3 **Formulate the trigonometric functions in terms of *x*, *y*, and *r*.**

Let $x = -1, y = 0, r = 1$.

$$\sin\theta = \frac{y}{r} = \frac{0}{1} \qquad \cos\theta = \frac{x}{r} = \frac{-1}{1} \qquad \tan\theta = \frac{y}{x} = \frac{0}{-1}$$

$$\csc\theta = \frac{r}{y} = \frac{1}{0} \qquad \sec\theta = \frac{r}{x} = \frac{1}{-1} \qquad \cot\theta = \frac{x}{y} = \frac{-1}{0}$$

■ **Answer:** $\sin\theta = -1$ $\quad \cos\theta = 0$ $\quad \csc\theta = -1$ $\quad \tan\theta$ is undefined $\quad \sec\theta$ is undefined $\quad \cot\theta = 0$

STEP 4 **Write the values of the six trigonometric functions for θ.**

| $\sin\theta = 0$ | $\cos\theta = -1$ | $\tan\theta = 0$ |

| $\csc\theta$ is undefined | $\sec = -1$ | $\cot\theta$ is undefined |

Note: In Example 5, we could have used the definitions of sine, cosine, and tangent along with the reciprocal identities to calculate cosecant, secant, and cotangent.

■ **YOUR TURN** 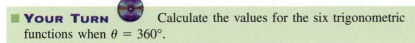 Calculate the values for the six trigonometric functions when $\theta = 360°$.

The following table summarizes the trigonometric function values for common quadrantal angles: 0°, 90°, 180°, 270°, and 360°.

θ	$\sin\theta$	$\cos\theta$	$\tan\theta$	$\cot\theta$	$\sec\theta$	$\csc\theta$
0°	0	1	0	undefined	1	undefined
90°	1	0	undefined	0	undefined	1
180°	0	−1	0	undefined	−1	undefined
270°	−1	0	undefined	0	undefined	−1
360°	0	1	0	undefined	1	Undefined

To confirm these values, use a calculator to evaluate each of these functions for the specified angles. Make sure the calculator is set in degrees (not radians) mode.

SECTION 2.2 SUMMARY

In this section, trigonometric functions were defined in the Cartesian plane as ratios of coordinates and distances. Right triangle trigonometric definitions learned in Chapter 1 are consistent with these definitions. We now have the ability to evaluate trigonometric functions for nonacute angles. In this section we learned that trigonometric functions are not always defined for quadrantal angles.

SECTION 2.2 EXERCISES

■ **SKILLS**

In Exercises 1–16, the terminal side of an angle, θ, in standard position passes through the indicated point. Calculate the values of the six trigonometric functions for angle θ.

1. $(1, 2)$ **2.** $(2, 3)$ **3.** $(3, 6)$ **4.** $(8, 4)$

5. $\left(\dfrac{1}{2}, \dfrac{2}{5}\right)$ **6.** $\left(\dfrac{4}{7}, \dfrac{2}{3}\right)$ **7.** $(-2, 4)$ **8.** $(-1, 3)$

■ **Answer:** $\sin\theta = 0$ $\cos\theta = 1$ $\cot\theta$ is undefined
$\csc\theta$ is undefined $\sec\theta = 1$ $\tan\theta = 0$

9. $(-4, -7)$ **10.** $(-9, -5)$ **11.** $(-\sqrt{2}, \sqrt{3})$ **12.** $(-\sqrt{3}, \sqrt{2})$

13. $(-\sqrt{5}, -\sqrt{3})$ **14.** $(-\sqrt{6}, -\sqrt{5})$ **15.** $\left(-\dfrac{10}{3}, -\dfrac{4}{3}\right)$ **16.** $\left(-\dfrac{2}{9}, -\dfrac{1}{3}\right)$

In Exercises 17–24, calculate the values for the six trigonometric functions of the angle θ, given in standard position, if the terminal side of θ is defined by the following lines.

17. $y = 2x$ $x \geq 0$ **18.** $y = 3x$ $x \geq 0$ **19.** $y = \dfrac{1}{2}x$ $x \geq 0$ **20.** $y = \dfrac{1}{2}x$ $x \leq 0$

21. $y = -\dfrac{1}{3}x$ $x \geq 0$ **22.** $y = -\dfrac{1}{3}x$ $x \leq 0$ **23.** $2x + 3y = 0$ $x \leq 0$ **24.** $2x + 3y = 0$ $x \geq 0$

In Exercises 25–36, calculate (if possible) the values for the six trigonometric functions of the angle, θ, given in standard position.

25. $\theta = 450°$ **26.** $\theta = 540°$ **27.** $\theta = 630°$ **28.** $\theta = 720°$

29. $\theta = -270°$ **30.** $\theta = -180°$ **31.** $\theta = -90°$ **32.** $\theta = -360°$

33. $\theta = -450°$ **34.** $\theta = -540°$ **35.** $\theta = -630°$ **36.** $\theta = -720°$

 ■ **APPLICATIONS**

In Exercises 37 and 38, refer to the following figure.

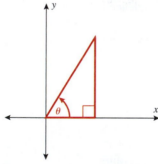

In Exercises 39 and 40, refer to the following figure.

37. Geometry. A right triangle is drawn in QI with one leg on the x-axis and its hypotenuse on the terminal side of $\angle\theta$ drawn in standard position. If $\sin\theta = \dfrac{7}{25}$, then what is $\tan\theta$?

38. Geometry. A right triangle is drawn in QI with one leg on the x-axis and its hypotenuse on the terminal side of $\angle\theta$ drawn in standard position. If $\tan\theta = \dfrac{84}{13}$, then what is $\cos\theta$?

39. Angle of Elevation. If $\angle\theta$ is the angle of elevation from a point on the ground to the top of a tree, and if $\cos\theta = \dfrac{11}{61}$ and the distance from the point on the ground to the base of the tree is 22 feet, then how high is the tree?

40. Angle of Elevation. If $\angle\theta$ is the angle of elevation from a point on the ground to the top of a tree, and if $\sin\theta = \dfrac{40}{41}$ and the tree is 20 feet high, then how far from the base of the tree is the point on the ground?

 ■ **CATCH THE MISTAKE**

In Exercises 41 and 42, explain the mistake that is made.

41. The terminal side of an angle θ, in standard position, passes through the point $(1, 2)$. Calculate $\sin\theta$.

Solution:

Label the coordinates. $x = 1, y = 2$

Calculate r. $r = 1^2 + 2^2 = 5$

Use the definition of sine. $\sin\theta = \dfrac{y}{r}$

Substitute $y = 2, r = 5$. $\sin\theta = \dfrac{2}{5}$

This is incorrect. What mistake was made?

42. Evaluate sec 810° exactly.

Solution:

Find a coterminal angle that lies between 0° and 360°.

$810° - 360° = 450°$
$450° - 360° = 90°$

810° has the same terminal side as 90°.

Evaluate cosine for 90°. $\cos 90° = 0$

Evaluate secant for 90°. $\sec 90° = \dfrac{1}{\cos 90°} = \dfrac{1}{0} = 0$

This is incorrect. What mistake was made?

■ CHALLENGE

43. T or F: $\sin 30° + \sin 60° = \sin 90°$

44. T or F: $\tan 0° + \tan 90° = \tan 90°$

45. T or F: $\cos\theta = \cos(\theta + 360n)$, n is an integer

46. T or F: $\sin\theta = \sin(\theta + 360n)$, n is an integer

47. If the terminal side of angle θ passes through the point $(-3a, 4a)$, find $\cos\theta$.

48. If the terminal side of angle θ passes through the point $(-3a, 4a)$, find $\sin\theta$.

49. If the line $y = mx$ makes an angle, θ, with the x-axis, find the slope, m, in terms of a single trigonometric function.

50. Find x if $(x, -2)$ is on the terminal side of angle θ and $\csc\theta = -\dfrac{\sqrt{29}}{2}$.

51. Find the equation of the line with *positive* slope that passes through the point $(a, 0)$ and makes an acute angle θ with the x-axis. The equation of the line will be in terms of x, a, and a trigonometric function of θ.

52. Find the equation of the line with *negative* slope that passes through the point $(a, 0)$ and makes an acute angle θ with the x-axis. The equation of the line will be in terms of x, a, and a trigonometric function of θ.

■ TECHNOLOGY

In Exercises 53–56, use a calculator to evaluate the following expressions. If you get an error, explain why.

53. $\sin 270°$

54. $\cos 270°$

55. $\tan 270°$

56. $\cot 270°$

SECTION 2.3 Evaluating Trigonometric Functions for Any Angle

Skills Objectives

■ Determine the reference angle of a nonacute angle.
■ Evaluate trigonometric functions exactly for common angles.
■ Approximate trigonometric functions of nonacute angles.

Conceptual Objectives

■ Determine algebraic signs of trigonometric functions for all four quadrants.
■ Determine values for trigonometric functions for quadrantal angles.
■ Determine ranges for trigonometric functions.

In Section 2.2, we defined trigonometric functions in the Cartesian plane as ratios of x, y, and r. We calculated the trigonometric function values, given a point on the terminal side of the angle in standard position. We also discussed the special case of a quadrantal angle. In this section, we now evaluate trigonometric functions for special angles exactly. Additionally, we will approximate trigonometric functions for nonacute angles using an acute reference angle and knowledge of algebraic signs of trigonometric functions in particular quadrants.

Algebraic Signs of Trigonometric Functions

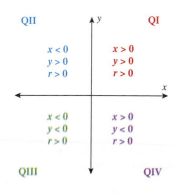

In Section 2.2, we defined trigonometric functions as ratios of x, y, and r. Since r is the distance from the origin to the point (x, y) and distance is never negative, r is always taken as the positive solution to $r^2 = \sqrt{x^2 + y^2}$, so $r > 0$.

The x-coordinate is positive in quadrants **I** and **IV** and negative in quadrants **II** and **III**. The y-coordinate is positive in quadrants **I** and **II** and negative in quadrants **III** and **IV**. Recall the definition of the six trigonometric functions in the Cartesian plane:

$$\sin\theta = \frac{y}{r} \qquad\qquad \cos\theta = \frac{x}{r} \qquad\qquad \tan\theta = \frac{y}{x} \quad (x \neq 0)$$

$$\csc\theta = \frac{r}{y} \quad (y \neq 0) \qquad \sec\theta = \frac{r}{x} \quad (x \neq 0) \qquad \cot\theta = \frac{x}{y} \quad (y \neq 0)$$

Therefore, the algebraic sign, $+$ or $-$, of each trigonometric function will depend on which quadrant contains the terminal side of angle θ. Let us look at the three main trigonometric functions: sine, cosine, and tangent. In quadrant I, all three functions are positive since x, y, and r are all positive. However, in quadrant II, only sine is positive since y and r are both positive. In quadrant III, only tangent is positive and in quadrant IV, only cosine is positive. The phrase "**A**ll **S**tudents **T**ake **C**alculus" helps us remember which of the three trigonometric functions is positive in each quadrant.

STUDY TIP

All **S**tudents **T**ake **C**alculus is a phrase that helps us remember which of the three (sine, cosine, tangent) functions are positive in quadrants I, II, III, and IV.

PHRASE	QUADRANT	POSITIVE TRIGONOMETRIC FUNCTION
All	I	**A**ll three: sine, cosine, and tangent
Students	II	**S**ine
Take	III	**T**angent
Calculus	IV	**C**osine

The following table indicates the algebraic sign of all six trigonometric functions when the terminal side of the angle, θ, lies in each quadrant.

TERMINAL SIDE OF θ IN QUADRANT	$\sin\theta$	$\cos\theta$	$\tan\theta$	$\cot\theta$	$\sec\theta$	$\csc\theta$
I	$+$	$+$	$+$	$+$	$+$	$+$
II	$+$	$-$	$-$	$-$	$-$	$+$
III	$-$	$-$	$+$	$+$	$-$	$-$
IV	$-$	$+$	$-$	$-$	$+$	$-$

EXAMPLE 1 Using the Algebraic Sign of Trigonometric Functions

If $\cos\theta = -\dfrac{3}{5}$, and the terminal side of θ lies in quadrant III, find $\sin\theta$.

Solution:

STEP 1 Draw some angle θ in QIII.

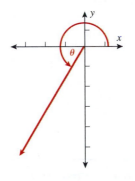

STEP 2 Identify known quantities from information given.

Recall that $\cos\theta = \dfrac{x}{r}$ and $r > 0$.

$\cos\theta = -\dfrac{3}{5} = \dfrac{-3}{5} = \dfrac{x}{r}$

Identify x and r.

$x = -3$ and $r = 5$

STEP 3 Since x and r are known, find y.

Substitute $x = -3$ and $r = 5$
into $x^2 + y^2 = r^2$.

$(-3)^2 + y^2 = 5^2$

Solve for y.

$9 + y^2 = 25$
$y^2 = 16$
$y = \pm 4$

STEP 4 Select the sign of y based on quadrant information.

Since the terminal side of θ lies in
quadrant III, $y < 0$.

$y = -4$

STEP 5 Find $\sin\theta$.

$\sin\theta = \dfrac{y}{r} = \dfrac{-4}{5}$

$\sin\theta = -\dfrac{4}{5}$

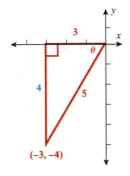

■ **YOUR TURN**   If $\sin\theta = -\frac{3}{4}$, and the terminal side of θ lies in quadrant III, find $\cos\theta$.

A table can also be made when the terminal side of θ lies along one of the axes (quadrantal angle). This table was calculated for specific angles in Section 2.2, and we will develop it again here using algebraic signs of trigonometric functions in quadrants.

When the terminal side lies along the x-axis, then $y = 0$. When $y = 0$, notice that $r = \sqrt{x^2 + y^2} = \sqrt{x^2} = |x|$. When the terminal side lies along the positive x-axis, $x > 0$, and when the terminal side lies along the negative x-axis, $x < 0$. Therefore, when the terminal side of the angle lies on the positive x-axis, then $y = 0$, $x > 0$, and

■ **Answer:** $\cos\theta = -\dfrac{\sqrt{7}}{4}$

$r = x$, and when the terminal side lies along the negative x-axis, then $y = 0$, $x < 0$, and $r = |x|$. A similar argument can be made for the y-axis that results in $r = |y|$.

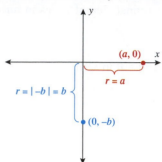

TERMINAL SIDE OF θ LIES ALONG THE	$\sin\theta$	$\cos\theta$	$\tan\theta$	$\cot\theta$	$\sec\theta$	$\csc\theta$
Positive x-axis (0° or 360°)	0	1	0	undefined	1	undefined
Positive y-axis (90°)	1	0	undefined	0	undefined	1
Negative x-axis (180°)	0	−1	0	undefined	−1	undefined
Negative y-axis (270°)	−1	0	undefined	0	Undefined	−1

EXAMPLE 2 Using the Value of Trigonometric Functions for Quadrantal Angles

Evaluate the following expressions (if possible):

a. $\cos 540° + \sin 270°$ **b.** $\cot 90° + \tan(-90°)$

Solution (a):

The terminal side of an angle with measure 540° lies along the negative x-axis. $540° - 360° = 180°$

Evaluate cosine for an angle with the terminal side along the negative x-axis. $\cos 540° = -1$

Evaluate sine for an angle with the terminal side along the negative y-axis. $\sin 270° = -1$

Sum the sine and cosine terms. $\cos 540° + \sin 270° = -1 + (-1)$

$$\cos 540° + \sin 270° = -2$$

Check: Evaluate this expression using a calculator.

Solution (b):

Evaluate cotangent for an angle with the terminal side along the positive y-axis. $\cot 90° = 0$

The terminal side of an angle with measure $-90°$ lies along the negative y-axis. $\tan(-90°) = \tan 270°$

The tangent function is undefined for an angle with the terminal side along the negative y-axis. $\tan(-90°)$ is undefined

Even though $\cot 90°$ is defined, since $\tan(-90°)$ is undefined, the sum of the two expressions is also undefined.

TECHNOLOGY TIP

Use a TI/scientific calculator to check the value of the expression:
(a) $\cos 540° + \sin 270°$

```
cos(540)+sin(270
)
            -2
```

(b) $\cot 90° + \tan(-90°)$

```
(tan(90))-1+tan(-
90)
```

```
ERR:DOMAIN
1:Quit
2:Goto
```

✅**CONCEPT CHECK** Along what axis would the terminal side of an angle in standard position lie if the angle has measure $-630°$?

◾ **YOUR TURN** Evaluate the following expressions if possible:

a. $\csc 90° + \sec 180°$ **b.** $\csc(-630°) + \sec(-630°)$

Ranges of Trigonometric Functions

Thus far, we have discussed what the algebraic sign of a trigonometric function is in a particular quadrant, but we haven't discussed how to find actual values of the trigonometric functions for nonacute angles. We will shortly be using reference angles and triangles. However, before we proceed let's get a feel for the ranges (values of the functions) we will expect. Although formal definitions of domain and range of trigonometric functions will occur in Chapter 4, we will limit our discussion here to typical values (ranges) of trigonometric functions.

Let us start with an angle, θ, in quadrant I and the sine function defined as the ratio

$$\sin\theta = \frac{y}{r}$$

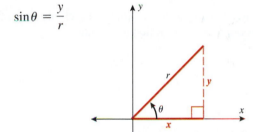

If the measure of θ increases toward $90°$, then y increases. Notice that even though the value of y approaches the value of r, they will only be equal when $\theta = 90°$, and y can never be larger than r.

$$y \leq r$$

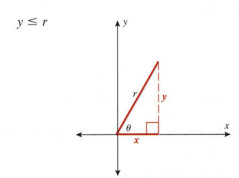

A similar analysis can be conducted in quadrant IV when θ approaches $-90°$ (note that y is negative in quadrant IV). A result that is valid in all four quadrants is $|y| \leq r$.

Write the absolute value inequality as a double inequality. $\qquad -r \leq y \leq r$

Divide both sides by r. $\qquad\qquad\qquad\qquad\qquad\qquad -1 \leq \dfrac{y}{r} \leq 1$

Let $\sin\theta = \dfrac{y}{r}$. $\qquad\qquad\qquad\qquad\qquad\qquad\qquad -1 \leq \sin\theta \leq 1$

Similarly, by allowing θ to approach $0°$ and $180°$, we can show that $|x| \leq r$, which leads to the range of the cosine function: $-1 \leq \cos\theta \leq 1$. Sine and cosine range between -1 and 1, and since secant and cosecant are reciprocal functions of cosine and sine functions, their ranges are determined by the following:

$$\sec\theta \leq -1 \text{ or } \sec\theta \geq 1 \qquad \csc\theta \leq -1 \text{ or } \csc\theta \geq 1$$

Since $\tan\theta = \dfrac{y}{x}$ and $\cot\theta = \dfrac{x}{y}$ and since it is possible for $x < y, x = y$, or $x > y$, values of the tangent and cotangent functions can be any real numbers. The following box summarizes the ranges of the trigonometric functions.

RANGES OF TRIGONOMETRIC FUNCTIONS

For any angle, θ, for which the trigonometric functions are defined, the six trigonometric functions have the following ranges:

- $-1 \leq \sin\theta \leq 1$
- $-1 \leq \cos\theta \leq 1$
- $\sec\theta \leq -1$ or $\sec\theta \geq 1$
- $\csc\theta \leq -1$ or $\csc\theta \geq 1$
- $\tan\theta$ can equal any real number.
- $\cot\theta$ can equal any real number.

EXAMPLE 3 Determining If a Value Is Within the Range of a Trigonometric Function

Determine if each statement is possible or impossible.

a. $\cos\theta = 1.001$ $\qquad\qquad$ **b.** $\cot\theta = 0$ $\qquad\qquad$ **c.** $\sec\theta = \dfrac{\sqrt{3}}{2}$

Solution (a): It is not possible, because $1.001 > 1$.

Solution (b): It is possible, because $\cot 90° = 0$.

Solution (c): It is not possible, because $\dfrac{\sqrt{3}}{2} \approx 0.866 < 1$.

YOUR TURN Determine if each statement is possible or impossible.

a. $\sin\theta = -1.1$ $\qquad$ **b.** $\tan\theta = 2$ $\qquad$ **c.** $\csc\theta = \sqrt{3}$

Reference Triangle and Angle

Now that we know what ranges the trigonometric functions have and what algebraic signs they each have in the four quadrants, we can evaluate trigonometric functions for any nonacute angle. But before we do that, we first must discuss *reference angles* and *reference triangles*.

Recall the story in the chapter opener about the small-town basketball team that played the state championship in the enormous arena. As it turned out, their small-town gymnasium court had the same measurements as the court in the large arena. It was as if their hometown court, had been picked up and placed inside the arena. Right triangle trigonometry (Chapter 1) is our hometown court, and the Cartesian plane with nonacute angles is the large arena.

Every nonquadrantal angle in standard position has a corresponding *reference angle* and *reference triangle*. We have already calculated the trigonometric function values for quadrantal angles.

DEFINITION   **REFERENCE ANGLE**

For any angle θ in standard position whose terminal side lies in one of the four quadrants, there exists a **reference angle**, α, which is an acute angle with positive measure that is formed by the terminal side of θ and the x-axis.

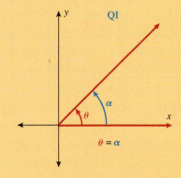

$\theta = \alpha$

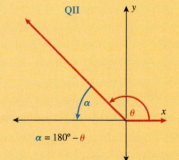

$\alpha = 180° - \theta$

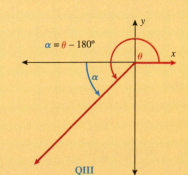

$\alpha = \theta - 180°$

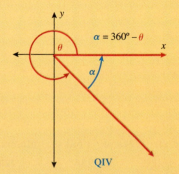

$\alpha = 360° - \theta$

CAUTION

The reference angle is the acute angle that the terminal side makes with the x-axis, *not* the y-axis.

The reference angle is the acute angle with positive measure that the terminal side makes with the x-axis.

EXAMPLE 4 **Finding Reference Angles**

Find the reference angle for each angle given.

a. 210°   **b.** 135°   **c.** 422°

Solution (a):

The terminal side of θ lies in quadrant III.

The reference angle is made with the terminal side and the x-axis.

$210° - 180° = \boxed{30°}$

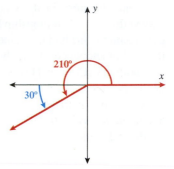

Solution (b):

The terminal side of θ lies in quadrant II.

The reference angle is made with the terminal side and the x-axis.

$180° - 135° = \boxed{45°}$

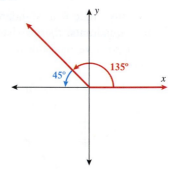

Solution (c):

The terminal side of θ lies in quadrant I.

The reference angle is made with the terminal side and the x-axis.

$422° - 360° = \boxed{62°}$

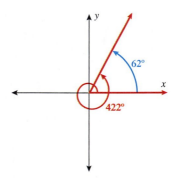

■ **YOUR TURN**   Find the reference angle for each angle given.

a. 160°   **b.** 285°   **c.** 600°

■ **Answers: a.** 20°   **b.** 75°   **c.** 60°

DEFINITION   REFERENCE TRIANGLE

To form a **reference triangle** for angle θ, drop a perpendicular from the terminal side of the angle to the x-axis. The right triangle now has the reference angle, α, as one of its angles.

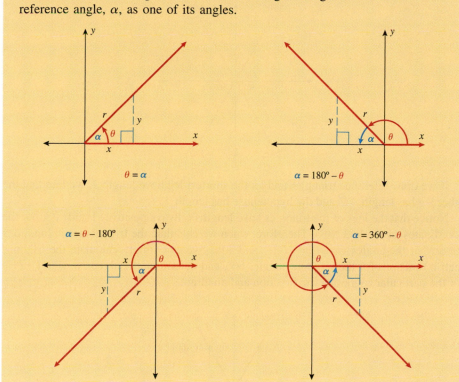

$\theta = \alpha$

$\alpha = 180° - \theta$

$\alpha = \theta - 180°$

$\alpha = 360° - \theta$

In Section 1.3, we first defined the trigonometric functions of an acute angle as ratios of lengths of sides of a right triangle. For example, $\sin\theta = \dfrac{\text{opposite}}{\text{hypotenuse}}$. The lengths of triangles are always positive.

In Section 2.2, we defined the sine function of any angle as $\sin\theta = \dfrac{y}{r}$. Notice in the above box that for a nonacute angle, θ, $\sin\theta = \dfrac{y}{r}$ and for the acute reference angle, α, $\sin\alpha = \dfrac{|y|}{r}$. The only difference between these two expressions is the algebraic sign, since r is always positive and y is positive or negative depending on the quadrant.

Therefore, to calculate the trigonometric function values for a nonacute angle, simply find the trigonometric values for the reference angle and determine the correct algebraic sign according to the quadrant the terminal side lies in.

Evaluating Trigonometric Functions for Nonacute Angles

Let's look at a specific example before we generalize a procedure for evaluating nonacute angles.

STUDY TIP

To find the trigonometric function values of nonacute angles, first find the trigonometric values for the reference angle, and then use the algebraic sign information to determine the sign.

Suppose we have the angles in standard position with measure 60°, 120°, 240°, and 300°. Notice that the reference angle for all these angles is 60°.

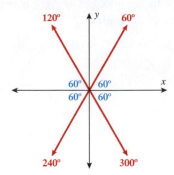

If we draw reference triangles and let the shortest leg have length **1**, we find that the other leg has length $\sqrt{3}$ and the hypotenuse has length **2**.

Notice that the legs of the triangles have lengths (always positive) 1 and $\sqrt{3}$, but the coordinates are $(\pm 1, \pm \sqrt{3})$. Therefore, when we calculate the trigonometric functions for any of the angles 60°, 120°, 240°, and 300°, we can simply calculate the trigonometric functions for the reference angle, 60°, and determine the algebraic sign $(+/-)$ for the particular trigonometric function and quadrant.

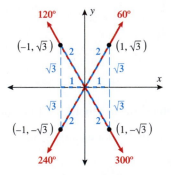

To find the value of cos 120°, we first recognize that the terminal side of an angle with 120° measure lies in quadrant II. We also know that cosine is negative in quadrant II. We then calculate the cosine of the reference angle, 60°.

$$\cos 60° = \frac{\text{adjacent}}{\text{hypotenuse}} = \frac{1}{2}$$

Since we know that cos 120° is negative because it lies in quadrant II, we know that

$$\cos 120° = -\frac{1}{2}$$

Similarly, we know that $\cos 240° = -\frac{1}{2}$ and $\cos 300° = \frac{1}{2}$.

For any angle whose terminal side lies along one of the axes, we consult the table in this section for the values of the trigonometric functions for quadrantal angles. If the terminal side lies in one of the four quadrants, the angle is said to be nonquadrantal, and the following procedure can be used.

PROCEDURE FOR EVALUATING FUNCTION VALUES FOR ANY NONQUADRANTAL ANGLE, θ

Step 1: If $\theta < 0°$, then add 360° as many times as needed to get a coterminal angle with measure between 0° and 360°.

If $\theta > 360°$, then subtract 360° as many times as needed to get a coterminal angle with measure between 0° and 360°.

Step 2: Find the quadrant in which the terminal side of the angle in Step 1 lies.

Step 3: Find the reference angle, α, of the coterminal angle found in Step 1.

Step 4: Find the trigonometric function values for the reference angle, α.

Step 5: Determine the correct algebraic signs $(+/-)$ for the trigonometric functions based on the quadrant identified in Step 2.

Step 6: Combine the trigonometric values found in Step 4 with the algebraic signs in Step 5 to evaluate the trigonometric function values of θ.

It is important to note that the same attention must be given to the difference between evaluating trigonometric functions exactly and approximating with a calculator for non-acute angles as was done with acute angles. We follow the above procedure for all angles except when we get to Step 4. In Step 4, we evaluate exactly if possible (30°, 45°, 60°); otherwise we use a calculator to approximate.

EXAMPLE 5 Exact Evaluation for Special Angles

Find the exact value of cos 210°.

Solution:

The terminal side of $\theta = 210°$ lies in quadrant III.

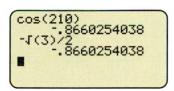

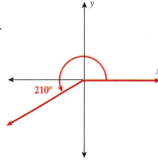

Find the reference angle for $\theta = 210°$.	$210° - 180° = 30°$
Find the value of the cosine of the reference angle.	$\cos 30° = \dfrac{\sqrt{3}}{2}$
Determine the algebraic sign for cosine in quadrant III.	Negative $(-)$
Combine the algebraic sign of cosine in quadrant III with the value of cosine of the reference angle.	$\cos 210° = -\dfrac{\sqrt{3}}{2}$

✓**CONCEPT CHECK**   Is sine positive or negative in quadrant IV?

■ **YOUR TURN**   Find the exact value of sin 330°.

■ **Answer:** $-\dfrac{1}{2}$

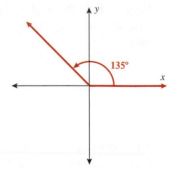

tan(495)
                    -1

EXAMPLE 6 **Exact Evaluation for Special Angles**

Find the exact value of tan 495°.

Solution:

Subtract 360° to get a coterminal angle
between 0° and 360°. $495° - 360° = 135°$

The terminal side of the angle lies in quadrant II.

Find the reference angle for 135°. $180° - 135° = 45°$

Find the value of the tangent of the reference angle. $\tan 45° = 1$

Determine the algebraic sign for tangent in quadrant II.   Negative $(-)$

Combine the algebraic sign of tangent in quadrant II
with the value of tangent of the reference angle. $\tan 495° = -1$

✓CONCEPT CHECK   Is tangent positive or negative in quadrant IV?

■ **YOUR TURN**   Find the exact value of tan 660°.

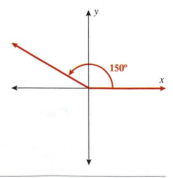

(sin(-210))⁻¹
                    2
■

EXAMPLE 7 **Exact Evaluation for Special Angles**

Find the exact value of $\csc(-210°)$.

Solution:

Add 360° to get a coterminal angle
between 0° and 360°. $-210° + 360° = 150°$

The terminal side of the angle lies in quadrant II.

■ **Answer:** $-\sqrt{3}$

Find the reference angle for 150°.	$180° - 150° = 30°$
Find the value of the cosecant of the reference angle.	$\csc 30° = \dfrac{1}{\sin 30°} = \dfrac{1}{\dfrac{1}{2}} = 2$
Determine the algebraic sign for cosecant in quadrant II.	Positive $(+)$
Combine the algebraic sign of cosecant in quadrant II with the value of cosecant of the reference angle.	$\csc(-210°) = 2$

■ **YOUR TURN**   Find the exact value of $\sec(-330°)$.

EXAMPLE 8 Finding Exact Angle Measures Given Trigonometric Function Values

Find all values of θ, where $0° \leq \theta \leq 360°$, when $\sin\theta = -\dfrac{\sqrt{3}}{2}$.

Solution:

Determine in what quadrants sine is negative.	QIII and QIV
Since the absolute value of $\sin\theta$ is $\dfrac{\sqrt{3}}{2}$, the reference angle is 60°.	$\sin 60° = \dfrac{\sqrt{3}}{2}$

Determine angles in QIII and QIV with reference angle 60°.

Quadrant III:	$180° + 60° = 240°$
Quadrant IV:	$360° - 60° = 300°$

The two angles are   240°   and   300° .

■ **YOUR TURN**   Find all values of θ, where $0° \leq \theta \leq 360°$, when $\cos\theta = -\dfrac{\sqrt{3}}{2}$.

We can evaluate trigonometric functions exactly for quadrantal angles and any angle that has a special angle (30°, 45°, or 60°) for a reference angle. For all other angles, we can approximate the trigonometric function value with a calculator. Let us now evaluate $\sin 337°$ using the procedure for evaluating function values for nonquadrantal angles.

■ **Answer:** $\dfrac{2\sqrt{3}}{3}$ ■ **Answer:** 150° and 210°

Words	Math

The terminal side of $\theta = 337°$ lies in quadrant IV.

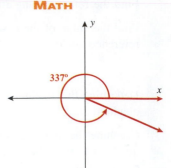
337°

Find the reference angle for θ. $\qquad$ $360° - 337° = 23°$

Find the value of sign for the reference angle. $\qquad$ $\sin 23° = 0.390731$

Determine the algebraic sign of sine in quadrant IV. $\qquad$ $-$ (negative)

Combine the value for the reference angle and the algebraic sign. $\qquad$ $\sin 337° = -0.3907$

We could have simply evaluated the trigonometric function directly with our calculator. Make sure the calculator is set in degrees mode.

337 $\quad$ +/− $\quad$ sin $\qquad\qquad$ -0.3907

In the following examples we will use the direct calculator approach. Approximations will be rounded to the nearest ten thousandth (four decimal places).

TECHNOLOGY TIP

```
tan(466)
        -3.487414444
(sin(-313))-1
         1.367327461
■
```

EXAMPLE 9 Using a Calculator to Evaluate Trigonometric Values of Angles

Use a calculator to evaluate the following:

a. $\tan 466°$ $\qquad\qquad$ **b.** $\csc(-313°)$

Solution (a): $\tan 466 = \qquad$ -3.4874

Solution (b): $\sin +/- \ 313 \qquad 1/x \qquad 1.3673$

YOUR TURN Use a calculator to evaluate the following:

a. $\sec(-118°)$ $\qquad$ **b.** $\cot 226°$

TECHNOLOGY TIP

When $\sin \theta = -0.6293$, use the $\boxed{\sin^{-1}}$ key and the absolute value to find the reference angle.

```
sin-1(0.6293)
       38.99849667
```

EXAMPLE 10 Finding Approximate Angle Measures Given Trigonometric Function Values

Find the measure of θ (rounded to the nearest degree) if $\sin \theta = -0.6293$ and the terminal side of θ (in standard position) lies in quadrant III.

Solution:

Sine of the reference angle is 0.6293. $\qquad$ $\sin \alpha = 0.6293$

Find the reference angle. $\qquad$ $\alpha = \sin^{-1}(0.6293) = 38.998°$

■ **Answers: a.** -2.1301 $\quad$ **b.** 0.9657

Round the reference angle to the nearest degree.

$\alpha = 39°$

Find θ, which lies in quadrant III.

$180° + 39° = 219°$

$\theta = 219°$

☐ Check with a calculator.

$\sin 219° = -0.6293$

■ **YOUR TURN**   Find the measure of θ (rounded to the nearest degree) if $\cos\theta = -0.5299$ and the terminal side of θ (in standard position) lies in quadrant II.

SECTION 2.3 SUMMARY

In this section, the algebraic signs of the trigonometric functions in each quadrant were identified. The trigonometric functions can be evaluated for nonquadrantal angles using reference angles and knowledge of algebraic signs in each quadrant. When reference angles are 30°, 45°, or 60°, nonacute angles can be evaluated exactly; otherwise we use a calculator to evaluate trigonometric values for any angles.

SECTION 2.3 EXERCISES

■ **SKILLS**

In Exercises 1–6, indicate the quadrants in which the terminal side of θ must lie in order for the following to be true.

1. $\cos\theta$ is positive and $\sin\theta$ is negative.

2. $\cos\theta$ is negative and $\sin\theta$ is positive.

3. $\tan\theta$ is negative and $\sin\theta$ is positive.

4. $\tan\theta$ is positive and $\cos\theta$ is negative.

5. $\sec\theta$ and $\csc\theta$ are both positive.

6. $\sec\theta$ and $\csc\theta$ are both negative.

In Exercises 7–16, find the indicated trigonometric function values.

7. If $\cos\theta = -\dfrac{3}{5}$, and the terminal side of θ lies in quadrant III, find $\sin\theta$.

8. If $\tan\theta = -\dfrac{5}{12}$, and the terminal side of θ lies in quadrant II, find $\cos\theta$.

9. If $\sin\theta = \dfrac{60}{61}$, and the terminal side of θ lies in quadrant II, find $\tan\theta$.

10. If $\cos\theta = \dfrac{40}{41}$, and the terminal side of θ lies in quadrant IV, find $\tan\theta$.

11. If $\tan\theta = \dfrac{84}{13}$, and the terminal side of θ lies in quadrant III, find $\sin\theta$.

■ **Answer:** 122°

12. If $\sin\theta = -\dfrac{7}{25}$, and the terminal side of θ lies in quadrant IV, find $\cos\theta$.

13. If $\sec\theta = -2$, and the terminal side of θ lies in quadrant III, find $\tan\theta$.

14. If $\cot\theta = 1$, and the terminal side of θ lies in quadrant I, find $\sin\theta$.

15. If $\sin\theta = \dfrac{\sqrt{11}}{6}$, and the terminal side of θ lies in quadrant II, find $\tan\theta$.

16. If $\tan\theta = -\dfrac{1}{2}$, and the terminal side of θ lies in quadrant II, find $\cos\theta$.

In Exercises 17–26, evaluate the following expressions if possible:

17. $\cos(-270°) + \sin 450°$

18. $\sin(-270°) + \cos 450°$

19. $\sin 630° + \tan(-540°)$

20. $\cos(-720°) + \tan 720°$

21. $\cos 540° - \sec(-540°)$

22. $\sin(-450°) + \csc 270°$

23. $\csc(-630°) - \cot 630°$

24. $\sec(-540°) + \tan 540°$

25. $\tan 720° + \sec 720°$

26. $\cot 450° - \cos(-450°)$

In Exercises 27–34, determine if each statement is possible or impossible.

27. $\sin\theta = -0.999$

28. $\cos\theta = 1.0001$

29. $\cos\theta = \dfrac{2\sqrt{6}}{3}$

30. $\sin\theta = \dfrac{\sqrt{2}}{10}$

31. $\tan\theta = 4\sqrt{5}$

32. $\cot\theta = -\dfrac{\sqrt{6}}{7}$

33. $\sec\theta = -\dfrac{4}{\sqrt{7}}$

34. $\csc\theta = \dfrac{\pi}{2}$

In Exercises 35–44, evaluate the following expressions *exactly*.

35. $\cos 240°$

36. $\cos 120°$

37. $\sin 300°$

38. $\sin 315°$

39. $\tan 210°$

40. $\sec 135°$

41. $\tan(-315°)$

42. $\sec(-330°)$

43. $\csc 330°$

44. $\csc(-240°)$

In Exercises 45–52, find all values of θ, where $0° \leq \theta \leq 360°$, when the following are true.

45. $\cos\theta = \dfrac{\sqrt{3}}{2}$

46. $\sin\theta = \dfrac{\sqrt{3}}{2}$

47. $\sin\theta = -\dfrac{1}{2}$

48. $\cos\theta = -\dfrac{1}{2}$

49. $\cos\theta = 0$

50. $\sin\theta = 0$

51. $\sin\theta = -1$

52. $\cos\theta = -1$

In Exercises 53–62, evaluate the trigonometric expressions with a calculator. Round to four decimal places.

53. $\sin 237°$

54. $\cos 317°$

55. $\tan(-265°)$

56. $\tan 622°$

57. $\sec 421°$

58. $\sec(-222°)$

59. $\csc(-111°)$

60. $\csc 211°$

61. $\cot 159°$

62. $\cot(-82°)$

In Exercises 63–72, find the positive measure of θ (rounded to the nearest degree) if the indicated information is true.

63. $\sin\theta = 0.9397$ and the terminal side of θ lies in quadrant II.

64. $\cos\theta = 0.7071$ and the terminal side of θ lies in quadrant IV.

65. $\cos\theta = -0.7986$ and the terminal side of θ lies in quadrant II.

66. $\sin\theta = -0.1746$ and the terminal side of θ lies in quadrant III.

67. $\tan\theta = -0.7813$ and the terminal side of θ lies in quadrant IV.

68. $\cos\theta = -0.3420$ and the terminal side of θ lies in quadrant III.

69. $\tan\theta = -0.8391$ and the terminal side of θ lies in quadrant II.

70. $\tan\theta = 11.4301$ and the terminal side of θ lies in quadrant III.

71. $\sin\theta = -0.3420$ and the terminal side of θ lies in quadrant IV.

72. $\sin\theta = -0.4226$ and the terminal side of θ lies in quadrant III.

 ■ APPLICATIONS

When light passes from one substance to another, such as from air to water, its path bends. This is called *refraction* and is what is seen in eyeglass lenses, camera lenses, and gems. The rule governing the change in the path is called *Snell's Law*, named after a Dutch astronomer. Snell's Law: $n_1 \sin\theta_1 = n_2 \sin\theta_2$, where the n's are the indices of refraction of the different substances and the θs are the respective angles that light makes with a line perpendicular to the surface, where the substances change from one to the other. The figure shows the path of light rays going from air to water. Assume that the index of refraction in air is 1.

In Exercises 73–76, use the following figure.

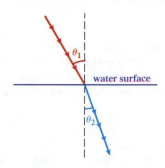

73. Optics. If light rays hit the water's surface at an angle 30° from the perpendicular and are refracted to an angle of 22° from the perpendicular, then what is the refraction index for water if you can assume that the refraction index for air is 1?

74. Optics. If light rays hit a glass surface at an angle 30° from the perpendicular and are refracted to an angle of 18° from the perpendicular, then what is the refraction index for glass if you can assume that the refraction index for air is 1?

75. Optics. If the refraction index for a diamond is 2.4, then what angle is light refracted to if it enters the diamond at an angle of 30°?

76. Optics. If the refraction index for a rhinestone is 1.9, then what angle is light refracted to if it enters the rhinestone at an angle of 30°?

 ■ CATCH THE MISTAKE

In Exercises 77 and 78, explain the mistake that is made.

77. Evaluate the expression $\sec 120°$ exactly.

Solution:

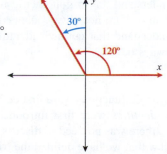

120° lies in quadrant II.
The reference angle is 30°.

Find the cosine of the reference angle.	$\cos 30° = \dfrac{\sqrt{3}}{2}$
Cosine is negative in quadrant II.	$\cos 120° = -\dfrac{\sqrt{3}}{2}$
Secant is the reciprocal of cosine.	$\sec 120° = -\dfrac{2}{\sqrt{3}} = -\dfrac{2\sqrt{3}}{3}$

This is incorrect. What mistake was made?

78. Find the measure of θ (rounded to the nearest degree) if $\cos\theta = -0.2388$ and the terminal side of θ (in standard position) lies in quadrant III.

Solution:

Evaluate with a calculator.

$$\theta = \cos^{-1}(-0.2388) = 103.8157°$$

Approximate to the nearest degree.　　　$\theta \approx 104°$

This is incorrect. What mistake was made?

■ CHALLENGE

79. T or F: It is possible for all six trigonometric functions of the same angle to have positive values.

80. T or F: It is possible for all six trigonometric functions of the same angle to have negative values.

81. T or F: The trigonometric function value for any angle with negative measure must be negative.

82. T or F: The trigonometric function value for any angle with positive measure must be positive.

83. If $\tan\theta = \dfrac{a}{b}$ where a and b are positive, and θ lies in quadrant III, find $\sin\theta$.

84. If $\tan\theta = -\dfrac{a}{b}$ where a and b are positive, and θ lies in quadrant II, find $\cos\theta$.

■ TECHNOLOGY

85. Use a calculator to evaluate $\sin 80°$ and $\sin 100°$. Now use the calculator to evaluate $\sin^{-1}(0.9848)$. When sine is positive, in which of the quadrants, I or II, does the calculator assume the terminal side of the angle lies?

86. Use a calculator to evaluate $\sin 260°$ and $\sin 280°$. Now use the calculator to evaluate $\sin^{-1}(-0.9848)$. When sine is negative, in which of the quadrants, III or IV, does the calculator assume the terminal side of the angle lies?

87. Use a calculator to evaluate $\cos 75°$ and $\cos(-75°)$. Now use the calculator to evaluate $\cos^{-1}(0.2588)$. When cosine is positive, in which of the quadrants, I or IV, does the calculator assume the terminal side of the angle lies?

88. Use a calculator to evaluate $\cos 105°$ and $\cos 255°$. Now use the calculator to evaluate $\cos^{-1}(-0.2588)$. When cosine is negative, in which of the quadrants, II or III, does the calculator assume the terminal side of the angle lies?

SECTION 2.4 Basic Trigonometric Identities

Skills Objectives

- Learn the reciprocal identities.
- Learn the quotient identities.
- Learn the Pythagorean identities.
- Use the basic identities to simplify expressions.

Conceptual Objectives

- Understand that trigonometric reciprocal identities are not always defined.
- Understand that quotient identities are not always defined.

In Section 1.3, when trigonometric functions were first defined as ratios of side lengths of triangles, the *reciprocal identities* were first introduced. Since sides of triangles always have positive lengths, there was no need to discuss the possibility of a denominator being equal to zero. Now that we have defined the trigonometric functions in the

Cartesian plane, we have found that for quadrantal angles some trigonometric functions are undefined. In this section, we revisit the *reciprocal* and *quotient identities* and develop the Pythagorean identities. These identities will allow us to simplify expressions. In mathematics, an **identity** is a statement that is true for all values of the variable (for which the expressions are defined). For example, $x^2 = x \cdot x$ is an identity because the statement is true no matter what values are selected for x. Similarly, $\dfrac{2x}{5x} = \dfrac{2}{5}$ is an identity, but it is important to note that this identity holds for all x except 0 ($x \neq 0$), since the expression on the left is not defined when $x = 0$. In our discussions on identities, it is important to note that the identities hold for all values of θ for which the expressions are defined.

Reciprocal Identities

Recall that the **reciprocal** of a nonzero number, x, is $\dfrac{1}{x}$. For example, the reciprocal of 5 is $\dfrac{1}{5}$ and the reciprocal of $\dfrac{9}{17}$ is $\dfrac{17}{9}$. There is no reciprocal for 0. In mathematics the expression $\dfrac{1}{0}$ is undefined. Calculators have a reciprocal key typically labeled $1/x$ or x^{-1}. This can be used to find the reciprocal for any number (except 0).

In Section 2.2, the trigonometric functions were defined as

$$\sin\theta = \frac{y}{r} \qquad\qquad \cos\theta = \frac{x}{r} \qquad\qquad \tan\theta = \frac{y}{x} \quad (x \neq 0)$$

$$\csc\theta = \frac{r}{y} \quad (y \neq 0) \qquad \sec\theta = \frac{r}{x} \quad (x \neq 0) \qquad \cot\theta = \frac{x}{y} \quad (y \neq 0)$$

WORDS	MATH
Write the definition of the cosecant function.	$\csc\theta = \dfrac{r}{y} \qquad (y \neq 0)$
The reciprocal identity holds for $y \neq 0$.	$\csc\theta = \dfrac{r}{y} = \dfrac{1}{\dfrac{y}{r}} \qquad y \neq 0$
Substitute the definition of sine, $\sin\theta = \dfrac{y}{r}$.	$\csc\theta = \dfrac{r}{y} = \dfrac{1}{\dfrac{y}{r}} = \dfrac{1}{\sin\theta}$
Write the $y \neq 0$ in terms of θ.	$\sin\theta \neq 0$
State the reciprocal identity.	$\csc\theta = \dfrac{1}{\sin\theta} \qquad \sin\theta \neq 0$

An equivalent reciprocal identity can be written as $\sin\theta = \dfrac{1}{\csc\theta}$. There is no need to make the restriction that $\csc\theta \neq 0$, since $\csc\theta = \dfrac{r}{y}$ is always greater than or equal to 1.

The following box summarizes the *reciprocal identities* that are true for all values of θ that do not correspond with the trigonometric function in the denominator having a zero value.

RECIPROCAL IDENTITIES

Reciprocal Identities	Equivalent Forms	Domain Restriction
$\csc\theta = \dfrac{1}{\sin\theta}$	$\sin\theta = \dfrac{1}{\csc\theta}$	$\theta = n\pi \quad n = \text{integer}$
$\sec\theta = \dfrac{1}{\cos\theta}$	$\cos\theta = \dfrac{1}{\sec\theta}$	$\theta \neq \dfrac{n\pi}{2} \quad n = \text{odd integer}$
$\cot\theta = \dfrac{1}{\tan\theta}$	$\tan\theta = \dfrac{1}{\cot\theta}$	$\theta \neq \dfrac{n\pi}{2} \quad n = \text{integer}$

STUDY TIP

Reciprocals always have the same algebraic sign.

Reciprocals always have the same algebraic sign. Therefore, if sine is positive, then cosecant is positive. If cosine is negative, then secant is negative.

EXAMPLE 1 Using Reciprocal Identities

a. If $\cos\theta = -\dfrac{1}{2}$, find $\sec\theta$.

b. If $\sin\theta = \dfrac{\sqrt{3}}{2}$, find $\csc\theta$.

c. If $\tan\theta = c$, find $\cot\theta$.

Solution (a):

Secant is the reciprocal of cosine. $\sec\theta = \dfrac{1}{\cos\theta}$

Substitute $\cos\theta = -\dfrac{1}{2}$ into the secant expression. $\sec\theta = \dfrac{1}{-\dfrac{1}{2}}$

Simplify. $\boxed{\sec\theta = -2}$

Solution (b):

Cosecant is the reciprocal of sine. $\csc\theta = \dfrac{1}{\sin\theta}$

Substitute $\sin\theta = \dfrac{\sqrt{3}}{2}$ into the cosecant expression. $\csc\theta = \dfrac{1}{\dfrac{\sqrt{3}}{2}}$

Simplify. $\csc\theta = \dfrac{2}{\sqrt{3}}$

Rationalize the radical in the denominator. $\csc\theta = \dfrac{2}{\sqrt{3}} \cdot \dfrac{\sqrt{3}}{\sqrt{3}}$

$\boxed{\csc\theta = \dfrac{2\sqrt{3}}{3}}$

Solution (c):

Cotangent is the reciprocal of tangent. $\cot\theta = \dfrac{1}{\tan\theta}$

Substitute $\tan\theta = c$ into the cotangent expression. $\cot\theta = \dfrac{1}{c}$

State any restrictions on c. $\cot\theta = \dfrac{1}{c} \qquad c \neq 0$

YOUR TURN **a.** If $\cos\theta = -1$, find $\sec\theta$.

b. If $\sin\theta = x$, find $\csc\theta$.

Quotient Identities

Let us now use the trigonometric definitions to derive the *quotient identities.*

WORDS	MATH
Write the definition of the tangent function.	$\tan\theta = \dfrac{y}{x} \qquad (x \neq 0)$
Multiply the two sides of the equation by $\dfrac{1/r}{1/r}$.	$\tan\theta = \dfrac{\dfrac{y}{r}}{\dfrac{x}{r}} \qquad (x \neq 0)$
Substitute $\sin\theta = \dfrac{y}{r}$ and $\cos\theta = \dfrac{x}{r}$.	$\tan\theta = \dfrac{\dfrac{y}{r}}{\dfrac{x}{r}} = \dfrac{\sin\theta}{\cos\theta}$
Write the $x \neq 0$ in terms of θ.	$\cos\theta \neq 0$
State the quotient identity.	$\tan\theta = \dfrac{\sin\theta}{\cos\theta} \qquad \cos\theta \neq 0$

A similar quotient identity can be written as
$\cot\theta = \dfrac{\cos\theta}{\sin\theta}$ where $\sin\theta \neq 0$.

The following box summarizes the quotient identities that are true for all values of θ that do not correspond to the trigonometric function in the denominator having a zero value.

QUOTIENT IDENTITIES

$$\tan\theta = \frac{\sin\theta}{\cos\theta} \qquad \cot\theta = \frac{\cos\theta}{\sin\theta}$$

Answers a. $\sec\theta = -1$ **b.** $\csc\theta = \dfrac{1}{x} \qquad x \neq 0$

EXAMPLE 2 Using the Quotient Identities

If $\sin\theta = \dfrac{3}{5}$ and $\cos\theta = -\dfrac{4}{5}$, find $\tan\theta$ and $\cot\theta$.

Solution:

Write the quotient identity involving tangent.

$$\tan\theta = \frac{\sin\theta}{\cos\theta}$$

Write the quotient identity involving cotangent.

$$\cot\theta = \frac{\cos\theta}{\sin\theta}$$

Substitute $\sin\theta = \dfrac{3}{5}$ and

$\cos\theta = -\dfrac{4}{5}$.

$$\tan\theta = \frac{\left(\dfrac{3}{5}\right)}{\left(-\dfrac{4}{5}\right)} \quad \text{and} \quad \cot\theta = \frac{\left(-\dfrac{4}{5}\right)}{\left(\dfrac{3}{5}\right)}$$

Simplify.

$$\tan\theta = -\frac{3}{4} \quad \text{and} \quad \cot\theta = -\frac{4}{3}$$

Note: We could have found $\tan\theta$ and then used the reciprocal identity, $\cot\theta = \dfrac{1}{\tan\theta}$, to find the cotangent function value.

■ **YOUR TURN**   If $\sin\theta = \dfrac{4}{5}$ and $\cos\theta = \dfrac{3}{5}$, find $\tan\theta$ and $\cot\theta$.

Pythagorean Identities

Before we develop the *Pythagorean identities*, we first need to discuss notation.

OPERATION	SHORTHAND NOTATION	WORDS	EXAMPLE
$(\sin\theta)^2$	$\sin^2\theta$	Sine squared of theta	$\sin\theta = \dfrac{\sqrt{3}}{2}$, then $\sin^2\theta = \left(\dfrac{\sqrt{3}}{2}\right)^2 = \dfrac{3}{4}$
$(\cos\theta)^3$	$\cos^3\theta$	Cosine cubed of theta	$\cos\theta = -\dfrac{1}{2}$, then $\cos^3\theta = \left(-\dfrac{1}{2}\right)^3 = -\dfrac{1}{8}$

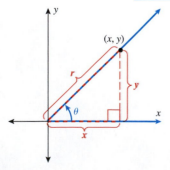

Let us now derive our first *Pythagorean identity*. Recall that in the Cartesian plane, the relationship between the coordinates x and y and the distance, r, from the origin to the point (x, y) is given by $x^2 + y^2 = r^2$.

■ **Answer:** $\tan\theta = \dfrac{4}{3}$ and $\cot\theta = \dfrac{3}{4}$

WORDS	MATH
Write the distance relationship in the Cartesian plane.	$x^2 + y^2 = r^2$
Divide the equation by r^2.	$\dfrac{x^2 + y^2}{r^2} = \dfrac{r^2}{r^2}$
Simplify.	$\dfrac{x^2}{r^2} + \dfrac{y^2}{r^2} = 1$
Use the power rule of exponents, $\dfrac{a^2}{b^2} = \left(\dfrac{a}{b}\right)^2$.	$\left(\dfrac{x}{r}\right)^2 + \left(\dfrac{y}{r}\right)^2 = 1$
Substitute the sine and cosine definitions, $\sin\theta = \dfrac{y}{r}$ and $\cos\theta = \dfrac{x}{r}$ into the left side of the equation.	$(\cos\theta)^2 + (\sin\theta)^2 = 1$
Use shorthand notation.	$\cos^2\theta + \sin^2\theta = 1$

Note: $\cos^2\theta + \sin^2\theta = 1$ and $\sin^2\theta + \cos^2\theta = 1$ are equivalent identities.

Two important things to note at this time are:

- The angle must be the same for this identity to hold.
- The sum of the squares of sine and cosine of the same angle is always 1.

Let's verify both with special angle values and a calculator:

θ	$\sin\theta$	$\cos\theta$	$\sin^2\theta + \cos^2\theta = 1$
30°	$\dfrac{1}{2}$	$\dfrac{\sqrt{3}}{2}$	$\left(\dfrac{1}{2}\right)^2 + \left(\dfrac{\sqrt{3}}{2}\right)^2 = \dfrac{1}{4} + \dfrac{3}{4} = 1$
27°	0.4540	0.8910	$(0.4540)^2 + (0.8910)^2 = 0.999997 \approx 1$

The *second Pythagorean identity* can be derived from the first.

WORDS	MATH
Write the first Pythagorean identity.	$\sin^2\theta + \cos^2\theta = 1$
Divide both sides by $\cos^2\theta$.	$\dfrac{\sin^2\theta}{\cos^2\theta} + \dfrac{\cos^2\theta}{\cos^2\theta} = \dfrac{1}{\cos^2\theta}$
Use longhand notation.	$\dfrac{(\sin\theta)^2}{(\cos\theta)^2} + \dfrac{(\cos\theta)^2}{(\cos\theta)^2} = \dfrac{1}{(\cos\theta)^2}$
Use the property of exponents.	$\left(\dfrac{\sin\theta}{\cos\theta}\right)^2 + \left(\dfrac{\cos\theta}{\cos\theta}\right)^2 = \left(\dfrac{1}{\cos\theta}\right)^2$
Use the quotient and reciprocal identities $\tan\theta = \dfrac{\sin\theta}{\cos\theta}$ and $\sec\theta = \dfrac{1}{\cos\theta}$ to write the expressions in terms of tangent and secant functions.	$(\tan\theta)^2 + 1 = (\sec\theta)^2$
Use shorthand notation.	$\tan^2\theta + 1 = \sec^2\theta$

In this derivation, had we divided by $\sin^2\theta$ instead of $\cos^2\theta$, we would have arrived at the *third Pythagorean identity*, $1 + \cot^2\theta = \csc^2\theta$. It is important to note in the second and third identities that we divided by the squares of cosine and sine functions. Therefore, the second and third identities only hold for angles for which the functions are defined.

PYTHAGOREAN IDENTITIES

$$\sin^2\theta + \cos^2\theta = 1 \qquad \tan^2\theta + 1 = \sec^2\theta \qquad 1 + \cot^2\theta = \csc^2\theta$$

TECHNOLOGY TIP

To check the answers

$\cos\theta = -\dfrac{3}{5}$ and $\tan\theta = -\dfrac{4}{3}$,

use the Pythagorean identity

$1 + \tan^2\theta = \sec^2\theta$ [1] [+] [(]

[(-)] [4] [÷] [3] [)] [x²] [ENTER]

Use the key [Frac] to change a decimal number to a fraction. Type

[MATH] [1: ► Frac] [ENTER] [ENTER]

[(] [(-)] [5] [÷] [3] [)] [x²]

[MATH] [1: ► Frac] [ENTER] [ENTER]

```
1+(-4/3)²
        2.777777778
Ans►Frac
             25/9
(-5/3)²►Frac
             25/9
```

EXAMPLE 3 Using the Identities to Find Trigonometric Function Values

Find $\cos\theta$ and $\tan\theta$ if $\sin\theta = \dfrac{4}{5}$ and the terminal side of θ lies in quadrant II.

Solution:

Write the first Pythagorean identity. $\sin^2\theta + \cos^2\theta = 1$

Substitute $\sin\theta = \dfrac{4}{5}$ into the equation. $\left(\dfrac{4}{5}\right)^2 + \cos^2\theta = 1$

Eliminate parentheses. $\dfrac{16}{25} + \cos^2\theta = 1$

Subtract $\dfrac{16}{25}$ from both sides of the equation. $\cos^2\theta = \dfrac{9}{25}$

Use the square root property. $\cos\theta = \pm\sqrt{\dfrac{9}{25}}$

Simplify. $\cos\theta = \pm\dfrac{3}{5}$

Cosine is negative in quadrant II. $\cos\theta = -\dfrac{3}{5}$

Write the quotient identity involving tangent. $\tan\theta = \dfrac{\sin\theta}{\cos\theta}$

Substitute $\sin\theta = \dfrac{4}{5}$ and $\cos\theta = -\dfrac{3}{5}$. $\tan\theta = \dfrac{\dfrac{4}{5}}{-\dfrac{3}{5}}$

Simplify. $\tan\theta = -\dfrac{4}{3}$

CONCEPT CHECK Is the sine function positive or negative in quadrant III?

■ **YOUR TURN** Find $\sin\theta$ and $\tan\theta$ if $\cos\theta = -\dfrac{4}{5}$ and the terminal side of θ lies in quadrant III.

Answer: $\sin\theta = -\dfrac{3}{5}$ and $\tan\theta = \dfrac{3}{4}$ ■

EXAMPLE 4 Using the Identities to Find Trigonometric Function Values

Find $\sin\theta$ and $\cos\theta$ if $\tan\theta = \dfrac{3}{4}$ and if the terminal side of θ lies in quadrant III.

COMMON MISTAKE

A common mistake when given the tangent function is assuming that the numerator is the value of sine and the denominator is the value of cosine.

 CORRECT

Write the second Pythagorean identity.

$$\tan^2\theta + 1 = \sec^2\theta$$

Substitute $\tan\theta = \dfrac{3}{4}$ into the equation.

$$\left(\frac{3}{4}\right)^2 + 1 = \sec^2\theta$$

Eliminate parentheses.

$$\frac{9}{16} + 1 = \sec^2\theta$$

Simplify.

$$\sec^2\theta = \frac{25}{16}$$

Apply the square root property.

$$\sec\theta = \pm\sqrt{\frac{25}{16}} = \pm\frac{5}{4}$$

Substitute the reciprocal identity, $\sec\theta = \dfrac{1}{\cos\theta}$.

$$\frac{1}{\cos\theta} = \pm\frac{5}{4}$$

Solve for $\cos\theta$.

$$\cos\theta = \pm\frac{4}{5}$$

Cosine is negative in quadrant III.

$$\cos\theta = -\frac{4}{5}$$

Write the first Pythagorean identity.

$$\sin^2\theta + \cos^2\theta = 1$$

 INCORRECT

Write the quotient identity involving tangent.

$$\tan\theta = \frac{\sin\theta}{\cos\theta}$$

Substitute $\tan\theta = \dfrac{3}{4}$ into the equation.

$$\frac{3}{4} = \frac{\sin\theta}{\cos\theta}$$

Identify $\sin\theta$ and $\cos\theta$.

$\sin\theta = 3$, $\cos\theta = 4$ **ERROR**

Recall the ranges of sine and cosine.

$-1 \le \sin\theta \le 1$ and
$-1 \le \cos\theta \le 1$

Substitute $\cos\theta = -\dfrac{4}{5}$.

$$\sin^2\theta + \left(-\frac{4}{5}\right)^2 = 1$$

Simplify.

$$\sin^2\theta = \frac{9}{25}$$

Apply the square root property.

$$\sin\theta = \pm\sqrt{\frac{9}{25}} = \pm\frac{3}{5}$$

Sine is negative in quadrant III.

$$\sin\theta = -\frac{3}{5}$$

In Chapter 5 we will cover trigonometric identities in more detail. For now, we can use these basic identities to simplify expressions.

EXAMPLE 5 Using Identities to Simplify Expressions

Multiply $(1 - \cos\theta)(1 + \cos\theta)$ and simplify.

Solution:

Multiply using the FOIL method.

$$(1 - \cos\theta)(1 + \cos\theta) = 1 - \cos\theta + \cos\theta - \cos^2\theta$$

Combine like terms.

$$= 1 - \cos\theta + \cos\theta - \cos^2\theta$$

$$= 1 - \cos^2\theta$$

Rewrite the Pythagorean identity, $\sin^2\theta + \cos^2\theta = 1$.

$$\sin^2\theta = 1 - \cos^2\theta$$

$$(1 - \cos\theta)(1 + \cos\theta) = \sin^2\theta$$

■ **YOUR TURN** Multiply $(1 - \sin\theta)(1 + \sin\theta)$ and simplify.

SECTION 2.4 SUMMARY

In this section, basic trigonometric identities (reciprocal, quotient, and Pythagorean) were used to calculate trigonometric function values given other trigonometric function values and information regarding which quadrant the terminal side of the angle lies in. Basic trigonometric identities can be used to simplify expressions.

■ **Answer:** $\cos^2\theta$

SECTION 2.4 EXERCISES

■ SKILLS

In Exercises 1–10, use a reciprocal identity to find the function value indicated. Rationalize denominators if necessary.

1. If $\cos\theta = \dfrac{7}{8}$, find $\sec\theta$.

2. If $\sin\theta = -\dfrac{3}{7}$, find $\csc\theta$.

3. If $\sin\theta = -0.6$, find $\csc\theta$.

4. If $\cos\theta = 0.8$, find $\sec\theta$.

5. If $\tan\theta = -5$, find $\cot\theta$.

6. If $\tan\theta = 0.5$, find $\cot\theta$.

7. If $\csc\theta = \dfrac{\sqrt{5}}{2}$, find $\sin\theta$.

8. If $\sec\theta = \dfrac{\sqrt{11}}{2}$, find $\cos\theta$.

9. If $\cot\theta = -\dfrac{\sqrt{7}}{5}$, find $\tan\theta$.

10. If $\cot\theta = 3.5$, find $\tan\theta$.

In Exercises 11–18, use a quotient identity to find the function value indicated. Rationalize denominators if necessary.

11. If $\sin\theta = -\dfrac{1}{2}$ and $\cos\theta = \dfrac{\sqrt{3}}{2}$, find $\tan\theta$.

12. If $\sin\theta = -\dfrac{1}{2}$ and $\cos\theta = \dfrac{\sqrt{3}}{2}$, find $\cot\theta$.

13. If $\sin\theta = -0.6$ and $\cos\theta = -0.8$, find $\cot\theta$.

14. If $\sin\theta = -0.6$ and $\cos\theta = -0.8$, find $\tan\theta$.

15. If $\sin\theta = -\dfrac{\sqrt{11}}{6}$ and $\cos\theta = -\dfrac{5}{6}$, find $\tan\theta$.

16. If $\sin\theta = -\dfrac{\sqrt{11}}{6}$ and $\cos\theta = -\dfrac{5}{6}$, find $\cot\theta$.

17. If $\sin\theta = a$ and $\cos\theta = b$, find $\cot\theta$ and state any restrictions on a or b.

18. If $\sin\theta = a$ and $\cos\theta = b$, find $\tan\theta$ and state any restrictions on a or b.

In Exercises 19–22, find the indicated expression.

19. If $\sin\theta = \dfrac{\sqrt{5}}{8}$, find $\sin^2\theta$.

20. If $\cos\theta = 0.1$ find $\cos^3\theta$.

21. If $\csc\theta = -2$ find $\csc^3\theta$.

22. If $\tan\theta = -5$ find $\tan^2\theta$.

In Exercises 23–36, use a Pythagorean identity to find the function value indicated. Rationalize denominators if necessary.

23. If $\sin\theta = -\dfrac{1}{2}$ and the terminal side of θ lies in quadrant III, find $\cos\theta$.

24. If $\sin\theta = -\dfrac{3}{5}$ and the terminal side of θ lies in quadrant III, find $\cos\theta$.

25. If $\cos\theta = \dfrac{2}{5}$ and the terminal side of θ lies in quadrant IV, find $\sin\theta$.

26. If $\cos\theta = \dfrac{2}{7}$ and the terminal side of θ lies in quadrant IV, find $\sin\theta$.

27. If $\tan\theta = 4$ and the terminal side of θ lies in quadrant III, find $\sec\theta$.

28. If $\tan\theta = -5$ and the terminal side of θ lies in quadrant II, find $\sec\theta$.

29. If $\cot\theta = 2$ and the terminal side of θ lies in quadrant III, find $\csc\theta$.

30. If $\cot\theta = -3$ and the terminal side of θ lies in quadrant II, find $\csc\theta$.

31. If $\sin\theta = \dfrac{8}{15}$ and the terminal side of θ lies in quadrant II, find $\tan\theta$.

32. If $\sin\theta = -\dfrac{7}{15}$ and the terminal side of θ lies in quadrant III, find $\tan\theta$.

33. If $\cos\theta = -\dfrac{7}{15}$ and the terminal side of θ lies in quadrant III, find $\csc\theta$.

34. If $\cos\theta = -\dfrac{8}{15}$ and the terminal side of θ lies in quadrant II, find $\csc\theta$.

35. If $\sec\theta = \dfrac{\sqrt{13}}{3}$ and the terminal side of θ lies in quadrant IV, find $\sin\theta$.

36. If $\sec\theta = -\dfrac{\sqrt{61}}{5}$ and the terminal side of θ lies in quadrant II, find $\sin\theta$.

In Exercises 37–42, use identities to find the function value indicated. Rationalize denominators if necessary.

37. Find $\sin\theta$ and $\cos\theta$ if $\tan\theta = -\dfrac{4}{3}$ and the terminal side of θ lies in quadrant II.

38. Find $\sin\theta$ and $\cos\theta$ if $\tan\theta = -\dfrac{3}{4}$ and the terminal side of θ lies in quadrant II.

39. Find $\sin\theta$ and $\cos\theta$ if $\tan\theta = 2$ and the terminal side of θ lies in quadrant III.

40. Find $\sin\theta$ and $\cos\theta$ if $\tan\theta = 5$ and the terminal side of θ lies in quadrant III.

41. Find $\sin\theta$ and $\cos\theta$ if $\tan\theta = 0.6$ and the terminal side of θ lies in quadrant III.

42. Find $\sin\theta$ and $\cos\theta$ if $\tan\theta = 0.8$ and the terminal side of θ lies in quadrant III.

In Exercises 43–52, perform the indicated operation and simplify your answers if possible. Leave all answers in terms of $\sin\theta$ and $\cos\theta$.

43. $\sec\theta\cot\theta$

44. $\csc\theta\tan\theta$

45. $\tan^2\theta - \sec^2\theta$

46. $\cot^2\theta - \csc^2\theta$

47. $\csc\theta - \sin\theta$

48. $\sec\theta - \cos\theta$

49. $\dfrac{\sec\theta}{\tan\theta}$

50. $\dfrac{\csc\theta}{\cot\theta}$

51. $(\sin\theta + \cos\theta)^2$

52. $(\sin\theta - \cos\theta)^2$

 ■ **APPLICATIONS**

The *bifolium* is a curve that can be drawn using either an algebraic equation or an equation using trigonometric functions. Even though the trigonometric equation uses polar

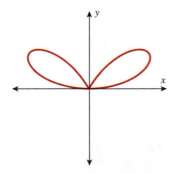

coordinates, it's much easier to solve the trigonometric equation for function values than to solve the algebraic equation. The graph of $r = 8\sin\theta\cos^2\theta$ is shown below. You'll see more on polar coordinates and graphs of polar equations in Chapter 8. The following graph is in polar coordinates.

53. **Bifolium.** The equation for the bifolium on the left is $r = 8\sin\theta\cos^2\theta$. Use a Pythagorean identity to rewrite the equation using just the function $\sin\theta$. Then find r if $\theta = 30°, 60°,$ and $90°$.

54. **Bifolium.** The equation for the bifolium on the left is $r = 8\sin\theta\cos^2\theta$. Use a Pythagorean identity to rewrite the equation using just the function $\sin\theta$. Then find r if $\theta = -30°, -60°,$ and $-90°$.

■ CATCH THE MISTAKE

In Exercises 55 and 56, explain the mistake that is made.

55. If $\sin\theta = -\dfrac{1}{3}$ and the terminal side of θ lies in quadrant III, find $\cos\theta$.

Solution:

Write the first Pythagorean identity.	$\sin^2\theta + \cos^2\theta = 1$
Substitute $\sin\theta = -\dfrac{1}{3}$ into the identity.	$\left(-\dfrac{1}{3}\right)^2 + \cos^2\theta = 1$
Eliminate parentheses.	$\dfrac{1}{9} + \cos^2\theta = 1$
Subtract $\dfrac{1}{9}$ from both sides.	$\cos^2\theta = \dfrac{8}{9}$
Take the square root of both sides.	$\cos\theta = \sqrt{\dfrac{8}{9}}$
Simplify.	$\cos\theta = \dfrac{2\sqrt{2}}{3}$

This is incorrect. What mistake was made?

56. Find $\sin\theta$ and $\cos\theta$ if $\tan\theta = -\dfrac{1}{4}$ and the terminal side of θ lies in quadrant II.

Solution:

Write the quotient identity.	$\tan\theta = \dfrac{\sin\theta}{\cos\theta}$
Substitute $\tan\theta = -\dfrac{1}{4}$ into the identity.	$-\dfrac{1}{4} = \dfrac{\sin\theta}{\cos\theta}$
In quadrant II, sine is positive and cosine is negative.	$\dfrac{1}{-4} = \dfrac{\sin\theta}{\cos\theta}$
Identify sine and cosine.	$\sin\theta = 1$ and $\cos\theta = -4$

This is incorrect. What mistake was made?

■ CHALLENGE

57. T or F: It is possible for $\sin\theta > 0$ and $\csc\theta < 0$.

58. T or F: It is possible for $\cos\theta < 0$ and $\sec\theta > 0$.

59. Find the measure of an angle θ, $0° < \theta \le 180°$, that satisfies $\sin\theta = \csc\theta$.

60. Find the measure of an angle θ, $0° < \theta \le 180°$, that satisfies $\cos\theta = \sec\theta$.

61. Write $\cot\theta$ in terms of only $\sin\theta$.

62. Write $\csc\theta$ in terms of only $\cos\theta$.

63. Let $x = 8\sin\theta$ in the expression $\sqrt{64 - x^2}$ and simplify.

64. Let $x = 6\cos\theta$ in the expression $\sqrt{36 - x^2}$ and simplify.

■ TECHNOLOGY

65. Verify the quotient $\tan\theta = \dfrac{\sin\theta}{\cos\theta}$ numerically for $\theta = 22°$. Use a calculator to find

a. $\sin 22°$ **b.** $\cos 22°$ **c.** $\dfrac{\sin 22°}{\cos 22°}$ **d.** $\tan 22°$

Are the results in (c) and (d) the same?

66. Is $\tan 10° = \dfrac{\sin 160°}{\cos 16°}$? Use a calculator to find

a. $\sin 160°$ **b.** $\cos 16°$ **c.** $\dfrac{\sin 160°}{\cos 16°}$ **d.** $\tan 10°$

Are the results in (c) and (d) the same?

Gary is installing a wheelchair ramp leading up the front porch of his family's home for his brother. Since the ADA (Americans with Disabilities Act) recommends an incline of about 5 degrees for wheelchair ramps, Gary uses the planned horizontal length (13 feet) and height (1 foot) of the ramp to estimate its angle of inclination (θ). Here are his calculations:

$$\tan\theta = \frac{\text{opp}}{\text{adj}} = \frac{1}{13}$$

$$\theta = \tan^{-1}\left(\frac{1}{13}\right) \approx 0.077°$$

However, from his scaled drawing Gary can tell that the angle is steeper than 0.077°. What is wrong with his calculations?

A: Gary's calculator was in radian mode rather than degree mode. Thus, his calculated value of θ is in radians, not degrees. Once this has been corrected, he calculates

$$\theta = \tan^{-1}\left(\frac{1}{13}\right) \approx 4.4°$$

So his angle of inclination is acceptable.

Now estimate the measures of the following angles in degrees on your calculator.

$$\theta = \cos^{-1}(0.52)$$
$$\alpha = \sin^{-1}(0.95)$$
$$\gamma = \tan^{-1}(-27)$$

TYING IT ALL TOGETHER

A plane is flying on a heading of N 50° E. There is a monument with a bearing of N 35° E that is 100 feet tall and lies 300 feet from the plane's current position (at the origin of the coordinate system). The airport is 6 miles away in the direction of the plane's flight. Call the reference angle α that is made with the terminal side of the angle as the flight path, and call the reference angle β with the terminal side of the angle as the path to the monument. Find α and β, and state the coordinates of the monument (the plane's current position is still at the origin of the coordinate system).

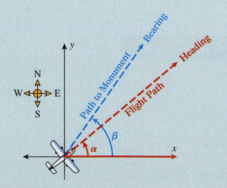

SECTION	TOPIC	PAGES	REVIEW EXERCISES	KEY CONCEPTS
2.1	Angles in the Cartesian plane	72–77	1–12	
	Angles in standard position	72–74	1–8	

$$
\begin{array}{c}
90^\circ \\
\uparrow y
\end{array}
$$

90° < θ < 180°   0° < θ < 90°

QII   QI

180° ←————————→ 0° or 360°   (x)

180° < θ < 270°   270° < θ < 360°

QIII   QIV

270°

SECTION	TOPIC	PAGES	REVIEW EXERCISES	KEY CONCEPTS
	Coterminal angles	74–75	9–12	Two angles are coterminal if they share the same terminal side. The measures of coterminal angles differ by multiples of 360°.
	Common angles	76–77		

$\left(-\frac{1}{2}, \frac{\sqrt{3}}{2}\right)$ $\left(\frac{1}{2}, \frac{\sqrt{3}}{2}\right)$

$\left(-\frac{\sqrt{2}}{2}, \frac{\sqrt{2}}{2}\right)$ 120°   60° $\left(\frac{\sqrt{2}}{2}, \frac{\sqrt{2}}{2}\right)$

135"   45°

$\left(-\frac{\sqrt{3}}{2}, \frac{1}{2}\right)$ 150°   30° $\left(\frac{\sqrt{3}}{2}, \frac{1}{2}\right)$

−1   1   x

$\left(-\frac{\sqrt{3}}{2}, -\frac{1}{2}\right)$ 210°   330° $\left(\frac{\sqrt{3}}{2}, -\frac{1}{2}\right)$

$\left(-\frac{\sqrt{2}}{2}, -\frac{\sqrt{2}}{2}\right)$ 225°   315° $\left(\frac{\sqrt{2}}{2}, -\frac{\sqrt{2}}{2}\right)$

240°   −1   300°

$\left(-\frac{1}{2}, -\frac{\sqrt{3}}{2}\right)$ $\left(\frac{1}{2}, -\frac{\sqrt{3}}{2}\right)$

SECTION	TOPIC	PAGES	REVIEW EXERCISES	KEY CONCEPTS
2.2	Definition 2 of trigonometric functions: Cartesian plane	80–87	13–22	$\sin\theta = \dfrac{y}{r}$ $\cos\theta = \dfrac{x}{r}$ $\tan\theta = \dfrac{y}{x}$ $\csc\theta = \dfrac{r}{y}$ $\sec\theta = \dfrac{r}{x}$ $\cot\theta = \dfrac{x}{y}$ where $x^2 + y^2 = r^2$. The distance, r, is positive: $r > 0$.
	Trigonometric function values for quadrantal angles	85–87	23–26	U—Undefined

θ	0°	90°	180°	270°
$\sin\theta$	0	1	0	−1
$\cos\theta$	1	0	−1	0
$\tan\theta$	0	U	0	U
$\cot\theta$	U	0	U	0
$\sec\theta$	1	U	−1	U
$\csc\theta$	U	1	U	−1

SECTION	TOPIC	PAGES	REVIEW EXERCISES	KEY CONCEPTS
2.3	Evaluating trigonometric functions for any angle	89–103	27–48	**Step 1:** Find the quadrant in which the terminal side of the angle lies. **Step 2:** Find the reference angle. **Step 3:** Find the trigonometric value of the reference angle. **Step 4:** Determine the correct algebraic sign $(+/-)$ based on the quadrant found in Step 1. **Step 5:** Combine Steps 3 and 4.
	Algebraic signs	90–93	27–34	(see table below)

θ	QI	QII	QIII	QIV
$\sin\theta$	+	+	−	−
$\cos\theta$	+	−	−	+
$\tan\theta$	+	−	+	−

When $x > 0$ cosine is positive
When $x < 0$ cosine is negative
When $y > 0$ sine is positive
When $y < 0$ sine is negative

SECTION	TOPIC	PAGES	REVIEW EXERCISES	KEY CONCEPTS
	Trigonometric ranges	93–94	39–42	$-1 \leq \sin\theta \leq 1$ and $-1 \leq \cos\theta \leq 1$
	Reference angles	95–103	43–48	The reference angle, α, for angle θ is given by ■ QI: $\alpha = \theta$ ■ QII: $\alpha = 180° - \theta$ ■ QIII: $\alpha = \theta - 180°$ ■ QIV: $\alpha = 360° - \theta$
2.4	Basic trigonometric identities	106–114	57–76	Reciprocal identities Quotient identities Pythagorean identities
	Reciprocal identities	107–109	57–70	$\csc\theta = \dfrac{1}{\sin\theta}$ $\sec\theta = \dfrac{1}{\cos\theta}$ $\cot\theta = \dfrac{1}{\tan\theta}$
	Quotient identities	109–110	57–70	$\tan\theta = \dfrac{\sin\theta}{\cos\theta}$ $\cot\theta = \dfrac{\cos\theta}{\sin\theta}$
	Pythagorean identities	110–114	57–70	$\sin^2\theta + \cos^2\theta = 1$ $\tan^2\theta + 1 = \sec^2\theta$ $1 + \cot^2\theta = \csc^2\theta$

2.1 Angles in the Cartesian Plane

State in what quadrant or on what axes the following angles with given measure in standard position would lie.

1. $300°$ 2. $280°$ 3. $150°$ 4. $-180°$

Sketch the angles with given measure in standard position.

5. $120°$ 6. $-30°$ 7. $450°$ 8. $-185°$

Determine the angles of the smallest possible positive measure that are coterminal with the following angles.

9. $1000°$ 10. $-800°$ 11. $480°$ 12. $-10°$

2.2 Definition 2 of Trigonometric Functions: Cartesian Plane

The terminal side of an angle, θ, in standard position passes through the indicated point. Calculate the values of the six trigonometric functions for angle θ.

13. $(6, -8)$ 14. $(-24, -7)$ 15. $(-6, 2)$

16. $(-40, 9)$ 17. $(\sqrt{3}, 1)$ 18. $(-9, -9)$

Calculate the values for the six trigonometric functions of the angle θ, given in standard position, if the terminal side of θ is defined by the following lines.

19. $y = x, x \le 0$ 20. $y = -2x, x \le 0$

21. $3x + 4y = 0, x \ge 0$ 22. $5x - 12y = 0, x \ge 0$

Calculate (if possible) the values for the six trigonometric functions of the angle, θ, given in standard position.

23. $\theta = 270°$ 24. $\theta = 1080°$

25. $\theta = -1080°$ 26. $\theta = -360n°$, n is an integer

2.3 Evaluating Trigonometric Functions for Any Angle

Indicate the quadrants in which the terminal side of θ must lie in order for the following to be true.

27. $\sin\theta$ is negative and $\tan\theta$ is positive.

28. $\cos\theta$ is negative and $\csc\theta$ is positive.

29. $\tan\theta$ is negative and $\sec\theta$ is positive.

30. $\tan\theta$ and $\sin\theta$ are both negative.

Find the indicated trigonometric function values.

31. If $\tan\theta = -\dfrac{4}{3}$ and the terminal side of θ lies in QII, find $\sin\theta$.

32. If $\sin\theta = -\dfrac{8}{17}$ and the terminal side of θ lies in QIV, find $\cos\theta$.

33. If $\cos\theta = \dfrac{1}{3}$ and the terminal side of θ lies in QIV, find $\tan\theta$.

34. If $\cot\theta = \sqrt{3}$ and the terminal side of θ lies in QIII, find $\sin\theta$.

Evaluate the following expressions if possible.

35. $\cos 270° + \sin(-270°)$

36. $\tan 450° - \cot 540°$

37. $\sec(-720°) + \csc(-450°)$

38. $\tan 1080° - \sin(-1080°)$

Determine if each statement is possible or impossible.

39. $\sin\theta = -\sqrt{2}$ 40. $\cos\theta = 0.0004$

41. $\tan\theta = -5.4321$ 42. $\csc\theta = \dfrac{15}{17}$

Evaluate the following expressions exactly.

43. $\sin 330°$ 44. $\cos(-300°)$

45. $\tan 150°$ 46. $\cot 315°$

47. $\sec(-150°)$ 48. $\csc 210°$

Evaluate the trigonometric expressions with a calculator. Round to four decimal places.

49. $\sin(-14°)$ 50. $\cos 275°$

51. $\tan 46°$ 52. $\cot 500°$

53. $\sec 111°$ 54. $\csc 215°$

55. $\cot(-57°)$ 56. $\csc(-59°)$

2.4 Basic Trigonometric Identities

Use trigonometric identities to find the indicated value(s).

57. If $\sin\theta = -\dfrac{7}{11}$, find $\csc\theta$.

58. If $\cot\theta = \dfrac{3\sqrt{5}}{5}$, find $\tan\theta$.

59. If $\sin\theta = \dfrac{8}{17}$ and $\cos\theta = -\dfrac{15}{17}$, find $\tan\theta$.

60. If $\sin\theta = -\dfrac{12}{13}$ and $\cos\theta = -\dfrac{5}{13}$, find $\cot\theta$.

61. If $\sin\theta = \dfrac{2\sqrt{3}}{9}$ find $\sin^2\theta$.

62. If $\cos\theta = -0.45$ find $\cos^2\theta$.

63. If $\tan\theta = \sqrt{3}$ find $\tan^3\theta$.

64. If $\sin\theta = \dfrac{11}{61}$ and the terminal side of θ lies in quadrant II, find $\tan\theta$.

65. If $\cos\theta = \dfrac{7}{25}$ and the terminal side of θ lies in quadrant IV, find $\cot\theta$.

66. If $\tan\theta = -1$ and the terminal side of θ lies in quadrant IV, find $\sec\theta$.

67. If $\csc\theta = \dfrac{13}{12}$ and the terminal side of θ lies in quadrant II, find $\cot\theta$.

68. Find $\sin\theta$ and $\cos\theta$ if $\tan\theta = -\dfrac{24}{7}$ and the terminal side of θ lies in quadrant II.

69. Find $\sin\theta$ and $\cos\theta$ if $\cot\theta = \dfrac{\sqrt{3}}{3}$ and the terminal side of θ lies in quadrant III.

70. Find $\sin\theta$ and $\cos\theta$ if $\cot\theta = \sqrt{3}$ and the terminal side of θ lies in quadrant III.

Perform the operations and simplify. Write answers in terms of $\sin\theta$ and $\cos\theta$.

71. $\sec\theta\cot^2\theta$

72. $\sin\theta\cot\theta$

73. $\dfrac{\tan\theta}{\sec\theta}$

74. $(\cos\theta + \sec\theta)^2$

75. $\cot\theta(\sec\theta + \sin\theta)$

76. $\dfrac{\sin^2\theta}{\csc\theta}$

1. Fill in the table with exact values for the quadrantal angles and algebraic signs for the nonquadrantal angles.

	0°	QI	90°	QII	180°	QIII	270°	QIV	360°
$\sin\theta$									
$\cos\theta$									

2. If $\cot\theta < 0$ and $\sec\theta > 0$, in which quadrant does the terminal side of θ lie?

3. Find the values of θ, $0° \leq \theta \leq 360°$ for which $\tan\theta$ is undefined.

4. Are two angles with measures $-170°$ and $890°$ coterminal?

5. Find the smallest angle with positive measure that is coterminal with an angle that has measure $-500°$.

6. Find $\cos\theta$ if the terminal side of angle θ passes through the point $(-2, -5)$.

7. Find the value of $\sin\theta$ if the terminal side of angle θ lies along the graph of $y = |x|$.

8. Evaluate $\sin 210°$ exactly.

9. Evaluate $\tan(-315°)$ exactly.

10. Evaluate $\sec 135°$ exactly.

11. Evaluate $\cot 222°$ with a calculator. Round to four decimal places.

12. Find $\sin\theta$ and $\cos\theta$ if $\tan\theta = \dfrac{4}{3}$ and the terminal side of angle θ lies in quadrant III.

13. Find $\sin\theta$ and $\tan\theta$ if $\cos\theta = \dfrac{1}{6}$ and the terminal side of angle θ lies in quadrant IV.

14. Simplify the expression $\dfrac{(1 - \cos\theta)(1 + \cos\theta)}{\tan^2\theta}$ and write the answer only in terms of $\cos\theta$.

15. Which of the following is not possible?

   a. $\cos\theta = -\sqrt{3}$ b. $\tan\theta = -\sqrt{3}$

Radian Measure and the Unit Circle Approach

Courtesy Ford Motor Company

How does an odometer or speedometer on an automobile work? The transmission counts how many times the tires rotate (how many full revolutions take place) per second. A computer then calculates how far the car has traveled in that second by multiplying the number of revolutions by the tire circumference. Distance is given by the odometer, and the speedometer takes the distance per second and converts to miles per hour (or km/h). Realize that the computer chip is programmed to the tire designed for the vehicle. If a person were to change the tire size (smaller or larger than the original specifications), then the odometer and speedometer would need to be adjusted.

Suppose you bought a Ford Expedition Eddie Bauer Edition, which comes standard with 17″ rims (corresponding to a tire with 25.7″ diameter), and you decide to later upgrade these tires for 19″ rims (corresponding to a tire with 28.2″ diameter). If the onboard computer is not adjusted, is the actual speed faster or slower than the speedometer indicator?

In this case, the speedometer would read 9.6% too low. For example, if your speedometer read 60 mph, your actual speed would be 65.8 mph. You will see in this chapter that the *angular speed* (rotations of tires per second), *radius* (of the tires), and *linear speed* (speed of the automobile) are related.

In this chapter you will learn a second way to measure angles using radians. You will convert between degrees and radians. You will calculate arc lengths, areas of circular sectors, and angular and linear speeds. Finally, the third definition of trigonometric functions using the unit circle approach will be given. You will work with the trigonometric functions in the context of a unit circle.

Radian Measure and Circular Functions

Angle Measure

- Radians
- Relationship Between Radians and Degrees

Circles

- Arc Length
- Area of a Sector
- Linear and Angular Speeds

Trigonometric Functions

- Unit Circle Approach

CHAPTER OBJECTIVES

- Understand that radians are another measure of angles.
- Understand that radians are unitless real numbers.
- Convert between degrees and radians.
- Calculate the arc length and area of a circular sector.
- Relate angular and linear speeds.
- Define trigonometric functions using the unit circle approach.
- Draw the unit circle, and label sine and cosine values for special angles (in both degrees and radians).

NAVIGATION THROUGH SUPPLEMENTS

DIGITAL VIDEO SERIES #3

STUDENT SOLUTIONS MANUAL CHAPTER 3

BOOK COMPANION SITE
www.wiley.com/college/young

Radian Measure

In geometry and most applications, angles are measured in degrees. However, *radian measure* is another way to measure angles. Using radian measure allows us to write trigonometric functions as functions not only of angles but also of real numbers in general.

Recall that in Section 1.1 we defined one full rotation as an angle having measure 360°. Now we think of the angle in the context of a circle. A **central angle** is an angle that has its vertex at the center of a circle.

The measure of the central angle whose intercepted arc is equal in length to the radius is equal to 1 **radian**.

From geometry we know that the ratio of the measures of two angles is equal to the ratio of the lengths of the arcs subtended by those angles.

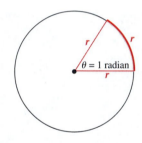

$$\frac{\theta_1}{\theta_2} = \frac{s_1}{s_2}$$

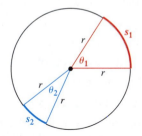

If $\theta_2 = 1$ radian, then the length of the subtended arc is equal to the radius, $s_2 = r$. This leads to a general definition of *radian measure*.

CAUTION

To correctly calculate radians from the formula $\theta = s/r$, the radius and arc length must be expressed in the same units.

DEFINITION   RADIAN MEASURE

If a central angle θ in a circle with radius r intercepts an arc on the circle of length s, then the measure of θ, in **radians**, is given by

$$\theta \text{ (in radians)} = \frac{s}{r}$$

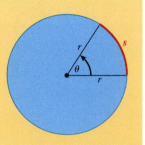

Note: The formula is valid only if s (arc length) and r (radius) are expressed in the same units.

Note that both s and r are measured in units of length. When both are given in the same units, the units cancel, giving the number of radians as a *dimensionless* (unitless) number.

One full rotation corresponds to an arc length equal to the circumference, $2\pi r$, of the circle with radius r. We find that one full rotation is equal to 2π radians.

$$\theta_{\text{full rotation}} = \frac{2\pi r}{r} = 2\pi$$

 EXAMPLE 1 Finding the Radian Measure of an Angle

What is the measure (in radians) of a central angle, θ, that intercepts an arc of length 6 centimeters on a circle with radius 2 meters?

COMMON MISTAKE

A common mistake is forgetting to first put the radius and arc length in the same units.

 CORRECT

Write the formula relating radian measure to arc length and radius.

$$\theta \text{ (in radians)} = \frac{s}{r}$$

Substitute $s = 6$ cm and $r = 2$ m into the radian expression.

$$\theta = \frac{6 \text{ cm}}{2 \text{ m}}$$

Convert the radius (2) meters to centimeters: 2 m = 200 cm

$$\theta = \frac{6 \text{ cm}}{200 \text{ cm}}$$

The units, cm, cancel and the result is a unitless real number.

$$\theta = 0.03 \text{ rad}$$

 INCORRECT

Write the formula relating radian measure to arc length and radius.

$$\theta \text{ (in radians)} = \frac{s}{r}$$

Substitute $s = 6$ cm and $r = 2$ m into the radian expression.

$$\theta = \frac{6 \text{ cm}}{2 \text{ m}}$$

Simplify.

ERROR $\theta = 3$ rad

YOUR TURN   What is the measure (in radians) of a central angle, θ, that intercepts an arc of length 12 mm on a circle with radius 4 cm?

In the above example the units, cm, canceled, therefore correctly giving *radians* as a unitless real number. Because radians are unitless, the word radians (or rad) is often omitted. If an angle measure is given simply as a real number, then radians are implied.

CAUTION

If no unit of angle measure is specified, then radian measure is implied.

WORDS	MATH
The measure of θ is 4 degrees.	$\theta = 4°$
The measure of θ is 4 radians.	$\theta = 4$

Converting Between Degrees and Radians

In order to convert between degrees and radians, we must first look for a relationship between them. We start by considering one full rotation around the circle. An angle corresponding to one full rotation is said to have measure 360°. Radians are defined as the ratio of the arc length that the angle sides and circle make to the radius of the circle. One full rotation corresponds to an arc length equal to the circumference of the circle.

WORDS	MATH
Write the angle measure (in degrees) that corresponds to one full rotation.	$\theta = 360°$
Write the angle measure (in radians) that corresponds to one full rotation.	
Arc length is the circumference of the circle.	$s = 2\pi r$
Substitute $s = 2\pi r$ into θ (in radians) $= \dfrac{s}{r}$.	$\theta = \dfrac{2\pi r}{r} = 2\pi$ radians
Equate the measures corresponding to one full rotation.	$360° = 2\pi$ radians
Divide by 2.	$180° = \pi$ radians
Divide 180°.	$1 = \dfrac{\pi \text{ radians}}{180°}$
The conversion factor:	$1 = \dfrac{\pi \text{ rad}}{180°}$ or $1 = \dfrac{180°}{\pi}$

This leads us to formulas that convert between degrees and radians. Let θ_d represent an angle measure given in degrees and θ_r represent an angle measure given in radians.

CONVERTING DEGREES TO RADIANS

To convert degrees to radians, multiply the degree measure by $\dfrac{\pi}{180°}$.

$$\theta_r = \theta_d\left(\frac{\pi}{180°}\right)$$

CONVERTING RADIANS TO DEGREES

To convert radians to degrees, multiply the radian measure by $\dfrac{180°}{\pi}$.

$$\theta_d = \theta_r\left(\frac{180°}{\pi}\right)$$

Before we begin converting between degrees and radians, let's first get a feel for radians. How many degrees is 1 radian? Use the technique outlined in the box above.

WORDS	**MATH**
Multiply 1 radian by $\dfrac{180°}{\pi}$.	$1\left(\dfrac{180°}{\pi}\right)$
Approximate π by 3.14.	$1\left(\dfrac{180°}{3.14}\right)$
Use a calculator to evaluate and round to the nearest degree.	$\approx 57°$

$$1 \text{ radian} \approx 57°$$

A radian is much larger than a degree. Let's compare two angles, one measuring 30 radians and the other measuring 30°. Note that $\dfrac{30 \text{ rad}}{2\pi \text{ rad}} \approx 4.77$ revolutions.

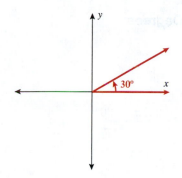
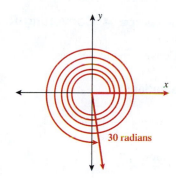

EXAMPLE 2 Converting Degrees to Radians

Convert 45° to radians.

Solution:

Multiply 45° by $\dfrac{\pi}{180°}$.

$$(45°)\left(\dfrac{\pi}{180°}\right) = \dfrac{45°\pi}{180°}$$

Simplify.

$$= \dfrac{\pi}{4} \text{ radians}$$

Note: $\dfrac{\pi}{4}$ is the exact value. A calculator can be used to evaluate this expression. Scientific and graphing calculators have a π button (on most scientific calculators it requires using a shift or second command). The decimal approximation rounded to three decimal places is 0.785.

Exact Value: $\dfrac{\pi}{4}$

Approximate Value: 0.785

■ **YOUR TURN** Convert 60° to radians.

■ **Answer:** $\dfrac{\pi}{3}$ or 1.047

EXAMPLE 3 Converting Degrees to Radians

Convert 472° to radians.

Solution:

Multiply 472° by $\dfrac{\pi}{180°}$.

$$472° \left(\dfrac{\pi}{180°} \right)$$

Simplify (factor out the common 4).

$$\dfrac{118}{45}\pi$$

Use a calculator to approximate.

$$\approx 8.238 \text{ radians}$$

■ **YOUR TURN** Convert 460° to radians.

EXAMPLE 4 Converting Radians to Degrees

Convert $\dfrac{2\pi}{3}$ to degrees.

Solution:

Multiply $\dfrac{2\pi}{3}$ by $\dfrac{180°}{\pi}$.

$$\dfrac{2\pi}{3} \cdot \dfrac{180°}{\pi}$$

Simplify.

$$120°$$

■ **YOUR TURN** Convert $\dfrac{3\pi}{2}$ to degrees.

We can now draw the unit circle with special angles in both **degrees** and **radians**.

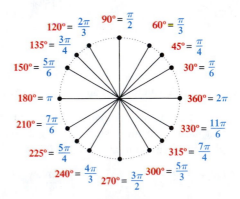

In order to evaluate trigonometric functions for angles measured in radians, we first convert from radians to degrees and then recall the exact trigonometric values for special angles. The following table lists sine and cosine values for special angles in both degrees and radians. Tangent, secant, cosecant, and cotangent can all be found from sine and cosine values using ratio and reciprocal identities. The table only lists quadrant I and quadrantal special angles. Values in quadrants II, III, and IV can be found using reference angles and knowledge of the algebraic sign $(+/-)$ of sine and cosine in each quadrant.

ANGLE, θ		VALUE OF TRIGONOMETRIC FUNCTION	
RADIANS	**DEGREES**	$\sin\theta$	$\cos\theta$
0	0°	0	1
$\dfrac{\pi}{6}$	30°	$\dfrac{1}{2}$	$\dfrac{\sqrt{3}}{2}$
$\dfrac{\pi}{4}$	45°	$\dfrac{\sqrt{2}}{2}$	$\dfrac{\sqrt{2}}{2}$
$\dfrac{\pi}{3}$	60°	$\dfrac{\sqrt{3}}{2}$	$\dfrac{1}{2}$
$\dfrac{\pi}{2}$	90°	1	0
π	180°	0	-1
$\dfrac{3\pi}{2}$	270°	-1	0
2π	360°	0	1

EXAMPLE 5 Evaluating Trigonometric Functions for Angles in Radian Measure

Evaluate $\sin\dfrac{\pi}{3}$ exactly.

Solution:

Convert $\dfrac{\pi}{3}$ to degrees. $\dfrac{\pi}{3}\cdot\dfrac{180°}{\pi}=60°$

Find the value of $\sin 60°$. $\sin 60°=\dfrac{\sqrt{3}}{2}$

Equate the sine of 60° and $\dfrac{\pi}{3}$. $\boxed{\sin\dfrac{\pi}{3}=\dfrac{\sqrt{3}}{2}}$

To check the answer with a calculator, first make sure that the calculator is in radian mode, then find $\sin\dfrac{\pi}{3}$. The result is 0.866 (rounded), which is also the approximate value of $\dfrac{\sqrt{3}}{2}$.

■ **YOUR TURN** Evaluate $\cos\dfrac{\pi}{3}$ exactly.

TECHNOLOGY TIP

Set a TI/scientific calculator to radian mode by typing

MODE ▼ Radian ENTER

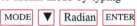

Use a TI/scientific calculator to check the value for $\sin\dfrac{\pi}{3}$ and $\dfrac{\sqrt{3}}{2}$. Press 2nd ^ for π.

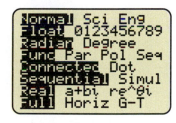

■ **Answer:** $\dfrac{1}{2}$

If the angle the trigonometric function is to be evaluated for has its terminal side in quadrant II, III, or IV, then we use reference angles and knowledge of the algebraic sign $(+/-)$ in that quadrant. Let θ be the angle and α be the reference angle; then the following chart represents their relationships in both degrees and radians.

Terminal side lies in . . .	Degrees	Radians
QI	$\alpha = \theta$	$\alpha = \theta$
QII	$\alpha = 180° - \theta$	$\alpha = \pi - \theta$
QIII	$\alpha = \theta - 180°$	$\alpha = \theta - \pi$
QIV	$\alpha = 360° - \theta$	$\alpha = 2\pi - \theta$

Example 6 Finding Reference Angles in Radians

Find the reference angle for each angle given.

a. $\dfrac{3\pi}{4}$ **b.** $\dfrac{11\pi}{6}$

Solution (a):

The terminal side of θ lies in quadrant II.

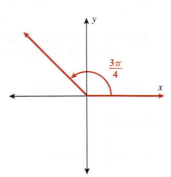

The reference angle is made with the terminal side and the x-axis.

$$\pi - \frac{3\pi}{4} = \boxed{\frac{\pi}{4}}$$

Solution (b):

The terminal side of θ lies in quadrant IV.

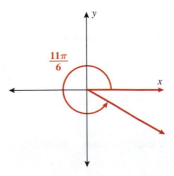

The reference angle is made with the terminal side and the x-axis.

$$2\pi - \frac{11\pi}{6} = \boxed{\frac{\pi}{6}}$$

EXAMPLE 7 **Evaluating Trigonometric Functions for Angles in Radian Measure Using Reference Angles**

Evaluate $\cos \dfrac{5\pi}{4}$ exactly.

TECHNOLOGY TIP

Use the TI/scientific calculator to check the value for $\cos \dfrac{5\pi}{4}$ and $-\dfrac{\sqrt{2}}{2}$.

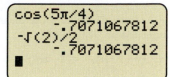

Solution:

The terminal side of angle $\dfrac{5\pi}{4}$ lies in quadrant III.

The reference angle is $\dfrac{\pi}{4} = 45°$.

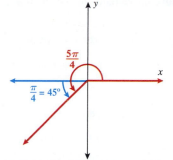

Find the cosine value for the reference angle.

$$\cos \dfrac{\pi}{4} = \cos 45° = \dfrac{\sqrt{2}}{2}$$

Determine the algebraic sign for cosine in quadrant III.

Negative $(-)$

Combine the algebraic sign of cosine in quadrant III with the value of cosine of the reference angle.

$$\cos \dfrac{5\pi}{4} = -\dfrac{\sqrt{2}}{2}$$

Check with a calculator (make sure the calculator is in radian mode).

$$-0.707 = -0.707$$

✓**CONCEPT CHECK** What is the reference angle (in both radian and degree measure) of $\theta = \dfrac{7\pi}{4}$?

■ **YOUR TURN** Evaluate $\sin \dfrac{7\pi}{4}$ exactly.

SECTION 3.1 **SUMMARY**

In this section, a second measure of angles was introduced. A central angle of a circle has radian measure equal to the ratio of the arc length to the radius of the circle. Radians and degrees are related by the relation that $\pi = 180°$. To convert from radians to degrees, multiply the radian measure by $\dfrac{180°}{\pi}$. Similarly, to convert from degrees to radians, multiply the degree measure by $\dfrac{\pi}{180°}$. One radian is approximately equal to 57°. Careful attention must be paid to what mode (degrees or radians) calculators are set in when evaluating trigonometric functions. To evaluate a trigonometric function exactly for radian measure, simply convert radian measure to degrees and evaluate the trigonometric function for special angles. To evaluate a trigonometric function for nonacute angles, we use reference angles and knowledge of the algebraic sign.

■ **Answer:** $-\dfrac{\sqrt{2}}{2}$

SECTION 3.1 EXERCISES

■ **SKILLS**

In Exercises 1–10, find the measure (in radians) of a central angle, θ, that intercepts an arc on a circle of radius r with indicated arc length, s.

1. $r = 10$ cm, $s = 2$ cm
2. $r = 20$ cm, $s = 2$ cm
3. $r = 22$ in., $s = 4$ in.
4. $r = 6$ in., $s = 1$ in.
5. $r = 100$ cm, $s = 20$ mm
6. $r = 1$ m, $s = 2$ cm
7. $r = \frac{1}{4}$ in., $s = \frac{1}{32}$ in.
8. $r = \frac{3}{4}$ cm, $s = \frac{3}{14}$ cm
9. $r = 2.5$ cm, $s = 5$ mm
10. $r = 1.6$ cm, $s = 0.2$ mm

In Exercises 11–24, convert from degrees to radians. Leave answers in terms of π.

11. $30°$
12. $60°$
13. $45°$
14. $90°$
15. $315°$

16. $270°$
17. $75°$
18. $100°$
19. $170°$
20. $340°$

21. $780°$
22. $540°$
23. $-210°$
24. $-320°$

In Exercises 25–38, convert from radians to degrees.

25. $\dfrac{\pi}{6}$
26. $\dfrac{\pi}{4}$
27. $\dfrac{3\pi}{4}$
28. $\dfrac{7\pi}{6}$
29. $\dfrac{3\pi}{8}$

30. $\dfrac{11\pi}{9}$
31. $\dfrac{5\pi}{12}$
32. $\dfrac{7\pi}{3}$
33. 9π
34. -6π

35. $\dfrac{19\pi}{20}$
36. $\dfrac{13\pi}{36}$
37. $-\dfrac{7\pi}{15}$
38. $-\dfrac{8\pi}{9}$

In Exercises 39–44, convert from radian measure to degrees. Round answers to the nearest hundredth of a degree.

39. 4
40. 3
41. 0.85
42. 3.27
43. -2.7989
44. -5.9841

In Exercises 45–50, convert from degrees to radians. Round answers to three significant digits.

45. $47°$
46. $65°$
47. $112°$
48. $172°$
49. $56.5°$
50. $298.7°$

In Exercises 51–58, find the reference angle in terms of both radians and degrees.

51. $\dfrac{2\pi}{3}$
52. $\dfrac{3\pi}{4}$
53. $\dfrac{7\pi}{4}$
54. $\dfrac{5\pi}{4}$

55. $\dfrac{5\pi}{12}$
56. $\dfrac{7\pi}{12}$
57. $\dfrac{4\pi}{3}$
58. $\dfrac{9\pi}{4}$

In Exercises 59–70, find the *exact* value of the following expressions.

59. $\sin\dfrac{\pi}{4}$
60. $\cos\dfrac{\pi}{6}$
61. $\sin\dfrac{7\pi}{4}$
62. $\cos\dfrac{2\pi}{3}$

63. $\tan\dfrac{11\pi}{6}$
64. $\tan\dfrac{5\pi}{3}$
65. $\tan\dfrac{\pi}{6}$
66. $\tan\dfrac{5\pi}{6}$

67. $\cot\dfrac{3\pi}{2}$
68. $\csc\left(-\dfrac{\pi}{2}\right)$
69. $\sec 5\pi$
70. $\cot\left(-\dfrac{3\pi}{2}\right)$

■ APPLICATIONS

71. Electronic Signals. Two electronic signals that are not cophased are called out of phase. Two signals that cancel each other out are said to be 180° out of phase, or the difference in their phases is 180°. How many radians out of phase are two signals whose phase difference is 270°?

72. Electronic Signals. Two electronic signals that are not cophased are called out of phase. Two signals that cancel each other out are said to be 180° out of phase, or the difference in their phases is 180°. How many radians out of phase are two signals whose phase difference is 110°?

73. Construction. In China, you find circular clan homes called *tulou*. Some tulou are three or four stories high and exceed 70 meters in diameter. If a wedge or section on the third floor of such a building has a central angle measuring 36°, how many radians is this?

Kin Cheung/Reuters/Landov

74. Construction. In China, you find circular clan homes called *tulou*. Some tulou are three or four stories high and exceed 70 meters in diameter. If a wedge or section on the third floor of such a building has a central angle measuring 72°, how many radians is this?

75. Clock. How many radians does the second hand of a clock turn in $2\frac{1}{2}$ minutes?

76. Clock. How many radians does the second hand of a clock turn in 3 minutes and 15 seconds?

77. London Eye. The London Eye has 32 capsules (each capable of holding 25 passengers with an unobstructed view of London). What is the radian measure of the angle made between the center of the wheel and the spokes aligning with each capsule?

78. Space Needle. The space needle in Seattle has a restaurant that offers views of Mount Rainier and Puget Sound. The restaurant completes one full rotation in approximately 45 minutes. How many radians will the restaurant have rotated in 25 minutes?

■ CATCH THE MISTAKE

In Exercises 79–82, explain the mistake that is made.

79. What is the measure (in radians) of a central angle, θ, that intercepts an arc of length 6 cm on a circle with radius 2 m?

Solution:

Write the formula for radians. $\theta = \dfrac{s}{r}$

Substitute $s = 6, r = 2$. $\theta = \dfrac{6}{2}$

Write the angle in terms of radians. $\theta = 3$ radians

This is incorrect. What mistake was made?

80. What is the measure (in radians) of a central angle, θ, that intercepts an arc of length 2 inches on a circle with radius 1 foot?

Solution:

Write the formula for radians. $\theta = \dfrac{s}{r}$

Substitute $s = 2, r = 1$. $\theta = \dfrac{2}{1}$

Write the angle in terms of radians. $\theta = 2$ radians

This is incorrect. What mistake was made?

81. Evaluate $6 \tan 45 + 5 \sec \dfrac{\pi}{3}$.

Solution:

Evaluate $\tan 45$ and $\sec \dfrac{\pi}{3}$. $\tan 45 = 1 \quad \sec \dfrac{\pi}{3} = 2$

Substitute the values of the trigonometric functions.

$$6 \tan 45 + 5 \sec \dfrac{\pi}{3} = 6(1) + 5(2)$$

Simplify. $6 \tan 45 + 5 \sec \dfrac{\pi}{3} = 16$

This is incorrect. What mistake was made?

82. Approximate with a calculator:
$\cos(42) + \tan(65) - \sin(12)$. Round to three decimal places.

Solution:

Evaluate the trigonometric functions individually.

$$\cos(42) = 0.743, \tan(65) = 2.145, \sin(12) = 0.208$$

Substitute the values into the expression.

$$\cos(42) + \tan(65) - \sin(12) = 0.743 + 2.145 - 0.208$$

Simplify.

$$\cos(42) + \tan(65) - \sin(12) = 2.680$$

This is incorrect. What mistake was made?

■ **CHALLENGE**

83. T or F: For an angle with positive measure, it is possible for the numerical values of the degree and radian measures to be equal.

84. T or F: The sum of the angles with radian measure in a triangle is π.

85. Find the sum of complementary angles in radian measure.

86. How many complete revolutions does an angle with measure 92 radians make?

87. The distance between Atlanta, Georgia, and Boston, Massachusetts, is approximately 900 miles along the surface of the Earth. The radius of the Earth is approximately 4000 miles. What is the central angle with vertex at the center of the Earth and sides of the angles intersecting the surface of the Earth in Atlanta and Boston?

88. The radius of the Earth is approximately 6400 km. If a central angle, with vertex at the center of the Earth, intersects the surface of the Earth in London (UK) and Rome (Italy) with a central angle of 0.22 radians, what is the distance between London and Rome? Round to the nearest hundredth kilometer.

89. At 8:20, what is the radian measure of the smaller angle between the hour hand and minute hand?

90. At 9:05, what is the radian measure of the larger angle between the hour hand and minute hand?

■ **TECHNOLOGY**

91. With a calculator set in radian mode, find $\sin 42$. With a calculator set in degree mode, find $\sin\left(42\dfrac{180°}{\pi}\right)$. Why do your results make sense?

92. With a calculator set in radian mode, find $\cos 5$. With a calculator set in degree mode, find $\cos\left(5\dfrac{180°}{\pi}\right)$. Why do your results make sense?

Skills Objectives

- Calculate arc length of a circle.
- Calculate the area of a circular sector.
- Solve application problems involving circle arc lengths and sectors.

Conceptual Objective

- Understand why the angle in the arc length formula must be in radian measure.

In Section 3.1, radian measure was defined in terms of the ratio of a circular arc of length s and length of the circle's radius, r.

$$\theta \text{ (in radians)} = \frac{s}{r}$$

In this section (3.2) and in the next section (3.3) we look at applications of radian measure that involve calculating *arc lengths* and *areas of circular sectors* and calculating *angular and linear speeds*. All of these applications are related to the definition of radians.

Arc Length

From geometry we know the length of an arc of a circle is proportional to its central angle. In Section 3.1 we learned that for the special case when the arc is equal to the circumference of the circle, the angle measure corresponding to one full rotation is 2π. Let us now assume we are given the central angle and we want to find the arc length.

WORDS	MATH
Write the definition of radian measure.	$\theta = \dfrac{s}{r}$
Multiply both sides of the equation by r.	$r \cdot \theta = \dfrac{s}{r} \cdot r$
Simplify.	$r\theta = s$

This formula, $s = r\theta$, is only true when θ is in radians. To develop a formula when θ is in degrees, we multiply θ by $\dfrac{\pi}{180°}$.

DEFINITION **ARC LENGTH**

If a central angle, θ, in a circle with radius r intercepts an arc on the circle of length s, then the **arc length**, s, is given by

$$s = r\theta_r \qquad \theta_r \text{ is in radians.}$$

or

$$s = r\theta_d\left(\frac{\pi}{180°}\right) \qquad \theta_d \text{ is in degrees.}$$

STUDY TIP

To use the relationship

$$s = r\theta$$

the angle θ must be in radians.

EXAMPLE 1 Finding Arc Length When the Angle Has Radian Measure

In a circle with radius 10 centimeters, an arc is intercepted by a central angle with measure $\dfrac{7\pi}{4}$. Find the arc length.

Solution:

Write the formula for arc length when the angle has radian measure.

$$s = r\,\theta_r$$

Substitute $r = 10$ cm and $\theta_r = \dfrac{7\pi}{4}$.

$$s = (10\text{ cm})\left(\dfrac{7\pi}{4}\right)$$

Simplify.

$$s = \dfrac{35\pi}{2}\text{ cm}$$

■ **YOUR TURN**   In a circle with radius 15 inches, an arc is intercepted by a central angle with measure $\dfrac{\pi}{3}$. Find the arc length.

EXAMPLE 2 Finding Arc Length When the Angle Has Degree Measure

In a circle with radius 7.5 centimeters, an arc is intercepted by a central angle with measure 76°. Find the arc length. Approximate the arc length to the nearest centimeter.

Solution:

Write the formula for arc length when the angle has degree measure.

$$s = r\theta_d\left(\dfrac{\pi}{180°}\right)$$

Substitute $r = 7.5$ cm and $\theta_d = 76°$.

$$s = (7.5\text{ cm})(76°)\left(\dfrac{\pi}{180°}\right)$$

Evaluate with a calculator.

$$s = 9.948\text{ cm}$$

Round to the nearest tenth.

$$s \approx 9.9\text{ cm}$$

■ **YOUR TURN**   In a circle with radius 20 meters, an arc is intercepted by a central angle with measure 113°. Find the arc length. Approximate the arc length to the nearest meter.

EXAMPLE 3 Path of International Space Station

The International Space Station is in an approximate circular orbit 400 kilometers above the surface of the Earth. If the ground station tracks the space station when it is within a 45° cone above the tracking antenna, how many kilometers

■ **Answer:** 5π inches   ■ **Answer:** 39.44 m

does the ISS cover during the ground station track? Assume that the radius of the Earth is 6400 kilometers. Round to the nearest kilometer.

Solution:

Write the formula for arc length when the angle has degree measure.

$$s = r\theta_d\left(\frac{\pi}{180°}\right)$$

Substitute $r = 6400 + 400 = 6800$ km and $\theta_d = 45°$.

$$s = (6800 \text{ km})(45°)\left(\frac{\pi}{180°}\right)$$

Evaluate with a calculator.

$$s = 5340.708 \text{ km}$$

Round to the nearest km.

$$s \approx 5341 \text{ km}$$

The ISS travels approximately 5341 kilometers during the ground station tracking.

■ **YOUR TURN** If the ground station in Example 3 could track the ISS within a 60° cone, how far would the ISS travel during the ground station tracking?

■ **EXAMPLE 4** Gears

Gears are inside many devices: automobiles, power meters, and VHS players are a few. When the smaller gear drives the larger gear, then typically the driving gear is rotated faster than a larger gear would be if it were the drive gear. In general, smaller ratios of radius of the driving gear to the driven gear are called for when machines are expected to yield more power. The smaller gear has a radius of 3 centimeters and the larger gear has a radius of 6.4 centimeters. If the smaller gear rotates 170°, how many degrees has the larger gear rotated? Round the answer to the nearest degree.

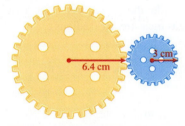

■ **Answer:** 7121 km

TECHNOLOGY TIP
When solving for θ_d, be sure to use a pair of parentheses for the product in the denominator.

$$\theta_d = \frac{180°}{\pi} \cdot \frac{17\pi \text{ cm}}{6(6.4 \text{ cm})}$$

$$= \frac{180° \cdot 17}{6(6.4)}$$

```
180*17/(6*6.4)
             79.6875
■
```

Solution:

The small gear arc length = the large gear arc length.

Smaller Gear

Write the formula for arc length when the angle has degree measure.

$$s = r\theta_d\left(\frac{\pi}{180°}\right)$$

Substitute the values for the smaller gear:
$r = 3$ cm and $\theta_d = 170°$.

$$s_{\text{smaller}} = (3 \text{ cm})(170°)\left(\frac{\pi}{180°}\right)$$

Simplify.

$$s_{\text{smaller}} = \left(\frac{17\pi}{6}\right) \text{ cm}$$

Larger Gear

The larger gear's arc length is equal to the smaller gear's arc length.

$$s = \left(\frac{17\pi}{6}\right) \text{ cm}$$

Write the formula for arc length when the angle has degree measure.

$$s = r\theta_d\left(\frac{\pi}{180°}\right)$$

Substitute $s = \left(\frac{17\pi}{6}\right)$ cm and $r = 6.4$ cm.

$$\left(\frac{17\pi}{6} \text{ cm}\right) = (6.4 \text{ cm})\theta_d\left(\frac{\pi}{180°}\right)$$

Solve for θ_d.

$$\theta_d = \frac{180°}{\pi} \cdot \frac{17\pi \text{ cm}}{6(6.4 \text{ cm})}$$

Simplify.

$$\theta_d = 79.6875°$$

Round to the nearest degree.

$$\theta_d = 80°$$

The larger gear rotates approximately 80°.

Area of a Circular Sector

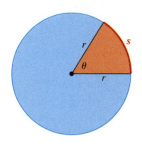

A restaurant lists a piece of French silk pie as having 400 calories. How does the chef arrive at that number? She calculates the calories of all the ingredients that went into making the entire pie, and then divides by the number of slices the pie yields. For example, if an entire pie has 3200 calories and it is sliced into 8 equal pieces, then each piece has 400 calories. Although that example involves volume, the idea is the same with areas of sectors of circles. *Circular sectors* can be thought of as "pieces of a pie."

Recall that arc lengths of a circle are proportional to the central angle (in radians) and the radius. Similarly, a circular sector is a portion of the entire circle. Let A represent the area of the *sector of the circle* and θ_r represent the central angle (in radians) that forms the sector. Then, let us consider the entire circle. The area is πr^2 and the angle that represents one full rotation has measure 2π (radians).

WORDS	MATH
Write the ratio of the area of the sector to the area of the entire circle.	$\dfrac{A}{\pi r^2}$
Write the ratio of the central angle, θ_r, to the measure of one full rotation.	$\dfrac{\theta_r}{2\pi}$

The ratios must be equal (proportionality of sector to circle).

$$\frac{A}{\pi r^2} = \frac{\theta_r}{2\pi}$$

Multiply both sides of the equation by πr^2.

$$\pi r^2 \cdot \frac{A}{\pi r^2} = \frac{\theta_r}{2\pi} \cdot \pi r^2$$

Simplify.

$$A = \frac{1}{2} r^2 \theta_r$$

DEFINITION   AREA OF A CIRCULAR SECTOR

The **area of a sector of a circle** with radius r and central angle θ is given by

$$A = \frac{1}{2} r^2 \theta_r \qquad \theta_r \text{ is in radians.}$$

or

$$A = \frac{1}{2} r^2 \theta_d \left(\frac{\pi}{180°} \right) \qquad \theta_d \text{ is in degrees.}$$

STUDY TIP

To use the relationship

$$A = \frac{1}{2} r^2 \theta$$

the angle θ must be in radians.

EXAMPLE 5 Finding the Area of a Sector When the Angle Has Radian Measure

Find the area of the sector associated with a single slice of pizza if the entire pizza has a 14 inch diameter and the pizza is cut into 8 equal pieces.

Solution:

The radius is half the diameter.

$$r = \frac{14}{2} = 7 \text{ in.}$$

Find the angle of each slice if a pizza is cut into 8 pieces.

$$\theta_r = \frac{2\pi}{8} = \frac{\pi}{4}$$

Write the formula for circular sector area in radians.

$$A = \frac{1}{2} r^2 \theta_r$$

Substitute $r = 7$ inches and $\theta_r = \frac{\pi}{4}$ into the area equation.

$$A = \frac{1}{2} (7 \text{ in.})^2 \left(\frac{\pi}{4} \right)$$

Simplify.

$$A = \frac{49\pi}{8} \text{ sq. in.}$$

Approximate with a calculator (round to the nearest square inch).

19 sq. in.

■ **YOUR TURN**   Approximate the area of a slice of pizza if the entire pizza has a 16 inch diameter.

■ **Answer:** 25 in.^{2}

EXAMPLE 6 Finding the Area of a Sector When the Angle Has Degree Measure

Sprinkler heads come in all different sizes depending on the angle of rotation desired. If a sprinkler head rotates 90° and has enough pressure to keep a constant 25 foot spray, what is the area of the sector of the lawn that gets watered? Round to the nearest square foot.

Solution:

Write the formula for circular sector area in degrees.

$$A = \frac{1}{2}r^2\theta_d\left(\frac{\pi}{180°}\right)$$

Substitute $r = 25$ ft and $\theta_d = 90°$ into the area equation.

$$A = \frac{1}{2}(25 \text{ ft})^2(90°)\left(\frac{\pi}{180°}\right)$$

Simplify.

$$A = \left(\frac{625\pi}{4}\right)\text{ft}^2 \approx 490.87 \text{ ft}^2$$

Round to the nearest foot.

$$A \approx 491 \text{ sq. ft.}$$

■ **YOUR TURN** If a sprinkler head rotates 180° and has enough pressure to keep a constant 30 foot spray, what is the area of the sector of the lawn that gets watered? Round to the nearest square foot.

SECTION 3.2   SUMMARY

In this section, we used the proportionality concept (both the arc length and area of a sector are proportional to the central angle of a circle). The definition of radian measure was used to develop formulas for the arc length of a circle when the central angle is given in either radians or degrees.

$$s = r\theta_r \qquad\qquad \theta_r \text{ is in radians.}$$

or

$$s = r\theta_d\left(\frac{\pi}{180°}\right) \qquad \theta_d \text{ is in degrees.}$$

The formula for the area of a sector of a circle was also developed for the cases in which the central angle is given in either radians or degrees.

$$A = \frac{1}{2}r^2\theta_r \qquad\qquad \theta_r \text{ is in radians.}$$

or

$$A = \frac{1}{2}r^2\theta_d\left(\frac{\pi}{180°}\right) \qquad \theta_d \text{ is in degrees.}$$

Arc length and sector area are rich with applications, as you will see in the exercises.

■ **Answer:** 1414 ft²

SECTION 3.2 EXERCISES

■ **SKILLS**

In Exercises 1–8, find the exact length of each arc made by the indicated central angle and radius of each circle.

1. $\theta = 3$, $r = 4$ mm

2. $\theta = 4$, $r = 5$ cm

3. $\theta = \dfrac{\pi}{12}$, $r = 8$ ft

4. $\theta = \dfrac{\pi}{8}$, $r = 6$ yd

5. $\theta = 22°$, $r = 18$ μm

6. $\theta = 14°$, $r = 15$ μm

7. $\theta = 8°$, $r = 1500$ km

8. $\theta = 3°$, $r = 1800$ km

In Exercises 9–16, find the exact length of each radius given the arc length and central angle of each circle.

9. $s = \dfrac{5\pi}{2}$ ft, $\theta = \dfrac{\pi}{10}$

10. $s = \dfrac{5\pi}{6}$ m, $\theta = \dfrac{\pi}{12}$

11. $s = \dfrac{24\pi}{5}$ in., $\theta = \dfrac{3\pi}{5}$

12. $s = \dfrac{5\pi}{9}$ km, $\theta = \dfrac{\pi}{180}$

13. $s = \dfrac{4\pi}{9}$ yd, $\theta = 20°$

14. $s = \dfrac{11\pi}{6}$ cm, $\theta = 15°$

15. $s = \dfrac{8\pi}{3}$ mi, $\theta = 40°$

16. $s = \dfrac{\pi}{4}$ μm, $\theta = 30°$

In Exercises 17–24, use a calculator to approximate the length of each arc made by the indicated central angle and radius of each circle. Round answers to two significant digits.

17. $\theta = 3.3$, $r = 0.4$ mm

18. $\theta = 2.4$, $r = 5.5$ cm

19. $\theta = \dfrac{\pi}{15}$, $r = 8$ yd

20. $\theta = \dfrac{\pi}{10}$, $r = 6$ ft

21. $\theta = 79.5°$, $r = 1.55$ μm

22. $\theta = 19.7°$, $r = 0.63$ μm

23. $\theta = 29°$, $r = 2500$ km

24. $\theta = 11°$, $r = 2200$ km

In Exercises 25–32, find the area of the circular sector given the indicated radius and central angle. Round answers to three significant digits.

25. $\theta = \dfrac{\pi}{6}$, $r = 7$ ft

26. $\theta = \dfrac{\pi}{5}$, $r = 3$ in.

27. $\theta = \dfrac{3\pi}{8}$, $r = 2.2$ km

28. $\theta = \dfrac{5\pi}{6}$, $r = 13$ mi

29. $\theta = 56°$, $r = 4.2$ cm

30. $\theta = 27°$, $r = 2.5$ mm

31. $\theta = 1.2°$, $r = 1.5$ ft

32. $\theta = 14°$, $r = 3.0$ ft

■ **APPLICATIONS**

33. Low Earth Orbit Satellites. A low Earth orbit (LEO) satellite is in an approximate circular orbit 300 kilometers above the surface of the Earth. If the ground station tracks the satellite when it is within a 45° cone above the tracking antenna (directly overhead), how many kilometers does the satellite cover during the ground station track? Assume the radius of the Earth is 6400 kilometers. Round to the nearest kilometer.

34. Low Earth Orbit Satellites. A low Earth orbit (LEO) satellite is in an approximate circular orbit 250 kilometers above the surface of the Earth. If the ground station tracks the satellite when it is within a 30° cone above the tracking antenna (directly overhead), how many kilometers does the satellite cover during the ground station track? Assume the radius of the Earth is 6400 kilometers. Round to the nearest kilometer.

35. Big Ben. The famous clock tower in London has a minute hand that is 14 feet long. How far does the tip

of the minute hand of Big Ben travel in 25 minutes? Round to two decimal places.

Getty Images, Inc.

36. **Big Ben.** The famous clock tower in London has a minute hand that is 14 feet long. How far does the tip of the minute hand of Big Ben travel in 35 minutes? Round to two decimal places.

37. **London Eye.** The London Eye has 32 capsules and a diameter of 400 feet. What is the distance someone has traveled once they reach the highest point for the first time?

Index Stock

38. **London Eye.** Assuming every capsule in Exercise 37 is loaded and unloaded, what is the distance someone has traveled from the point they first get in the capsule to the point at which the eye stops for the 6th time (not including the stop made for them)?

39. **Gears.** The smaller gear has a radius of 5 centimeters and the larger gear has a radius of 12.1 centimeters. If the smaller gear rotates 120°, how many degrees has the larger gear rotated? Round the answer to the nearest degree.

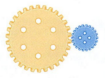

40. **Gears.** The smaller gear has a radius of 3 inches and the larger gear has a radius of 15 inches (see above figure). If the smaller gear rotates 420°, how many degrees has the larger gear rotated? Round the answer to the nearest degree.

41. **Bicycle Low Gear.** If a bicycle has 26 inch diameter wheels, the front chain drive has a radius of 2.2 inches, and the back drive has a radius of 3 inches, how far does the bicycle travel every one rotation of the cranks (pedals)?

42. **Bicycle High Gear.** If a bicycle has 26 inch diameter wheels, the front chain drive has a radius of 4 inches, and the back drive has a radius of 1 inch, how far does the bicycle travel every one rotation of the cranks (pedals)?

Getty Images, Inc.

43. **Odometer.** A Ford Expedition Eddie Bauer Edition comes standard with 17″ rims (which corresponds to a tire with 25.7″ diameter). Suppose you decide to later upgrade these tires for 19″ rims (corresponding to a tire with 28.2″ diameter). If you do not get your onboard computer reset for the new tires, the odometer will not be accurate. After your tires have actually driven 1000 miles, how many miles will the odometer report the Expedition has been driven? Round to the nearest mile.

Courtesy Ford Motor Company

44. **Odometer.** For the same Ford Expedition Eddie Bauer Edition in Exercise 43, after you have driven 50,000 miles, how many miles will the odometer report the Expedition has been driven if the computer is not reset to account for the new oversized tires? Round to the nearest mile.

45. **Sprinkler Coverage.** A sprinkler has a 20 foot spray and covers an angle of 45°. What is the area that the sprinkler waters?

46. **Sprinkler Coverage.** A sprinkler has a 22 foot spray and covers an angle of 60°. What is the area that the sprinkler waters?

47. **Windshield Wiper.** A windshield wiper that is 12 inches long (blade and arm) rotates 70°. If the rubber part is 8 inches long, what is the area cleared by the wiper? Round to the nearest inch.

48. **Windshield Wiper.** A windshield wiper that is 11 inches long (blade and arm) rotates 65°. If the rubber part is 7 inches long, what is the area cleared by the wiper? Round to the nearest inch.

■ CATCH THE MISTAKE

In Exercises 49 and 50, explain the mistake that is made.

49. A circle with radius 5 cm has an arc that is made from a central angle with measure 65°. Approximate the arc length to the nearest millimeter.

Solution:

Write the formula for arc length. $s = r\theta$

Substitute $r = 5$ cm and $\theta = 65°$ into the formula. $s = (5 \text{ cm})(65)$

Simplify. $s = 325$ cm

This is incorrect. What mistake was made?

50. For a circle with radius $r = 2.2$ cm, find the area of the circular sector with central angle measuring $\theta = 25°$. Round the answer to three significant digits.

Solution:

Write the formula for area of a circular sector. $A = \dfrac{1}{2}r^2\theta_r$

Substitute $r = 2.2$ cm and $\theta = 25°$ into the formula. $A = \dfrac{1}{2}(2.2 \text{ cm})^2(25°)$

Simplify. $A = 60.5$ cm^2

This is incorrect. What mistake was made?

■ CHALLENGE

51. T or F: If the radius of a circle doubles, then the arc length (associated with a fixed central angle) doubles.

52. T or F: If the radius of a circle doubles, then the area of the sector (associated with a fixed central angle) doubles.

53. If a smaller gear has radius r_1 and a larger gear has radius r_2 and the smaller gear rotates $\theta°_1$ what is the degree measure of the angle the larger gear rotates?

54. If a circle with radius r_1 has an arc length s_1 associated with a particular central angle, write the formula for the area of the sector of the circle formed by that central angle, in terms of the radius and arc length.

You may think that a baseball field is a circular sector but it is not. If it were, the distances from home plate to left field,

center field, and right field would all be the same (the radius). Where the infield dirt meets the outfield grass and along the fence in the outfield are arc lengths associated with a circle of radius 95 ft and with a vertex located at the pitcher's mound (not home plate).

55. What is the area enclosed in the circular sector with radius 95 feet and central angle 150°? Round to the nearest hundred square feet.

56. Approximate the area of the infield by adding the area in blue to the result in Exercise 55. Neglect the area near first and third bases and the foul line. Round to the nearest hundred square feet.

57. If a batter wants to bunt a ball so that it is fair (in front of home plate and between the foul lines) but keep it in the dirt (in the sector in front of home plate), within how small of an area is the batter trying to keep his bunt? Round to the nearest square foot.

58. Most bunts would fall within the blue triangle in the diagram on the left. Assume the catcher only fields bunts that fall in the sector calculated in Exercise 57 and the pitcher only fields bunts that fall on the pitcher's mound. Approximately how much area do the first baseman and third baseman *each* need to cover? Round to the nearest square foot.

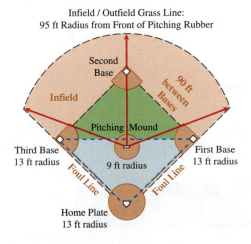

Infield / Outfield Grass Line:
95 ft Radius from Front of Pitching Rubber

Second Base

90 ft between Bases

Infield

Pitching Mound

Third Base
13 ft radius

9 ft radius

First Base
13 ft radius

Foul Line

Foul Line

Home Plate
13 ft radius

Skills Objectives
- Calculate angular speed.
- Calculate linear speed.
- Solve application problems involving angular and linear speeds.

Conceptual Objective
- Relate angular speed to linear speed.

In the chapter opener about a Ford Expedition with standard 17″ rims, we learned that the onboard computer that determines distance (odometer reading) and speed (speedometer) combines the number of tire rotations and the size of the tire. Because the onboard computer is set for 17″ rims (which corresponds to a tire with 25.7″ diameter), if the owner decided to upgrade to 19″ rims (corresponding to a tire with 28.2″ diameter), the computer must be updated with this new information. If the computer is not updated with the new tire size, both the odometer and speedometer readings will be incorrect.

Courtesy Ford Motor Company

You will see in this section that the *angular speed* (rotations of tires per second), *radius* (of the tires), and *linear speed* (speed of the automobile) are related. In the context of a circle, we will first define *linear speed*, then *angular speed*, and then relate them using the *radius*.

Linear Speed

It is important to note that although **velocity** and **speed** are often used as synonyms, in physics the difference is that velocity has direction and is written as a vector (Chapter 7), and speed is the magnitude of the velocity vector, which results in a real number (Chapter 7). In this chapter *speed* will be used.

Recall the relationship between distance, rate, and time: $d = rt$. Rate is speed, and in words this formula can be rewritten as:

$$\text{speed} = \frac{\text{distance}}{\text{time}}$$

If we think of a car driving around a circular track, the distance it travels is the arc length, s, and if we let v represent speed and t represent time we have the formula for speed around a circle (*linear speed*):

$$v = \frac{s}{t}$$

DEFINITION   **LINEAR SPEED**

If a point, P, moves along the circumference of a circle at a constant speed, then the **linear speed**, v, is given by

$$v = \frac{s}{t}$$

where s is the arc length and t is the time.

EXAMPLE 1   Linear Speed

A car travels at a constant speed around a circular track with circumference equal to 2 miles. If the car records a time of 15 minutes for 9 laps, what is the linear speed of the car in miles per hour?

Solution:

Calculate the distance traveled around the circular track.

$$s = (9 \text{ laps})\left(\frac{2 \text{ miles}}{\text{lap}}\right) = 18 \text{ miles}$$

Substitute $t = 15$ minutes and $s = 18$ miles into $v = \frac{s}{t}$.

$$v = \frac{18 \text{ miles}}{15 \text{ minutes}}$$

Convert the linear speed from miles/minute to miles/hour.

$$v = \left(\frac{18 \text{ miles}}{15 \text{ minutes}}\right)\left(\frac{60 \text{ minutes}}{1 \text{ hour}}\right)$$

Simplify.

$$v = 72 \text{ miles per hour}$$

■ **YOUR TURN**   A car travels at a constant speed around a circular track with circumference equal to 3 miles. If the car records a time of 12 minutes for 7 laps, what is the linear speed of the car in miles per hour?

Angular Speed

In linear speed we find how fast a position along the circumference of a circle is changing. In **angular speed** we find how fast the central angle is changing.

DEFINITION   **ANGULAR SPEED**

If a point, P, moves along the circumference of a circle at a constant speed, then the central angle, θ, that is formed with the terminal side passing through the point P also changes over some time, t, at a constant speed. The **angular speed**, ω (omega), is given by

$$\omega = \frac{\theta}{t} \qquad \text{where } \theta \text{ is given in radians.}$$

EXAMPLE 2 Angular Speed

A lighthouse in the middle of a channel rotates its light in a circular motion with constant speed. If the beacon of light completes one rotation every 10 seconds, what is the angular speed of the beacon in radians per minute?

Solution:

Calculate the angle measure in radians associated with one rotation.

$$\theta = 2\pi$$

Substitute $\theta = 2\pi$ and $t = 10$ seconds into $\omega = \dfrac{\theta}{t}$.

$$\omega = \frac{2\pi \text{ (radians)}}{10 \text{ seconds}}$$

Convert the angular speed from rad/sec to rad/min.

$$\omega = \frac{2\pi \text{ (radians)}}{10 \text{ seconds}} \cdot \frac{60 \text{ seconds}}{1 \text{ minutes}}$$

Simplify.

$$\omega = 12\pi \text{ rad/min}$$

■ **YOUR TURN** If the lighthouse in Example 2 is adjusted so that the beacon rotates 1 time every 40 seconds, what is the angular speed of the beacon in radians per minute?

Relationship Between Linear and Angular Speeds

In the chapter opener, we discussed the Ford Expedition with 17″ standard rims that would have odometer and speedometer errors if the owner decided to upgrade to 19″ rims. That is because *angular speed* (rotations of tires per second), *radius* (of the tires), and *linear speed* (speed of the automobile) are related. To see how, let us start with the definition of arc length (Section 3.2), which comes from the definition of radian measure (Section 3.1).

WORDS	**MATH**
Write the definition of radian measure.	$\theta = \dfrac{s}{r}$
Write the definition of arc length (θ in radians).	$s = r\theta$
Divide both sides by t.	$\dfrac{s}{t} = \dfrac{r\theta}{t}$
Rewrite the right side of the equation.	$\dfrac{s}{t} = r\dfrac{\theta}{t}$
Recall the definitions of **linear** and **angular** speeds.	$v = \dfrac{s}{t}$ and $\omega = \dfrac{\theta}{t}$
Substitute $v = \dfrac{s}{t}$ and $\omega = \dfrac{\theta}{t}$ into $\dfrac{s}{t} = r\dfrac{\theta}{t}$.	$v = r\omega$

■ **Answer:** $\omega = 3\pi$ rad/min

RELATING LINEAR AND ANGULAR SPEEDS

If a point, P, moves at a constant speed along the circumference of a circle with radius r, then the **linear speed**, v, and the **angular speed**, ω, are related by

$$v = r\omega \quad \text{or} \quad \omega = \frac{v}{r}$$

Note: This relationship is only true when θ is given in radians.

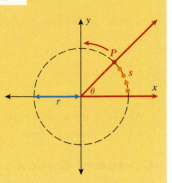

STUDY TIP

The relationship between linear and angular speed assumes the angle is given in radians.

We now will investigate the Ford Expedition scenario with upgraded tires. Notice that tires of 2 different radii with the same angular speed have different linear speed. The larger tire has the faster linear speed.

 EXAMPLE 3 Relating Linear and Angular Speeds

A Ford Expedition comes standard with tires that have a diameter of 25.7 inches. If the owner decided to upgrade to tires with a diameter of 28.2 inches without having the onboard computer updated, how fast will the Expedition *actually* be traveling when the speedometer reads 75 mph?

Solution:

The computer onboard the Expedition "thinks" the tires are 25.7 inches in diameter and knows the angular speed. Use the programmed tire diameter and speedometer reading to calculate the angular speed. Then use that angular speed and the upgraded tire diameter to get the actual speed (linear speed).

Write the formula for the angular speed.

$$\omega = \frac{v}{r}$$

Substitute $v = 75 \dfrac{\text{miles}}{\text{hour}}$ and

$r = \dfrac{25.7}{2} = 12.85$ inches into the formula.

$$\omega = \frac{75 \dfrac{\text{miles}}{\text{hour}}}{12.85 \text{ inches}}$$

1 mile = 5280 feet = 63,360 inches.

$$\omega = \frac{75(63{,}360) \dfrac{\text{inches}}{\text{hour}}}{12.85 \text{ inches}}$$

Simplify.

$$\omega \approx 369{,}805 \frac{\text{radians}}{\text{hour}}$$

Write the linear speed formula.

$$v = r\omega$$

Substitute $r = \dfrac{28.2}{2} = 14.1$ inches

and $\omega \approx 369{,}805 \dfrac{\text{radians}}{\text{hour}}$.

$$v = (14.1 \text{ inches})\left(369{,}805 \frac{\text{radians}}{\text{hour}} \right)$$

Simplify.

$$v \approx 5{,}214{,}251 \frac{\text{inches}}{\text{hour}}$$

1 mile = 5280 feet = 63,360 inches.

$$v \approx \frac{5,214,251 \dfrac{\text{inches}}{\text{hour}}}{63,360 \dfrac{\text{inches}}{\text{mile}}}$$

$$v \approx 82.296 \frac{\text{miles}}{\text{hour}}$$

Although the speedometer indicates a speed of 75 mph, the actual speed is approximately 82 mph.

CONCEPT CHECK   If you downsized the tires on your car from the original specifications, do you expect the speedometer to overreport or underreport your actual speed?

■ **YOUR TURN**   Suppose the owner of the Expedition in Example 3 decides to downsize the tires from their original 25.7 inch diameter to a 24.4 inch diameter. If the speedometer indicates a speed of 65 mph, what is the actual speed of the Expedition?

SECTION 3.3   # SUMMARY

In this section, circular motion was defined in terms of linear speed (speed along the circumference of a circle), v, and angular speed (speed of angle rotation), ω.

$$\text{Linear speed: } v = \frac{s}{t}$$

$$\text{Angular speed: } \omega = \frac{\theta}{t} \text{ where } \theta \text{ is given in radians.}$$

Linear and angular speeds associated with circular motion are related through the radius, r, of the circle.

$$v = r\omega \quad \text{or} \quad \omega = \frac{v}{r}$$

It is important to note that these formulas hold true only when angular speed is given in radians per unit of time.

SECTION 3.3 EXERCISES

SKILLS

In Exercises 1–8, find the linear speed of a point that moves with constant speed in a circular motion if the point travels along the circle an arc length s in time t.

1. $s = 2$ m, $t = 5$ sec

2. $s = 12$ ft, $t = 3$ min

3. $s = 68,000$ km, $t = 250$ hr

4. $s = 7,524$ mi, $t = 12$ days

■ **Answer:** Approximately 62 mph

5. $s = 1.75$ nm (nanometers), $t = 0.25$ milliseconds

6. $s = 3.6$ μm (microns), $t = 9$ nanoseconds

7. $s = \dfrac{1}{16}$ in., $t = 4$ min

8. $s = \dfrac{2}{5}$ cm, $t = 8$ hr

In Exercises 9–16, find the distance traveled (arc length) of a point that moves with constant speed v along a circle in time t.

9. $v = 2.8$ m/sec, $t = 3.5$ sec

10. $v = 6.2$ km/hr, $t = 4.5$ hr

11. $v = 4.5$ mi/hr, $t = 20$ min

12. $v = 5.6$ ft/sec, $t = 2$ min

13. $v = 60$ mi/hr, $t = 15$ min

14. $v = 72$ km/hr, $t = 10$ min

15. $v = 750$ km/min, $t = 4$ days

16. $v = 120$ ft/sec, $t = 27$ min

In Exercises 17–24, find the angular speed (radians/second) associated with rotating a central angle θ in time t.

17. $\theta = 25\pi$, $t = 10$ sec

18. $\theta = \dfrac{3\pi}{4}$, $t = \dfrac{1}{6}$ sec

19. $\theta = 100\pi$, $t = 5$ sec

20. $\theta = \dfrac{\pi}{2}$, $t = \dfrac{1}{10}$ sec

21. $\theta = 200°$, $t = 5$ sec

22. $\theta = 60°$, $t = 0.2$ sec

23. $\theta = 780°$, $t = 3$ sec

24. $\theta = 420°$, $t = 6$ sec

In Exercises 25–30, find the linear speed of a point traveling at a constant speed along the circumference of a circle with radius r and angular speed ω.

25. $\omega = \dfrac{2\pi \text{ rad}}{3 \text{ sec}}$, $r = 9$ in.

26. $\omega = \dfrac{3\pi \text{ rad}}{4 \text{ sec}}$, $r = 8$ cm

27. $\omega = \dfrac{\pi \text{ rad}}{20 \text{ sec}}$, $r = 5$ mm

28. $\omega = \dfrac{5\pi \text{ rad}}{16 \text{ sec}}$, $r = 24$ ft

29. $\omega = \dfrac{4\pi \text{ rad}}{15 \text{ sec}}$, $r = 2.5$ in.

30. $\omega = \dfrac{8\pi \text{ rad}}{15 \text{ sec}}$, $r = 4.5$ cm

In Exercises 31–36, find the distance a point travels along a circle, s, over a time t, given the angular speed, ω, and radius of the circle, r. Round to three significant digits.

31. $r = 5$ cm, $\omega = \dfrac{\pi \text{ rad}}{6 \text{ sec}}$, $t = 10$ sec

32. $r = 2$ mm, $\omega = 6\pi \dfrac{\text{rad}}{\text{sec}}$, $t = 11$ sec

33. $r = 5.2$ in., $\omega = \dfrac{\pi \text{ rad}}{15 \text{ sec}}$, $t = 10$ min

34. $r = 3.2$ ft, $\omega = \dfrac{\pi \text{ rad}}{4 \text{ sec}}$, $t = 3$ min

35. $r = 15$ in., $\omega = 5$ rotations per second, $t = 15$ min (express distance in miles*)

36. $r = 17$ in., $\omega = 6$ rotations per second, $t = 10$ min (express distance in miles*)

*1 mi = 5280 ft

■ APPLICATIONS

37. Tires. A car owner decides to upgrade from tires with a diameter of 24.3 inches to tires with a diameter of 26.1 inches. If she doesn't update the onboard computer, how fast will she actually be traveling when the speedometer reads 65 mph?

38. Tires. A car owner decides to upgrade from tires with a diameter of 24.8 inches to tires with a diameter of 27.0 inches. If she doesn't update the onboard computer, how fast will she actually be traveling when the speedometer reads 70 mph?

39. Planets. The Earth rotates every 24 hours (actually 23 hours, 56 minutes, and 4 seconds) and has a diameter of 7926 miles. If you're standing on the equator, how fast are you traveling (how fast is the Earth spinning)? Compute this using 24 hours and then with 23 hours, 56 minutes, 4 seconds.

40. Planets. The planet Jupiter rotates every 9.9 hours and has a diameter of 88,846 miles. If you're standing on its equator, how fast are you traveling (linear speed)?

41. Carousel. A boy wants to jump onto a moving carousel that is spinning at the rate of 5 revolutions per minute. If the carousel is 60 feet in diameter, how fast must the boy run, in feet per second, to match the speed of the carousel and jump on?

42. Carousel. A boy wants to jump onto a playground carousel that is spinning at the rate of 30 revolutions per minute. If the carousel is 6 feet in diameter, how fast must the boy run, in feet per second, to match the speed of the carousel and jump on?

43. Music. Some people still have their phonograph collections and play the records on turntables. A phonograph record is a vinyl disc that rotates on the turntable. If a 12 inch diameter record rotates at 33 1/3 revolutions per minute, what is the angular speed?

44. Music. Some people still have their phonograph collections and play the records on turntables. A phonograph record is a vinyl disc that rotates on the turntable. If a 12 inch diameter record rotates at 33 1/3 revolutions per minute, what is the linear speed of a point on the outer edge?

45. Bicycle. How fast is a bicyclist traveling in miles per hour if his tires are 27 inches in diameter and his angular speed is 5π radians per second?

46. Bicycle. How fast is a bicyclist traveling in miles per hour if his tires are 22 inches in diameter and his angular speed is 5π radians per second?

47. Electric Motor. If a 2 inch diameter pulley that's being driven by an electric motor and running at 1600 revolutions per minute is connected by a belt to a 5 inch diameter pulley to drive a saw, what is the speed of the saw in revolutions per minute?

48. Electric Motor. If a 2.5 inch diameter pulley that's being driven by an electric motor and running at 1800 revolutions per minute is connected by a belt to a 4 inch diameter pulley to drive a saw, what is the speed of the saw in revolutions per minute?

NASA explores artificial gravity as a way to counter the physiologic effects of extended weightlessness for future space exploration. NASA's centrifuge has a 58 foot diameter arm.

Courtesy NASA

49. NASA. If two humans are on opposite (red and blue) ends of the centrifuge and their linear speed is 200 miles per hour, how fast is the arm rotating?

50. NASA. If two humans are on opposite (red and blue) ends of the centrifuge and they rotate one full rotation every second, what is their linear speed in feet per second?

To achieve similar weightlessness as that on NASA's centrifuge, ride the *Gravitron* at a carnival or fair. The Gravitron has a diameter of 14 meters, and in the first 20 seconds it achieves zero gravity and the floor drops.

©David Burton

51. Gravitron. If the gravitron rotates 24 times per minute, find the linear speed of the people riding it in meters per second.

52. Gravitron. If the Gravitron rotates 30 times per minute, find the linear speed of the people riding in kilometers per hour.

■ CATCH THE MISTAKE

In Exercises 53 and 54, explain the mistake that is made.

53. If the radius of a set of tires on a car is 15 inches and the tires rotate 180° per second, how fast is the car traveling (linear speed) in miles per hour?

Solution:

Write the formula for
linear speed. $v = r\omega$

Let $r = 15$ inches and
$\omega = 180°$ per second. $v = (15 \text{ in.})(180°/\text{sec})$

Simplify. $v = 2700 \text{ in./sec}$

Let 1 mile $= 5280$
feet $= 63,360$ inches
and 1 hour $= 3600$ seconds. $v = \left(\dfrac{2700 \cdot 3600}{63,360}\right) \text{mph}$

Simplify. $v \approx 153.4 \text{ mph}$

This is incorrect. The correct answer is approximately 2.7 mph. What mistake was made?

54. If a bicycle has tires with radius 10 inches and the tires rotate 90° per $\frac{1}{2}$ second, how fast is the bicycle traveling (linear speed) in miles per hour?

Solution:

Write the formula for
linear speed. $v = r\omega$

Let $r = 10$ inches and
$\omega = 180°$ per second. $v = (10 \text{ in.})(180°/\text{sec})$

Simplify. $v = 1800 \text{ in./sec}$

Let 1 mile $= 5280$
feet $= 63,360$ inches and
1 hour $= 3600$ seconds. $v = \left(\dfrac{1800 \cdot 3600}{63,360}\right) \text{mph}$

Simplify. $v \approx 102.3 \text{ mph}$

This is incorrect. The correct answer is approximately 1.8 mph. What mistake was made?

■ CHALLENGE

55. T or F: Angular and linear speed are inversely proportional.

56. T or F: Angular and linear speed are directly proportional.

57. In the chapter opener about the Ford Expedition, if the standard tires have radius r_1 and the upgraded tires have radius r_2, assuming the owner does not get the onboard computer adjusted, find the actual speed the Ford is traveling, v_2, in terms of the indicated speed on the speedometer, v_1.

58. For the Ford in Exercise 57, find the actual mileage the Ford has traveled, s_2, in terms of the indicated mileage on the odometer, s_1.

In Exercises 59 and 60, use the diagram shown.

The large gear has a radius of 6 cm, the medium gear has a radius of 3 cm, and the small gear has a radius of 1 cm.

59. If the small gear rotates 1 revolution per second, what is the linear speed of a point traveling along the circumference of the large gear?

60. If the small gear rotates 1.5 revolutions per second, what is the linear speed of a point traveling along the circumference of the large gear?

Skills Objectives

- Draw a unit circle with special angles and label cosine and sine values.
- Determine the domain and range of circular functions.
- Classify circular functions as even or odd.

Conceptual Objectives

- Define trigonometric functions using the unit circle approach.
- Relate x-coordinates and y-coordinates of points on a unit circle to the values of cosine and sine functions.
- Visualize periodic properties of circular functions.

Trigonometric Functions and the Unit Circle

Recall that the first definition of trigonometric functions we developed was in terms of ratios of sides of right triangles (Section 1.3). Then, in Section 2.2 we superimposed right triangles in the Cartesian plane, which led to a second definition of trigonometric functions (for any angle) in terms of ratios of x and y coordinates of a point and the distance from the origin to that point. In this section we inscribe the right triangles into the unit circle in the Cartesian plane, which will yield a third definition of trigonometric functions. It is important to note that all three definitions are consistent with one another.

Recall that the equation for a unit circle centered at the origin is given by $x^2 + y^2 = 1$. A **unit circle** has radius equal to 1 and is centered at the origin. We will use the term *circular function* later in this section, but it is important to note that a circle is not a function (it does not pass the vertical line test).

If we form a central angle θ in the unit circle such that the terminal side lies in quadrant I, we can use the previous two definitions of sine and cosine when $r = 1$ (unit circle).

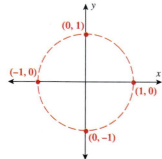

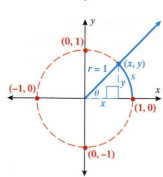

TRIGONOMETRIC FUNCTION	RIGHT TRIANGLE TRIGONOMETRY	CARTESIAN PLANE
$\sin\theta$	$\dfrac{\text{opposite}}{\text{hypotenuse}} = \dfrac{y}{1} = y$	$\dfrac{y}{r} = \dfrac{y}{1} = y$
$\cos\theta$	$\dfrac{\text{adjacent}}{\text{hypotenuse}} = \dfrac{x}{1} = x$	$\dfrac{x}{r} = \dfrac{x}{1} = x$

Notice that the point (x, y) on the unit circle can be written as $(\cos\theta, \sin\theta)$. If we recall the unit circle coordinate values for special angles (Section 2.1), we can now summarize the exact values for **sine** and **cosine** in the illustration on the following page.

The following observations are consistent with previous properties of trigonometric functions:

- $\sin > 0$ in QI and QII.

- $\cos > 0$ in QI and QIV.

- Unit circle equation $x^2 + y^2 = 1$ leads to the Pythagorean identity $\cos^2\theta + \sin^2\theta = 1$.

STUDY TIP

$(\cos\theta, \sin\theta)$ represents a point on the unit circle.

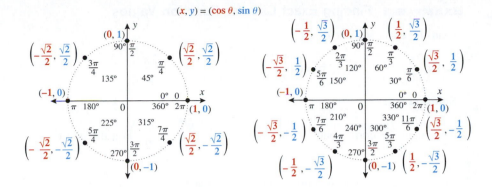

Circular Functions

Trigonometric functions thus far have been functions of angle measure (degrees or radians). Let s represent the distance from $(1, 0)$ to (x, y) along the circumference of the unit circle. The arc length, s, is related to the radius and central angle through the definition of radian measure: θ (in radians) $= \dfrac{s}{r}$.

A unit circle has $r = 1$ unit; therefore a central angle in a unit circle has the same measure as the arc length from the point $(1, 0)$ to the point (x, y) on the unit circle where the terminal side intersects. If we let $\theta = s$, we can define the trigonometric functions as *circular functions* of real numbers. Recall that once sine and cosine are found, then tangent can be found using the quotient identity and the remaining three can be found using reciprocal functions.

DEFINITION (3)   **TRIGONOMETRIC FUNCTIONS**

UNIT CIRCLE APPROACH

Let (x, y) be any point on the unit circle. If θ is a real number that represents the distance from the point $(1, 0)$ along the circumference to the point (x, y), then

$$\sin\theta = y \qquad\qquad \cos\theta = x \qquad\qquad \tan\theta = \frac{y}{x} \qquad x \neq 0$$

$$\cot\theta = \frac{x}{y} \qquad y \neq 0 \qquad \sec\theta = \frac{1}{x} \qquad x \neq 0 \qquad \csc\theta = \frac{1}{y} \qquad y \neq 0$$

The coordinates of the points along the unit circle can be written as $(\cos\theta, \sin\theta)$ and since θ is a real number, the **trigonometric functions** are often called **circular functions**.

EXAMPLE 1 Finding Exact Circular Function Values

Find the exact values for

a. $\sin\dfrac{7\pi}{4}$ **b.** $\cos\dfrac{5\pi}{6}$ **c.** $\tan\dfrac{3\pi}{2}$

Solution (a):

The angle $\dfrac{7\pi}{4}$ corresponds to the coordinates $\left(\dfrac{\sqrt{2}}{2}, -\dfrac{\sqrt{2}}{2}\right)$ on the unit circle.

The value of the sine function is the *y*-coordinate. $\boxed{\sin\dfrac{7\pi}{4} = -\dfrac{\sqrt{2}}{2}}$

Solution (b):

The angle $\dfrac{5\pi}{6}$ corresponds to the coordinates $\left(-\dfrac{\sqrt{3}}{2}, \dfrac{1}{2}\right)$ on the unit circle.

The value of the cosine function is the *x*-coordinate. $\boxed{\cos\dfrac{5\pi}{6} = -\dfrac{\sqrt{3}}{2}}$

Solution (c):

The angle $\dfrac{3\pi}{2}$ corresponds to the coordinates $(0, -1)$ on the unit circle.

The value of the cosine function is the *x*-coordinate. $\cos\dfrac{3\pi}{2} = 0$

The value of the sine function is the *y*-coordinate. $\sin\dfrac{3\pi}{2} = -1$

Tangent is the ratio of sine to cosine. $\tan\dfrac{3\pi}{2} = \dfrac{\sin\dfrac{3\pi}{2}}{\cos\dfrac{3\pi}{2}}$

Let $\cos\dfrac{3\pi}{2} = 0$ and $\sin\dfrac{3\pi}{2} = -1$. $\tan\dfrac{3\pi}{2} = \dfrac{-1}{0}$

$\boxed{\tan\dfrac{3\pi}{2} \text{ is undefined.}}$

■ **YOUR TURN** Find the exact values for

a. $\sin\dfrac{5\pi}{6}$ **b.** $\cos\dfrac{7\pi}{4}$ **c.** $\tan\dfrac{2\pi}{3}$

■ **Answer: a.** $\dfrac{1}{2}$ **b.** $\dfrac{\sqrt{2}}{2}$ **c.** $-\sqrt{3}$

EXAMPLE 2 Solving Equations Involving Circular Functions

Use the unit circle to find all values of θ, $0 \leq \theta \leq 2\pi$, for which $\sin\theta = -\dfrac{1}{2}$.

Solution:

The value of sine is the y-coordinate. The angles corresponding to $\sin\theta = -\dfrac{1}{2}$ are $\dfrac{7\pi}{6}$ and $\dfrac{11\pi}{6}$.

There are two values for θ that are greater than or equal to zero and less than or equal to 2π and correspond to $\sin\theta = -\dfrac{1}{2}$.

$$\theta = \frac{7\pi}{6}, \frac{11\pi}{6}$$

■ **YOUR TURN** Find all values of θ, $0 \leq \theta \leq 2\pi$, for which $\cos\theta = -\dfrac{1}{2}$.

Properties of Circular Functions

WORDS	MATH
A point (x, y) that lies on the unit circle, $x^2 + y^2 = 1$.	$-1 \leq x \leq 1$ and $-1 \leq y \leq 1$
Since $(x, y) = (\cos\theta, \sin\theta)$, the following holds.	$-1 \leq \cos\theta \leq 1$ and $-1 \leq \sin\theta \leq 1$
State the **domain of cosine and sine**.	$\theta \in \mathbb{R}$ or $(-\infty, \infty)$
Since $\cot\theta = \dfrac{\cos\theta}{\sin\theta}$ and $\csc\theta = \dfrac{1}{\sin\theta}$, values for θ that make $\sin\theta = 0$ must be eliminated from the **domain of cotangent and cosecant**.	$\theta \neq \pm\pi, \pm 2\pi, \pm 3\pi$ or $\theta \neq n\pi$, n any integer
Since $\tan\theta = \dfrac{\sin\theta}{\cos\theta}$ and $\sec\theta = \dfrac{1}{\cos\theta}$, values for θ that make $\cos\theta = 0$ must be eliminated from the **domain of tangent and secant**.	$\theta \neq \pm\dfrac{\pi}{2}, \pm\dfrac{3\pi}{2}, \pm\dfrac{5\pi}{2}$ or $\theta \neq \dfrac{(2n+1)\pi}{2}$, n any integer

■ **Answer:** $\theta = \dfrac{2\pi}{3}, \dfrac{4\pi}{3}$

The following box summarizes the domains and ranges of the circular functions.

DOMAINS AND RANGES OF THE CIRCULAR FUNCTIONS

For any real number, θ, and integer, n,

Function	Domain	Range
$\sin\theta$	$(-\infty, \infty)$	$[-1, 1]$
$\cos\theta$	$(-\infty, \infty)$	$[-1, 1]$
$\tan\theta$	all real numbers such that $\theta \neq \dfrac{(2n+1)\pi}{2}$	$(-\infty, \infty)$
$\cot\theta$	all real numbers such that $\theta \neq n\pi$	$(-\infty, \infty)$
$\sec\theta$	all real numbers such that $\theta \neq \dfrac{(2n+1)\pi}{2}$	$(-\infty, -1] \cup [1, \infty)$
$\csc\theta$	all real numbers such that $\theta \neq n\pi$	$(-\infty, -1] \cup [1, \infty)$

Recall from algebra that even and odd functions have both an algebraic and graphical interpretation. **Even functions** are functions for which $f(-x) = f(x)$, and they are symmetric about the y-axis. **Odd functions** are functions for which $f(-x) = -f(x)$, and they are symmetric about the origin.

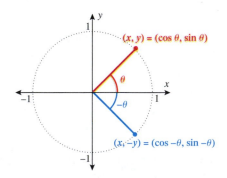

Cosine is an even function. $x = \cos\theta = \cos(-\theta)$

Sine is an odd function. $y = \sin\theta$ and $-y = \sin(-\theta)$

$\sin(-\theta) = -\sin\theta$

EXAMPLE 3 Using Properties of Circular Functions

Evaluate $\cos\left(-\dfrac{5\pi}{6}\right)$.

Solution:

Cosine is an even function.

$$\cos\left(-\frac{5\pi}{6}\right) = \cos\left(\frac{5\pi}{6}\right)$$

Use the unit circle to evaluate cosine.

$$\cos\left(\frac{5\pi}{6}\right) = -\frac{\sqrt{3}}{2}$$

$$\cos\left(-\frac{5\pi}{6}\right) = -\frac{\sqrt{3}}{2}$$

```
cos(-5π/6)
          -.8660254038
-√(3)/2
          -.8660254038
■
```

YOUR TURN Evaluate $\sin\left(-\dfrac{5\pi}{6}\right)$.

It is important to note that, although circular functions can be evaluated exactly for some special angles, a calculator can be used to approximate circular functions for any value. It is important to set the calculator to radian mode first.

 EXAMPLE 4 Evaluating Circular Functions with a Calculator

Use a calculator to evaluate $\sin\left(\dfrac{7\pi}{12}\right)$. Round the answer to four decimal places.

COMMON MISTAKE

CORRECT

Evaluate with a calculator.

0.965925826

Round to four decimal places.

$$\sin\left(\frac{7\pi}{12}\right) = 0.9659$$

INCORRECT

Evaluate with a calculator.

0.031979376 **ERROR**

(calculator in degree mode)

Round to four decimal places.

$$\sin\left(\frac{7\pi}{12}\right) = 0.0320 \quad \textbf{INCORRECT}$$

Many calculators automatically reset to degrees mode after every calculation, so be sure to always check what mode the calculator indicates.

YOUR TURN Use a calculator to evaluate $\tan\left(\dfrac{9\pi}{5}\right)$. Round the answer to four decimal places.

■ **Answer:** $-\dfrac{1}{2}$ ■ **Answer:** -0.7265

EXAMPLE 5 Even and Odd Circular Functions

Show that secant is an even function.

Solution:

Show that $\sec(-\theta) = \sec(\theta)$.

Secant is the reciprocal of cosine. $\sec(-\theta) = \dfrac{1}{\cos(-\theta)}$

Cosine is an even function, $\cos(-\theta) = \cos(\theta)$. $\sec(-\theta) = \dfrac{1}{\cos(\theta)}$

Secant is the reciprocal of cosine, $\sec\theta = \dfrac{1}{\cos\theta}$. $\sec(-\theta) = \dfrac{1}{\cos(\theta)} = \sec\theta$

☐ Since $\sec(-\theta) = \sec(\theta)$, secant is an even function.

SECTION 3.4 **SUMMARY**

In this section, we have defined trigonometric functions in terms of circular functions. Any point (x, y) that lies on a unit circle satisfies the equation $x^2 + y^2 = 1$. The Pythagorean identity, $\cos^2\theta + \sin^2\theta = 1$, can also be interpreted as a unit circle if $(x, y) = (\cos\theta, \sin\theta)$, where θ is the central angle made when the terminal side intersects the unit circle at the point (x, y). Cosine is an even function, $\cos(-\theta) = \cos\theta$, and sine is an odd function, $\sin(-\theta) = -\sin\theta$.

SECTION 3.4 EXERCISES

SKILLS

In Exercises 1–12, find the exact values of the indicated trigonometric functions using the unit circle.

1. $\sin\left(\dfrac{5\pi}{3}\right)$

2. $\cos\left(\dfrac{5\pi}{3}\right)$

3. $\cos\left(\dfrac{7\pi}{6}\right)$

4. $\sin\left(\dfrac{7\pi}{6}\right)$

5. $\sin\left(\dfrac{3\pi}{4}\right)$

6. $\cos\left(\dfrac{3\pi}{4}\right)$

7. $\tan\left(\dfrac{7\pi}{4}\right)$

8. $\cot\left(\dfrac{7\pi}{4}\right)$

9. $\sec 225°$

10. $\csc 300°$

11. $\tan 240°$

12. $\cot 330°$

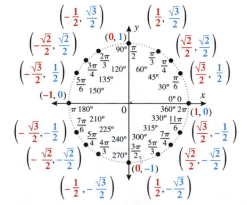

In Exercises 13–28, use the unit circle and the fact that sine is an odd function and cosine is an even function to find the exact values of the indicated functions.

13. $\sin\left(-\dfrac{2\pi}{3}\right)$

14. $\sin\left(-\dfrac{5\pi}{4}\right)$

15. $\sin\left(-\dfrac{\pi}{3}\right)$

16. $\sin\left(-\dfrac{7\pi}{6}\right)$

17. $\cos\left(-\dfrac{3\pi}{4}\right)$

18. $\cos\left(-\dfrac{5\pi}{3}\right)$

19. $\cos\left(-\dfrac{5\pi}{6}\right)$

20. $\cos\left(-\dfrac{7\pi}{4}\right)$

21. $\sin\left(-225°\right)$

22. $\sin\left(-180°\right)$

23. $\sin\left(-270°\right)$

24. $\sin\left(-60°\right)$

25. $\cos\left(-45°\right)$

26. $\cos\left(-135°\right)$

27. $\cos\left(-90°\right)$

28. $\cos\left(-210°\right)$

In Exercises 29–44, use the unit circle to find all of the exact values of θ that make the equation true in the indicated interval.

29. $\cos\theta = \dfrac{\sqrt{3}}{2}, \ 0 \le \theta \le 2\pi$

30. $\cos\theta = -\dfrac{\sqrt{3}}{2}, \ 0 \le \theta \le 2\pi$

31. $\sin\theta = -\dfrac{\sqrt{3}}{2}, \ 0 \le \theta \le 2\pi$

32. $\sin\theta = \dfrac{\sqrt{3}}{2}, \ 0 \le \theta \le 2\pi$

33. $\sin\theta = 0, \ 0 \le \theta \le 4\pi$

34. $\sin\theta = -1, \ 0 \le \theta \le 4\pi$

35. $\cos\theta = -1, \ 0 \le \theta \le 4\pi$

36. $\cos\theta = 0, \ 0 \le \theta \le 4\pi$

37. $\tan\theta = -1, \ 0 \le \theta \le 2\pi$

38. $\cot\theta = 1, \ 0 \le \theta \le 2\pi$

39. $\sec\theta = -\sqrt{2}, \ 0 \le \theta \le 2\pi$

40. $\csc\theta = \sqrt{2}, \ 0 \le \theta \le 2\pi$

41. $\csc\theta$ is undefined, $0 \le \theta \le 2\pi$

42. $\sec\theta$ is undefined, $0 \le \theta \le 2\pi$

43. $\tan\theta$ is undefined, $0 \le \theta \le 2\pi$

44. $\cot\theta$ is undefined, $0 \le \theta \le 2\pi$

 ■ **APPLICATIONS**

45. Atmospheric Temperature. The average daily temperature in Peoria, Illinois, can be predicted with $T = 50 - 28\cos\dfrac{2\pi(x - 31)}{365}$, where x is the number of the day in the year (January 1 = 1, February 1 = 32, etc.) and T is in degrees Fahrenheit. What is the expected temperature on February 15?

46. Atmospheric Temperature. The average daily temperature in Peoria, Illinois, can be predicted with $T = 50 - 28\cos\dfrac{2\pi(x - 31)}{365}$, where x is the number of the day in the year (January 1 = 1, February 1 = 32, etc.) and T is in degrees Fahrenheit. What is the expected temperature on August 15? (Assume it's not a leap year.)

47. Body Temperature. The human body temperature normally fluctuates during the day. If a person's body temperature can be predicted with $T = 99.1 - 0.5\sin\left(x + \dfrac{\pi}{12}\right)$, where x is the number of hours since midnight and T is in degrees Fahrenheit, then what is the person's temperature at 6:00 A.M.?

48. Body Temperature. The human body temperature normally fluctuates during the day. If a person's body temperature can be predicted with $T = 99.1 - 0.5\sin\left(x + \dfrac{\pi}{12}\right)$, where x is the number of hours since midnight and T is in degrees Fahrenheit, then what is the person's temperature at 9:00 P.M.?

49. Tides. The height of the water in a harbor changes with the tides. If the height of the water on a particular day can be determined with $h(x) = 5 + 4.8\sin\dfrac{\pi}{6}(x + 4)$, where x is the number of hours since midnight and h is the height of the tide in feet, then what is the height of the tide at 3:00 P.M.?

Alamy Ltd.

50. Tides. The height of the water in a harbor changes with the tides. If the height of the water on a particular day can be determined with $h(x) = 5 + 4.8\sin\dfrac{\pi}{6}(x + 4)$, where x is the number of hours since midnight and h is the height of the tide in feet, then what is the height of the tide at 5:00 A.M.?

51. Yo-Yo Dieting. A woman has been yo-yo dieting for years. Her weight changes throughout the year as she gains and loses weight. Her weight in a particular month can be determined with $w(x) = 145 + 10\cos\left(\dfrac{\pi}{6}x\right)$, where x is the month and w is in pounds. If $x = 1$ corresponds to January, how much does she weigh in June?

52. Yo-Yo Dieting. How much does the woman in Exercise 51 weigh in December?

53. Seasonal Sales. The average number of guests visiting the Magic Kingdom at Walt Disney World per day is given by $n(x) = 30,000 + 20,000\sin\left(\dfrac{\pi}{2}(x + 1)\right)$, where n is the number of guests and x is the month. If January corresponds to $x = 1$, how many people on average are visiting the Magic Kingdom per day in February?

54. Seasonal Sales. How many guests are visiting the Magic Kingdom in Exercise 53 in December?

■ CATCH THE MISTAKE

In Exercises 55 and 56, explain the mistake that is made.

55. Use the unit circle to evaluate $\tan\left(\dfrac{5\pi}{6}\right)$ exactly.

Solution:

Tangent is the ratio of sine to cosine.
$$\tan\left(\frac{5\pi}{6}\right) = \frac{\sin\left(\dfrac{5\pi}{6}\right)}{\cos\left(\dfrac{5\pi}{6}\right)}$$

Use the unit circle to identify sine and cosine.
$$\sin\left(\frac{5\pi}{6}\right) = -\frac{\sqrt{3}}{2} \text{ and } \cos\left(\frac{5\pi}{6}\right) = \frac{1}{2}$$

Substitute values for sine and cosine.
$$\tan\left(\frac{5\pi}{6}\right) = \frac{-\dfrac{\sqrt{3}}{2}}{\dfrac{1}{2}}$$

Simplify.
$$\tan\left(\frac{5\pi}{6}\right) = -\sqrt{3}$$

This is incorrect. What mistake was made?

56. Use the unit circle to evaluate $\sec\left(\dfrac{11\pi}{6}\right)$ exactly.

Solution:

Secant is the reciprocal of cosine.

$$\sec\left(\frac{11\pi}{6}\right) = \frac{1}{\cos\left(\dfrac{11\pi}{6}\right)}$$

Use the unit circle to evaluate cosine.

$$\cos\left(\frac{11\pi}{6}\right) = -\frac{1}{2}$$

Substitute the value for cosine.

$$\sec\left(\frac{11\pi}{6}\right) = \frac{1}{-\dfrac{1}{2}}$$

Simplify.

$$\sec\left(\frac{11\pi}{6}\right) = -2$$

This is incorrect. What mistake was made?

■ **CHALLENGE**

57. T or F: $\sin(2n\pi + \theta) = \sin\theta$, for n an integer.

58. T or F: $\cos(2n\pi + \theta) = \cos\theta$, for n an integer.

59. T or F: $\sin\theta = 1$ when $\theta = \dfrac{(2n+1)\pi}{2}$, for n an integer.

60. T or F: $\cos\theta = 1$ when $\theta = n\pi$, for n an integer.

61. Is cosecant an even or odd function? Justify your answer.

62. Is tangent an even or odd function? Justify your answer.

63. Find all the values of θ, $0 \leq \theta \leq 2\pi$, for which the equation is true: $\sin\theta = \cos\theta$.

64. Find all the values of θ (θ is any real number) for which the equation is true: $\sin\theta = \cos\theta$.

■ **TECHNOLOGY**

65. Use a calculator to approximate $\sin(423°)$. What do you expect $\sin(-423°)$ to be? Verify your answer with a calculator.

66. Use a calculator to approximate $\cos(227°)$. What do you expect $\cos(-227°)$ to be? Verify your answer with a calculator.

67. To approximate $\cos\left(\dfrac{\pi}{3}\right)$, take 5 steps and read the x coordinate.

68. To approximate $\sin\left(\dfrac{\pi}{3}\right)$, take 5 steps and read the y coordinate.

A graphing calculator can be used to graph the unit circle with parametric equations (these will be covered in more detail in Section 8.5). For now, set the calculator in parametric and radian modes and let

$$X_1 = \cos T$$
$$Y_1 = \sin T$$

Set the window so that $0 \leq t \leq 2\pi$, step $= \dfrac{\pi}{15}$, $-2 \leq X \leq 2$, and $-2 \leq Y \leq 2$.

Q: Charles has been taking flying lessons and is approaching the airport with his instructor on his first landing. He is circling the airport at a rate of 80 knots (about 92 miles per hour) at a radius of 8 miles from the airport. Completing one rotation of this circle took about 33 minutes. Air traffic control has directed him to move in so that he is circling the airport at a radius of 2 miles as he waits to land. He is expecting to complete only three-quarters of one rotation about this circle before beginning his landing, so Charles estimates that he will begin landing in about 25 minutes (three-quarters of 33 minutes). However, his instructor tells him that they will begin their landing in about 6 minutes. Why is Charles's calculated time so far off?

A: Charles assumed that it would take the same amount of time to complete three-quarters of one rotation about a circle of radius 2 miles as it did to complete three-quarters of one rotation about a circle of radius 8 miles. However, these arc lengths are not the same. The measure of the central angle to complete three-quarters of one rotation about a circle is $\theta = \dfrac{3\pi}{2}$ radians. Thus, the arc lengths are:

$$s = r\theta = 8\left(\frac{3\pi}{2}\right) = 12\pi \text{ miles (three-quarters of the distance about a circle of radius 8 miles)}$$

$$s = r\theta = 2\left(\frac{3\pi}{2}\right) = 3\pi \text{ miles (three-quarters of the distance about a circle of radius 2 miles)}$$

Since Charles maintains the same linear velocity of 92 miles per hour, it will take him less time to cover the lesser distance of 3π miles.

Time to fly 12π miles: $(12\pi \text{ miles})\left(\dfrac{1 \text{ hour}}{92 \text{ miles}}\right)\left(\dfrac{60 \text{ minutes}}{1 \text{ hour}}\right) \approx 25 \text{ minutes}$

Time to fly 3π miles: $(3\pi \text{ miles})\left(\dfrac{1 \text{ hour}}{92 \text{ miles}}\right)\left(\dfrac{60 \text{ minutes}}{1 \text{ hour}}\right) \approx 6 \text{ minutes}$

Now estimate the time it would take Charles to complete two-thirds of one rotation about a circle of radius 6 miles if he maintains his current linear velocity of 92 miles per hour. Round your answer to the nearest minute.

TYING IT ALL TOGETHER

A compact disk player reads a CD at a rate of 1.2 meters of track per second and can play for up to 74 minutes. The tracks begin at an inner radius of 25 mm and end at an outer radius of 58 mm, and the disk has a thickness of 1.2 mm. The wavelength of the laser is 780 nm. If a track begins at a radius of 30 mm, what is the angular speed at which the disk is being read as the song begins? If a track begins at a radius of 42 mm, what is the angular speed at which the disk is being read as the song begins? How do these angular speeds compare? What is the total length of the tracks if the CD is full (contains 74 minutes of music)?

SECTION	TOPIC	PAGES	REVIEW EXERCISES	KEY CONCEPTS
3.1	Radian measure	126–133	1–20	θ (in radians) $= \dfrac{s}{r}$ s (arc length) and r (radius) must have the same units.
	Converting between radian and degree measures	128–130	1–20	Degrees to radians: $\theta_r = \theta_d\left(\dfrac{\pi}{180°}\right)$ Radians to degrees: $\theta_d = \theta_r\left(\dfrac{180°}{\pi}\right)$
3.2	Arc length and area of a circular sector	137–142	21–40	
	Arc length	137–140	21–36	$s = r\theta_r$, θ_r is in radians or $s = r\theta_d\left(\dfrac{\pi}{180°}\right)$, θ_d is in degrees
	Area of circular sector	140–142	37–40	$A = \dfrac{1}{2}r^2\theta_r$, θ_r is in radians or $A = \dfrac{1}{2}r^2\theta_d\left(\dfrac{\pi}{180°}\right)$, θ_d is in degrees
3.3	Linear and angular speeds	146–150	41–60	Uniform circular motion ■ Linear speed: speed around the circumference of a circle ■ Angular speed: rotation speed of angle
	Linear speed	146–147	41–48	Linear speed, v, is given by $$v = \dfrac{s}{t}$$ where s is the arc length and t is time.
	Angular speed	147–148	49–54	Angular speed, ω, is given by $$\omega = \dfrac{\theta}{t}$$ where θ is given in radians.

SECTION	TOPIC	PAGES	REVIEW EXERCISES	KEY CONCEPTS
	Relationship between angular and linear speeds	148–150	55–60	Related through the radius of the circle $$v = r\omega \quad \text{or} \quad \omega = \frac{v}{r}$$ It is important to note that these formulas hold true only when angular speed is given in *radians* per unit of time.
3.4	Definition 3 of trigonometric functions: Unit circle approach	154–160	61–80	
	Values for sine and cosine as coordinates of points lying on the unit circle	154–155	61–72	
	Properties of circular functions	157–160	73–80	Cosine is an even function. $$\cos(-\theta) = \cos\theta$$ Sine is an odd function. $$\sin(-\theta) = -\sin\theta$$

CHAPTER 3 REVIEW EXERCISES

3.1 Radian Measure

Convert from degrees to radians. Leave your answers in terms of π.

1. $135°$ 2. $240°$ 3. $330°$ 4. $180°$

5. $216°$ 6. $108°$ 7. $504°$ 8. $600°$

9. $-150°$ 10. $-15°$

Convert from radians to degrees.

11. $\dfrac{\pi}{3}$ 12. $\dfrac{11\pi}{6}$ 13. $\dfrac{5\pi}{4}$ 14. $\dfrac{2\pi}{3}$

15. $\dfrac{5\pi}{9}$ 16. $\dfrac{17\pi}{10}$ 17. $\dfrac{13\pi}{4}$ 18. $\dfrac{11\pi}{3}$

19. $-\dfrac{5\pi}{18}$ 20. $-\dfrac{13\pi}{36}$

3.2 Arc Length and Area of a Circular Sector

Find the arc length made by the indicated central angle and radius of each circle. Round to two decimal places.

21. $\theta = \dfrac{\pi}{3}, r = 5$ cm   22. $\theta = \dfrac{5\pi}{6}, r = 10$ in.

23. $\theta = 100°, r = 5$ in.   24. $\theta = 36°, r = 12$ ft

Find the measure of the angle whose intercepted arc and radius of circle are given.

25. $r = 12$ in., $s = 6$ in.   26. $r = 10$ ft, $s = 27$ in.

27. $r = 6$ ft, $s = 4\pi$ ft   28. $r = 8$ m, $s = 2\pi$ m

29. $r = 5$ ft, $s = 10$ in.   30. $r = 4$ km, $s = 4$ m

Find the measure of each radius given the arc length and central angle of each circle.

31. $\theta = \dfrac{5\pi}{8}, s = \pi$ in.   32. $\theta = \dfrac{2\pi}{3}, s = 3\pi$ km

33. $\theta = 150°, s = 14\pi$ m   34. $\theta = 63°, s = 14\pi$ in.

35. $\theta = 10°, s = \dfrac{5\pi}{18}$ yd   36. $\theta = 80°, s = \dfrac{5\pi}{3}$ ft

Find the area of the circular sector given the indicated radius and central angle.

37. $\theta = \dfrac{\pi}{3}, r = 24$ mi   38. $\theta = \dfrac{5\pi}{12}, r = 9$ in.

39. $\theta = 60°, r = 60$ m   40. $\theta = 81°, r = 36$ cm

3.3 Linear and Angular Speeds

Find the linear speed of a point that moves with constant speed in a circular motion if the point travels arc length s in time t.

41. $s = 3$ ft, $t = 9$ sec   42. $s = 5280$ ft, $t = 4$ min

43. $s = 15$ mi, $t = 3$ min   44. $s = 12$ cm, $t = 0.25$ sec

Find the distance traveled of a point that moves with constant speed v along a circle in time t.

45. $v = 15$ mi/hr, $t = 1$ day

46. $v = 16$ ft/sec, $t = 1$ min

47. $v = 80$ mi/hr, $t = 15$ min

48. $v = 1.5$ cm/hr, $t = 6$ sec

Find the angular speed (radians/second) associated with rotating a central angle, θ, in time t.

49. $\theta = 6\pi, t = 9$ sec   50. $\theta = \pi, t = 0.05$ sec

51. $\theta = 225°, t = 20$ sec   52. $\theta = 330°, t = 22$ sec

Find the linear speed of a point traveling at a constant speed along the circumference of a circle with radius r and angular speed ω.

53. $\omega = \dfrac{5\pi\,\text{rad}}{6\ \text{sec}}, r = 12$ m   54. $\omega = \dfrac{\pi\ \text{rad}}{20\,\text{sec}}, r = 30$ in.

Find the distance a point travels along a circle, s, over a time t, given the angular speed, ω, and radius of the circle, r.

55. $r = 10$ ft, $\omega = \dfrac{\pi\,\text{rad}}{4\ \text{sec}}, t = 30$ sec

56. $r = 6$ in., $\omega = \dfrac{3\pi\,\text{rad}}{4\ \text{sec}}, t = 6$ sec

57. $r = 12$ yd, $\omega = \dfrac{2\pi\,\text{rad}}{3\ \text{sec}}, t = 30$ sec

58. $r = 100$ in., $\omega = \dfrac{\pi\ \text{rad}}{18\,\text{sec}}, t = 3$ min

Applications

59. A ladybug is clinging to the outer edge of a child's spinning disk. The disk is 4 inches in diameter and is spinning at 60 revolutions per minute. How fast is the ladybug traveling?

60. How fast is a motorcyclist traveling in miles per hour if his tires are 30 inches in diameter and his angular speed is 10π radians per second?

3.4 Definition 3 of Trigonometric Functions: Unit Circle Approach

Find the exact values of the indicated trigonometric functions.

61. $\tan\left(\dfrac{5\pi}{6}\right)$ 62. $\cos\left(\dfrac{5\pi}{6}\right)$ 63. $\sin\left(\dfrac{11\pi}{6}\right)$

64. $\sec\left(\dfrac{11\pi}{6}\right)$ 65. $\cot\left(\dfrac{5\pi}{4}\right)$ 66. $\csc\left(\dfrac{5\pi}{4}\right)$

67. $\sin\left(\dfrac{3\pi}{2}\right)$ 68. $\cos\left(\dfrac{3\pi}{2}\right)$ 69. $\cos\pi$

70. $\tan 315°$ 71. $\cos 60°$ 72. $\sin 330°$

Find the exact values of the indicated functions.

73. $\sin\left(-\dfrac{5\pi}{6}\right)$ 74. $\cos\left(-\dfrac{5\pi}{4}\right)$

75. $\cos(-240°)$ 76. $\sin(-135°)$

Find all of the exact values of θ that make the equation true in the indicated interval.

77. $\sin\theta = -\frac{1}{2}, 0 \le \theta \le 2\pi$

78. $\cos\theta = -\frac{1}{2}, 0 \le \theta \le 2\pi$

79. $\tan\theta = 0, 0 \le \theta \le 4\pi$

80. $\sin\theta = -1, 0 \le \theta \le 4\pi$

CHAPTER 3 PRACTICE TEST

1. Find the measure (in radians) of a central angle, θ, that intercepts an arc on a circle with radius $r = 20$ cm with arc length $s = 4$ mm.

2. Convert $\dfrac{13\pi}{4}$ to degree measure.

3. Convert $260°$ to radian measure. Leave the answer in terms of π.

4. Convert $217°$ to radians. Round to two decimal places.

5. What is the reference angle to $\theta = \dfrac{7\pi}{12}$?

6. Calculate the arc length on a circle with central angle $\theta = \dfrac{\pi}{15}$ and radius $r = 8$ yards.

7. A sprinkler has a 25 foot spray and it covers an angle of $30°$. What is the area that the sprinkler waters? Round to the nearest square foot.

8. A bicycle with tires of radius $r = 15$ inches is being ridden by a boy at constant speed (the tires are making 5 rotations per second). How many miles will he have ridden in 15 minutes? (1 mile = 5280 ft)

9. The smaller gear has a radius of 2 cm and the larger gear has a radius of 5.2 cm. If the smaller gear rotates $135°$, how many degrees has the larger gear rotated? Round answer to the nearest degree.

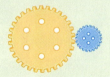

If possible, find the exact value of the indicated trigonometric function using the unit circle.

10. $\sin\left(-\dfrac{7\pi}{6}\right)$

11. $\tan\left(\dfrac{7\pi}{4}\right)$

12. $\csc\left(-\dfrac{3\pi}{4}\right)$

13. $\cot\left(-\dfrac{3\pi}{2}\right)$

14. $\sec\left(-\dfrac{7\pi}{2}\right)$

15. What is the measure in radians of the smaller angle between the hour and minute hands at 10:10?

Graphing Trigonometric Functions

John Stuart/Iconica/Getty Images, Inc.

Sine and cosine functions are used to represent periodic phenomena. Orbits, tide levels, the biological clock in animals and plants, and radio signals are all periodic (repetitive).

When you are standing on the shore of a placid lake and a motor boat goes by, the waves lap up to the shore at regular intervals and for a while the height of each wave appears to be constant.

If we graph the height of the water as a function of time, the result is a *sine wave*. The duration of time between each wave hitting the shore is called the *period*, and the height of the wave is called the *amplitude*.

In this chapter we will graph the trigonometric functions. We will start with simple sine and cosine functions and then proceed to translations of sine and cosine functions. We will then use the basic graphs of sine and cosine to determine graphs of tangent, secant, cosecant, and cotangent. Many periodic phenomena can be represented with sine and cosine functions, such as orbits, tides, biological clocks, and electromagnetic waves (radio, cell phones, light).

Graphing Trigonometric Functions

Graphs of Sine and Cosine
$y = A \sin Bx$
$y = A \cos Bx$

Translations of Sine and Cosine
$y = k + A \sin(Bx + C)$
$y = k + A \cos(Bx + C)$

Graphs of Other Trigonometric Functions: Tangent/Cotangent Secant/Cosecant

- Amplitude
- Period
- Domain/Range

- Horizontal (Phase) Shifts
- Vertical Shifts
- Domain/Range
- Curve-Fitting

- Period
- Domain/Range

CHAPTER OBJECTIVES

- Graph basic sine and cosine functions.
- Determine the amplitude and period of sine and cosine functions.
- Graph general sine and cosine functions using translations.
- Model periodic phenomena with trigonometric functions.
- Graph tangent, cotangent, secant, and cosecant functions.
- Determine domain and range of the trigonometric functions.

NAVIGATION THROUGH SUPPLEMENTS

DIGITAL VIDEO SERIES #4

STUDENT SOLUTIONS MANUAL CHAPTER 4

BOOK COMPANION SITE
www.wiley.com/college/young

Skills Objectives

- Graph sine and cosine functions.
- Determine the domain and range of sine and cosine functions.
- Determine the amplitude and period of sinusoidal graphs.
- Solve harmonic motion problems.

Conceptual Objectives

- Relate angles to the x coordinates and values of the trigonometric functions to the y coordinates.
- Understand why the graphs of sine and cosine are called sinusoidal graphs.

Periodic Functions

The following are examples of things that repeat in a predictable way (are roughly periodic):

- a heartbeat
- tide levels
- time of sunrise
- average outdoor temperature for the time of year

The trigonometric functions are *strictly* periodic. In the unit circle, the value of any of the trigonometric functions is the same for any angle with the same initial and terminal sides (no matter how many full rotations the angle makes). For example, if we add (or subtract) multiples of 2π to the angle θ, the values for sine and cosine are unchanged.

$$\sin(\theta + 2n\pi) = \sin\theta \quad \text{or} \quad \cos(\theta + 2n\pi) = \cos\theta \ (n \text{ is any integer})$$

DEFINITION   PERIODIC FUNCTION

A function, f, is called **periodic** if there is a positive number p such that

$$f(x + p) = f(x) \qquad \text{for all } x \text{ in the domain of } f$$

If p is the smallest such number for which this equation holds, then p is called the **fundamental period**.

You will see in this chapter that sine, cosine, secant, and cosecant have fundamental period 2π, but that tangent and cotangent have fundamental period π.

The Graph of $y = \sin x$

Let us start by point-plotting the sine function. We select values for the sine function corresponding to "special angles" we have covered in the previous chapters.

x	$y = \sin x$	(x, y)
0	$\sin 0 = 0$	$(0, 0)$
$\dfrac{\pi}{4}$	$\sin \dfrac{\pi}{4} = \dfrac{\sqrt{2}}{2}$	$\left(\dfrac{\pi}{4}, \dfrac{\sqrt{2}}{2}\right)$
$\dfrac{\pi}{2}$	$\sin \dfrac{\pi}{2} = 1$	$\left(\dfrac{\pi}{2}, 1\right)$
$\dfrac{3\pi}{4}$	$\sin \dfrac{3\pi}{4} = \dfrac{\sqrt{2}}{2}$	$\left(\dfrac{3\pi}{4}, \dfrac{\sqrt{2}}{2}\right)$
π	$\sin \pi = 0$	$(\pi, 0)$
$\dfrac{5\pi}{4}$	$\sin \dfrac{5\pi}{4} = -\dfrac{\sqrt{2}}{2}$	$\left(\dfrac{5\pi}{4}, -\dfrac{\sqrt{2}}{2}\right)$
$\dfrac{3\pi}{2}$	$\sin \dfrac{3\pi}{2} = -1$	$\left(\dfrac{3\pi}{2}, -1\right)$
$\dfrac{7\pi}{4}$	$\sin \dfrac{7\pi}{4} = -\dfrac{\sqrt{2}}{2}$	$\left(\dfrac{7\pi}{4}, -\dfrac{\sqrt{2}}{2}\right)$
2π	$\sin 2\pi = 0$	$(2\pi, 0)$

Plotting these coordinates, (x, y), we obtain the graph of one **period**, or **cycle**, of the graph of $y = \sin x$. Note that $\dfrac{\sqrt{2}}{2} \approx 0.7$.

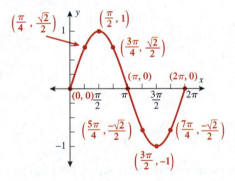

We can extend the graph horizontally in both directions (left and right).

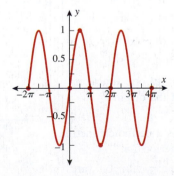

In this chapter, we are no longer showing angles on the unit circle but are now showing an angle as a *real number* in radians on the x-axis of the *Cartesian* graph. Therefore, we no longer illustrate a terminal side to an angle—the physical arcs and angles no longer exist; only their measures exist, as values of the x coordinate.

Set a TI/scientific calculator to radian mode by typing ⬛MODE⬛.

Set the window at Xmin at -2π, Xmax at 4π, Xscl at $\frac{\pi}{2}$, Ymin at -1, Ymax at 1, and Yscl at 1. Setting Xscl at $\frac{\pi}{2}$ will mark the labels on the x-axis in terms of multiples of $\frac{\pi}{2}$.

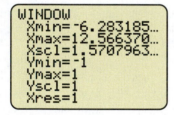

Use ⬛Y=⬛ to enter the function $\sin(X)$.

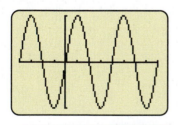

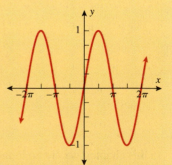

If we graph the function $y = \sin x$, the x-intercepts correspond to values of x when the sine function is equal to zero.

x	$y = \sin x$	(x, y)
0	$\sin 0 = 0$	$(0, 0)$
π	$\sin \pi = 0$	$(\pi, 0)$
2π	$\sin 2\pi = 0$	$(2\pi, 0)$
3π	$\sin 3\pi = 0$	$(3\pi, 0)$
4π	$\sin 4\pi = 0$	$(4\pi, 0)$
$\cdots$		
$n\pi$	$\sin n\pi = 0$	$(n\pi, 0)$ where n is an integer.

Notice that the point $(0, 0)$ is both a y-intercept and an x-intercept. The maximum value of the sine function is 1, and the minimum value of the sine function is -1, which occurs at odd multiples of $\frac{\pi}{2}$.

x	$y = \sin x$	(x, y)
$\frac{\pi}{2}$	$\sin \frac{\pi}{2} = 1$	$\left(\frac{\pi}{2}, 1\right)$
$\frac{3\pi}{2}$	$\sin \frac{3\pi}{2} = -1$	$\left(\frac{3\pi}{2}, -1\right)$
$\frac{5\pi}{2}$	$\sin \frac{5\pi}{2} = 1$	$\left(\frac{5\pi}{2}, 1\right)$
$\frac{7\pi}{2}$	$\sin \frac{7\pi}{2} = -1$	$\left(\frac{7\pi}{2}, -1\right)$
$\cdots$		
$\frac{(2n+1)}{\pi}$	$\sin\left(\frac{2n+1}{\pi}\right) = \pm 1$	$\left(\frac{2n+1}{\pi}, \pm 1\right)$

The following box summarizes the sine function. Note that we can use either $y = \sin x$ or $f(x) = \sin x$ notation.

SINE FUNCTION $f(x) = \sin x$

- Domain: $(-\infty, \infty)$ or $-\infty < x < \infty$
- Range: $[-1, 1]$ or $-1 \leq y \leq 1$
- The sine function is an odd function.
  - Symmetric about the origin
  - $f(-x) = -f(x)$
- The sine function is a periodic function with fundamental period 2π.
- The x-intercepts, $0, \pm\pi, \pm 2\pi, \ldots$, are of the form $n\pi$, where n is an integer.
- The maximum (1) and minimum (-1) values of sine correspond to x values of the form $\frac{(2n+1)\pi}{2}$, such as $\pm\frac{\pi}{2}, \pm\frac{3\pi}{2}, \pm\frac{5\pi}{2}, \ldots$.

The Graph of $y = \cos x$

Let us start by point-plotting the cosine function.

x	$y = \cos x$	(x, y)
0	$\cos 0 = 1$	$(0, 1)$
$\dfrac{\pi}{4}$	$\cos \dfrac{\pi}{4} = \dfrac{\sqrt{2}}{2}$	$\left(\dfrac{\pi}{4}, \dfrac{\sqrt{2}}{2}\right)$
$\dfrac{\pi}{2}$	$\cos \dfrac{\pi}{2} = 0$	$\left(\dfrac{\pi}{2}, 0\right)$
$\dfrac{3\pi}{4}$	$\cos \dfrac{3\pi}{4} = -\dfrac{\sqrt{2}}{2}$	$\left(\dfrac{3\pi}{4}, -\dfrac{\sqrt{2}}{2}\right)$
π	$\cos \pi = -1$	$(\pi, -1)$
$\dfrac{5\pi}{4}$	$\cos \dfrac{5\pi}{4} = -\dfrac{\sqrt{2}}{2}$	$\left(\dfrac{5\pi}{4}, -\dfrac{\sqrt{2}}{2}\right)$
$\dfrac{3\pi}{2}$	$\cos \dfrac{3\pi}{2} = 0$	$\left(\dfrac{3\pi}{2}, 0\right)$
$\dfrac{7\pi}{4}$	$\cos \dfrac{7\pi}{4} = \dfrac{\sqrt{2}}{2}$	$\left(\dfrac{7\pi}{4}, \dfrac{\sqrt{2}}{2}\right)$
2π	$\cos 2\pi = 1$	$(2\pi, 1)$

Plotting these coordinates, (x, y), we obtain the graph of one period, or cycle, of the graph of $y = \cos x$. Note that $\dfrac{\sqrt{2}}{2} \approx 0.7$.

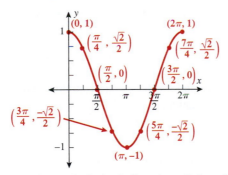

We can extend the graph horizontally in both directions (left and right).

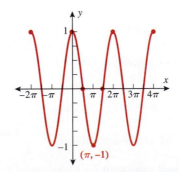

If we graph the function $y = \cos x$, the x-intercepts correspond to values of x when the cosine function is equal to zero, where n is an integer.

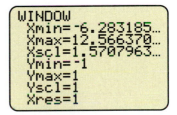
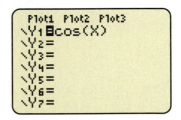
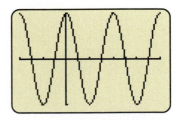

x	$y = \cos x$	(x, y)
$\dfrac{\pi}{2}$	$\cos\dfrac{\pi}{2} = 0$	$\left(\dfrac{\pi}{2}, 0\right)$
$\dfrac{3\pi}{2}$	$\cos\dfrac{3\pi}{2} = 0$	$\left(\dfrac{3\pi}{2}, 0\right)$
$\dfrac{5\pi}{2}$	$\cos\dfrac{5\pi}{2} = 0$	$\left(\dfrac{5\pi}{2}, 0\right)$
$\dfrac{7\pi}{2}$	$\cos\dfrac{7\pi}{2} = 0$	$\left(\dfrac{7\pi}{2}, 0\right)$
$\ldots$		
$\dfrac{(2n+1)}{\pi}$	$\cos\left(\dfrac{2n+1}{\pi}\right) = 0$	$\left(\dfrac{2n+1}{\pi}, 0\right)$

The point $(0, 1)$ is the y-intercept, and there are several x-intercepts of the form $\left(\dfrac{2n+1}{\pi}, 0\right)$. The maximum value of the cosine function is 1, and the minimum value of the cosine function is -1, which occurs at integer multiples of π.

x	$y = \cos x$	(x, y)
0	$\cos 0 = 1$	$(0, 1)$
π	$\cos \pi = -1$	$(\pi, -1)$
2π	$\cos 2\pi = 1$	$(2\pi, 1)$
3π	$\cos 3\pi = -1$	$(3\pi, -1)$
4π	$\cos 4\pi = 1$	$(4\pi, 1)$
$\ldots$		
$n\pi$	$\cos n\pi = \pm 1$	$(n\pi, \pm 1)$

The following box summarizes the cosine function. Note that we can use either $y = \cos x$ or $f(x) = \cos x$ notation.

COSINE FUNCTION $f(x) = \cos x$

- Domain: $(-\infty, \infty)$ or $-\infty < x < \infty$
- Range: $[-1, 1]$ or $-1 \leq y \leq 1$
- The cosine function is an even function.
  - Symmetric about the y-axis
  - $f(-x) = f(x)$
- The cosine function is a periodic function with fundamental period 2π.
- The x-intercepts, $\pm\dfrac{\pi}{2}, \pm\dfrac{3\pi}{2}, \pm\dfrac{5\pi}{2}, \ldots$, are odd multiples of $\dfrac{\pi}{2}$, which have the form $\dfrac{(2n+1)\pi}{2}$, where n is an integer.
- The maximum (1) and minimum (-1) values of cosine correspond to x values of the form $n\pi$, such as $0, \pm\pi, \pm2\pi, \ldots$.

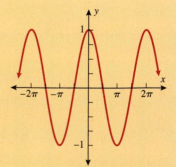

The Amplitude and Period of Sinusoidal Graphs

In mathematics, the word **sinusoidal** means "resembling the sine function." Let us start by graphing $f(x) = \sin x$ and $f(x) = \cos x$ on the same graph. Notice that they have similar characteristics (domain, range, period, and shape).

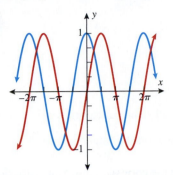

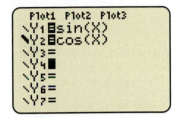

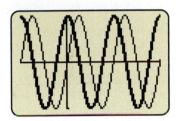

In fact, if we were to shift the cosine graph to the right $\dfrac{\pi}{2}$ units the two graphs would be identical. For that reason we refer to any graphs that involve either sine or cosine functions as **sinusoidal functions**.

We now turn our attention to graphs of the form $y = A \sin Bx$ and $y = A \cos Bx$, which are graphs like $y = \sin x$ and $y = \cos x$ that have been stretched or compressed either vertically or horizontally.

EXAMPLE 1 Vertical Stretching and Compressing

Plot the functions $y = 2 \sin x$ and $y = \dfrac{1}{2} \sin x$ on the same graph with $y = \sin x$ on the interval $-4\pi \le x \le 4\pi$.

Solution:

STEP 1 Make a table with the coordinate values of the graphs:

x	0	$\dfrac{\pi}{2}$	π	$\dfrac{3\pi}{2}$	2π
$\sin x$	0	1	0	-1	0
$2 \sin x$	0	2	0	-2	0
$\dfrac{1}{2} \sin x$	0	$\dfrac{1}{2}$	0	$-\dfrac{1}{2}$	0

STEP 2 Label the points on the graph and connect with a smooth curve over one period, $0 \le x \le 2\pi$.

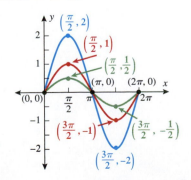

STEP 3 Extend the graph in both directions (repeats every 2π).

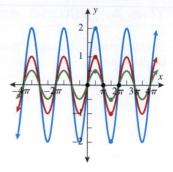

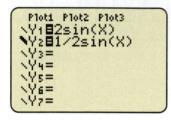
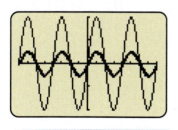
YOUR TURN   Plot the functions $y = 3\cos x$ and $y = \frac{1}{3}\cos x$ on the same graph with $y = \cos x$ on the interval $-2\pi \le x \le 2\pi$.

Notice in Example 1 and the corresponding Your Turn that:

■ $y = 2\sin x$ has the shape and period of $y = \sin x$ but stretched vertically.

■ $y = \frac{1}{2}\sin x$ has the shape and period of $y = \sin x$ but compressed vertically.

■ $y = 3\cos x$ has the shape and period of $y = \cos x$ but stretched vertically.

■ $y = \frac{1}{3}\cos x$ has the shape and period of $y = \cos x$ but compressed vertically.

In general, functions of the form $y = A\sin x$ and $y = A\cos x$ are stretched in the vertical direction when $|A| > 1$ and compressed in the vertical direction when $|A| < 1$.

The **amplitude** of a periodic function is half the difference between the maximum value of the function and the minimum value of the function. For the functions $y = \sin x$ and $y = \cos x$, the maximum value is 1 and the minimum value is -1. Therefore the amplitude of these two functions is $A = \frac{1}{2}|1 - (-1)| = 1$.

AMPLITUDE OF SINUSOIDAL FUNCTIONS

For sinusoidal functions of the form $y = A\sin Bx$ and $y = A\cos Bx$, the **amplitude** is $|A|$. When $|A| < 1$, the graph is compressed in the vertical direction, and when $|A| > 1$ the graph is stretched in the vertical direction.

EXAMPLE 2 Finding the Amplitude of Sinusoidal Functions

State the amplitude of (a) $f(x) = -4\cos x$ and (b) $g(x) = \frac{1}{5}\sin x$.

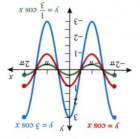

Answer: $y = 3\cos x$

$y = \frac{1}{3}\cos x$

$y = \cos x$

Solution (a): The amplitude is the magnitude of -4. $\qquad A = |-4| = 4$

Solution (b): The amplitude is the magnitude of $\dfrac{1}{5}$. $\qquad A = \left|\dfrac{1}{5}\right| = \dfrac{1}{5}$

EXAMPLE 3 Horizontal Stretching and Compressing

TECHNOLOGY TIP

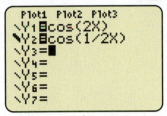

Plot the functions $y = \cos 2x$ and $y = \cos\left(\dfrac{1}{2}x\right)$ on the same graph with $y = \cos x$ on the interval $-2\pi \le x \le 2\pi$.

Solution:

When stretching or compressing in the horizontal direction, careful attention must be paid to the x coordinates in the point-plotting table.

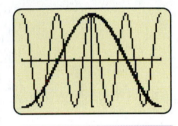

STEP 1 Make a table with the coordinate values of the graphs. It is only necessary to select the points that correspond to x-intercepts, $(y = 0)$, and maximum and minimum points, $(y = \pm 1)$:

x	0	$\dfrac{\pi}{4}$	$\dfrac{\pi}{2}$	$\dfrac{3\pi}{4}$	π	$\dfrac{5\pi}{4}$	$\dfrac{3\pi}{2}$	$\dfrac{7\pi}{4}$	2π
$\cos x$	1		0		-1		0		1
$\cos 2x$	1	0	-1	0	1	0	-1	0	1
$\cos\left(\dfrac{1}{2}x\right)$	1				0				-1

STEP 2 Label the points on the graph and connect with a smooth curve.

$y = \cos x$

$\cos 2x$

$\cos\left(\dfrac{1}{2}x\right)$

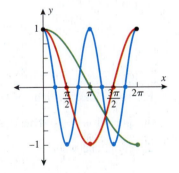

STEP 3 Extend the graph to cover the entire interval: $-2\pi \le x \le 2\pi$.

$y = \cos x$

$\cos 2x$

$\cos\left(\dfrac{1}{2}x\right)$

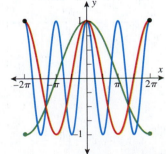

■ **YOUR TURN**   Plot the functions $y = \sin 2x$ and $y = \sin\left(\dfrac{1}{2}x\right)$ on the same graph with $y = \sin x$ on the interval $-2\pi \leq x \leq 2\pi$.

Notice in Example 3 and the corresponding Your Turn that:

- $y = \cos 2x$ has the shape and amplitude of $y = \cos x$ but compressed horizontally.

- $y = \cos\left(\dfrac{1}{2}x\right)$ has the shape and amplitude of $y = \cos x$ but stretched horizontally.

- $y = \sin 2x$ has the shape and amplitude of $y = \sin x$ but compressed horizontally.

- $y = \sin\left(\dfrac{1}{2}x\right)$ has the shape and amplitude of $y = \sin x$ but stretched horizontally.

In general, functions of the form $y = \sin Bx$ and $y = \cos Bx$, with $B > 0$, are compressed in the horizontal direction when $B > 1$ and stretched in the horizontal direction when $B < 1$.

Since sine is an odd function, $\sin(-Bx) = -\sin Bx$, and since cosine is an even function, $\cos(-Bx) = \cos Bx$, all functions can be rewritten with $B > 0$. We will discuss negative arguments in the next section in the context of *reflections*.

The period of functions $y = \sin x$ and $y = \cos x$ is 2π. To find the period of a function of the form $y = A\sin Bx$ or $y = A\cos Bx$, set Bx equal to 2π and solve for x.

$$Bx = 2\pi$$

$$x = \frac{2\pi}{B}$$

PERIOD OF SINUSOIDAL FUNCTIONS

For sinusoidal functions of the form $y = A\sin Bx$ and $y = A\cos Bx$, with $B > 0$, the **period** is $\dfrac{2\pi}{B}$. When $B < 1$, the graph is stretched in the horizontal direction, and when $B > 1$, the graph is compressed in the horizontal direction.

■ **Answer:** $y = \sin x$

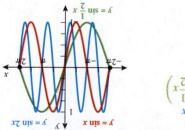

For the remainder of this section we will address only positive arguments, and so we will omit the absolute value signs.

EXAMPLE 4 Finding the Period of Sinusoidal Functions

State the period of (a) $f(x) = \cos 4x$ and (b) $g(x) = \sin\left(\frac{1}{3}x\right)$.

Solution (a):

Compare $\cos 4x$ with $\cos Bx$ to identify B. $\qquad\qquad B = 4$

Calculate the period of $\cos 4x$, using $p = \dfrac{2\pi}{B}$. $\qquad p = \dfrac{2\pi}{4} = \dfrac{\pi}{2}$

The period of $\cos 4x$ is $\boxed{p = \dfrac{\pi}{2}}$.

Solution (b):

Compare $\sin\left(\dfrac{1}{3}x\right)$ with $\sin Bx$ to identify B. $\qquad B = \dfrac{1}{3}$

Calculate the period of $\sin\left(\dfrac{1}{3}x\right)$, using $p = \dfrac{2\pi}{B}$. $\qquad p = \dfrac{2\pi}{\dfrac{1}{3}} = 6\pi$

The period of $\sin\left(\dfrac{1}{3}x\right)$ is $\boxed{p = 6\pi}$.

■ **YOUR TURN** State the period of (a) $f(x) = \sin 3x$ and (b) $g(x) = \cos\left(\dfrac{1}{2}x\right)$.

Now that you know the basic graphs of $y = \sin x$ and $y = \cos x$, you can sketch one cycle (period) of these graphs with the following x values: $0, \dfrac{\pi}{2}, \pi, \dfrac{3\pi}{2}, 2\pi$. For a period of 2π, we used steps of $\dfrac{\pi}{2}$. Therefore when we start at the origin, as long as we have 4 steps during one period we are able to sketch the graphs.

STUDY TIP

Divide the period by 4 to get the values along the x-axis.

STRATEGY FOR SKETCHING GRAPHS OF SINUSOIDAL FUNCTIONS

To graph $y = A \sin Bx$ or $y = A \cos Bx$ with $B > 0$:

Step 1: Find the amplitude, $|A|$, and period, $\dfrac{2\pi}{B}$.

Step 2: Divide the period into four equal parts (steps).

Step 3: Make a table and evaluate the function for x values from Step 2.

Step 4: Draw the xy plane (label the y-axis up to $\pm A$) and plot points found in Step 3.

Step 5: Connect the points with a sinusoidal curve (with amplitude $|A|$).

Step 6: Repeat the graph over several periods in both directions (left and right).

■ **Answer:** (a) $p = \dfrac{2\pi}{3}$, (b) $p = 4\pi$

TECHNOLOGY TIP

Use a TI calculator to check the graph of $y = 3\sin 2x$.

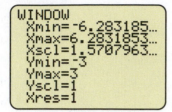

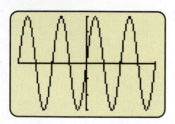

EXAMPLE 5 Graphing Sinusoidal Functions of the Form $y = A\sin Bx$

Use the strategy for graphing sinusoidal functions to graph $y = 3\sin 2x$.

Solution:

STEP 1 Find the amplitude and period ($A = 3$ and $B = 2$).

$$A = |3| = 3 \quad \text{and} \quad p = \frac{2\pi}{B} = \frac{2\pi}{2} = \pi$$

STEP 2 Divide the period, π, into four equal steps. $\dfrac{\pi}{4}$

STEP 3 Make a table, starting at $x = 0$ to the period π in steps of $\dfrac{\pi}{4}$.

x	$y = 3\sin 2x$	(x, y)
0	$3[\sin 0] = 3[0] = 0$	$(0, 0)$
$\dfrac{\pi}{4}$	$3\left[\sin\left(\dfrac{\pi}{2}\right)\right] = 3[1] = 3$	$\left(\dfrac{\pi}{4}, 3\right)$
$\dfrac{\pi}{2}$	$3[\sin \pi] = 3[0] = 0$	$\left(\dfrac{\pi}{2}, 0\right)$
$\dfrac{3\pi}{4}$	$3\left[\sin\left(\dfrac{3\pi}{2}\right)\right] = 3[-1] = -3$	$\left(\dfrac{3\pi}{4}, -3\right)$
π	$3[\sin 2\pi] = 3[0] = 0$	$(\pi, 0)$

STEP 4 Draw the xy plane and label the points in the table.

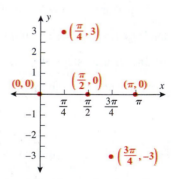

STEP 5 Connect the points with a sinusoidal curve.

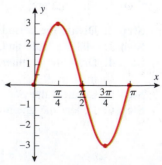

STEP 6 Repeat over several periods (to the left and right).

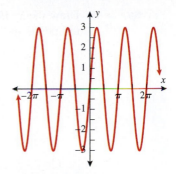

YOUR TURN Use the strategy for graphing sinusoidal functions to graph $y = 2 \sin 3x$.

EXAMPLE 6 Graphing Sinusoidal Functions of the Form $y = A \cos Bx$

Use the strategy for graphing sinusoidal functions to graph $y = -2\cos\left(\frac{1}{3}x\right)$.

Solution:

STEP 1 Find the amplitude and period,

$$A = -2 \text{ and } B = \frac{1}{3}. \qquad A = |-2| = 2 \text{ and } p = \frac{2\pi}{B} = \frac{2\pi}{\frac{1}{3}} = 6\pi$$

STEP 2 Divide the period, 6π, into four equal steps. $\quad \dfrac{6\pi}{4} = \dfrac{3\pi}{2}$

STEP 3 Make a table, starting at $x = 0$ to the period 6π in steps of $\dfrac{3\pi}{2}$.

x	$y = -2\cos\left(\dfrac{1}{3}x\right)$	(x, y)
0	$-2[\cos 0] = -2[1] = -2$	$(0, -2)$
$\dfrac{3\pi}{2}$	$-2\left[\cos\left(\dfrac{\pi}{2}\right)\right] = -2[0] = 0$	$\left(\dfrac{3\pi}{2}, 0\right)$
3π	$-2[\cos(\pi)] = -2[-1] = 2$	$(3\pi, 2)$
$\dfrac{9\pi}{2}$	$-2\left[\cos\left(\dfrac{3\pi}{2}\right)\right] = -2[0] = 0$	$\left(\dfrac{9\pi}{2}, 0\right)$
6π	$-2[\cos 2\pi] = -2[1] = -2$	$(6\pi, -2)$

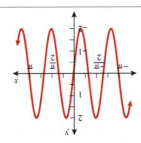

Answer:

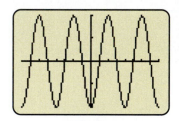
STEP 4 Draw the xy plane and label the points in the table.

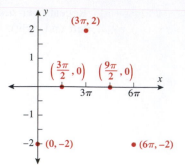

STEP 5 Connect the points with a sinusoidal curve.

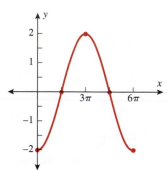

STEP 6 Repeat over several periods (to the left and right).

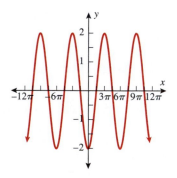

YOUR TURN Use the strategy for graphing sinusoidal functions to graph $y = -3\cos\left(\dfrac{1}{2}x\right)$.

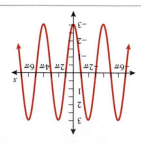

Answer:

Notice in Example 6 and the corresponding Your Turn that when A is negative, the result is rotation of the original function around the x-axis.

EXAMPLE 7 Finding an Equation for a Sinusoidal Graph

Find an equation for the following graph:

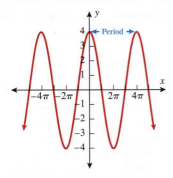

TECHNOLOGY TIP

Use a TI calculator to check the graph of $y = 4\cos\left(\frac{1}{2}x\right)$.

Set the window at Xmin at -6π, Xmax at 6π, Xscl at π, Ymin at -4, Ymax at 4, and Yscl at 1. Setting Xscl at π will mark the labels on the x-axis in terms of multiples of π.

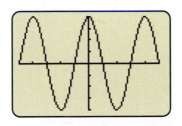

Solution:

This graph is similar to the cosine function. $y = A\cos Bx$

The amplitude is 4 (half the maximum spread). $A = 4$

The period, $\dfrac{2\pi}{B}$, is equal to 4π. $\dfrac{2\pi}{B} = 4\pi$

Solve for B. $B = \dfrac{1}{2}$

Substitute $A = 4$ and $B = \dfrac{1}{2}$ into $y = A\cos Bx$. $y = 4\cos\left(\dfrac{1}{2}x\right)$

■ **YOUR TURN** Find an equation of the following graph:

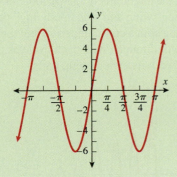

Harmonic Motion

One of the most important applications of sinusoidal functions is in describing *harmonic motion*, which we define as the symmetric periodic movement of an object or quantity about a center (equilibrium) position or value. The oscillation of a pendulum is a form of harmonic motion; other examples are the recoil of a spring-balance scale when a weight is placed on the tray, or the variation of current or voltage within an AC circuit.

There are three types of harmonic motion: *simple harmonic motion, damped harmonic motion*, and *resonance*.

Simple Harmonic Motion

Simple harmonic motion is the kind of *unvarying* periodic motion that would occur in an ideal situation in which no resistive forces, such as friction, caused the amplitude of oscillation to decrease over time: the amplitude stays in exactly the same range in each period as time—the variable on the horizontal axis—increases. This type of motion will also occur if energy is being supplied at the correct rate to overcome resistive forces. Simple harmonic motion occurs, for example, in an AC electric circuit when a power source is consistently supplying energy. When you are swinging on a swing and "pumping" energy into the swing to keep it in motion at a constant period and amplitude, you are sustaining simple harmonic motion.

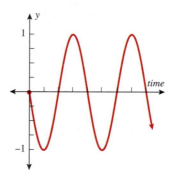

Damped Harmonic Motion

In damped harmonic motion, the amplitude of the periodic motion decreases as time increases. If you are on a moving swing and stop "pumping" new energy into the swing, the swing will continue moving with a constant period, but the amplitude—the height to which the swing will rise—will diminish with each cycle as the swing is slowed down by friction with the air or between its own moving parts.

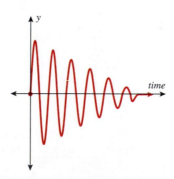

Hulton Archive/Getty Images, Inc.   Topham/The Image Works

Resonance

Resonance is what occurs when the amplitude of periodic motion increases as time increases. It is caused when the energy applied to an oscillating object or system is more than what is needed to oppose friction or other forces and sustain simple harmonic motion; instead, the applied energy *increases* the amplitude of harmonic motion with each cycle. With resonance, eventually, the amplitude becomes unbounded and the result is disastrous. Bridges have collapsed because of resonance. To the right is a picture of the Tacoma Narrows Bridge (near Seattle, Washington) that collapsed due to high winds resulting in resonance. Soldiers know that when they march across a bridge they must break cadence to prevent a regular oscillation from building up.

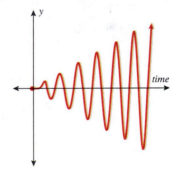

If we hang a weight from a spring, while at rest, we say it is in the equilibrium position.

If we then pull down on the weight, the elasticity in the spring causes the weight to oscillate up and down.

If we neglect friction and air resistance, we can imagine that the combination of the weight and the spring will oscillate indefinitely; the height of the weight with respect to the equilibrium position is then modeled by a simple sinusoidal function. This is called **simple harmonic motion**.

SIMPLE HARMONIC MOTION

The position of a point oscillating around an equilibrium position at time t is modeled by the sinusoidal function:

$$y = A \sin \omega t \qquad \text{or} \qquad y = A \cos \omega t$$

where $|A|$ is the amplitude, and the period is $\dfrac{2\pi}{\omega}$ where $\omega > 0$.

EXAMPLE 8 Simple Harmonic Motion

Let the height of the seat of a swing be equal to zero when the swing is at rest. Assume that a child starts swinging until she reaches the highest she can swing and keeps that effort constant. The height of the seat is given by

$$h(t) = 8 \sin\left(\frac{\pi}{2} t\right)$$

where t is time in seconds and h is the height in feet. Note that positive h indicates height reached swinging forward and negative h indicates height reached swinging backwards, assuming that $t = 0$ when the child passes through the equilibrium position swinging forward.

a. Graph the height function, $h(t)$, for $0 \le t \le 4$.
b. What is the maximum height above ground reached by the seat of the swing?
c. What is the period of the swinging child?

Solution (a):

Make a table with integer values of t. $\qquad 0 \le t \le 4$

t (seconds)	$y = h(t) = 8 \sin\left(\dfrac{\pi}{2} t\right)$ (feet)	(x, y)
0	$8 \sin 0 = 0$	$(0, 0)$
1	$8 \sin\left(\dfrac{\pi}{2}\right) = 8$	$(1, 8)$
2	$8 \sin(\pi) = 0$	$(2, 0)$
3	$8 \sin\left(\dfrac{3\pi}{2}\right) = -8$	$(3, -8)$
4	$8 \sin(2\pi) = 0$	$(4, 0)$

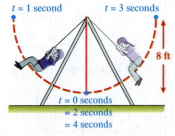

$(1, 8)$

$(2, 0)$ $(4, 0)$

$(0, 0)$

$(3, -8)$

Labeling the time and height on the original diagram, we see that the maximum height is 8 feet and the period is 4 seconds.

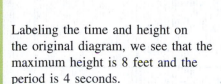

$t = 1$ second $t = 3$ seconds

8 ft

$t = 0$ seconds
$= 2$ seconds
$= 4$ seconds

Solutions (b) and (c):

$$h(t) = 8\sin\left(\frac{\pi}{2}t\right) = \underset{A}{8}\sin\left(\frac{2\pi}{\underset{\omega}{4}}t\right)$$

The maximum height above the equilibrium is the amplitude.

$$A = 8 \text{ ft}$$

The period is $\dfrac{2\pi}{\omega}$.

$$p = \frac{2\pi}{\dfrac{2\pi}{4}} = 4 \text{ sec}$$

Damped harmonic motion is any sinusoidal function whose amplitude decreases as time increases. If we again hang a weight from a spring so that it is suspended at rest, and then pull down on the weight and release, the weight will oscillate about the equilibrium point. This time, if we do not neglect friction and air resistance the weight will oscillate closer and closer to the equilibrium point over time until the weight eventually comes to a rest at the equilibrium point. This is an example of damped harmonic motion.

Mathematically, any function that decreases can be used to damp oscillatory motion. Here are two examples of functions that describe damped harmonic motion.

$$y = \frac{1}{t}\sin\omega t \qquad y = e^{-t}\cos\omega t$$

e represents an exponential function.

EXAMPLE 9 Damped Harmonic Motion

Assume that the child in Example 8 decides to stop pumping and allows the swing to continue moving until he eventually comes to rest. Assume that

$$h(t) = \frac{8}{t}\cos\left(\frac{\pi}{2}t\right)$$

where t is time in seconds and h is the height in feet above the resting position. Note that positive h indicates height reached swinging forward and negative h indicates height reached swinging backwards, assuming that $t = 1$ when the child passes through the equilibrium position swinging backward and stops "pumping."

a. Graph the height function, $h(t)$, for $1 \le t \le 8$.
b. What is the height above ground at 4 seconds? At 8 seconds? After 1 minute?

Solution (a):

Make a table with integer values of t. $\qquad 1 \le t \le 8$

t (seconds)	$y = h(t) = \dfrac{8}{t}\cos\left(\dfrac{\pi}{2}t\right)$ (feet)	(x, y)
1	$\dfrac{8}{1}\cos\left(\dfrac{\pi}{2}\right) = 0$	$(1, 0)$
2	$\dfrac{8}{2}\cos(\pi) = -4$	$(2, -4)$
3	$\dfrac{8}{3}\cos\left(\dfrac{3\pi}{2}\right) = 0$	$(3, 0)$
4	$\dfrac{8}{4}\cos(2\pi) = 2$	$(4, 2)$
5	$\dfrac{8}{5}\cos\left(\dfrac{5\pi}{2}\right) = 0$	$(5, 0)$
6	$\dfrac{8}{6}\cos(3\pi) = -\dfrac{4}{3}$	$\left(6, -\dfrac{4}{3}\right)$
7	$\dfrac{8}{7}\cos\left(\dfrac{7\pi}{2}\right) = 0$	$(7, 0)$
8	$\dfrac{8}{8}\cos(4\pi) = 1$	$(8, 1)$

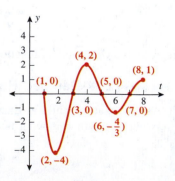

Solution (b):

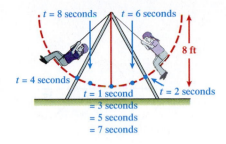

The height is 2 feet when t is 4 seconds.

$$\frac{8}{4}\cos(2\pi) = 2$$

The height is 1 foot when t is 8 seconds.

$$\frac{8}{8}\cos(4\pi) = 1$$

The height is 0.13 feet when t is
1 minute (60 seconds).

$$\frac{8}{60}\cos(30\pi) = 0.1333$$

Resonance is any sinusoidal function whose amplitude increases as time increases. Mathematically, any function that increases can be used to cause resonance. Here are two examples of functions that result in resonance as time increases.

$$y = t\cos\omega t \qquad y = e^t\sin\omega t$$

SECTION 4.1 SUMMARY

The sine function is an odd function and its graph is symmetric about the origin. The cosine function is an even function and its graph is symmetric about the y-axis.

Graphs of the form $y = A\sin Bx$ and $y = A\cos Bx$ have amplitude $|A|$ and period $\dfrac{2\pi}{B}$.

In order to graph sinusoidal functions, you can start by point-plotting. A more efficient way is to first determine the amplitude and period. Divide the period into four equal parts and choose those values for x. Make a table of those four points and graph (this is the graph of one period). Extend the graph to the left and right. To find the equation of a sinusoidal function, given its graph, start by first finding the amplitude (half the distance between the maximum and minimum values). Then determine the period. Harmonic motion is one of the major applications of sinusoidal functions.

SECTION 4.1 EXERCISES

■ **SKILLS**

In Exercises 1–10, match each function with its graph.

1. $y = -\sin x$

2. $y = \sin x$

3. $y = \cos x$

4. $y = -\cos x$

5. $y = 2\sin x$

6. $y = 2\cos x$

7. $y = \sin\dfrac{1}{2}x$

8. $y = \cos\dfrac{1}{2}x$

9. $y = -2\cos\dfrac{1}{2}x$

10. $y = -2\sin\dfrac{1}{2}x$

a.
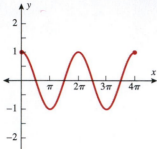

b.
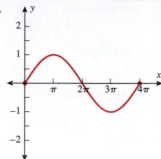

c.
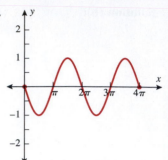

d.
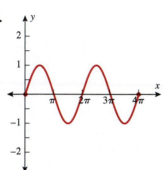

e.
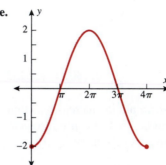

f.
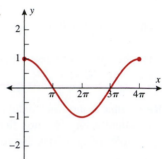

g.
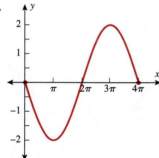

h.
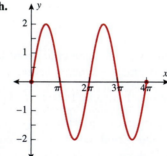

i.
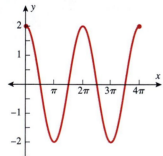

j.
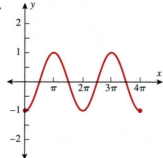

In Exercises 11–20, state the amplitude and period of each function.

11. $y = \dfrac{3}{2}\cos 3x$

12. $y = \dfrac{2}{3}\sin 4x$

13. $y = -\sin 5x$

14. $y = -\cos 7x$

15. $y = \dfrac{2}{3}\cos\left(\dfrac{3}{2}x\right)$

16. $y = \dfrac{3}{2}\sin\left(\dfrac{2}{3}x\right)$

17. $y = -3\cos(\pi x)$

18. $y = -2\sin(\pi x)$

19. $y = 5\sin\left(\dfrac{\pi}{3}x\right)$

20. $y = 4\cos\left(\dfrac{\pi}{4}x\right)$

In Exercises 21–32, graph the given functions over one period.

21. $y = 8\cos x$

22. $y = 7\sin x$

23. $y = \sin 4x$

24. $y = \cos 3x$

25. $y = -3\cos\left(\dfrac{1}{2}x\right)$

26. $y = -2\sin\left(\dfrac{1}{4}x\right)$

27. $y = -3\sin(\pi x)$

28. $y = -2\cos(\pi x)$

29. $y = 5\cos(2\pi x)$

30. $y = 4\sin(2\pi x)$

31. $y = -3\sin\left(\dfrac{\pi}{4}x\right)$

32. $y = -4\sin\left(\dfrac{\pi}{2}x\right)$

In Exercises 33–40, graph the given function over the interval $[-2p, 2p]$, where p is the period of the function.

33. $y = -4\cos\left(\dfrac{1}{2}x\right)$

34. $y = -5\sin\left(\dfrac{1}{2}x\right)$

35. $y = -\sin 6x$

36. $y = -\cos 4x$

37. $y = 3\cos\left(\dfrac{\pi}{4}x\right)$

38. $y = 4\sin\left(\dfrac{\pi}{4}x\right)$

39. $y = \sin(4\pi x)$

40. $y = \cos(6\pi x)$

In Exercises 41–48, find an equation for each graph.

41.

42.

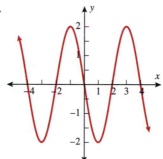

43.

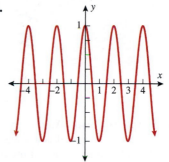

44.

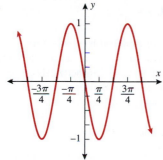

45.

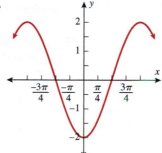

46.

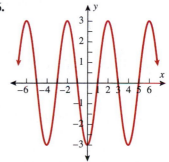

47.

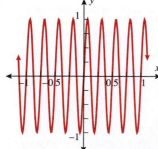

48.

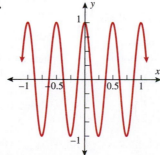

■ APPLICATIONS

49. Simple Harmonic Motion. A weight hanging on a spring will oscillate up and down about its equilibrium position after it's pulled down and released. This is an example of simple harmonic motion. This motion would continue forever if there were not any friction or air resistance. Simple harmonic motion can be described with the function $y = A \cos\left(t\sqrt{\dfrac{k}{m}}\right)$, where A is the amplitude, t is the time in seconds, m is the mass, and k is a constant particular to that spring. If a spring is measured in centimeters and the weight in grams, then what are the amplitude and mass if $y = 4\cos\left(\dfrac{t\sqrt{k}}{2}\right)$?

50. Simple Harmonic Motion. A weight hanging on a spring will oscillate up and down about its equilibrium position after it's pulled down and released. This is an example of simple harmonic motion. This motion would continue forever if there were not any friction or air resistance. Simple harmonic motion can be described with the function $y = A\cos\left(t\sqrt{\dfrac{k}{m}}\right)$, where A is the amplitude, t is the time in seconds, m is the mass, and k is a constant particular to that spring. If a spring is measured in centimeters and the weight in grams, then what are the amplitude and mass if $y = 3\cos\left(3t\sqrt{k}\right)$?

51. Frequency of Oscillations. A weight hanging on a spring will oscillate up and down about its equilibrium position after it's pulled down and released. This is an example of simple harmonic motion. This motion would continue forever if there were not any friction or air resistance. Simple harmonic motion can be described with the function $y = A\cos\left(t\sqrt{\dfrac{k}{m}}\right)$, where A is the amplitude, t is the time in seconds, m is the mass, and k is a constant particular to that spring. The frequency of oscillation, f, is given by $f = \dfrac{1}{p}$, where p is the period. What is the frequency of oscillation modeled by $y = 3\cos\left(\dfrac{t}{2}\right)$?

52. Frequency of Oscillations. A weight hanging on a spring will oscillate up and down about its equilibrium position after it's pulled down and released. This is an example of simple harmonic motion. This motion would continue forever if there were not any friction or air resistance. Simple harmonic motion can be described with the function $y = A\cos\left(t\sqrt{\dfrac{k}{m}}\right)$ where A is the amplitude, t is the time in seconds, m is the mass, and k is a constant particular to that spring. The frequency of oscillation, f, is given by $f = \dfrac{1}{p}$, where p is the period. What is the frequency of oscillation modeled by $y = 3.5\cos\left(3t\right)$?

In Exercises 53–56 the term frequency is used. Frequency is the reciprocal of period, $f = \dfrac{1}{p}$.

53. Sound Waves. A pure tone created by a vibrating tuning fork shows up as a sine wave on an oscilloscope's screen. A tuning fork vibrating at 256 hertz (Hz) gives the tone middle C and can have the equation $y = 0.005 \sin 2\pi(256t)$, where the amplitude is in centimeters (cm) and the time in seconds. What are the amplitude and frequency of the wave?

54. Sound Waves. A pure tone created by a vibrating tuning fork shows up as a sine wave on an oscilloscope's screen. A tuning fork vibrating at 288 hertz gives the tone D and can have the equation $y = 0.005 \sin 2\pi(288t)$, where the amplitude is in centimeters (cm) and the time in seconds. What are the amplitude and frequency of the wave?

55. Sound Waves. If a sound wave is represented by $y = 0.008 \sin 750\pi t$ cm, what are its amplitude and frequency?

56. Sound Waves. If a sound wave is represented by $y = 0.006 \cos 1000\pi t$ cm, what are its amplitude and frequency?

57. Sonic Booms. When an airplane flies faster than the speed of sound, the sound waves that are formed take on a cone shape, and where the cone hits the ground, a sonic boom is heard. If θ is the angle of the vertex of the cone, then $\sin\dfrac{\theta}{2} = \dfrac{330 \text{ m/s}}{V} = \dfrac{1}{M}$, where V is the speed of the plane and M is the mach number. What is the speed of the plane if the plane is flying at mach 2?

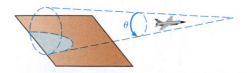

58. Sonic Booms. When an airplane flies faster than the speed of sound, the sound waves that are formed take on a cone shape, and where the cone hits the ground, a sonic boom is heard. If θ is the angle of the vertex of the cone, then $\sin\dfrac{\theta}{2} = \dfrac{330 \text{ m/s}}{V} = \dfrac{1}{M}$, where V is the speed of the plane and M is the mach number. What is the mach number if the plane is flying at 990 m/s?

59. Sonic Booms. When an airplane flies faster than the speed of sound, the sound waves that are formed take on a cone shape, and where the cone hits the ground, a sonic boom is heard. If θ is the angle of the vertex of the cone, then $\sin\dfrac{\theta}{2} = \dfrac{330 \text{ m/s}}{V} = \dfrac{1}{M}$, where V is the speed of the plane and M is the mach number. What is the speed of the plane if the cone angle is 60°?

60. Sonic Booms. When an airplane flies faster than the speed of sound, the sound waves that are formed take on a cone shape, and where the cone hits the ground, a sonic boom is heard. If θ is the angle of the vertex of the cone, then $\sin\dfrac{\theta}{2} = \dfrac{330 \text{ m/s}}{V} = \dfrac{1}{M}$, where V is the speed of the plane and M is the mach number. What is the speed of the plane if the cone angle is 30°?

 ■ CATCH THE MISTAKE

In Exercises 61 and 62, explain the mistake that is made.

61. Graph the function $y = -2\cos x$.

Solution:

Find the amplitude. $A = |-2| = 2$

The graph of $y = -2\cos x$ is similar to the graph of $y = \cos x$ with amplitude 2.

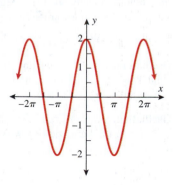

This is incorrect. What mistake was made?

62. Graph the function $y = -\sin(2x)$.

Solution:

Make a table with values:

x	$y = -\sin(2x)$	(x, y)
0	$y = -\sin(0) = 0$	$(0, 0)$
$\dfrac{\pi}{2}$	$y = -\sin(\pi) = 0$	$(0, 0)$
π	$y = -\sin(2\pi) = 0$	$(0, 0)$
$\dfrac{3\pi}{2}$	$y = -\sin(3\pi) = 0$	$(0, 0)$
2π	$y = -\sin(4\pi) = 0$	$(0, 0)$

Graph the function by plotting these points and connecting with a sinusoidal curve.

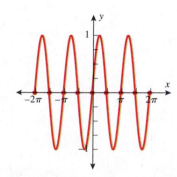

This is incorrect. What mistake was made?

■ CHALLENGE

In Exercises 63–70, A and B are positive real numbers.

63. T or F: The graph of $y = -A\cos Bx$ is the graph of $y = A\cos Bx$ reflected about the x-axis.

64. T or F: The graph of $y = A\sin(-Bx)$ is the graph of $y = A\sin(Bx)$ reflected about the x-axis.

65. T or F: The graph of $y = -A\cos(-Bx)$ is the graph of $y = A\cos Bx$.

66. T or F: The graph of $y = -A\sin(-Bx)$ is the graph of $y = A\sin Bx$.

67. Find the y-intercept of the function $y = A\cos Bx$.

68. Find the y-intercept of the function $y = A\sin Bx$.

69. Find the x-intercepts of the function $y = A\sin Bx$.

70. Find the x-intercepts of the function $y = A\cos Bx$.

■ TECHNOLOGY

71. Use a graphing calculator to graph $Y_1 = 5\sin x$ and $Y_2 = \sin(5x)$. Is the following statement true based on what you see? $y = \sin cx$ has the same graph as $y = c\sin x$.

72. Use a graphing calculator to graph $Y_1 = 3\cos x$ and $Y_2 = \cos(3x)$. Is the following statement true based on what you see? $y = \cos cx$ has the same graph as $y = c\cos x$.

73. Use a graphing calculator to graph $Y_1 = \sin x$ and $Y_2 = \cos\left(x - \dfrac{\pi}{2}\right)$. What do you notice?

74. Use a graphing calculator to graph $Y_1 = \cos x$ and $Y_2 = \sin\left(x + \dfrac{\pi}{2}\right)$. What do you notice?

75. Use a graphing calculator to graph $Y_1 = \cos x$ and $Y_2 = \cos(x + c)$, where

a. $c = \dfrac{\pi}{3}$, and explain the relationship between Y_2 and Y_1.

b. $c = -\dfrac{\pi}{3}$, and explain the relationship between Y_2 and Y_1.

76. Use a graphing calculator to graph $Y_1 = \sin x$ and $Y_2 = \sin(x + c)$, where

a. $c = \dfrac{\pi}{3}$, and explain the relationship between Y_2 and Y_1.

b. $c = -\dfrac{\pi}{3}$, and explain the relationship between Y_2 and Y_1.

Damped oscillatory motion, or **damped oscillation**, occurs when things in oscillatory motion experience friction or resistance. The friction causes the amplitude to decrease as a function of time. Mathematically, we use a negative exponential to damp the oscillations in the form of

$$f(t) = e^{-t}\sin t$$

77. Damped Oscillation. Graph the functions $Y_1 = e^{-t}$, $Y_2 = \sin t$, and $Y_3 = e^{-t}\sin t$ in the same viewing window (let t range from 0 to 2π). What happens as t increases?

78. Damped Oscillation. Graph $Y_1 = e^{-t}\sin t$, $Y_2 = e^{-2t}\sin t$, and $Y_3 = e^{-4t}\sin t$ in the same viewing window. What happens to $Y = e^{-kt}\sin t$ as k increases?

Skills Objectives

- Graph reflections of sine and cosine functions.
- Graph vertical translations of sine and cosine functions.
- Graph horizontal translations of sine and cosine functions.
- Graph combinations of translations and reflections of sine and cosine functions.
- Calculate the amplitude, period, and phase shift of a sinusoidal function.

Conceptual Objectives

- Relate a horizontal translation to a phase shift.
- Find a sinusoidal function from data.

In Appendix 0.4 there is a discussion on translations of functions. A shift inside the function, $f(x \pm c)$, results in a horizontal shift opposite the sign of c. A shift outside the function, $f(x) \pm c$, results in a vertical shift with the sign of c. A negative sign inside the function, $f(-x)$, results in reflection of $f(x)$ about the y-axis, and a negative sign outside the function, $-f(x)$, results in reflection of $f(x)$ about the x-axis. These rules are for all functions; therefore they apply to trigonometric functions.

In Section 4.1 we graphed functions of the form $y = A \sin Bx$ and $y = A \cos Bx$. Using the properties of translations and reflections of functions in this chapter, we will graph functions of the form $y = k + A \sin(Bx + C)$ or $y = k + A \cos(Bx + C)$.

Reflections

In Section 4.1, when graphing the functions $f(x) = A \sin Bx$ and $f(x) = A \cos Bx$, we determined the period to be $\dfrac{2\pi}{B}$, but we only discussed cases when B was positive, $B > 0$. Realize that even if the argument is negative, $f(-x) = A \sin(-Bx)$ or $f(-x) = A \cos(-Bx)$, we can use properties of even and odd functions to rewrite the expression with a positive argument.

	TYPE OF FUNCTION	PROPERTY	REWRITE WITH POSITIVE ARGUMENT
$f(-x) = A \sin(-Bx)$	odd	$f(-x) = -f(x)$	$f(-x) = -A \sin(Bx)$
$f(-x) = A \cos(-Bx)$	even	$f(-x) = f(x)$	$f(-x) = A \cos(Bx)$

Since we do not need to consider negative values of B (because of the properties of even and odd functions) there is no need to consider reflection about the y-axis. We do, however, need to consider reflection about the x-axis.

Recall that a negative sign outside the function, $-f(x)$, results in reflection of $f(x)$ about the x-axis. The following are graphs of $y = \sin x$ and $y = -\sin x$:

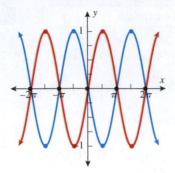

Table of values for one period:

x	$y = \sin x$	(x, y)	$y = -\sin x$	(x, y)
0	$y = \sin 0 = 0$	$(0, 0)$	$y = -\sin 0 = 0$	$(0, 0)$
$\dfrac{\pi}{2}$	$y = \sin\left(\dfrac{\pi}{2}\right) = 1$	$\left(\dfrac{\pi}{2}, 1\right)$	$y = -\sin\left(\dfrac{\pi}{2}\right) = -1$	$\left(\dfrac{\pi}{2}, -1\right)$
π	$y = \sin \pi = 0$	$(\pi, 0)$	$y = -\sin \pi = 0$	$(\pi, 0)$
$\dfrac{3\pi}{2}$	$y = \sin\left(\dfrac{3\pi}{2}\right) = -1$	$\left(\dfrac{3\pi}{2}, -1\right)$	$y = -\sin\left(\dfrac{3\pi}{2}\right) = 1$	$\left(\dfrac{3\pi}{2}, 1\right)$
2π	$y = \sin 2\pi = 0$	$(2\pi, 0)$	$y = -\sin 2\pi = 0$	$(2\pi, 0)$

Suppose we are asked to graph $y = -2\cos(\pi x)$. We can proceed one of two ways.

- Make a table for the values of $y = -2\cos(\pi x)$ and point-plot, or

- Graph the function $y = 2\cos(\pi x)$ and reflect that graph about the x-axis to arrive at the graph of $y = -2\cos(\pi x)$.

We will proceed with the second option.

WORDS

Determine the amplitude and period of $y = 2\cos(\pi x)$.

Divide the period, 2, into 4 equal parts and make a table.

Graph $y = 2\cos(\pi x)$.

MATH

$A = 2,\ p = \dfrac{2\pi}{\pi} = 2$

x	0	$\frac{1}{2}$	1	$\frac{3}{2}$	2
$y = 2\cos(\pi x)$	2	0	-2	0	2

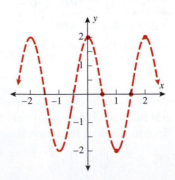

Reflect around the x-axis.

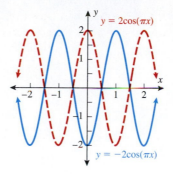

$y = 2\cos(\pi x)$

$y = -2\cos(\pi x)$

Vertical Translations

Recall that graphing functions using vertical translations (shifts) occur in the following way:

- To graph $f(x) + c$, shift $f(x)$ *up c* units.

- To graph $f(x) - c$, shift $f(x)$ *down c* units.

Therefore, functions like $y = k + A \sin Bx$ or $y = k + A \cos Bx$ are graphed by shifting the graphs of $y = A \sin Bx$ or $y = A \cos Bx$ in the vertical direction (up or down) k units. Although we found with reflection that it was just as easy to plot points and graph as opposed to first graphing one function and then reflecting about the x-axis, in this case it is much easier to first graph the simpler function and then perform a vertical shift.

EXAMPLE 1 Graphing Functions of the Form $y = k + A \sin Bx$

Graph $y = -3 + 2 \sin(\pi x)$, $-2 \le x \le 2$.

Solution:

STEP 1 Graph $y = 2 \sin(\pi x)$ over one period.

The amplitude is 2, the period is $\dfrac{2\pi}{\pi} = 2$.

Divide 0 to 2 into four equal parts: each $\dfrac{1}{2}$.

x	$y = 2 \sin(\pi x)$	(x, y)
0	$y = 2 \sin(0) = 0$	$(0, 0)$
$\dfrac{1}{2}$	$y = 2 \sin\left(\dfrac{\pi}{2}\right) = 2(1) = 2$	$\left(\dfrac{1}{2}, 2\right)$
1	$y = 2 \sin(\pi) = 0$	$(\pi, 0)$
$\dfrac{3}{2}$	$y = 2 \sin\left(\dfrac{3\pi}{2}\right) = 2(-1) = -2$	$\left(\dfrac{3}{2}, -2\right)$
2	$y = 2 \sin(2\pi) = 0$	$(2\pi, 0)$

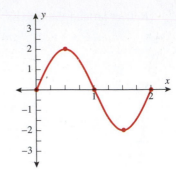

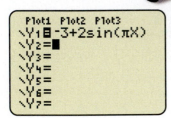

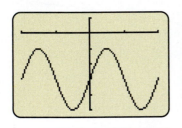

STEP 2 Extend the graph over two periods, $-2 \leq x \leq 2$.

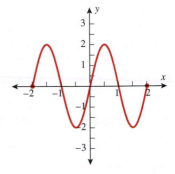

STEP 3 $y = -3 + 2\sin(\pi x)$ is
$y = 2\sin(\pi x)$ shifted down 3 units.

Note: The axis of symmetry (*x*-axis) is shifted down 3 units ($y = -3$).

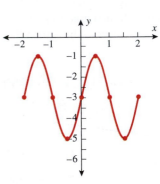

■ **YOUR TURN** Graph $y = -2 + 3\sin(2\pi x)$, $-1 \leq x \leq 1$.

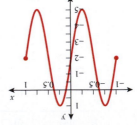

■ **Answer:**

EXAMPLE 2 **Graphing Functions of the Form** $y = k + A\cos Bx$

Graph $y = 1 - \cos\left(\dfrac{1}{2}x\right)$, $-4\pi \le x \le 4\pi$.

Solution:

STEP 1 Graph $y = -\cos\left(\dfrac{1}{2}x\right)$ over one period.

The amplitude is 1, the period is $\dfrac{2\pi}{\dfrac{1}{2}} = 4\pi$.

Divide 0 to 4π into four equal parts: each π.

x	$y = -\cos\left(\dfrac{1}{2}x\right)$	(x, y)
0	$y = -\cos(0) = -1$	$(0, -1)$
π	$y = -\cos\left(\dfrac{\pi}{2}\right) = 0$	$(\pi, 0)$
2π	$y = -\cos(\pi) = -(-1) = 1$	$(2\pi, 1)$
3π	$y = -\cos\left(\dfrac{3\pi}{2}\right) = 0$	$(3\pi, 0)$
4π	$y = -\cos(2\pi) = -1$	$(4\pi, -1)$

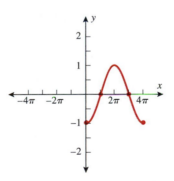

STEP 2 Extend the graph over two periods,
$-4\pi \le x \le 4\pi$.

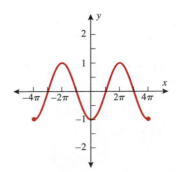

TECHNOLOGY TIP
Use a TI calculator to check

the graph of $y = 1 - \cos\left(\dfrac{1}{2}x\right)$.

Set the window at Xmin at -4π, Xmax at 4π, Xscl at π, Ymin at -2, Ymax at 1, and Yscl at 1.

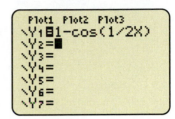

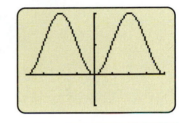

STEP 3 $y = 1 - \cos\left(\frac{1}{2}x\right)$ is $y = -\cos\left(\frac{1}{2}x\right)$ shifted up 1 unit.

Note: The axis of symmetry (x-axis) is shifted up 1 unit ($y = 1$).

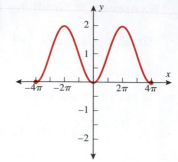

YOUR TURN Graph $y = -1 + \cos(2x)$, $-\pi \le x \le \pi$.

Horizontal Translations: Phase Shift

Recall that graphing functions using horizontal translations (shifts) occur in the following way:

■ To graph $f(x + c)$, shift $f(x)$ to the left c units.

■ To graph $f(x - c)$, shift $f(x)$ to the right c units.

Therefore, to graph functions such as $y = A\sin(x \pm C)$ or $y = A\cos(x \pm C)$, we simply shift the graphs of $y = A\sin x$ and $y = A\cos x$ horizontally (left or right) C units.

For example, to graph $y = \cos\left(x - \frac{\pi}{2}\right)$, we first graph $y = \cos x$ and then shift that graph $\frac{\pi}{2}$ units to the right.

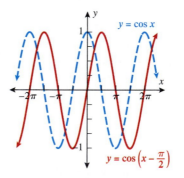

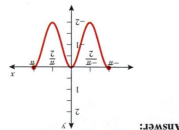

Answer:

Notice that the cosine function shifted to the right $\frac{\pi}{2}$ units coincides with the graph

of $y = \sin x$. This leads to the statement $\cos\left(x - \frac{\pi}{2}\right) = \sin x$. The shift of $\frac{\pi}{2}$ units is of-

ten called the *phase shift*. For this reason we say that sine and cosine are 90°, or $\frac{\pi}{2}$, out

of phase.

If we want to graph functions such as $y = A\sin(Bx \pm C)$ or $y = A\cos(Bx \pm C)$,

we must first factor the common B term into **standard form** $y = A\sin\left[B\left(x \pm \frac{C}{B}\right)\right]$ or

$y = A\sin\left[B\left(x \pm \frac{C}{B}\right)\right]$. The period of functions such as $y = A\sin(Bx \pm C)$ or

$y = A\cos(Bx \pm C)$ is still $\frac{2\pi}{B}$, only now there is an additional phase shift. These func-

tions can be obtained by shifting the function $y = A\sin Bx$ or $y = A\cos Bx$ horizontally

$\frac{C}{B}$ units. This horizontal shift, $\frac{C}{B}$, is called the **phase shift**. The sign of the phase shift

determines if the shift is to the left or to the right.

A second way of interpreting this is as follows. Since the graph of $y = A\sin x$ or
$y = A\cos x$ sketched over one period goes from $x = 0$ to $x = 2\pi$, then the graph of
$y = A\sin(Bx + C)$ or $y = A\cos(Bx + C)$ sketched over one period goes from:

$$Bx + C = 0 \qquad \text{to} \qquad Bx + C = 2\pi$$

$$x = -\frac{C}{B} \qquad \text{to} \qquad x = -\frac{C}{B} + \frac{2\pi}{B}$$

The phase shift, $-\frac{C}{B}$, is $\frac{C}{B}$ units to the *left* and the period is $\frac{2\pi}{B}$.

Note: Had we used the function $y = A\sin(Bx - C)$ or $y = A\cos(Bx - C)$, the

phase shift would be $\frac{C}{B}$ units to the *right* and the period is still $\frac{2\pi}{B}$.

STUDY TIP

Rewriting in standard form

$$y = A\sin\left[B\left(x \pm \frac{C}{B}\right)\right]$$

makes identifying the phase shift easier.

EXAMPLE 3 Graphing Functions of the Form $y = A\cos(Bx \pm C)$

Graph $y = 5\cos(4x + \pi)$ over one period.

Solution:

STEP 1 State the amplitude. $\qquad\qquad A = |5| = 5$

STEP 2 Calculate the period and phase shift.

The interval for one period
is from 0 to 2π. $\qquad\qquad 4x + \pi = 0 \quad \text{to} \quad 4x + \pi = 2\pi$

Solve for x. $\qquad\qquad x = -\frac{\pi}{4} \quad \text{to} \quad x = -\frac{\pi}{4} + \frac{\pi}{2}$

Identify the **phase shift**. $\qquad\qquad -\frac{C}{B} = -\frac{\pi}{4}$

Identify the **period**. $\qquad\qquad \frac{2\pi}{B} = \frac{\pi}{2}$

STUDY TIP

An alternative method for finding the period and phase shift is to first write the function in standard form.

$$y = 5\cos\left[4\left(x + \frac{\pi}{4}\right)\right]$$

$B = 4$

$$\text{Period} = \frac{2\pi}{B} = \frac{2\pi}{4} = \frac{\pi}{2}$$

Phase shift $= \frac{\pi}{4}$ units to the left.

STEP 3 Graph.

Draw a cosine function starting at $x = -\dfrac{\pi}{4}$ with period $x = \dfrac{\pi}{2}$ and amplitude 5.

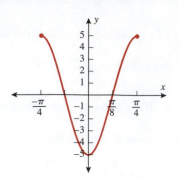

■ **YOUR TURN** Graph $y = 3\cos(2x - \pi)$ over one period.

STUDY TIP

An alternative method for finding the period and phase shift is to first write the function in standard form.

$$3\sin\left[-\frac{1}{2}(x - \pi)\right]$$

$$= -3\sin\left[\frac{1}{2}(x - \pi)\right]$$

$$B = \frac{1}{2}$$

$$\text{Period} = \frac{2\pi}{B} = \frac{2\pi}{\frac{1}{2}} = 4\pi$$

Phase shift = π units to the right.

EXAMPLE 4 Graphing Functions of the Form $y = A\sin(Bx \pm C)$

Graph $y = 3\sin\left(-\dfrac{1}{2}(x - \pi)\right)$.

STEP 1 Use the property of odd functions:
$f(-x) = -f(x)$ $y = -3\sin\left(\dfrac{1}{2}(x - \pi)\right)$

STEP 2 State the amplitude. $A = |-3| = 3$

The graph of $y = -3\sin\left(\dfrac{1}{2}(x - \pi)\right)$ will be a reflection of

$y = 3\sin\left(\dfrac{1}{2}(x - \pi)\right)$ about the x-axis.

STEP 3 Calculate the period and phase shift.

The interval for one period is from 0 to 2π. $\dfrac{1}{2}(x - \pi) = 0$ to $\dfrac{1}{2}(x - \pi) = 2\pi$

Solve for x. $x = \pi$ to $x = \pi + 4\pi$

Identify the **phase shift**. $-\dfrac{C}{B} = \pi$

Identify the **period**. $\dfrac{2\pi}{B} = 4\pi$

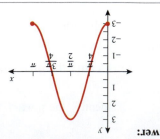

■ **Answer:**

STEP 4 Graph.
Draw a sine function starting at $x = \pi$ with period $x = 4\pi$ and amplitude 3 with a dashed curve. Then reflect this curve about the x-axis (solid).

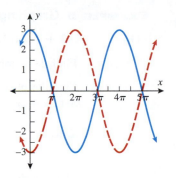

Use a TI calculator to check the graph of $y = 3\sin\left(-\dfrac{1}{2}(x - \pi)\right)$.

Set the window at Xmin at $-\pi$, Xmax at 4π, Xscl at π, Ymin at -3, Ymax at 3, and Yscl at 1.

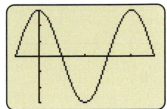

YOUR TURN Graph $y = -4\sin(-2x + \pi)$.

Graphing $y = k + A\sin(Bx + C)$ and $y = k + A\cos(Bx + C)$

We now put everything (reflections, vertical shifts, and horizontal shifts) together to graph functions of the form $y = k + A\sin(Bx + C)$ and $y = k + A\cos(Bx + C)$. We summarize with the following statements.

STRATEGY FOR GRAPHING $y = k + A\sin(Bx + C)$
AND $y = k + A\cos(Bx + C)$

A strategy for graphing $y = k + A\sin(Bx + C)$ is outlined below. The same strategy can be used to graph $y = k + A\cos(Bx + C)$.

Step 1: Find the amplitude, $|A|$.

Step 2: Find the period, $\dfrac{2\pi}{B}$, and phase shift, $\dfrac{C}{B}$ $\left(\text{if } \dfrac{C}{B} < 0 \text{ shift right; if}\right.$

$\dfrac{C}{B} > 0 \left.\text{shift left}\right)$.

Step 3: Graph $y = A\sin(Bx + C)$ over one period $\left(\text{from } -\dfrac{C}{B} \text{ to } -\dfrac{C}{B} + \dfrac{2\pi}{B}\right)$.

Step 4: Extend the graph over several periods.

Step 5: Shift the graph of $y = A\sin(Bx + C)$ vertically k units $(k > 0 \text{ up}; k < 0 \text{ down})$.

Note: If we rewrite in standard form

$$y = k + A\sin\left(B\left(x + \dfrac{C}{B}\right)\right)$$

if $B < 0$, we can use properties of even and odd functions:

$$\sin(-x) = -\sin x$$
$$\cos(-x) = \cos x$$

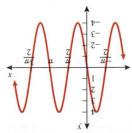

■ **Answer:**

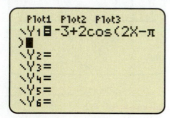

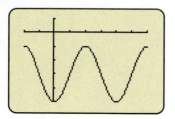

EXAMPLE 5 Graphing Sinusoidal Functions

Graph $y = -3 + 2\cos(2x - \pi)$.

STEP 1 Find the amplitude. $\qquad$ $A = |2| = 2$

STEP 2 Find the phase shift and period.
Set $2x - \pi$ equal to 0 and 2π. $\quad 2x - \pi = 0 \quad$ to $\quad 2x - \pi = 2\pi$

Solve for x. $\qquad\qquad x = \dfrac{\pi}{2} \quad$ to $\quad x = \dfrac{\pi}{2} + \pi$

Identify phase shift and period. $\quad \dfrac{\pi}{2}$ (Phase shift)

π (Period)

STEP 3 Graph $y = 2\cos(2x - \pi)$ starting
at $x = \dfrac{\pi}{2}$ over one period, π.

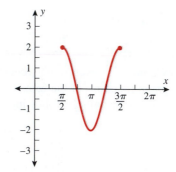

STEP 4 Extend the graph of
$y = 2\cos(2x - \pi)$
over several periods.

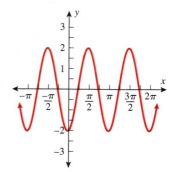

STEP 5 Shift the graph of $y = 2\cos(2x - \pi)$
down 3 units to arrive at the graph
of $y = -3 + 2\cos(2x - \pi)$.

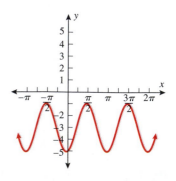

■ **YOUR TURN**   Graph $y = 2 - 3\sin(2x + \pi)$.

Finding Sinusoidal Functions from Data

Often things that are periodic in nature can be modeled with sinusoidal functions—for example, the temperature in a particular city. It is said that Orlando, Florida, has an average temperature of 72.3°. If we look at the average temperature by month, we see that it is warmest in the summer (June, July, and August) and cooler in the winter (December, January, and February). We expect this pattern to repeat every year (12 months).

Average Temperature by Month (°F)
Orlando, FL

Jan	Feb	Mar	Apr	May	Jun	Jul	Aug	Sep	Oct	Nov	Dec
59.7	61.2	66.7	71.2	76.9	81.1	82.3	82.5	81.0	75.2	68.0	62.1

If we let the *x*-axis represent the months ($x = 1$ corresponds to January and $x = 12$ corresponds to December) and the *y*-axis represent the temperature in degrees Fahrenheit, we can find a sinusoidal function that represents the data.

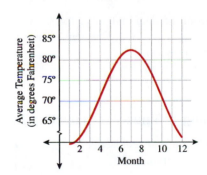

The most general sinusoidal function is given by $y = k + A\sin(Bx + C)$. Since the graphs of sine and cosine only differ by a horizontal shift, we can assume the general form of the sine function without loss of generality.

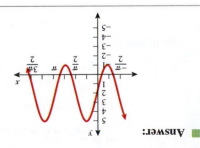

■ **Answer:**

The following procedure summarizes how to find a sinusoidal function that fits the data we have.

Assume a general sinusoidal function: $y = k + A \sin(Bx + C)$.

Step 1: Calculate the amplitude, A.

$$A = \frac{\text{largest data value} - \text{smallest data value}}{2}$$

Step 2: Calculate the vertical shift, k.

$$k = \frac{\text{largest data value} + \text{smallest data value}}{2}$$

Step 3: Calculate B. Assume that the period, $p = \dfrac{2\pi}{B}$, is equal to the time it takes for the data to repeat.

Step 4: Calculate the horizontal shift, $\dfrac{C}{B}$, by solving the equation for C:

$$y = k + A \sin(Bx + C)$$

Select a point (x, y) from the data and solve for C. Answers will vary depending on which point is selected. For consistency, we will always choose the simplest value for x ($x = 1$).

EXAMPLE 6 Finding a Sinusoidal Function from Data

Find a sinusoidal function that represents the monthly average temperature in Orlando, Florida.

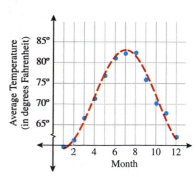

Solution:

$y = k + A \sin(Bx + C)$, where x represents the month ($x = 1$ corresponds to January and $x = 12$ corresponds to December).

STEP 1 Find the amplitude.

$$A = \frac{\text{largest data value} - \text{smallest data value}}{2} = \frac{82.5 - 59.7}{2} = 11.4$$

STEP 2 Find the vertical shift.

$$k = \frac{\text{largest data value} + \text{smallest data value}}{2} = \frac{82.5 + 59.7}{3} = 71.1$$

STEP 3 Calculate B.

The period is 12 months. $12 = \dfrac{2\pi}{B}$

$$B = \frac{\pi}{6}$$

STEP 4 Solve for C.

Substitute

$A = 12, k = 71.1, B = \dfrac{\pi}{6}.$ $y = 71.1 + 11.4 \sin\left(\dfrac{\pi}{6}x + C\right)$

Select the point,
$x = 1, y = 59.7.$ $59.7 = 71.1 + 11.4 \sin\left(\dfrac{\pi}{6} + C\right)$

Subtract 71.1. $-11.4 = 11.4 \sin\left(\dfrac{\pi}{6} + C\right)$

Divide by 11.4. $-1 = \sin\left(\dfrac{\pi}{6} + C\right)$

$\sin\theta = -1$ when $\theta = \dfrac{3\pi}{2}.$ $\dfrac{\pi}{6} + C = \dfrac{3\pi}{2}$

Subtract $\dfrac{\pi}{6}.$ $C = \dfrac{4\pi}{3}$

The function $y = 71.1 + 11.4 \sin\left(\dfrac{\pi}{6}x + \dfrac{4\pi}{3}\right)$ models the average monthly temperature in Orlando when $x = 1$ corresponds to January.

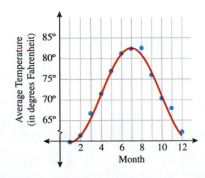

Circadian rhythms are the many different patterns that repeat "daily" in several living organisms arising from the so-called biological clock. Some of these cycles last 24 hours, while others are somewhat longer or shorter. In fact, experimentation has shown that a given Circadian rhythm can be altered if one controls the environment correctly. This is often done with space shuttle experiments related to germination in order to speed up the process.

EXAMPLE 7 **Circadian Rhythms in Photosynthesis***

It is known that photosynthesis (the process by which plants convert carbon dioxide and water into sugar and water) is a Circadian process exhibited by some plants. Assuming that a normal cycle takes 24 hours, the following is a typical graph of photosynthesis for a given plant. Here, photosynthesis is measured in terms of carbon assimilation, the units of which are micromoles of carbon per m^2/sec. Find its equation.

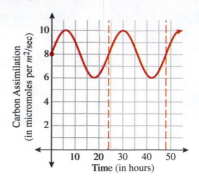

Solution:

$y = k + A \sin Bt$, where t represents time in hours.

STEP 1 Find the amplitude. $\qquad A = \dfrac{10 - 6}{2} = 2$

STEP 2 Find the vertical shift. $\qquad k = \dfrac{10 + 6}{2} = 8$

STEP 3 Calculate B.

The period is 24 hours. $\qquad 24 = \dfrac{2\pi}{B}$

$$B = \dfrac{2\pi}{24} = \dfrac{\pi}{12}$$

$y = 2 \sin\left(\dfrac{\pi}{12}t\right) + 8$ micromoles of carbon per m^2/sec, where t is in hours.

SECTION 4.2 **SUMMARY**

Graphs of the form $y = -A \sin Bx$ are just reflections of graphs of $y = A \sin Bx$ about the x-axis. Graphs of the form $y = A \sin Bx \pm k$ are obtained by first graphing $y = A \sin Bx$ and shifting up ($+$) or down ($-$), k units. Graphs of the form $y = A \sin(Bx \pm C)$ have the shape of a sine graph but with period $\dfrac{2\pi}{B}$ and horizontal shift to the left ($+$) or the right ($-$), $\dfrac{C}{B}$ units. To graph $y = k + A \sin(Bx + C)$, we first graph $y = A \sin(Bx + C)$ and then shift up or down k units. Many physical phenomena can be modeled by sinusoidal graphs. A sinusoidal function can be determined from data using the 4-step procedure.

---

*Example 7 is courtesy of Dr. Mark McKibben, Goucher College.

SECTION 4.2 EXERCISES

■ SKILLS

In Exercises 1–8, match the graph to the function.

1. $\sin\left(x - \dfrac{\pi}{2}\right)$ **2.** $\cos\left(x + \dfrac{\pi}{2}\right)$ **3.** $-\cos\left(x + \dfrac{\pi}{2}\right)$ **4.** $-\sin\left(x - \dfrac{\pi}{2}\right)$

5. $\cos x + 1$ **6.** $1 - \sin x$ **7.** $-1 + \sin x$ **8.** $1 - \cos x$

a.
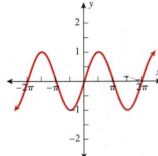

b.
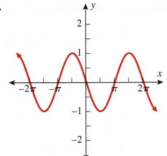

c.
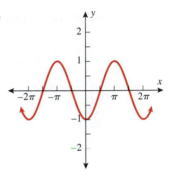

d.
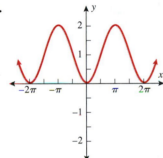

e.
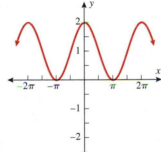

f.
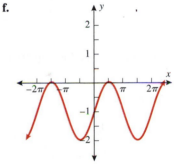

g.
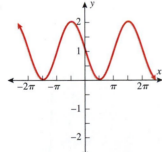

h.
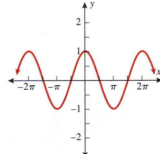

In Exercises 9–20, sketch the graph of the function over one period.

9. $y = -3 + \sin x$ **10.** $y = 3 - \cos x$ **11.** $y = 1 - \sin(x - \pi)$

12. $y = -1 + \cos(x + \pi)$ **13.** $y = 2 + \cos(2x)$ **14.** $y = 3 + \sin(\pi x)$

15. $y = 2 - \sin(4x)$ **16.** $y = 1 - \cos(2x)$ **17.** $y = 3 - 2\cos\left(\frac{1}{2}x\right)$

18. $y = 2 - 3\sin\left(\frac{1}{2}x\right)$ **19.** $y = -1 + 2\sin\left(\frac{\pi}{2}x\right)$ **20.** $y = -2 + \cos\left(\frac{\pi}{2}x\right)$

In Exercises 21–26, state the amplitude, period, and phase shift of each function.

21. $y = 2\sin(\pi x - 1)$ **22.** $y = 4\cos(x + \pi)$ **23.** $y = -5\cos(3x + 2)$

24. $y = -7\sin(4x - 3)$ **25.** $y = 6\sin[-\pi(x + 2)]$ **26.** $y = 3\sin\left[-\frac{\pi}{2}(x - 1)\right]$

In Exercises 27–36, sketch the graph of the function over the indicated interval.

27. $y = \frac{1}{2} + \frac{3}{2}\cos[2x + \pi],\ \left[-\frac{3\pi}{2}, \frac{3\pi}{2}\right]$ **28.** $y = \frac{1}{3} + \frac{2}{3}\sin[2x - \pi],\ \left[-\frac{3\pi}{2}, \frac{3\pi}{2}\right]$

29. $y = \frac{1}{2} - \frac{1}{2}\sin\left[\frac{1}{2}x - \frac{\pi}{4}\right],\ \left[-\frac{7\pi}{2}, \frac{9\pi}{2}\right]$ **30.** $y = -\frac{1}{2} + \frac{1}{2}\cos\left[\frac{1}{2}x + \frac{\pi}{4}\right],\ \left[-\frac{9\pi}{2}, \frac{7\pi}{2}\right]$

31. $y = -3 + 4\sin[\pi(x - 2)],\ [0, 4]$ **32.** $y = 4 - 3\cos[\pi(x + 1)],\ [-1, 3]$

33. $y = 2 - 3\cos\left[3x - \frac{\pi}{2}\right],\ \left[-\frac{\pi}{2}, \frac{5\pi}{6}\right]$ **34.** $y = 1 - 2\sin\left[3x + \frac{\pi}{2}\right],\ \left[-\frac{5\pi}{6}, \frac{\pi}{2}\right]$

35. $y = 2 - \sin\left[-\frac{\pi}{2}\left(x - \frac{1}{2}\right)\right],\ \left[-\frac{7}{2}, \frac{9}{2}\right]$ **36.** $y = 1 - \cos\left[-\frac{\pi}{2}\left(x + \frac{1}{2}\right)\right],\ \left[-\frac{9}{2}, \frac{7}{2}\right]$

In Exercises 37–40, find an equation that corresponds to the graph.

37.

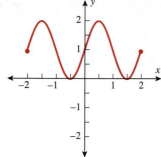

38.

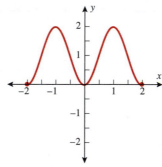

39.

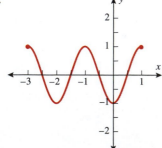

40.

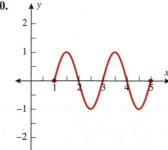

■ APPLICATIONS*

41. Electrical Current. The current, in amperes (A), flowing through an alternating current (AC) circuit at time t is

$$I = 220 \sin\left[20\pi\left(t - \frac{1}{100}\right)\right], \, t \geq 0$$

What is the maximum current? The minimum current? The period? The phase shift?

42. Electrical Current. The current, in amperes (amps), flowing through an alternating current (AC) circuit at time t is

$$I = 120 \sin\left[30\pi\left(t - \frac{1}{200}\right)\right], \, t \geq 0$$

What is the maximum current? The minimum current? The period? The phase shift?

For Exercises 43 and 44, find a sinusoidal function of the form $y = k + A \sin(Bx + C)$ that fits the data.

43. Temperature. The average temperature in Charlotte, North Carolina, is 60.1°F. The average monthly temperatures are given by

Average Temperature by Month (°F)
Charlotte, NC

Jan Feb Mar Apr May Jun Jul Aug Sep Oct Nov Dec

39.3 42.5 50.9 59.4 67.4 75.7 79.3 78.3 72.4 61.3 52.1 42.6

44. Temperature. The average temperature in Houston, Texas, is 67.9°F. The average monthly temperatures are given by

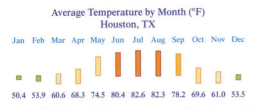

Average Temperature by Month (°F)
Houston, TX

Jan Feb Mar Apr May Jun Jul Aug Sep Oct Nov Dec

50.4 53.9 60.6 68.3 74.5 80.4 82.6 82.3 78.2 69.6 61.0 53.5

45. Roller Coaster. If a roller coaster at an amusement park is built using the sine curve determined by

$$y = 20 \sin\left(\frac{\pi}{400} x\right) + 30,$$ where x is the distance from the

beginning of the roller coaster in feet, then how high does the roller coaster go, and what distance does the roller coaster travel if it goes through three complete sine cycles?

46. Roller Coaster. If a roller coaster at an amusement park is built using the sine curve determined by

$$y = 25 \sin\left(\frac{\pi}{500} x\right) + 30,$$ where x is the distance from the

beginning of the roller coaster in feet, then how high does the roller coaster go, and what distance does the roller coaster travel if it goes through four complete sine cycles?

47. Deer Population. The number of deer on an island varies over time because of the amount of available food on the island. If the number of deer is determined by

$$y = 100 \sin\left(\frac{\pi t}{4}\right) + 500$$ where t is in years, then what are

the highest and lowest numbers of deer on the island, and how long is the cycle (how long between two different years when the number is the highest)?

48. Deer Population. The number of deer on an island varies over time because of the amount of available food on the island. If the number of deer is determined by

$$y = 500 \sin\left(\frac{\pi t}{2}\right) + 1000$$ where t is in years, then what

are the highest and lowest numbers of deer on the island, and how long is the cycle (how long between two different years when the number is the highest)?

49. Photosynthesis.* Suppose the cycle of a Circadian rhythm in photosynthesis is 22 hours and the maximum and minimum values obtained are 12 and 6 micromoles of carbon per m²/sec. Find an equation that models this photosynthesis.

50. Photosynthesis.* For the Circadian rhythm given in Exercise 49, what would the carbon assimilation be at (a) 46 hours and (b) 68 hours?

For Exercises 51 and 52, refer to the following:

With the advent of summer comes fireflies. They are intriguing because they emit a flashing luminescence that beckons their mate to them. It is known that the speed and intensity of the flashing are related to the temperature—the higher the temperature, the quicker and more intense the flashing becomes. If you ever watch a single firefly, you will see that the intensity of the flashing is periodic with time. The intensity of light emitted is measured in *candelas per square meter (of firefly)*. To give an idea of this unit of measure, the intensity of a picture on a typical TV screen is about 450 candelas. The measurement for the intensity of the light emitted by a typical firefly at its brightest moment is about 50 candelas. Assume that a typical cycle of this flashing is 4 seconds and that the intensity is essentially zero candelas at the beginning and ending of a cycle.

*Exercises 49–52 are courtesy of Dr. Mark McKibben, Goucher College.

51. Bioluminescence in Fireflies.* Find an equation that describes this flashing. What is the intensity of the flashing at 4 minutes?

■ CATCH THE MISTAKE

In Exercises 53 and 54, explain the mistake that is made.

53. Find the amplitude, phase shift, and period of

$$y = -\frac{2}{3}\sin[2x - \pi].$$

Solution:

Find the amplitude. $A = \left| -\dfrac{2}{3} \right| = \dfrac{2}{3}$

Find the phase shift. π units to the right

Find the period. $\dfrac{2\pi}{2} = \pi$

One of these is incorrect. Which is it, and what mistake was made?

54. Graph $y = 1 + \sin[2x]$.

First, graph $y = \sin[2x]$.

Find the amplitude. $A = |-1| = 1$

Make a table.

x	$y = -\sin[2x]$	(x, y)
0	$y = -\sin[0] = 0$	$(0, 0)$
π	$y = -\sin[2\pi] = 0$	$(\pi, 0)$
2π	$y = -\sin[4\pi] = 0$	$(2\pi, 0)$

52. Bioluminescence in Fireflies.* Graph this equation for $\frac{1}{2}$ minute of time.

Plot the points.

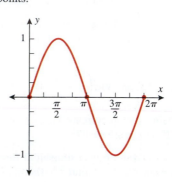

Shift the graph of $y = \sin[2x]$ up one unit to arrive at the graph of $y = 1 + \sin[2x]$.

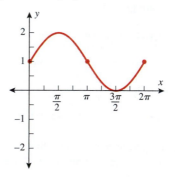

This is incorrect. What mistake was made?

■ CHALLENGE

55. T or F: The period of $y = k + A\sin(\omega t - \phi)$ increases as ω increases.

56. T or F: The phase shift of $y = k + A\sin(\omega t - \phi)$ is ϕ units.

57. T or F: $A\sin\left(\omega t + \dfrac{\omega\pi}{2}\right) = A\cos(\omega t)$

58. T or F: $A\sin\left(\omega t - \dfrac{\omega\pi}{2}\right) = A\cos(\omega t)$

59. If asked to graph $y = k + A\cos(\omega t - \phi)$ over one period, state the interval for x over which the graph would be drawn [phase shift, phase shift + period].

60. If asked to graph $y = k + A\cos(\omega t + \phi)$ over one period, state the interval for x over which the graph would be drawn [phase shift, phase shift + period].

*Exercises 49–52 are courtesy of Dr. Mark McKibben, Goucher College.

61. Using a graphing calculator, graph $Y_1 = x - \dfrac{x^3}{6}$ and

$Y_2 = \sin x$ for x in $[0, 0.2]$. Is the polynomial in Y_1 a good approximation of the sine function?

62. Using a graphing calculator, graph $Y_1 = 1 - \dfrac{x^2}{2}$ and

$Y_2 = \cos x$ for x in $[0, 0.2]$. Is the polynomial in Y_1 a good approximation of the cosine function?

63. Use a graphing calculator and the SINe REGression program to find the sine function that best fits the data in Exercise 43. Graph the function found by the graphing calculator and the function you found in Exercise 43 in the same viewing rectangle.

64. Use a graphing calculator and the SINe REGression program to find the sine function that best fits the data in Exercise 44. Graph the function found by the graphing calculator and the function you found in Exercise 44 in the same viewing rectangle.

SECTION 4.3 Graphs of Tangent, Cotangent, Secant, and Cosecant Functions

Skills Objectives

■ Graph basic tangent, cotangent, secant, and cosecant functions.
■ Determine the period of tangent, cotangent, secant, and cosecant functions.
■ Use translations to graph tangent, cotangent, secant, and cosecant functions.

Conceptual Objectives

■ Determine the domain and range of tangent, cotangent, secant, and cosecant functions.
■ Relate domain restrictions to vertical asymptotes.

The first two sections of this chapter focused on graphing sinusoidal functions (sine and cosine). We now turn our attention to graphing the other circular functions: tangent, cotangent, secant, and cosecant. We know the graphs of sine and cosine, and we can get the graphs of the other circular functions from the sinusoidal functions. Recall the reciprocal and quotient identities:

$$\tan x = \frac{\sin x}{\cos x} \qquad \cot x = \frac{\cos x}{\sin x} \qquad \sec x = \frac{1}{\cos x} \qquad \csc x = \frac{1}{\sin x}$$

Recall in algebra that when graphing rational functions, a *vertical asymptote* corresponds to the denominator being equal to zero. A **vertical asymptote** is a vertical (dashed) line that represents a value of x for which the function is not defined. As you will see in this section, tangent and secant functions have graphs with vertical asymptotes corresponding to when cosine is equal to zero, and cotangent and cosecant functions have graphs with vertical asymptotes corresponding to when sine is equal to zero.

One important difference between the sinusoidal functions ($y = \sin x$ and $y = \cos x$) and the other four ($y = \tan x$, $y = \sec x$, $y = \csc x$, $y = \cot x$) is that the sinusoidal functions have amplitude, whereas the other trigonometric functions do not (since they are unbounded).

The Tangent Function

Since the tangent function is a quotient that relies on sine and cosine, let us start with a table of values for the quadrantal angles.

x	sin x	cos x	$\tan x = \dfrac{\sin x}{\cos x}$	(x, y) OR ASYMPTOTE
0	0	1	0	(0, 0)
$\dfrac{\pi}{2}$	1	0	undefined	vertical asymptote: $x = \dfrac{\pi}{2}$
π	0	−1	0	$(\pi, 0)$
$\dfrac{3\pi}{2}$	−1	0	undefined	vertical asymptote: $x = \dfrac{3\pi}{2}$
2π	0	1	0	$(2\pi, 0)$

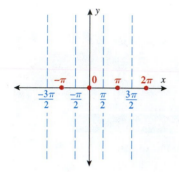

Notice that the x-intercepts correspond to integer multiples of π and vertical asymptotes correspond to odd multiples of $\dfrac{\pi}{2}$.

We know that the graph of the tangent function is undefined at the odd multiples of π, and therefore the graph cannot cross the vertical asymptotes. The question is what happens between the asymptotes? We know the x-intercepts; let us now make a table for special values of x.

x	sin x	cos x	$\tan x = \dfrac{\sin x}{\cos x}$	(x, y)
$\dfrac{\pi}{6}$	$\dfrac{1}{2}$	$\dfrac{\sqrt{3}}{2}$	$\dfrac{1}{\sqrt{3}} = \dfrac{\sqrt{3}}{3} \approx 0.577$	$\left(\dfrac{\pi}{6}, 0.577\right)$
$\dfrac{\pi}{4}$	$\dfrac{\sqrt{2}}{2}$	$\dfrac{\sqrt{2}}{2}$	1	$\left(\dfrac{\pi}{4}, 1\right)$
$\dfrac{\pi}{3}$	$\dfrac{\sqrt{3}}{2}$	$\dfrac{1}{2}$	$\sqrt{3} \approx 1.732$	$\left(\dfrac{\pi}{3}, 1.732\right)$
$\dfrac{2\pi}{3}$	$\dfrac{\sqrt{3}}{2}$	$-\dfrac{1}{2}$	$-\sqrt{3} \approx -1.732$	$\left(\dfrac{2\pi}{3}, -1.732\right)$
$\dfrac{3\pi}{4}$	$\dfrac{\sqrt{2}}{2}$	$-\dfrac{\sqrt{2}}{2}$	−1	$\left(\dfrac{3\pi}{4}, -1\right)$
$\dfrac{5\pi}{6}$	$\dfrac{1}{2}$	$-\dfrac{\sqrt{3}}{2}$	$\dfrac{1}{-\sqrt{3}} = -\dfrac{\sqrt{3}}{3} \approx -0.577$	$\left(\dfrac{5\pi}{6}, -0.577\right)$

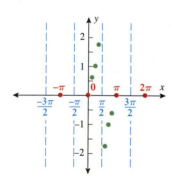

WORDS	MATH
What happens as x approaches $\dfrac{\pi}{2}$?	$\left(\lim\limits_{x \to \frac{\pi}{2}} \tan x\right)$
We must consider which way x approaches $\dfrac{\pi}{2}$:	
x approaches $\dfrac{\pi}{2}$ from the left.	$\left(\lim\limits_{x \to \frac{\pi}{2}^-} \tan x\right)$
x approaches $\dfrac{\pi}{2}$ from the right.	$\left(\lim\limits_{x \to \frac{\pi}{2}^+} \tan x\right)$
The numerical approximation to $\dfrac{\pi}{2}$ is 1.571.	$\dfrac{\pi}{2} \approx 1.571$

As x approaches $\dfrac{\pi}{2}$ from the left, the value of tangent increases toward a large, positive number.

$\displaystyle\lim_{x\to\frac{\pi}{2}^-}$	$(\tan x)$
1.5	14.1
1.55	48.1
1.57	1255.8

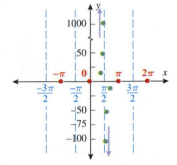

Thus, as x approaches $\dfrac{\pi}{2}$ from the left, $\tan x$ approaches infinity.

$$\left(\lim_{x\to\frac{\pi}{2}^-}\tan x\right)=\infty$$

As x approaches $\dfrac{\pi}{2}$ from the right, the value of tangent increases toward a large, negative number.

$\displaystyle\lim_{x\to\frac{\pi}{2}^+}$	$(\tan x)$
1.65	-12.6
1.59	-52.1
1.58	-108.6

TECHNOLOGY TIP

Thus, as x approaches $\dfrac{\pi}{2}$ from the right, $\tan x$ approaches negative infinity.

$$\left(\lim_{x\to\frac{\pi}{2}^+}\tan x\right)=-\infty$$

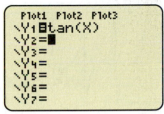

GRAPH OF $y = \tan x$

1. The x-intercepts occur at multiples of π: $x = n\pi$

2. Vertical asymptotes occur at odd multiples of $\dfrac{\pi}{2}$: $x = \dfrac{(2n+1)\pi}{2}$

3. The domain is the set of all real numbers except odd multiples of $\dfrac{\pi}{2}$. $x \neq \dfrac{(2n+1)\pi}{2}$

4. The range is the set of all real numbers.
5. $y = \tan x$ has period π.
6. $y = \tan x$ is an odd function (symmetric about the origin). $\tan(-x) = -\tan x$
7. The graph has no amplitude, since there are no maximum or minimum values.

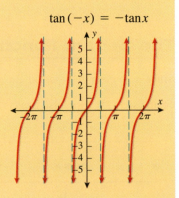

Note: n is an integer.

The Cotangent Function

The cotangent function is similar to the tangent function in that it is a quotient involving sine and cosine. The difference is that cotangent has cosine in the numerator and sine in the denominator: $\cot x = \dfrac{\cos x}{\sin x}$. The graph of $y = \tan x$ has x-intercepts corresponding to integer multiples of π and vertical asymptotes corresponding to odd integer multiples of $\frac{\pi}{2}$. The graph of cotangent is the reverse in that it has x-intercepts corresponding to odd integer multiples of $\frac{\pi}{2}$ and vertical asymptotes corresponding to integer multiples of π.

This is because the x-intercepts occur when the numerator, $\cos x$, is equal to 0 and the vertical asymptotes occur when the denominator, $\sin x$, is equal to 0.

TECHNOLOGY TIP

To graph cot x, use the reciprocal property to enter $(\tan (x))^{-1}$.

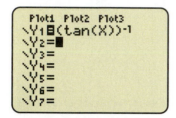

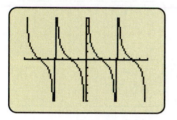

GRAPH OF $y = \cot x$

1. The x-intercepts occur at odd multiples of $\frac{\pi}{2}$: $x = \dfrac{(2n + 1)\pi}{2}$

2. Vertical asymptotes occur at multiples of π: $x = n\pi$

3. The domain is the set of all real numbers except multiples of π: $x \neq n\pi$

4. The range is the set of all real numbers.

5. $y = \cot x$ has period π.

6. $y = \cot x$ is an odd function (symmetric about the origin). $\cot (-x) = -\cot x$

7. The graph has no amplitude, since there are no maximum or minimum values.

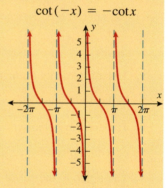

Note: n is an integer.

STUDY TIP

The graphs of $y = \tan x$ and $y = \cot x$ both have period π and neither has amplitude.

The Secant Function

Since $y = \cos x$ has period 2π, then the secant function, which is the reciprocal of cosine $\sec x = \dfrac{1}{\cos x}$, also has period 2π. We now illustrate values of secant with a table.

x	$\cos x$	$\sec x = \dfrac{1}{\cos x}$	(x, y) OR ASYMPTOTE
0	1	1	$(0, 1)$
$\dfrac{\pi}{2}$	0	undefined	vertical asymptote: $x = \dfrac{\pi}{2}$
π	-1	-1	$(\pi, -1)$
$\dfrac{3\pi}{2}$	0	undefined	vertical asymptote: $x = \dfrac{3\pi}{2}$
2π	1	1	$(2\pi, 1)$

Again, we ask the same question: what happens as x approaches the vertical asymptotes? The same asymptotic behavior occurs that we found with tangent.

If we graph $y = \cos x$ and $y = \sec x$ on the same graph we notice the following:

- x-intercepts of cosine correspond to the vertical asymptotes of secant.
- The range of cosine is $[-1, 1]$ and the range of secant is $(-\infty, -1] \cup [1, \infty)$.
- When cosine is positive, secant is positive, and vice versa.

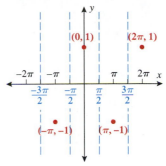

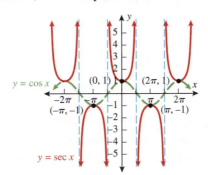

1. There are no x-intercepts.

2. Vertical asymptotes occur at odd multiples of $\dfrac{\pi}{2}$: $\qquad x = \dfrac{(2n+1)\pi}{2}$

3. The domain is the set of all real numbers except odd multiples of $\dfrac{\pi}{2}$. $\qquad x \neq \dfrac{(2n+1)\pi}{2}$

4. The range is $(-\infty, -1] \cup [1, \infty)$.

5. $y = \sec x$ has period 2π.

6. $y = \sec x$ is an even function (symmetric about the y-axis). $\qquad \sec(-x) = \sec x$

7. The graph has no amplitude, since there are no maximum or minimum values.

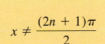

Note: n is an integer.

TECHNOLOGY TIP

To graph sec x, enter as $(\cos(x))^{-1}$.

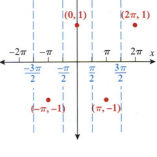

The Cosecant Function

Since $y = \sin x$ has period 2π, then the cosecant function, which is the reciprocal of sine $\csc x = \dfrac{1}{\sin x}$, also has period 2π. We now illustrate values of cosecant with a table.

x	$\sin x$	$\csc x = \dfrac{1}{\sin x}$	(x, y) OR ASYMPTOTE
0	0	undefined	vertical asymptote: $x = 0$
$\dfrac{\pi}{2}$	1	1	$\left(\dfrac{\pi}{2}, 1\right)$
π	0	undefined	vertical asymptote: $x = \pi$
$\dfrac{3\pi}{2}$	-1	-1	$\left(\dfrac{3\pi}{2}, -1\right)$
2π	0	undefined	vertical asymptote: $x = 2\pi$

Again, we ask the same question: what happens as x approaches the vertical asymptotes? The same asymptotic behavior occurs that we found with cotangent.

If we graph $y = \sin x$ and $y = \csc x$ on the same graph, we notice the following:

- x-intercepts of sine correspond to the vertical asymptotes of cosecant.
- The range of sine is $[-1, 1]$ and the range of cosecant is $(-\infty, -1] \cup [1, \infty)$.
- When sine is positive, cosecant is positive, and vice versa.

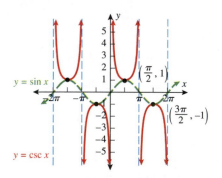

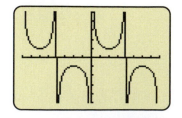
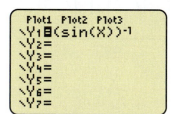
GRAPH OF $y = \csc x$

1. There are no x-intercepts.
2. Vertical asymptotes occur at integer multiples of π: $x = n\pi$
3. The domain is the set of all real numbers except integer multiples of π: $x \neq n\pi$
4. The range is $(-\infty, -1] \cup [1, \infty)$.
5. $y = \csc x$ has period 2π.
6. $y = \csc x$ is an odd function (symmetric about the origin). $\csc(-x) = -\csc x$
7. The graph has no amplitude, since there are no maximum or minimum values.

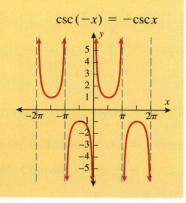

Note: n is an integer.

Graphing Tangent, Cotangent, Secant, and Cosecant Functions

FUNCTION	$y = \sin x$	$y = \cos x$	$y = \tan x$	$y = \cot x$	$y = \sec x$	$y = \csc x$
GRAPH OF ONE PERIOD						
DOMAIN	$\mathbb{R}$	$\mathbb{R}$	$x \neq \dfrac{(2n+1)\pi}{2}$	$x \neq n\pi$	$x \neq \dfrac{(2n+1)\pi}{2}$	$x \neq n\pi$
RANGE	$[-1, 1]$	$[-1, 1]$	$\mathbb{R}$	$\mathbb{R}$	$(-\infty, -1] \cup [1, \infty)$	$(-\infty, -1] \cup [1, \infty)$
AMPLITUDE	1	1	none	none	none	none
PERIOD	2π	2π	π	π	2π	2π
X-INTERCEPTS	$x = n\pi$	$x = \dfrac{(2n+1)\pi}{2}$	$x = n\pi$	$x = \dfrac{(2n+1)\pi}{2}$	none	none
VERTICAL ASYMPTOTES	none	none	$x = \dfrac{(2n+1)\pi}{2}$	$x = n\pi$	$x = \dfrac{(2n+1)\pi}{2}$	$x = n\pi$

Note: n is an integer.

We use these basic functions as the starting point for graphing general tangent, cotangent, secant, and cosecant functions.

GRAPHING TANGENT AND COTANGENT FUNCTIONS

Graphs of $y = A \tan Bx$ and $y = A \cot Bx$ can be obtained using the following steps:

Step 1: Calculate the period: $\dfrac{\pi}{B}$.

Step 2: Find two neighboring vertical asymptotes.

For $y = A \tan Bx$: $Bx = -\dfrac{\pi}{2}$ and $Bx = \dfrac{\pi}{2}$

For $y = A \cot Bx$: $Bx = 0$ and $Bx = \pi$

Step 3: Find the x-intercept between the two asymptotes.

For $y = A \tan Bx$: $Bx = 0$

For $y = A \cot Bx$: $Bx = \dfrac{\pi}{2}$

Step 4: Draw the vertical asymptotes and label the x-intercept.

Step 5: Divide the interval between the asymptotes into four equal parts. Set up a table with coordinates corresponding to the points in the interval.

Step 6: Connect the points with a smooth curve. Use arrows to indicate the behavior toward the asymptotes.

- If $A > 0$
  - $y = A \tan Bx$ increases from left to right.
  - $y = A \cot Bx$ decreases from left to right.
- If $A < 0$
  - $y = A \tan Bx$ decreases from left to right.
  - $y = A \cot Bx$ increases from left to right.

EXAMPLE 1 Graphing $y = A \tan Bx$

Graph $y = -3 \tan 2x$ on the interval $-\dfrac{\pi}{2} \le x \le \dfrac{\pi}{2}$.

Solution: $A = -3$, $B = 2$

STEP 1 Calculate the period. $\qquad\qquad\qquad \dfrac{\pi}{B} = \dfrac{\pi}{2}$

STEP 2 Find two vertical asymptotes. $\qquad Bx = -\dfrac{\pi}{2}$ and $Bx = \dfrac{\pi}{2}$

Substitute $B = 2$ and solve for x. $\qquad x = -\dfrac{\pi}{4}$ and $x = \dfrac{\pi}{4}$

STEP 3 Find the x-intercept between
the asymptotes. $\qquad\qquad\qquad\qquad Bx = 0$
$\qquad\qquad\qquad\qquad\qquad\qquad\qquad\qquad\quad x = 0$

STEP 4 Draw the vertical asymptotes,
$x = -\dfrac{\pi}{4}$ and $x = \dfrac{\pi}{4}$, and label the
x-intercept, $(0, 0)$.

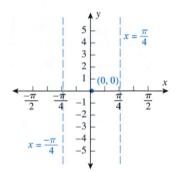

TECHNOLOGY TIP

Graph $y = -3 \tan 2x$ on the
interval $-\dfrac{\pi}{2} \le x \le \dfrac{\pi}{2}$.

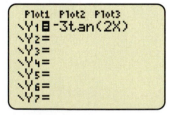

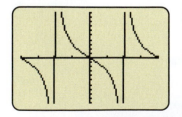

STEP 5 Divide the period, $\dfrac{\pi}{2}$, into four equal parts, $\dfrac{\pi}{8}$. Set up a table with
coordinates corresponding to values of $y = -3 \tan 2x$.

x	$y = -3 \tan 2x$	(x, y)
$-\dfrac{\pi}{4}$	undefined	vertical asymptote
$-\dfrac{\pi}{8}$	3	$\left(-\dfrac{\pi}{8}, 3\right)$
0	0	$(0, 0)$
$\dfrac{\pi}{8}$	-3	$\left(\dfrac{\pi}{8}, -3\right)$
$\dfrac{\pi}{4}$	undefined	vertical asymptote

STEP 6 Graph the points from the table and connect with smooth curve. Repeat to the right and left until reaching the interval endpoints.

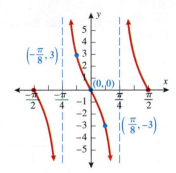

YOUR TURN Graph $y = \dfrac{1}{3}\tan\left(\dfrac{1}{2}x\right)$ on the interval $-\pi \le x \le \pi$.

EXAMPLE 2 Graphing $y = A \cot Bx$

Graph $y = 4\cot\left(\dfrac{1}{2}x\right)$ on the interval $-2\pi \le x \le 2\pi$.

Solution: $A = 4$, $B = \dfrac{1}{2}$

STEP 1 Calculate the period. $\dfrac{\pi}{B} = 2\pi$

STEP 2 Find two vertical asymptotes. $Bx = 0$ and $Bx = \pi$

Substitute $B = \dfrac{1}{2}$ and solve for x. $x = 0$ and $x = 2\pi$

STEP 3 Find the x-intercept between the asymptotes. $Bx = \dfrac{\pi}{2}$

$x = \pi$

STEP 4 Draw the vertical asymptotes, $x = 0$ and $x = 2\pi$, and label the x-intercept, $(\pi, 0)$.

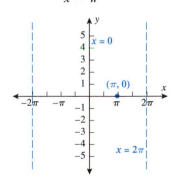

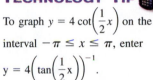
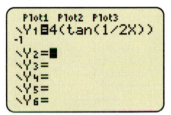
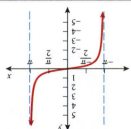

Answer:

STEP 5 Divide the period, 2π, into four equal parts, $\dfrac{\pi}{2}$. Set up a table with coordinates corresponding to values of $y = 4\cot\left(\dfrac{1}{2}x\right)$.

x	$y = 4\cot\left(\dfrac{1}{2}x\right)$	(x, y)
0	undefined	vertical asymptote
$\dfrac{\pi}{2}$	4	$\left(\dfrac{\pi}{2}, 4\right)$
π	0	$(\pi, 0)$
$\dfrac{3\pi}{2}$	-4	$\left(\dfrac{3\pi}{2}, -4\right)$
2π	undefined	vertical asymptote

STEP 6 Graph the points from the table and connect with smooth curve. Repeat to the right and left until reaching the interval endpoints.

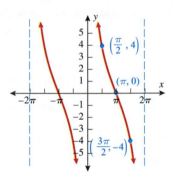

■ **YOUR TURN** Graph $y = 2\cot(2x)$ on the interval $-\pi \le x \le \pi$.

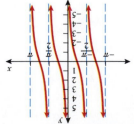

■ **Answer:**

GRAPHING SECANT AND COSECANT FUNCTIONS

Graphs of $y = A \sec Bx$ and $y = A \csc Bx$ can be obtained using the following steps:

Step 1: Graph the corresponding guide function with a dashed curve.

        For $y = A \sec Bx$, use $y = A \cos Bx$ as a guide.

        For $y = A \csc Bx$, use $y = A \sin Bx$ as a guide.

Step 2: Draw the asymptotes that correspond to the x-intercepts of the guide function.

Step 3: Draw the U shape between the asymptotes. If the guide function has a positive value between the asymptotes the U opens upward, and if the guide function has a negative value the U opens downward.

EXAMPLE 3 Graphing $y = A \sec Bx$

Graph $y = 2 \sec (\pi x)$ on the interval $-2 \le x \le 2$.

TECHNOLOGY TIP

To graph $y = 2 \sec (\pi x)$ on the interval $-2 \le x \le 2$, type $y = 2(\cos(\pi x))^{-1}$.

STEP 1 Graph the corresponding guide function with a dashed curve.

        For $y = 2 \sec (\pi x)$, use $y = 2 \cos (\pi x)$ as a guide.

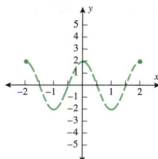

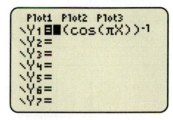

STEP 2 Draw the asymptotes that correspond to the x-intercepts of the guide function.

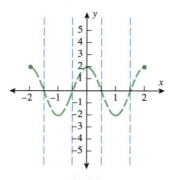

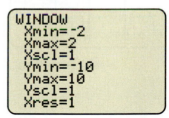

STEP 3 Draw the U shape between the asymptotes. If the guide function is positive the U opens upward, and if the guide function is negative the U opens downward.

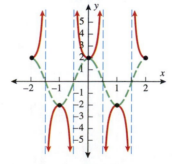

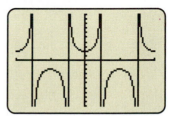

■ **YOUR TURN**   Graph $y = -\sec(2\pi x)$ on the interval $-1 \leq x \leq 1$.

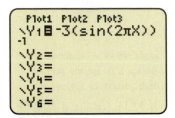
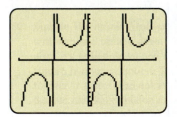
EXAMPLE 4 Graphing $y = A\csc Bx$

Graph $y = -3\csc(2\pi x)$ on the interval $-1 \leq x \leq 1$.

STEP 1   Graph the corresponding guide
function with a dashed curve.

For $y = -3\csc(2\pi x)$ use
$y = -3\sin(2\pi x)$ as a guide.

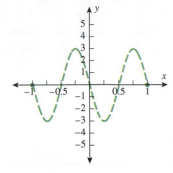

STEP 2   Draw the asymptotes that
correspond to the x-intercepts of
the guide function.

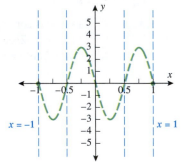

STEP 3   Draw the U shape between the
asymptotes. If the guide function is
positive the U opens upward, and if
the guide function is negative the U
opens downward.

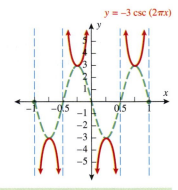

■ **YOUR TURN**   Graph $y = \dfrac{1}{2}\csc(\pi x)$ on the interval $-1 \leq x \leq 1$.

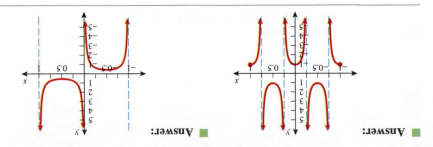

■ **Answer:**   ■ **Answer:**

Translations of Circular Functions

Vertical translations and horizontal translations (phase shifts) of tangent, cotangent, secant, and cosecant are graphed the same way as vertical and horizontal translations of sinusoidal graphs. For tangent and cotangent functions we follow the same procedure as we did with sinusoidal functions. For secant and cotangent, we graph the guide function first and then translate up or down depending on the sign of the vertical shift.

EXAMPLE 5 Graphing $y = k + A\tan(Bx + C)$

Graph $y = 1 - \tan\left(x - \dfrac{\pi}{2}\right)$ on $-\pi \le x \le \pi$. State the domain and range.

There are two ways to approach graphing this function. Both will be illustrated.

Solution (1): Plot $y = \tan x$, then do the following:

■ Shift to the right $\dfrac{\pi}{2}$ units:
$$y = \tan\left(x - \dfrac{\pi}{2}\right)$$

■ Reflect about the x-axis
(because of the negative):
$$y = -\tan\left(x - \dfrac{\pi}{2}\right)$$

■ Shift the entire graph up 1 unit.
$$y = 1 - \tan\left(x - \dfrac{\pi}{2}\right)$$

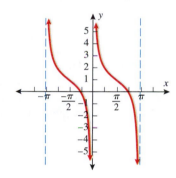

Solution (2): Graph $y = -\tan\left(x - \dfrac{\pi}{2}\right)$, then shift entire graph up 1 unit,

$B = 1$.

STEP 1 Calculate the period.
$$\dfrac{\pi}{B} = \pi$$

STEP 2 Find two vertical asymptotes.

Solve for x.
$$x - \dfrac{\pi}{2} = -\dfrac{\pi}{2} \text{ and } x - \dfrac{\pi}{2} = \dfrac{\pi}{2}$$
$$x = 0 \text{ and } x = \pi$$

STEP 3 Find the x-intercept between
the asymptotes.
$$x - \dfrac{\pi}{2} = 0$$
$$x = \dfrac{\pi}{2}$$

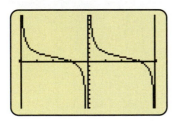
STEP 4 Draw the vertical asymptotes, $x = 0$ and $x = \pi$, and label the x-intercept, $\left(\dfrac{\pi}{2}, 0\right)$.

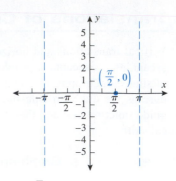

STEP 5 Divide the period, π, into four equal parts, $\dfrac{\pi}{4}$. Set up a table with coordinates corresponding to values of $y = -\tan\left(x - \dfrac{\pi}{2}\right)$ between the two asymptotes.

x	$y = -\tan\left(x - \dfrac{\pi}{2}\right)$	(x, y)
$x = 0$	undefined	vertical asymptote
$\dfrac{\pi}{4}$	1	$\left(\dfrac{\pi}{4}, 1\right)$
$\dfrac{\pi}{2}$	0	$\left(\dfrac{\pi}{2}, 0\right)$
$\dfrac{3\pi}{4}$	-1	$\left(\dfrac{3\pi}{4}, -1\right)$
$x = \pi$	undefined	vertical asymptote

STEP 6 Graph the points from the table and connect with smooth curve. Repeat to the right and left until reaching the interval endpoints.

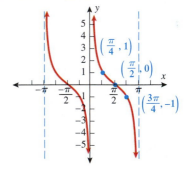

STEP 7 Shift entire graph up 1 unit to arrive at the graph of $y = 1 - \tan\left(x - \dfrac{\pi}{2}\right)$.

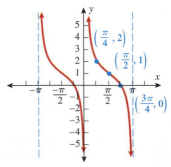

STEP 8 State the domain and range.

Domain: $(-\pi, 0) \cup (0, \pi)$
Range: $(-\infty, \infty)$

■ **YOUR TURN**   Graph $y = -1 + \cot\left(x + \dfrac{\pi}{2}\right)$ on $-\pi \le x \le \pi$. State the

domain and range.

EXAMPLE 6 Graphing $y = k + A\csc(Bx + C)$

Graph $y = 1 - \csc(2x - \pi)$ on $-\pi \le x \le \pi$. State the domain and range.

Solution: Graph $y = -\csc(2x - \pi)$ and shift the entire graph up 1 unit to arrive at the graph of $y = 1 - \csc(2x - \pi)$.

STEP 1 Draw the guide function,
$y = -\sin(2x - \pi)$.

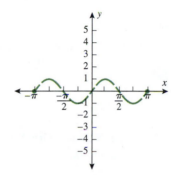

STEP 2 Draw the vertical asymptotes of
$y = -\csc(2x - \pi)$ that correspond
to the x-intercepts of
$y = -\sin(2x - \pi)$.

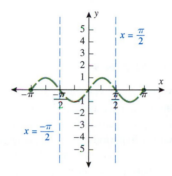

STEP 3 Draw the U shape between the
asymptotes. If the guide function is
positive the U opens upward, and if
the guide function is negative the U
opens downward.

$y = -\csc(2x - \pi)$

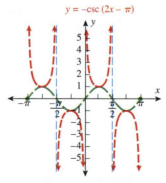

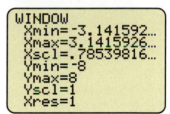

TECHNOLOGY TIP

To graph $y = 1 - \csc(2x - \pi)$
on the interval $-\pi \le x \le \pi$, type
$y = 1 - (\sin(2x - \pi))^{-1}$.

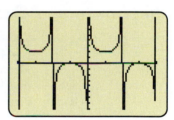

■

Answer: Domain: $\left[-\pi, -\dfrac{\pi}{2}\right) \cup \left(-\dfrac{\pi}{2}, \dfrac{\pi}{2}\right) \cup \left(\dfrac{\pi}{2}, \pi\right]$;

range: $(-\infty, \infty)$

STEP 4 Shift the entire graph up 1 unit to arrive at the graph of $y = 1 - \csc{(2x - \pi)}$.

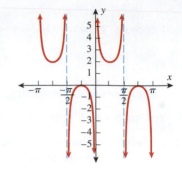

STEP 5 State the domain and range.

Domain: $\left(-\pi, -\dfrac{\pi}{2}\right) \cup \left(-\dfrac{\pi}{2}, \dfrac{\pi}{2}\right) \cup \left(\dfrac{\pi}{2}, \pi\right)$

Range: $(-\infty, 0] \cup [2, \infty)$

■ **YOUR TURN** Graph $y = -2 + \sec{(\pi x - \pi)}$ on $-1 \leq x \leq 1$. State the domain and range.

SECTION 4.3 ## SUMMARY

The tangent and cotangent functions have period π, whereas the secant and cosecant functions have period 2π. Tangent and cotangent functions are graphed by first identifying the vertical asymptotes and x-intercepts, then finding values of the function within a period. Graphs of secant and cosecant functions are found by first graphing their guide function (cosine and sine respectively) and then labeling vertical asymptotes that correspond to x-intercepts of the guide function. The graphs of secant and cosecant resemble the letter U. The secant and cosecant functions are positive when their guide function is positive and negative when their guide function is negative.

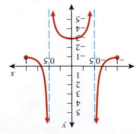

■ **Answer: Domain:** $\left[-1, -\dfrac{1}{2}\right) \cup \left(-\dfrac{1}{2}, \dfrac{1}{2}\right) \cup \left(\dfrac{1}{2}, 1\right]$; range: $(-\infty, -3] \cup [-1, \infty)$

SECTION 4.3 EXERCISES

■ SKILLS

In Exercises 1–8, match the graphs to the functions.

1. $y = -\tan x$ **2.** $y = -\csc x$ **3.** $y = \sec 2x$ **4.** $y = \csc 2x$

5. $y = \cot(\pi x)$ **6.** $y = -\cot(\pi x)$ **7.** $y = 3\sec x$ **8.** $y = 3\csc x$

a.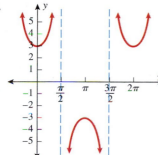

b.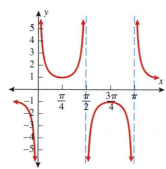

c.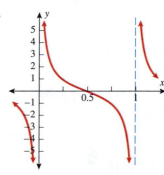

d.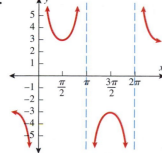

e.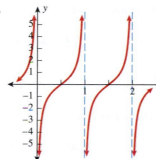

f.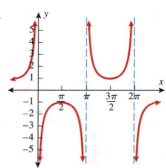

g.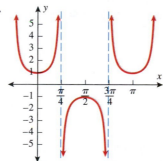

h.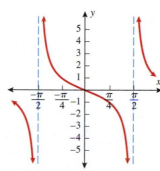

In Exercises 9–20, graph the functions over the indicated intervals.

9. $y = \tan\left(\frac{1}{2}x\right),\ -2\pi \le x \le 2\pi$

10. $y = \cot\left(\frac{1}{2}x\right),\ -2\pi \le x \le 2\pi$

11. $y = -\cot(2\pi x),\ -1 \le x \le 1$

12. $y = -\tan(2\pi x),\ -1 \le x \le 1$

13. $y = 2\tan(3x),\ -\pi \le x \le \pi$

14. $y = 2\tan\left(\frac{1}{3}x\right),\ -3\pi \le x \le 3\pi$

15. $y = -\tan\left(x - \dfrac{\pi}{2}\right)$, $-\pi \le x \le \pi$

16. $y = \tan\left(x + \dfrac{\pi}{4}\right)$, $-\pi \le x \le \pi$

17. $y = \cot\left(x - \dfrac{\pi}{4}\right)$, $-\pi \le x \le \pi$

18. $y = -\cot\left(x + \dfrac{\pi}{2}\right)$, $-\pi \le x \le \pi$

19. $y = \tan(2x - \pi)$, $-2\pi \le x \le 2\pi$

20. $y = \cot(2x - \pi)$, $-2\pi \le x \le 2\pi$

In Exercises 21–32, graph the functions over the indicated intervals.

21. $y = \sec\left(\dfrac{1}{2}x\right)$, $-2\pi \le x \le 2\pi$

22. $y = \csc\left(\dfrac{1}{2}x\right)$, $-2\pi \le x \le 2\pi$

23. $y = -\csc(2\pi x)$, $-1 \le x \le 1$

24. $y = -\sec(2\pi x)$, $-1 \le x \le 1$

25. $y = 2\sec(3x)$, $0 \le x \le 2\pi$

26. $y = 2\csc\left(\dfrac{1}{3}x\right)$, $-3\pi \le x \le 3\pi$

27. $y = -3\csc\left(x - \dfrac{\pi}{2}\right)$, over at least one period

28. $y = 5\sec\left(x + \dfrac{\pi}{4}\right)$, over at least one period

29. $y = \dfrac{1}{2}\sec(x - \pi)$, over at least one period

30. $y = -4\csc(x + \pi)$, over at least one period

31. $y = 2\sec(2x - \pi)$, $-2\pi \le x \le 2\pi$

32. $y = 2\csc(2x + \pi)$, $-2\pi \le x \le 2\pi$

In Exercises 33–40, graph the functions over at least one period.

33. $3 - 2\sec\left(x - \dfrac{\pi}{2}\right)$

34. $-3 + 2\csc\left(x + \dfrac{\pi}{2}\right)$

35. $y = \dfrac{1}{2} + \dfrac{1}{2}\tan\left(x - \dfrac{\pi}{2}\right)$

36. $y = \dfrac{3}{4} - \dfrac{1}{4}\cot\left(x + \dfrac{\pi}{2}\right)$

37. $y = -2 + 3\csc(2x - \pi)$

38. $y = -1 + 4\sec(2x + \pi)$

39. $y = -1 - \sec\left(\dfrac{1}{2}x - \dfrac{\pi}{4}\right)$

40. $y = -2 + \csc\left(\dfrac{1}{2}x + \dfrac{\pi}{4}\right)$

In Exercises 41–46, state the domain and range of the functions.

41. $y = \tan\left(\pi x - \dfrac{\pi}{2}\right)$

42. $y = \cot\left(x - \dfrac{\pi}{2}\right)$

43. $y = 2\sec(5x)$

44. $y = -4\sec(3x)$

45. $y = 2 - \csc\left(\dfrac{1}{2}x - \pi\right)$

46. $y = 1 - 2\sec\left(\dfrac{1}{2}x + \pi\right)$

■ APPLICATIONS

47. Height of a Mountain. When a chalet on the top of a mountain is viewed by a mountain climber from a point A, on the plain below, the measure of the angle of elevation is α degrees. When the mountain climber moves closer to the mountain, to point B, the measure of the angle of elevation is β degrees. If the distance between points A and B is d, then the height of the mountain can be found with $h = \dfrac{d}{\cot\alpha - \cot\beta}$. What is the height of the mountain if $\alpha = 20°$, $\beta = 25°$, and $d = 2$ miles?

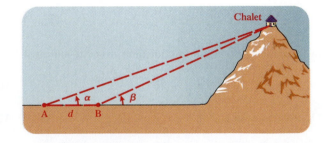

48. Height of a Mountain. When a chalet on the top of a mountain is viewed by a mountain climber from a point A, on the plain below, the measure of the angle of elevation is α degrees. When the mountain climber moves closer to the mountain, to point B, the measure of the angle of elevation is β degrees. If the distance between points A and B is d, then the height of the mountain can be found with $h = \dfrac{d}{\cot\alpha - \cot\beta}$. What is the height of the mountain if $\alpha = 15°, \beta = 35°$, and $d = 3$ miles?

49. Area of a Regular Polygon. The area of a regular polygon that is inscribed by a circle can be determined by $A = nr^2\tan\dfrac{180°}{n}$, where n is the number of sides of the polygon, and r is the radius of the inscribed circle. Find the area of a regular hexagon if the radius of the inscribed circle is 4 inches.

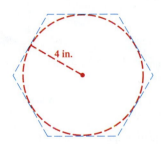

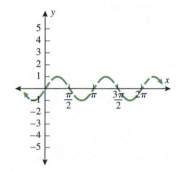

50. Area of a Regular Polygon. The area of a regular polygon that is inscribed by a circle can be determined by $A = nr^2\tan\dfrac{180°}{n}$, where n is the number of sides of the polygon, and r is the radius of the inscribed circle. Find the area of a regular octagon if the radius of the inscribed circle is 2 feet.

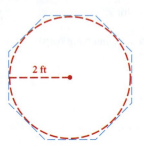

■ CATCH THE MISTAKE

In Exercises 51 and 52, explain the mistake that is made.

51. Graph $y = 3\csc(2x)$.

Solution:

Graph the guide function, $y = \sin 2x$.

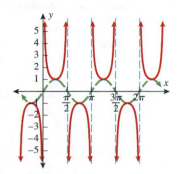

Draw vertical asymptotes at x values that correspond to x-intercepts of the guide function.

Draw the cosecant function.

This is incorrect. What mistake was made?

52. Graph $y = \tan(4x)$.

Solution:

Step 1: Calculate the period.
$$\frac{\pi}{B} = \frac{\pi}{4}$$

Step 2: Find two vertical asymptotes.
$$4x = 0 \text{ and } 4x = \pi$$

Solve for x.
$$x = 0 \text{ and } x = \frac{\pi}{4}$$

Step 3: Find the x-intercept between the asymptotes.
$$4x = \frac{\pi}{2}$$
$$x = \frac{\pi}{8}$$

Step 4: Draw the vertical asymptotes, $x = 0$ and $x = \frac{\pi}{4}$, and label the x-intercept, $\left(\frac{\pi}{8}, 0\right)$.

Step 5: Graph.

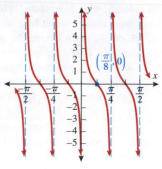

This is incorrect. What mistake was made?

■ CHALLENGE

53. T or F: $\sec\left(x - \frac{\pi}{2}\right) = \csc x$

54. T or F: $\csc\left(x - \frac{\pi}{2}\right) = \sec x$

55. For what values of n do $y = \tan x$ and $y = \tan(x - n\pi)$ have the same graph?

56. For what values of n do $y = \csc x$ and $y = \csc(x - n\pi)$ have the same graph?

57. Solve the equation $\tan(2x - \pi) = 0$ for x in the interval $[-\pi, \pi]$ by graphing.

58. Solve the equation $\csc(2x + \pi) = 0$ for x in the interval $[-\pi, \pi]$ by graphing.

■ TECHNOLOGY

59. What is the amplitude of the function $y = \cos x + \sin x$? Use a graphing calculator to graph $Y_1 = \cos x$, $Y_2 = \sin x$, and $Y_3 = \cos x + \sin x$ in the same viewing window.

60. Graph $Y_1 = \cos x + \sin x$ and $Y_2 = \sec x + \csc x$ in the same viewing window. Based on what you see, is $Y_1 = \cos x + \sin x$ the guide function for $Y_2 = \sec x + \csc x$?

Q: Anna is studying meteorology and is trying to graph the average number of daylight hours for her hometown of Vancouver, British Columbia. Here are her data:

JAN	FEB	MAR	APR	MAY	JUN
8.28	9.42	11.00	12.88	14.60	15.95

JUL	AUG	SEP	OCT	NOV	DEC
16.17	15.13	13.47	11.67	9.87	8.48

She lets x represent the months of the year ($x = 1$ corresponds to January, $x = 2$ corresponds to February, etc.) and lets y represent hours. She decides to use a piecewise-defined function to model the data by connecting the January and July data with a line and connecting the July and December data with a line. This yields the following function:

$$L(x) = \begin{cases} 1.315x + 6.965, & \text{if } 1 \leq x < 7 \\ -1.538x + 26.936, & \text{if } 7 \leq x \leq 12 \end{cases}$$

However, $L(6) = 14.855$, which varies more from the actual value of 15.95 than Anna would like. Is there a better way to model the data?

A: While $L(x)$ is not a bad representation of Anna's data, a sinusoidal function would be a better representation since the number of daylight hours is periodic (the pattern repeats every year). Find a function of the form $y = k + A \sin(Bx + C)$ to represent the data in the chart above.

TYING IT ALL TOGETHER

Emily attaches a 100 gram weight to the end of a 20 cm spring (see the illustration below) and hits the weight upward from below so that the spring and weight are set into motion. Let P be the point at the center of the weight. This point will oscillate back and forth between A and $-A$ as the weight oscillates. The origin (0) is at the equilibrium point, which is halfway between A and $-A$. We can represent the position of P at time t with the equation

$$x(t) = A \sin \omega t$$

The point P is undergoing simple harmonic motion. The amplitude of the motion is $|A|$. The time required for one complete oscillation of the point P from A to $-A$ and back to A again (or vice versa) is called the period. Its value is $\dfrac{2\pi}{|\omega|}$. The reciprocal of the period $\left(\dfrac{|\omega|}{2\pi}\right)$ is called the frequency, which is the number of oscillations per unit of time t. We will assume that there is no friction or air resistance.

Emily places a ruler beside the weight and notices that the distance from A to $-A$ is 10 cm. She watches the motion for 90 seconds and observes that it completes 45 oscillations and thus takes 2 seconds for the weight to complete one oscillation. Find the position function of P at time t (in other words, state the function $x(t)$ for the motion of the weight). What is the position of P at $t = \dfrac{1}{2}$ second?

CHAPTER 4 REVIEW

SECTION	TOPIC	PAGES	REVIEW EXERCISES	KEY CONCEPTS
4.1	Basic graphs of sine and cosine functions: Amplitude and period	172–191	1–20	
	$y = \sin x$	172–174	1–20	Odd function: $f(-x) = -f(x)$ Period: 2π Amplitude: 1
	$y = \cos x$	175–176	1–20	Even function: $f(-x) = f(x)$ Period: 2π Amplitude: 1
	$y = A \sin Bx$ and $y = A \cos Bx$ amplitude and period	177–186	1–20	Amplitude $= |A|$ ■ $|A| > 1$ stretch vertically ■ $|A| < 1$ compress vertically Period $= \dfrac{2\pi}{B}$ $B > 0$ ■ $B > 1$, compress horizontally ■ $B < 1$, stretch horizontally
4.2	Translations of the sine and cosine functions: Phase shift	197–210	21–40	$y = k + A \sin(Bx + C)$ or $y = k + A \cos(Bx + C)$
	Reflection	197–199	21–40	■ $y = -A \sin Bx$ is the reflection of $y = A \sin Bx$ about the x-axis. ■ $y = -A \cos Bx$ is the reflection of $y = A \cos Bx$ about the x-axis.
	Vertical shifts	199–202	21–40	■ $y = \pm k + A \sin Bx$ is found by shifting $y = A \sin Bx$ up $(+)$ or down $(-)$ k units. ■ $y = \pm k + A \cos Bx$ is found by shifting $y = A \cos Bx$ up $(+)$ or down $(-)$ k units.

SECTION	TOPIC	PAGES	REVIEW EXERCISES	KEY CONCEPTS
	Horizontal (phase) shifts	202–205	21–40	$y = A\sin(Bx \pm C) = A\sin\left[B\left(x \pm \dfrac{C}{B}\right)\right]$ has period $\dfrac{2\pi}{B}$ and a phase shift of $\dfrac{C}{B}$ unit to the left $(+)$ or the right $(-)$. $y = A\cos(Bx \pm C) = A\cos\left[B\left(x \pm \dfrac{C}{B}\right)\right]$ has period $\dfrac{2\pi}{B}$ and a phase shift of $\dfrac{C}{B}$ unit to the left $(+)$ or the right $(-)$.
	Combinations of translations	205–207	21–40	To graph $y = \pm k + A\sin(Bx + C)$ or $y = \pm k + A\cos(Bx + C)$, start with the graph of $y = A\sin(Bx + C)$ or $y = A\cos(Bx + C)$ and shift up or down k units.
4.3	Graphs of tangent, cotangent, secant, and cosecant functions	215–230	41–60	
	$y = \tan x$	215–217	41–60	x-intercepts: $n\pi$ Asymptotes: $x = \dfrac{(2n+1)}{2}\pi$ Period: π Amplitude: none
	$y = \cot x$	217–218	41–60	Asymptotes: $n\pi$ x-intercepts: $x = \dfrac{(2n+1)}{2}\pi$ Period: π Amplitude: none

SECTION	TOPIC	PAGES	REVIEW EXERCISES	KEY CONCEPTS
	$y = \sec x$	218–219	41–60	Asymptotes: $x = \dfrac{(2n+1)}{2}\pi$ Period: 2π Amplitude: None
	$y = \csc x$	219–220	41–60	Asymptotes: $n\pi$ Period: 2π Amplitude: None
	$y = A\tan(Bx+C)$ $y = A\cot(Bx+C)$	221–224	41–60	To find asymptotes, set $Bx+C$ equal to: • $-\dfrac{\pi}{2}$ and $\dfrac{\pi}{2}$ for tangent • 0 and π for cotangent To find the x-intercept, set $Bx+C$ equal to: • 0 for tangent • $\dfrac{\pi}{2}$ for cotangent
	$y = A\sec(Bx+C)$ $y = A\csc(Bx+C)$	225–227	41–60	• To graph $y = A\sec(Bx+C)$, use $y = A\cos(Bx+C)$ as the guide. • To graph $y = A\csc(Bx+C)$, use $y = A\sin(Bx+C)$ as the guide. Intercepts on the guide function correspond to vertical asymptotes of secant or cosecant functions.

4.1 Basic Graphs of Sine and Cosine Functions: Amplitude and Period

Refer to the graph of the cosine function to answer the questions.

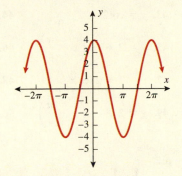

1. Determine the period of the function.

2. Determine the amplitude of the function.

3. Write an equation for the cosine function.

Refer to the graph of the sine function to answer the questions.

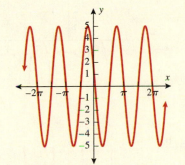

4. Determine the period of the function.

5. Determine the amplitude of the function.

6. Write an equation for the sine function.

Refer to the graph of the sine function to answer the questions.

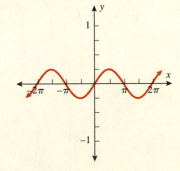

7. Determine the period of the function.

8. Determine the amplitude of the function.

9. Write an equation for the sine function.

Refer to the graph of the cosine function to answer the questions.

10. Determine the period of the function.

11. Determine the amplitude of the function.

12. Write an equation for the cosine function.

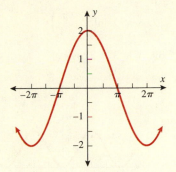

Determine the amplitude and period of each function.

13. $y = -2\cos 2\pi x$

14. $y = \dfrac{1}{3}\sin\dfrac{\pi}{2}x$

15. $y = \dfrac{1}{5}\sin 3x$

16. $y = -\dfrac{7}{6}\cos 6x$

Graph the function from -2π to 2π.

17. $y = -2\sin\dfrac{x}{2}$

18. $y = 3\sin 3x$

19. $y = \dfrac{1}{2}\cos 2x$

20. $y = -\dfrac{1}{4}\cos\dfrac{x}{2}$

4.2 Translations of the Sine and Cosine Functions: Phase Shift

State the amplitude, period, phase shift, and vertical shift of each function.

21. $y = 2 + 3\sin\left(x - \dfrac{\pi}{2}\right)$

22. $y = 3 - \dfrac{1}{2}\sin\left(x + \dfrac{\pi}{4}\right)$

23. $y = -2 - 4\cos 3\left(x + \dfrac{\pi}{4}\right)$

24. $y = -1 + 2\cos 2\left(x - \dfrac{\pi}{3}\right)$

25. $y = \dfrac{1}{2} - \cos(x - \pi)$

26. $y = \dfrac{3}{2} + \sin(x + \pi)$

27. $y = -\sin\left(\dfrac{1}{2}x + \pi\right)$

28. $y = \cos\left(\dfrac{1}{4}x - \pi\right)$

29. $y = 4 + 5\sin(4x - \pi)$

30. $y = -6\cos(3x + \pi)$

Sketch the graph of the function from -2π to 2π.

31. $y = 3 + \sin\left(x + \dfrac{\pi}{2}\right)$

32. $y = -2 + \sin\left(x - \dfrac{\pi}{2}\right)$

33. $y = -2 + \cos(x + \pi)$

34. $y = 3 + \cos(x - \pi)$

35. $y = 3 - \sin\left(\dfrac{\pi}{2}x\right)$

36. $y = 2\cos\left(\dfrac{\pi}{2}x\right)$

Match the graphs with their function equations.

37. $y = 2\sin\left(x - \dfrac{\pi}{4}\right)$

38. $y = -2 + 3\sin x$

39. $y = 3 - 3\cos x$

40. $y = 3\cos\left(x + \dfrac{\pi}{4}\right)$

a.

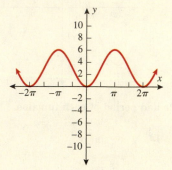

b.

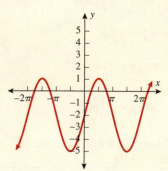

c.

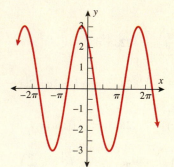

d.

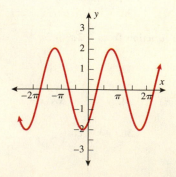

4.3 Graphs of Tangent, Cotangent, Secant, and Cosecant Functions

Match the graphs with their function equations.

41. $y = \dfrac{1}{2}\tan x$

42. $y = -\tan\left(\dfrac{1}{2}x\right)$

43. $y = \cot 2x$

44. $y = 2\cot x$

45. $y = 2 + \tan\left(x + \dfrac{\pi}{4}\right)$

46. $y = -\dfrac{1}{2} + \cot\left(x - \dfrac{\pi}{4}\right)$

47. $y = 2 + 2\sec x$

48. $y = -\sec\left(x - \dfrac{\pi}{2}\right)$

49. $y = 2 + \csc\left(x + \dfrac{\pi}{4}\right)$

50. $y = 2\csc x$

a.

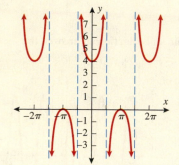

b.

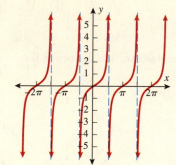

c.

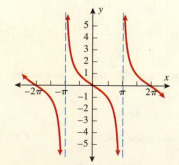

d.

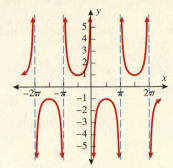

e.

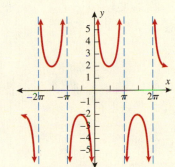

f.

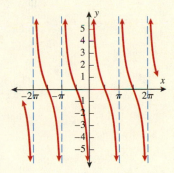

g.

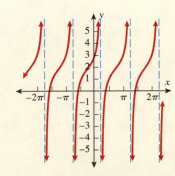

h.

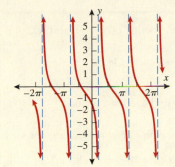

i.

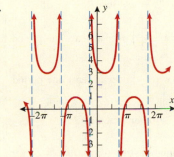

j.

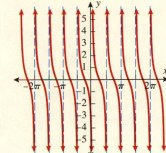

State the domain and range of the functions.

51. $y = 4\tan\left(x + \dfrac{\pi}{2}\right)$ **52.** $y = \cot 2\left(x - \dfrac{\pi}{2}\right)$

53. $y = 3\sec(2x)$ **54.** $y = 1 + 2\csc x$

Graph the function on $[-2\pi, 2\pi]$.

55. $y = -\tan\left(x - \dfrac{\pi}{4}\right)$ **56.** $y = 1 + \cot(2x)$

57. $y = 2 + \sec(x - \pi)$ **58.** $y = -\csc\left(x + \dfrac{\pi}{4}\right)$

59. $y = 2 + 3\sec\left(x - \dfrac{\pi}{3}\right)$ **60.** $y = 1 + \tan(2x + \pi)$

1. State the amplitude and period of $y = -5\sin(3x)$.

2. Graph $y = -2\cos\left(\frac{1}{2}x\right)$ over $-4\pi \le x \le 4\pi$.

3. Graph $y = \tan\left(\pi x - \frac{\pi}{2}\right)$ over two periods.

4. State the vertical asymptotes of $y = \csc(2\pi x)$ for all x.

5. Graph $y = -2\sec(\pi x)$ over two periods.

6. State the amplitude, period, and phase shift (if possible) of $y = -3\cot(4x - 2\pi)$.

7. Graph $y = -2 + 3\sin(2x - \pi)$ over $0 \le x \le 2\pi$.

8. Graph $y = 1 - \cos\left(\pi x - \frac{\pi}{2}\right)$ over at least one period.

9. Graph $y = 2 - \csc\left(\pi x - \frac{\pi}{2}\right)$ over at least one period.

10. Graph $y = \frac{1}{2} + \frac{1}{2}\cot(2x)$ over $0 \le x \le 2\pi$.

11. An electromagnetic signal has the form $E(t) = A\cos(\omega t + \phi)$. What are the period and the phase shift?

12. The vertical asymptotes of $y = 2\csc(3x - \pi)$ correspond to the _____ of $y = 2\sin(3x - \pi)$.

13. State the x-intercepts of $y = \tan(2x)$ for all x.

14. Determine the limit, $\lim\limits_{x \to \frac{\pi}{4}^-}[-\tan(2x)]$. That is, determine what $\tan(2x)$ approaches as x approaches $\frac{\pi}{4}$ from the left.

15. Select all of the following that are true statements.

   a. $\cos\left(x - \frac{\pi}{2}\right) = \sin x$

   b. $\cos\left(x + \frac{\pi}{2}\right) = \sin x$

   c. $\csc\left(x + \frac{\pi}{2}\right) = \sec x$

   d. $\sec\left(x + \frac{\pi}{2}\right) = \csc x$

   e. $\tan(x - \pi) = \tan x$

CHAPTER 5

Trigonometric Identities

Courtesy of Motorola

When you dial a phone number, how does the phone company know which phone to alert? Dual Tone Multi-Frequency (DTMF), also known as touch-tone dialing, was developed by Bell Labs in the 1960s. The touch-tone system also introduced a standardized keypad layout. After testing 18 different layouts, Bell Labs eventually chose the one familiar to us today, with 1 in the upper-left and 0 at the bottom between the star and the pound keys.

The keypad is laid out in a 4 × 3 matrix with each row representing a low frequency and each column representing a high frequency.

FREQUENCY	1209 Hz	1336 Hz	1477 Hz
697 Hz	1	2	3
770 Hz	4	5	6
852 Hz	7	8	9
941 Hz	*	0	#

When you press the number 8, the phone sends a sinusoidal tone that combines a low-frequency tone of 852 Hz and a high-frequency tone of 1336 Hz. The result can be found using sum-product *trigonometric identities*.

In this chapter we will review basic identities and use those to simplify trigonometric expressions. We will verify trigonometric identities. Specific identities that will be discussed are sum and difference, double-angle and half-angle, and product-to-sum and sum-to-product. Music, touch-tone key pads, and construction are applications of trigonometric identities. The trigonometric identities have useful applications, and they are used most frequently in calculus.

Trigonometric Identities

Identities	Sum and Difference	Double-Angle and Half-Angle	Product to Sum and Sum to Product
• Basic Identities • Simplify Trigonometric Expressions Using Identities • Verify Trigonometric Identities	• $\sin(A \pm B)$ • $\cos(A \pm B)$ • $\tan(A \pm B)$	• $\sin(2A)$ $\sin\left(\frac{1}{2}A\right)$ • $\cos(2A)$ $\cos\left(\frac{1}{2}A\right)$ • $\tan(2A)$ $\tan\left(\frac{1}{2}A\right)$	• $\sin A \pm \sin B$ • $\cos A \pm \cos B$ • $\cos A \cos B$ • $\sin A \sin B$ • $\cos A \sin B$

CHAPTER OBJECTIVES

- Review basic identities.
- Simplify a trigonometric expression using identities.
- Verify a trigonometric identity.
- Apply the sum and difference identities.
- Apply the double-angle and half-angle identities.
- Apply the product–sum identities.

NAVIGATION THROUGH SUPPLEMENTS

DIGITAL VIDEO SERIES #5

STUDENT SOLUTIONS MANUAL CHAPTER 5

BOOK COMPANION SITE
www.wiley.com/college/young

Skills Objectives

- Review basic trigonometric identities.
- Simplify trigonometric expressions using identities.
- Verify trigonometric identities.

Conceptual Objectives

- Understand that there is more than one way to verify an identity.
- Understand that identities must hold for all values in the domain of the functions.

Basic Identities

In Section 2.4 we discussed the *basic* (fundamental) trigonometric identities: reciprocal, quotient, and Pythagorean. In mathematics, an **identity** is an equation that is true for *all* values in the domain of the equation. If a statement is only true for *some* values of the variable, it is a **conditional equation**.

The following are **identities** (true for all x in the domain of the functions):

IDENTITY	TRUE FOR THESE VALUES OF x
$x^2 + 3x + 2 = (x + 2)(x + 1)$	All real numbers
$\tan x = \dfrac{\sin x}{\cos x}$	All real numbers except $x = \dfrac{n\pi}{2}$ where n is an odd integer.
$\sin^2 x + \cos^2 x = 1$	All real numbers

The following are **conditional equations** (only true for some x, not all x):

EQUATION	TRUE FOR THESE VALUES OF x
$x^2 + 3x + 2 = 0$	$x = -2$ and $x = -1$
$\tan x = 0$	$x = n\pi$ where n is an integer.
$\sin^2 x - \cos^2 x = 1$	$x = \dfrac{n\pi}{2}$ where n is an odd integer.

STUDY TIP

Just because an equation is true for *some* values of x does not make it an identity.

It is important to understand that just because an equation is true for some values of x does not make it an identity (true for all values of x).

The following boxes summarize the identities that were discussed in Section 2.4:

RECIPROCAL IDENTITIES

Reciprocal Identities	Equivalent Forms	Domain Restrictions	
$\csc x = \dfrac{1}{\sin x}$	$\sin x = \dfrac{1}{\csc x}$	$x \neq n\pi$	n is an integer
$\sec x = \dfrac{1}{\cos x}$	$\cos x = \dfrac{1}{\sec x}$	$x \neq \dfrac{n\pi}{2}$	n is an odd integer
$\cot x = \dfrac{1}{\tan x}$	$\tan x = \dfrac{1}{\cot x}$	$x \neq \dfrac{n\pi}{2}$	n is an integer

QUOTIENT IDENTITIES

$$\tan x = \frac{\sin x}{\cos x} \qquad \cot x = \frac{\cos x}{\sin x}$$

PYTHAGOREAN IDENTITIES

$$\sin^2 x + \cos^2 x = 1 \qquad \tan^2 x + 1 = \sec^2 x \qquad 1 + \cot^2 x = \csc^2 x$$

In the previous chapters we have discussed even and odd functions that have these properties:

TYPE OF FUNCTION	ALGEBRAIC IDENTITY	GRAPH
Even	$f(-x) = f(x)$	Symmetry about the y-axis
Odd	$f(-x) = -f(x)$	Symmetry about the origin

We have already pointed out in previous chapters that sine is an odd function and cosine is an even function. Combining this knowledge with the reciprocal and quotient identities, we arrive at the *even-odd identities*, which we can add to our categories of basic identities.

EVEN-ODD IDENTITIES

$$\sin(-x) = -\sin x \qquad \cos(-x) = \cos x \qquad \tan(-x) = -\tan x$$
$$\csc(-x) = -\csc x \qquad \sec(-x) = \sec x \qquad \cot(-x) = -\cot x$$

Simplifying Trigonometric Expressions Using Identities

In Section 2.4 we used the basic trigonometric identities to find values for functions, and we simplified trigonometric expressions using the identities. We now will use the basic identities and algebraic manipulation to simplify more complicated trigonometric expressions. In simplifying trigonometric expressions, one approach is to convert all expressions into sines and cosines first, then simplify.

EXAMPLE 1 Simplifying Trigonometric Expressions

Simplify $\tan x \sin x + \cos x$.

Solution:

STEP 1   Convert trigonometric functions to sine and cosine.

$$\tan x \sin x + \cos x = \left(\underbrace{\frac{\sin x}{\cos x}}_{\tan x}\right) \sin x + \cos x$$

STEP 2   Use algebraic techniques.

Simplify

$$= \frac{\sin^2 x}{\cos x} + \cos x$$

Write as a single quotient by getting a common denominator, $\cos x$.

$$= \frac{\sin^2 x}{\cos x} + \frac{\cos^2 x}{\cos x}$$

$$= \frac{\sin^2 x + \cos^2 x}{\cos x}$$

STEP 3   Use the Pythagorean identity: $\sin^2 x + \cos^2 x = 1$.

$$= \frac{1}{\cos x}$$

STEP 4   Use the reciprocal identity.

$$= \sec x$$

> ■ **YOUR TURN**   Simplify $\cot x \cos x + \sin x$.

In Example 1, $\tan x$ and $\sec x$ are not defined for odd multiples of $\frac{\pi}{2}$. In the Your Turn, $\cot x$ and $\csc x$ are not defined for integer multiples of π. Both the original expression and the simplified form have the same restrictions. There are times when the original expression has more domain restrictions than the simplified form and special care must be given to domain restrictions.

For example, the algebraic expression $\dfrac{x^2 - 1}{x + 1}$ has the domain restriction $x \neq -1$ because that value for x makes the value of the denominator equal to zero. If we forget to state the domain restrictions, we may have simplified the algebraic expression $\dfrac{x^2 - 1}{x + 1} = \dfrac{(x - 1)(x + 1)}{(x + 1)} = x - 1$ and assumed this was true for all values of x. The correct simplification is $\dfrac{x^2 - 1}{x + 1} = x - 1$ for $x \neq -1$. In fact, if we were to graph both the original expression $y = \dfrac{x^2 - 1}{x + 1}$ and the line $y = x - 1$, they would coincide, except the original expression would have a "hole" or discontinuity at $x = -1$. In this chapter it is assumed that the domain of the simplified expression is the same as the domain of the original expression.

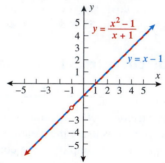

TECHNOLOGY TIP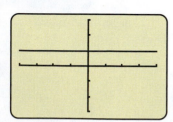

Graphs of

$$y = \frac{1}{\csc^2 x} + \frac{1}{\sec^2 x}$$

and $y = 1$ follow.

EXAMPLE 2 Simplifying Trigonometric Expressions

Simplify $\dfrac{1}{\csc^2 x} + \dfrac{1}{\sec^2 x}$.

Solution:

STEP 1 Rewrite the expression in terms of quotients squared.

$$\frac{1}{\csc^2 x} + \frac{1}{\sec^2 x} = \left(\frac{1}{\csc x}\right)^2 + \left(\frac{1}{\sec x}\right)^2$$

STEP 2 Use the reciprocal identities to write the cosecant and secant functions in terms of sines and cosines. $\sin x = \dfrac{1}{\csc x}$ and

$$\cos x = \frac{1}{\sec x}$$

$$= \sin^2 x + \cos^2 x$$

STEP 3 Use the Pythagorean identity: $\sin^2 x + \cos^2 x = 1$.

$$= 1$$

■ **Answer:** $\csc x$

■ **YOUR TURN**   Simplify $\dfrac{1}{\cos^2 x} - 1$.

Verifying Identities

We will now use the trigonometric identities to verify, or prove, other trigonometric identities for all values in the domain of the equation. For example,

$$(\sin x - \cos x)^2 - 1 = -2\sin x \cos x$$

The good news is that we will know we are done when we get there, since we know the desired identity. But how do we get there? How do we verify this is true? Remember that this must be true for all x, not just some x. Therefore, it is not enough to simply select values for x and show it is true for those.

WORDS	MATH
Start with one side of the equation (the more complicated side).	$(\sin x - \cos x)^2 - 1$
Expand the binomial squared: $(a - b)^2 = a^2 - 2ab + b^2$	$= \sin^2 x - 2\sin x \cos x + \cos^2 x - 1$
Group the $\sin^2 x$ and $\cos^2 x$ terms and use the Pythagorean identity.	$= -2\sin x \cos x + \underbrace{(\sin^2 x + \cos^2 x)}_{1} - 1$
Simplify.	$= -2\sin x \cos x$

When we arrive at the right side of the equation, then we have succeeded in verifying the identity. It is important to note that we are *working with only one side of the equation at a time*.

In verifying trigonometric identities, there is no one procedure that works for all identities. You manipulate one side of the equation until it looks like the other side. In general, there are two basic suggestions:

1. Convert all trigonometric functions to sines and cosines.

2. Write all sums or differences of quotients as a single quotient.

The following suggestions help guide the way to verifying trigonometric identities.

GUIDELINES FOR VERIFYING TRIGONOMETRIC IDENTITIES

- Start with the more complicated side of the equation.
- Sometimes it is helpful to convert all trigonometric functions into sines and cosines.
- Combine all sums and differences of quotients into a single quotient.
- Use basic trigonometric identities.
- Use algebraic techniques to manipulate one side of the equation until the other side of the equation is achieved.

It is important to note that trigonometric identities are valid for values of x (or of the independent variable) that are in the domain of the equation (or domains of the functions on both sides).

■ **Answer:** $\tan^2 x$

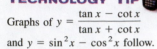

TECHNOLOGY TIP

Graphs of $y = \dfrac{\tan x - \cot x}{\tan x + \cot x}$
and $y = \sin^2 x - \cos^2 x$ follow.

```
Plot1 Plot2 Plot3
\Y1=(tan(X)-(tan
(X))⁻¹)/(tan(X)+(
tan(X))⁻¹)
\Y2◼(sin(X))²-(c
os(X))²
\Y3=
\Y4=
```

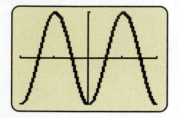

EXAMPLE 3 Verifying Trigonometric Identities

Verify the identity: $\dfrac{\tan x - \cot x}{\tan x + \cot x} = \sin^2 x - \cos^2 x$.

Solution:

STEP 1 Start with the more complicated side of the equation.

$$\frac{\tan x - \cot x}{\tan x + \cot x}$$

STEP 2 Use the quotient identities to write tangent and cotangent in terms of sine and cosine.

$$= \frac{\dfrac{\sin x}{\cos x} - \dfrac{\cos x}{\sin x}}{\dfrac{\sin x}{\cos x} + \dfrac{\cos x}{\sin x}}$$

STEP 3 Multiply by $\dfrac{\sin x \cos x}{\sin x \cos x}$.

$$= \left(\frac{\dfrac{\sin x}{\cos x} - \dfrac{\cos x}{\sin x}}{\dfrac{\sin x}{\cos x} + \dfrac{\cos x}{\sin x}} \right) \left(\frac{\sin x \cos x}{\sin x \cos x} \right)$$

Distribute

$$\frac{\sin x \cdot \cos x \left(\dfrac{\sin x}{\cos x} - \dfrac{\cos x}{\sin x} \right)}{\sin x \cdot \cos x \left(\dfrac{\sin x}{\cos x} + \dfrac{\cos x}{\sin x} \right)} = \frac{\dfrac{\sin^2 x \, \cancel{\cos x}}{\cancel{\cos x}} - \dfrac{\cancel{\sin x} \cos^2 x}{\cancel{\sin x}}}{\dfrac{\sin^2 x \, \cancel{\cos x}}{\cancel{\cos x}} + \dfrac{\cancel{\sin x} \cos^2 x}{\cancel{\sin x}}}$$

Simplify.

$$= \frac{\sin^2 x - \cos^2 x}{\sin^2 x + \cos^2 x}$$

STEP 4 Use the Pythagorean identity: $\sin^2 x + \cos^2 x = 1$.

$$= \sin^2 x - \cos^2 x$$

EXAMPLE 4 Determining If a Trigonometric Equation Is an Identity

Determine if $(1 - \cos^2 x)(1 + \cot^2 x) = 0$.

Solution:

STEP 1 Use the quotient identity to write cotangent in terms of sine and cosine.

$$(1 - \cos^2 x)(1 + \cot^2 x) = (1 - \cos^2 x)\left(1 + \frac{\cos^2 x}{\sin^2 x} \right)$$

STEP 2 Combine second parentheses into a single quotient.

$$= (1 - \cos^2 x)\left(\frac{\sin^2 x + \cos^2 x}{\sin^2 x} \right)$$

STEP 3 Use the Pythagorean identity.

$$= \underbrace{(1 - \cos^2 x)}_{\sin^2 x}\left(\frac{\overbrace{\sin^2 x + \cos^2 x}^{1}}{\sin^2 x} \right)$$

STEP 4 Eliminate parentheses.

$$= \frac{\sin^2 x}{\sin^2 x}$$

STEP 5 Simplify.

$$= 1$$

Since $1 \neq 0$, this is not an identity.

An alternative approach to Example 4
is to use two Pythagorean identities

$$\underbrace{(1 - \cos^2 x)}_{\sin^2 x}\underbrace{(1 + \cot^2 x)}_{\csc^2 x}$$

$$= \sin^2 x \cdot \frac{1}{\sin^2 x} = 1$$

EXAMPLE 5 Verifying Trigonometric Identities

Verify that $\dfrac{\sin(-x)}{\cos(-x)\tan(-x)} = 1.$

TECHNOLOGY TIP

Graphs of $y = \dfrac{\sin(-x)}{\cos(-x)\tan(-x)}$
and $y = 1$ follow.

Solution:

STEP 1	Start with the left side of the equation.	$\dfrac{\sin(-x)}{\cos(-x)\tan(-x)}$
STEP 2	Use the even-odd identities.	$= \dfrac{-\sin(x)}{-\cos(x)\tan(x)}$
STEP 3	Simplify.	$= \dfrac{\sin(x)}{\cos(x)\tan(x)}$
STEP 4	Use the quotient identity to write tangent in terms of sine and cosine.	$= \dfrac{\sin(x)}{\cos(x)\dfrac{\sin x}{\cos x}}$
STEP 5	Divide out the cosine term.	$= \dfrac{\sin(x)}{\sin(x)}$
STEP 6	Simplify.	$= 1$

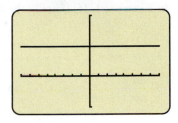

We have only discussed working with one side of the identity until arriving at the other side. Another method for verifying identities is to work with each side separately and use identities and algebraic techniques to arrive at the same result on both sides. It is important to note that the operations cannot cross the equals sign. Instead, each side must be treated independently.

EXAMPLE 6 Verifying an Identity by Working Both Sides Separately

Verify that $\dfrac{\sin x + 1}{\sin x} = -\dfrac{\cot^2 x}{1 - \csc x}.$

Solution:

Left-hand side: $\dfrac{\sin x + 1}{\sin x} = \dfrac{\sin x}{\sin x} + \dfrac{1}{\sin x} = 1 + \csc x$

Right-hand side: $\dfrac{-\cot^2 x}{1 - \csc x} = \dfrac{(1 - \csc^2 x)}{1 - \csc x}$

$= \dfrac{(1 - \csc x)(1 + \csc x)}{(1 - \csc x)} = 1 + \csc x$

Since the left-hand side equals the right-hand side, the equation is an identity.

SECTION 5.1 SUMMARY

We combined the basic trigonometric identities—reciprocal, quotient, Pythagorean, and even-odd—with algebraic techniques to simplify trigonometric expressions and verify more complex trigonometric identities. Two steps that are often used in both simplifying trigonometric expressions and verifying trigonometric identities are (1) writing all trigonometric functions in terms of sine and cosine functions, and (2) combining sums or differences of quotients into a single quotient.

When verifying trigonometric identities, typically we work with the more complicated side (keeping the other side in mind as our goal). Another approach to verifying trigonometric identities is to work on each side *separately* and arrive at the same result.

SECTION 5.1 EXERCISES

■ SKILLS

In Exercises 1–16, simplify the following trigonometric expressions.

1. $\sin x \csc x$

2. $\tan x \cot x$

3. $\sec(-x)\cot x$

4. $\tan(-x)\cos(-x)$

5. $\sin^2 x(\cot^2 x + 1)$

6. $\cos^2 x(\tan^2 x + 1)$

7. $(\sin x - \cos x)(\sin x + \cos x)$

8. $(\sin x + \cos x)^2$

9. $\dfrac{\csc x}{\cot x}$

10. $\dfrac{\sec x}{\tan x}$

11. $\dfrac{1 - \cos^4 x}{1 + \cos^2 x}$

12. $\dfrac{1 - \sin^4 x}{1 + \sin^2 x}$

13. $1 - \dfrac{\sin^2 x}{1 - \cos x}$

14. $1 - \dfrac{\cos^2 x}{1 + \sin x}$

15. $\dfrac{\tan x - \cot x}{\tan x + \cot x} + 2\cos^2 x$

16. $\dfrac{\tan x - \cot x}{\tan x + \cot x} + \cos^2 x$

In Exercises 17–38, verify the trigonometric identities.

17. $(\sin x + \cos x)^2 + (\sin x - \cos x)^2 = 2$

18. $(1 - \sin x)(1 + \sin x) = \cos^2 x$

19. $(\csc x + 1)(\csc x - 1) = \cot^2 x$

20. $(\sec x + 1)(\sec x - 1) = \tan^2 x$

21. $\tan x + \cot x = \csc x \sec x$

22. $\csc x - \sin x = \cot x \cos x$

23. $\dfrac{2 - \sin^2 x}{\cos x} = \sec x + \cos x$

24. $\dfrac{2 - \cos^2 x}{\sin x} = \csc x + \sin x$

25. $\dfrac{1}{\csc^2 x} + \dfrac{1}{\sec^2 x} = 1$

26. $\dfrac{1}{\cot^2 x} - \dfrac{1}{\tan^2 x} = \sec^2 x - \csc^2 x$

27. $\dfrac{1}{1 - \sin x} + \dfrac{1}{1 + \sin x} = 2\sec^2 x$

28. $\dfrac{1}{1 - \cos x} + \dfrac{1}{1 + \cos x} = 2\csc^2 x$

29. $\dfrac{\sin^2 x}{1 - \cos x} = 1 + \cos x$

30. $\dfrac{\cos^2 x}{1 - \sin x} = 1 + \sin x$

31. $\sec x + \tan x = \dfrac{1}{\sec x - \tan x}$

32. $\csc x + \cot x = \dfrac{1}{\csc x - \cot x}$

33. $\dfrac{\csc x - \tan x}{\sec x + \cot x} = \dfrac{\cos x - \sin^2 x}{\sin x + \cos^2 x}$

34. $\dfrac{\sec x + \tan x}{\csc x + 1} = \tan x$

35. $\dfrac{\cos^2 x + 1 + \sin x}{\cos^2 x + 3} = \dfrac{1 + \sin x}{2 + \sin x}$

36. $\dfrac{\sin x + 1 - \cos^2 x}{\cos^2 x} = \dfrac{\sin x}{1 - \sin x}$

37. $\sec x(\tan x + \cot x) = \dfrac{\csc x}{\cos^2 x}$

38. $\tan x(\csc x - \sin x) = \cos x$

In Exercises 39–50, determine if the following equations are conditional or identities.

39. $\cos^2 x(\tan x - \sec x)(\tan x + \sec x) = 1$

40. $\cos^2 x(\tan x - \sec x)(\tan x + \sec x) = \sin^2 x - 1$

41. $\dfrac{\csc x \cot x}{\sec x \tan x} = \cot^3 x$ **42.** $\sin x \cos x = 0$ **43.** $\sin x + \cos x = \sqrt{2}$ **44.** $\sin^2 x + \cos^2 x = 1$

45. $\tan^2 x - \sec^2 x = 1$ **46.** $\sec^2 x - \tan^2 x = 1$ **47.** $\sin x = \sqrt{1 - \cos^2 x}$ **48.** $\csc x = \sqrt{1 + \cot^2 x}$

49. $\sqrt{\sin^2 x + \cos^2 x} = 1$ **50.** $\sqrt{\sin^2 x + \cos^2 x} = \sin x + \cos x$

■ APPLICATIONS

Expression	Substitution		Trigonometric Identity
$\sqrt{a^2 - x^2}$	$x = a\sin\theta$	$-\dfrac{\pi}{2} \le \theta \le \dfrac{\pi}{2}$	$1 - \sin^2\theta = \cos^2\theta$
$\sqrt{a^2 + x^2}$	$x = a\tan\theta$	$-\dfrac{\pi}{2} < \theta < \dfrac{\pi}{2}$	$1 + \tan^2\theta = \sec^2\theta$
$\sqrt{x^2 - a^2}$	$x = a\sec\theta$	$0 \le \theta < \dfrac{\pi}{2}$ or $\pi \le \theta < \dfrac{3\pi}{2}$	$\sec^2\theta - 1 = \tan^2\theta$

In calculus, when integrating expressions such as $\sqrt{a^2 - x^2}$, $\sqrt{a^2 + x^2}$, and $\sqrt{x^2 - a^2}$, trigonometric functions are used as "dummy" functions to eliminate the radical. Once the integration is performed, the trigonometric function is "un-substituted." The following trigonometric substitutions (and corresponding trigonometric identities) are used to simplify these types of expressions.

When simplifying, it is important to remember that

$$|x| = \begin{cases} x & \text{if } x > 0 \\ -x & \text{if } x < 0 \end{cases}$$

51. Calculus (Trigonometric Substitution). Start with the expression $\sqrt{a^2 - x^2}$ and let $x = a \sin\theta$. Assuming $-\dfrac{\pi}{2} \le \theta \le \dfrac{\pi}{2}$, simplify the original expression so that it contains no radicals.

52. Calculus (Trigonometric Substitution). Start with the expression $\sqrt{a^2 + x^2}$ and let $x = a \tan\theta$, assuming $-\dfrac{\pi}{2} < \theta < \dfrac{\pi}{2}$, simplify the original expression so that it contains no radicals.

■ CATCH THE MISTAKE

In Exercises 53–56, explain the mistake that is made.

53. Verify the identity $\dfrac{\cos x}{1 - \tan x} + \dfrac{\sin x}{1 - \cot x} = \sin x + \cos x$.

Solution:

Start with the left side of the equation.

$$\dfrac{\cos x}{1 - \tan x} + \dfrac{\sin x}{1 - \cot x}$$

Write the tangent and cotangent functions in terms of sines and cosines.

$$\dfrac{\cos x}{1 - \dfrac{\sin x}{\cos x}} + \dfrac{\sin x}{1 - \dfrac{\cos x}{\sin x}}$$

Cancel the common cosine in the first term and sine in the second term.

$$\dfrac{1}{1 - \sin x} + \dfrac{1}{1 - \cos x}$$

This is not correct. What mistake was made?

54. Verify the identity $\dfrac{\cos^3 x \sec x}{1 - \sin x} = 1 + \sin x$.

Solution:

Start with the left side of the equation.

$$\dfrac{\cos^3 x \sec x}{1 - \sin x}$$

Rewrite secant in terms of sine.

$$\dfrac{\cos^3 x \dfrac{1}{\sin x}}{1 - \sin x}$$

Simplify.

$$\dfrac{\cos^3 x}{1 - \sin^2 x}$$

Use the Pythagorean identity.

$$\dfrac{\cos^3 x}{\cos^2 x}$$

Simplify. $\cos x$

This is not correct. What mistakes were made?

55. Determine if the equation is conditional or an identity.

$$\frac{\tan x}{\cot x} = 1$$

Solution:

Start with the left side of the equation. $\dfrac{\tan x}{\cot x}$

Rewrite the tangent and cotangent functions in terms of sines and cosines. $= \dfrac{\dfrac{\sin x}{\cos x}}{\dfrac{\cos x}{\sin x}}$

Simplify. $= \dfrac{\sin^2 x}{\cos^2 x} = \tan^2 x$

Let $x = \dfrac{\pi}{4}$. *Note*: $\tan\left(\dfrac{\pi}{4}\right) = 1$. $= 1$

Since $\dfrac{\tan x}{\cot x} = 1$, this equation is an identity.

This is incorrect. What mistake was made?

56. Determine if the equation is conditional or an identity.

$$|\sin x| - \cos x = 1$$

Solution:

Start with the left side of the equation. $|\sin x| - \cos x$

Let $x = \dfrac{(2n + 1)\pi}{2}$, where n is an integer.

$$\left|\sin\left(\frac{(2n + 1)\pi}{2}\right)\right| - \cos\left(\frac{(2n + 1)\pi}{2}\right)$$

Simplify. $|\pm 1| - 0 = 1$

Since $|\sin x| - \cos x = 1$, this is an identity.

This is incorrect. What mistake was made?

■ CHALLENGE

57. T or F: If an equation is true for some values (but not all values), then it is still an identity.

58. T or F: If an equation has an infinite number of solutions, then it is an identity.

59. In what quadrants is the equation $\cos \theta = \sqrt{1 - \sin^2 \theta}$ true?

60. In what quadrants is the equation $-\cos \theta = \sqrt{1 - \sin^2 \theta}$ true?

61. Simplify $(a \sin x + b \cos x)^2 + (b \sin x - a \cos x)^2$.

62. Simplify $\dfrac{1 + \cot^3 x}{1 + \cot x} + \cot x$.

63. Do you think $\sin (A + B) = \sin A + \sin B$? Why?

64. Do you think $\cos\left(\dfrac{1}{2}A\right) = \dfrac{1}{2}\cos A$? Why?

■ TECHNOLOGY

In the next section you will learn the sum and difference identities. In Exercises 65–68, we illustrate these identities with graphing calculators.

65. Determine the correct sign $(+/-)$ for
$\cos (A + B) = \cos A \cos B \overset{?}{\pm} \sin A \sin B$ by graphing
$Y_1 = \cos (A + B)$, $Y_2 = \cos A \cos B + \sin A \sin B$, and
$Y_2 = \cos A \cos B - \sin A \sin B$ in the same viewing rectangle for different values of A and B.

66. Determine the correct sign $(+/-)$ for
$\cos (A - B) = \cos A \cos B \overset{?}{\pm} \sin A \sin B$ by graphing
$Y_1 = \cos (A - B)$, $Y_2 = \cos A \cos B + \sin A \sin B$, and
$Y_2 = \cos A \cos B - \sin A \sin B$ in the same viewing rectangle for different values of A and B.

67. Determine the correct sign $(+/-)$ for
$\sin (A + B) = \sin A \cos B \overset{?}{\pm} \cos A \sin B$ by graphing
$Y_1 = \sin (A + B)$, $Y_2 = \sin A \cos B + \cos A \sin B$, and
$Y_2 = \sin A \cos B - \cos A \sin B$ in the same viewing rectangle for different values of A and B.

68. Determine the correct sign $(+/-)$ for
$\sin (A - B) = \sin A \cos B \overset{?}{\pm} \cos A \sin B$ by graphing
$Y_1 = \sin (A - B)$, $Y_2 = \sin A \cos B + \cos A \sin B$, and
$Y_2 = \sin A \cos B - \cos A \sin B$ in the same viewing rectangle for different values of A and B.

Skills Objectives

- Find exact values of functions of rational multiples of π by using sum and difference identities.
- Develop new identities from the sum and difference identities.

Conceptual Objectives

- Derive sum and difference identities for the cosine function by using the distance formula.
- Use the sum and difference identities for the cosine function to obtain the cofunction identities.
- Use the cofunction identities and the sum and difference identities for the cosine function to derive the sum and difference identities for the sine function.
- Use the cosine and sine sum and difference identities with the quotient identity to obtain the tangent sum and difference identities.
- Understand that a trigonometric function of a sum is not the sum of the trigonometric functions.

In the previous sections we discussed the more basic trigonometric identities. In this section we continue discussing trigonometric identities by deriving formulas for when the argument of the trigonometric function is a sum or difference. First, it is important to note that function notation is not distributive:

$$\cos(A + B) \neq \cos A + \cos B$$

It is easy to prove by counterexample. Let $A = \pi$ and $B = 0$:

$$\cos(A + B) = \cos(\pi + 0) = \cos(\pi) = -1$$

$$\cos A + \cos B = \cos \pi + \cos 0 = -1 + 1 = 0$$

In this section, we will derive some new and important identities.

- Sum and difference identities for cosine, sine, and tangent
- Cofunction identities

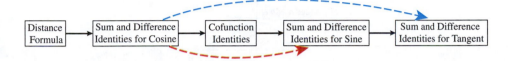

| Distance Formula | Sum and Difference Identities for Cosine | Cofunction Identities | Sum and Difference Identities for Sine | Sum and Difference Identities for Tangent |

Before we start deriving and working with trigonometric sum and difference identities, let us first discuss why these are important. Sum and difference identities (and later product-sum and sum-product identities), are important because they allow the calculation in functional (analytic) form and often lead to evaluating expressions exactly (as opposed to approximating with calculators). Functional form is important, as you will soon see in applications such as music, where the identities developed in this chapter allow the determination of the "beat" frequency. In calculus these identities will simplify the integration and differentiation processes.

Sum and Difference Identities for Cosine

Recall from Section 3.4 that the unit circle approach gave the relationship between the coordinates on the unit circle and the sine and cosine functions. Specifically, the x-coordinate corresponds to the value of the cosine function and the y-coordinate corresponds to the value of the sine function.

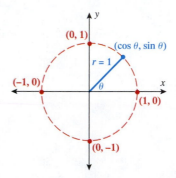

Let us now draw the unit circle with two angles, α and β, realizing that the two terminal sides of these angles form a third angle, $\alpha - \beta$.

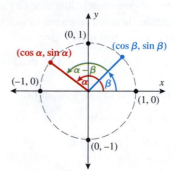

If we label the points $P_1 = (\cos \alpha, \sin \alpha)$ and $P_2 = (\cos \beta, \sin \beta)$ we can then draw a **segment** connecting points P_1 and P_2.

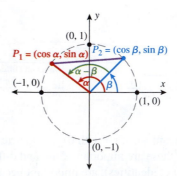

If we rotate the angle clockwise so the central angle, $\alpha - \beta$, is in standard position, then the two points where the initial and terminal sides intersect the unit circle are $P_3 = (\cos(\alpha - \beta), \sin(\alpha - \beta))$ and $P_4 = (1, 0)$.

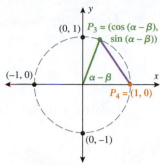

Notice that the **segment** has the same length because the angle $\alpha - \beta$ is the same and the two sides adjacent to $\alpha - \beta$ have the same length (1).

The distance from P_1 to P_2 is equal to the length of the **segment**. Similarly, the distance from P_3 to P_4 is equal to the length of the **segment**. Since the lengths of the **segments** are equal, we say that the distances are equal: $d(P_1, P_2) = d(P_3, P_4)$.

STUDY TIP

The distance from point $P_1 = (x_1, y_1)$ to $P_2 = (x_2, y_2)$ is given by the distance formula $d(P_1, P_2) = \sqrt{(x_2 - x_1)^2 + (y_2 - y_1)^2}$.

WORDS	**MATH**
Start with the distances (segment lengths) equal.	$d(P_1, P_2) = d(P_3, P_4)$
Write the distance formulas:	$\sqrt{(x_2 - x_1)^2 + (y_2 - y_1)^2} = \sqrt{(x_4 - x_3)^2 + (y_4 - y_3)^2}$

Substitute $P_1 = (x_1, y_1) = (\cos\alpha, \sin\alpha)$ and $P_2 = (x_2, y_2) = (\cos\beta, \sin\beta)$ into the left side of the equation and $P_3 = (x_3, y_3) = (\cos(\alpha - \beta), \sin(\alpha - \beta))$ and $P_4 = (x_4, y_4) = (1, 0)$ into the right side of the equation.

$$\sqrt{[\cos\beta - \cos\alpha]^2 + [\sin\beta - \sin\alpha]^2} = \sqrt{[1 - \cos(\alpha - \beta)]^2 + [0 - \sin(\alpha - \beta)]^2}$$

Square both sides of the equation.	$[\cos\beta - \cos\alpha]^2 + [\sin\beta - \sin\alpha]^2 = [1 - \cos(\alpha - \beta)]^2 + [0 - \sin(\alpha - \beta)]^2$
Eliminate the brackets.	

$$\cos^2\beta - 2\cos\alpha\cos\beta + \cos^2\alpha + \sin^2\beta - 2\sin\alpha\sin\beta + \sin^2\alpha$$
$$= 1 - 2\cos(\alpha - \beta) + \cos^2(\alpha - \beta) + \sin^2(\alpha - \beta)$$

Regroup terms on each side and use the Pythagorean identity.

$$\underbrace{\cos^2\alpha + \sin^2\alpha}_{1} - 2\cos\alpha\cos\beta - 2\sin\alpha\sin\beta + \underbrace{\cos^2\beta + \sin^2\beta}_{1}$$

$$= 1 - 2\cos(\alpha - \beta) + \underbrace{\cos^2(\alpha - \beta) + \sin^2(\alpha - \beta)}_{1}$$

Simplify.	$2 - 2\cos\alpha\cos\beta - 2\sin\alpha\sin\beta = 2 - 2\cos(\alpha - \beta)$
Subtract 2 from both sides.	$-2\cos\alpha\cos\beta - 2\sin\alpha\sin\beta = -2\cos(\alpha - \beta)$
Divide by -2.	$\cos\alpha\cos\beta + \sin\alpha\sin\beta = \cos(\alpha - \beta)$
Write the **difference identity for cosine**.	$\boxed{\cos(\alpha - \beta) = \cos\alpha\cos\beta + \sin\alpha\sin\beta}$

Now we can derive the sum identity for cosine from the difference identity for cosine and the properties of even and odd functions.

Replace β with $-\beta$ in the difference identity.

$$\cos(\alpha - (-\beta)) = \cos\alpha\cos(-\beta) + \sin\alpha\sin(-\beta)$$

Simplify the left side and use properties of even and odd functions on the right side.

$$\cos(\alpha + \beta) = \cos\alpha[\cos\beta] + \sin\alpha[-\sin\beta]$$

Write the **sum identity for cosine.**

$$\cos(\alpha + \beta) = \cos\alpha\cos\beta - \sin\alpha\sin\beta$$

SUM AND DIFFERENCE IDENTITIES FOR COSINE

Sum: $\quad\quad\quad\quad \cos(A + B) = \cos A\cos B - \sin A\sin B$

Difference: $\quad\quad \cos(A - B) = \cos A\cos B + \sin A\sin B$

EXAMPLE 1 Finding Exact Values for Cosine

Use the sum or difference identities for cosine to evaluate the cosine expressions exactly.

a. $\cos\left(\dfrac{7\pi}{12}\right)$ $\quad\quad\quad\quad$ **b.** $\cos(15°)$

Solution (a):

Write $\dfrac{7\pi}{12}$ as a sum.

$$\cos\left(\frac{7\pi}{12}\right) = \cos\left(\frac{4\pi}{12} + \frac{3\pi}{12}\right)$$

Simplify.

$$\cos\left(\frac{7\pi}{12}\right) = \cos\left(\frac{\pi}{3} + \frac{\pi}{4}\right)$$

Write the sum identity for cosine.

$$\cos(A + B) = \cos A\cos B - \sin A\sin B$$

Substitute $A = \dfrac{\pi}{3}$ and $B = \dfrac{\pi}{4}$.

$$\cos\left(\frac{7\pi}{12}\right) = \cos\left(\frac{\pi}{3}\right)\cos\left(\frac{\pi}{4}\right) - \sin\left(\frac{\pi}{3}\right)\sin\left(\frac{\pi}{4}\right)$$

Evaluate the expressions on the right exactly.

$$\cos\left(\frac{7\pi}{12}\right) = \frac{1}{2}\frac{\sqrt{2}}{2} - \frac{\sqrt{3}}{2}\frac{\sqrt{2}}{2}$$

Simplify.

$$\cos\left(\frac{7\pi}{12}\right) = \frac{\sqrt{2} - \sqrt{6}}{4}$$

Solution (b):

Write 15° as a sum.

$$\cos(15°) = \cos(45° - 30°)$$

Write the difference identity for cosine.

$$\cos(A - B) = \cos A\cos B + \sin A\sin B$$

Substitute $A = 45°$ and $B = 30°$.

$$\cos(15°) = \cos(45°)\cos(30°) + \sin(45°)\sin(30°)$$

Evaluate the expressions on the right exactly.

$$\cos(15°) = \frac{\sqrt{2}}{2}\frac{\sqrt{3}}{2} + \frac{\sqrt{2}}{2}\frac{1}{2}$$

Simplify.

$$\cos(15°) = \frac{\sqrt{6} + \sqrt{2}}{4}$$

TECHNOLOGY TIP

(a) Use a TI calculator to check the values for $\cos\left(\dfrac{7\pi}{12}\right)$ and $\dfrac{\sqrt{2} - \sqrt{6}}{4}$.

```
cos(7π/12)
          -.2588190451
(√(2)-√(6))/4
          -.2588190451
```

(b) Use a TI calculator to check the values of $\cos(15°)$ and $\dfrac{\sqrt{2} + \sqrt{6}}{4}$. Be sure the calculator is set in degree mode.

```
Normal Sci Eng
Float 0123456789
Radian Degrees
      Par Pol Seq
Connected Dot
Sequential Simul
Real a+bi re^θi
Full Horiz G-T
```

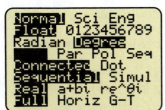

```
cos(15)
          .9659258263
(√(2)+√(6))/4
          .9659258263
■
```

Example 1 illustrates that an important characteristic of the sum and difference identities is that we can now find the exact value of angles that are multiples of 15° (or $\frac{\pi}{12}$).

■ **YOUR TURN** Use the sum or difference identities for cosine to evaluate the cosine expressions exactly.

a. $\cos\left(\dfrac{5\pi}{12}\right)$ **b.** $\cos(75°)$

EXAMPLE 2 Writing a Sum or Difference as a Single Cosine

Use the sum or difference identities for cosine to write the expressions as a single cosine.

a. $\sin 5x \sin 2x + \cos 5x \cos 2x$
b. $\cos x \cos 3x - \sin x \sin 3x$

Solution (a):

Because of the $+$ sign, this will be a cosine of a difference.

Write the formula reversed:	$\cos A \cos B + \sin A \sin B = \cos(A - B)$
Rewrite the expression.	$\cos 5x \cos 2x + \sin 5x \sin 2x$
Identify A and B.	$A = 5x$ and $B = 2x$
Substitute $A = 5x$ and $B = 2x$ into the difference identity.	$\cos 5x \cos 2x + \sin 5x \sin 2x = \cos(5x - 2x)$
Simplify.	$\cos 5x \cos 2x + \sin 5x \sin 2x = \cos(3x)$

Notice that if we had selected $A = 2x$ and $B = 5x$ instead, the result would have been $\cos(-3x)$, but since cosine is an even function this simplifies to $\cos(3x)$.

Solution (b):

Because of the $-$ sign, this will be a cosine of a sum.

Write the formula reversed:	$\cos A \cos B - \sin A \sin B = \cos(A + B)$
Compare the given expression.	$\cos x \cos 3x - \sin x \sin 3x$
Identify A and B.	$A = x$ and $B = 3x$
Substitute $A = x$ and $B = 3x$ into the sum identity.	$\cos x \cos 3x - \sin x \sin 3x = \cos(x + 3x)$
Simplify.	$\cos x \cos 3x - \sin x \sin 3x = \cos(4x)$

■ **YOUR TURN** Write the following expression as a single cosine:
$\cos 4x \cos 7x + \sin 4x \sin 7x$.

TECHNOLOGY TIP

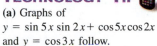

(a) Graphs of
$y = \sin 5x \sin 2x + \cos 5x \cos 2x$
and $y = \cos 3x$ follow.

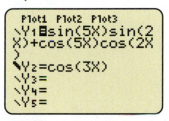

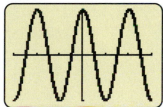

(b) Graphs of
$y = \cos x \cos 3x - \sin x \sin 3x$
and $y = \cos 4x$ follow.

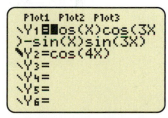

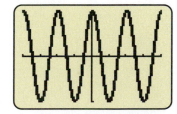

■ **Answer: a.** $\dfrac{\sqrt{6} - \sqrt{2}}{4}$ **b.** $\dfrac{\sqrt{6} - \sqrt{2}}{4}$ ■ **Answer:** $\cos(3x)$

Cofunction Identities

In Section 1.3, we discussed cofunction relationships for acute angles. Recall that the trigonometric function of an angle is equal to its cofunction value of the complementary angle. Now we use the sum and difference identities for cosine to develop the cofunction identities for any angle.

WORDS	**MATH**
Write the difference identity for cosine.	$\cos(A - B) = \cos A \cos B + \sin A \sin B$
Let $A = \dfrac{\pi}{2}$ and $B = \theta$.	$\cos\left(\dfrac{\pi}{2} - \theta\right) = \cos\left(\dfrac{\pi}{2}\right)\cos\theta + \sin\left(\dfrac{\pi}{2}\right)\sin\theta$
Evaluate known values for sine and cosine.	$\cos\left(\dfrac{\pi}{2} - \theta\right) = 0 \cdot \cos\theta + 1 \cdot \sin\theta$
Simplify.	$\cos\left(\dfrac{\pi}{2} - \theta\right) = \sin\theta$

Similarly, to determine the other cofunction identity:

Write the difference identity for cosine.	$\cos(A - B) = \cos A \cos B + \sin A \sin B$
Let $A = \dfrac{\pi}{2}$ and $B = \dfrac{\pi}{2} - \theta$.	

$$\cos\left[\dfrac{\pi}{2} - \left(\dfrac{\pi}{2} - \theta\right)\right] = \cos\left(\dfrac{\pi}{2}\right)\cos\left(\dfrac{\pi}{2} - \theta\right) + \sin\left(\dfrac{\pi}{2}\right)\sin\left(\dfrac{\pi}{2} - \theta\right)$$

Evaluate known values for sine and cosine.	$\cos(\theta) = 0 \cdot \cos\left(\dfrac{\pi}{2} - \theta\right) + 1 \cdot \sin\left(\dfrac{\pi}{2} - \theta\right)$
Simplify.	$\cos\theta = \sin\left(\dfrac{\pi}{2} - \theta\right)$

COFUNCTION IDENTITIES

$$\cos\left(\dfrac{\pi}{2} - \theta\right) = \sin\theta$$

$$\sin\left(\dfrac{\pi}{2} - \theta\right) = \cos\theta$$

Sum and Difference Identities for Sine

We can now use the cofunction identities with the sum and difference identities for cosine to develop the sum and difference identities for sine.

WORDS	**MATH**
Start with the cofunction identity.	$\sin\theta = \cos\left(\dfrac{\pi}{2} - \theta\right)$
Let $\theta = A + B$.	$\sin(A + B) = \cos\left(\dfrac{\pi}{2} - (A + B)\right)$

Regroup terms in the cosine expression.	$\sin(A + B) = \cos\left(\left(\dfrac{\pi}{2} - A\right) - B\right)$
Use the difference identity for cosine.	$\sin(A + B) = \cos\left(\dfrac{\pi}{2} - A\right)\cos B + \sin\left(\dfrac{\pi}{2} - A\right)\sin B$
Use the cofunction identities.	$\sin(A + B) = \underbrace{\cos\left(\dfrac{\pi}{2} - A\right)}_{\sin A}\cos B + \underbrace{\sin\left(\dfrac{\pi}{2} - A\right)}_{\cos A}\sin B$
Simplify.	$\sin(A + B) = \sin A \cos B + \cos A \sin B$

Now we can derive the difference identity for sine using the sum identity for sine and the properties of even and odd functions.

WORDS   **MATH**

Replace B with $-B$
in the sum identity. $\sin(A + (-B)) = \sin A \cos(-B) + \cos A \sin(-B)$

Simplify using even and
odd identities. $\sin(A - B) = \sin A \cos B - \cos A \sin B$

SUM AND DIFFERENCE IDENTITIES FOR SINE

Sum: $\sin(A + B) = \sin A \cos B + \cos A \sin B$
Difference: $\sin(A - B) = \sin A \cos B - \cos A \sin B$

EXAMPLE 3 Finding Exact Values for Sine

Use the sum or difference identities for sine to evaluate the sine expressions exactly.

a. $\sin\left(\dfrac{5\pi}{12}\right)$ **b.** $\sin(75°)$

Solution (a):

Write $\dfrac{5\pi}{12}$ as a sum.	$\sin\left(\dfrac{5\pi}{12}\right) = \sin\left(\dfrac{2\pi}{12} + \dfrac{3\pi}{12}\right)$
Simplify.	$\sin\left(\dfrac{5\pi}{12}\right) = \sin\left(\dfrac{\pi}{6} + \dfrac{\pi}{4}\right)$
Write the sum identity for sine.	$\sin(A + B) = \sin A \cos B + \cos A \sin B$
Substitute $A = \dfrac{\pi}{6}$ and $B = \dfrac{\pi}{4}$.	$\sin\left(\dfrac{5\pi}{12}\right) = \sin\left(\dfrac{\pi}{6}\right)\cos\left(\dfrac{\pi}{4}\right) + \cos\left(\dfrac{\pi}{6}\right)\sin\left(\dfrac{\pi}{4}\right)$

Evaluate the expressions on the right exactly.

$$\sin\left(\frac{5\pi}{12}\right) = \left(\frac{1}{2}\right)\left(\frac{\sqrt{2}}{2}\right) + \left(\frac{\sqrt{3}}{2}\right)\left(\frac{\sqrt{2}}{2}\right)$$

Simplify.

$$\sin\left(\frac{5\pi}{12}\right) = \frac{\sqrt{2} + \sqrt{6}}{4}$$

Solution (b):

Write 75° as a sum.

$$\sin(75°) = \sin(45° + 30°)$$

Write the sum identity for sine.

$$\sin(A + B) = \sin A \cos B + \cos A \sin B$$

Substitute $A = 45°$ and $B = 30°$.

$$\sin(75°) = \sin 45°\cos 30° + \cos 45°\sin 30°$$

Evaluate the expressions on the right exactly.

$$\sin(75°) = \left(\frac{\sqrt{2}}{2}\right)\left(\frac{\sqrt{3}}{2}\right) + \left(\frac{\sqrt{2}}{2}\right)\left(\frac{1}{2}\right)$$

Simplify.

$$\sin(75°) = \frac{\sqrt{6} + \sqrt{2}}{4}$$

YOUR TURN Use the sum or difference identities for sine to evaluate the sine expressions exactly.

a. $\sin\left(\dfrac{7\pi}{12}\right)$ **b.** $\sin(15°)$

We see in Example 3 that the sum and difference identities allow us to calculate exact values for trigonometric functions of angles that are multiples of 15° or $\frac{\pi}{12}$.

TECHNOLOGY TIP

Graphs of $y = 3\sin x \cos 3x + 3\cos x \sin 3x$ and $y = 3\sin 4x$ follow.

```
Plot1 Plot2 Plot3
\Y1=3sin(X)cos(3
X)+3cos(X)sin(3X
)
\Y2=3sin(4X)
\Y3=
\Y4=
\Y5=
```

EXAMPLE 4 Writing a Sum or Difference as a Single Sine

Graph $y = 3\sin x \cos 3x + 3\cos x \sin 3x$.

Solution:

Use the sum identity for sine to write the expression as a single sine.

STEP 1 Factor the common 3. $y = 3(\sin x \cos 3x + \cos x \sin 3x)$

STEP 2 Write the sum identity for sine. $\sin A \cos B + \cos A \sin B = \sin(A + B)$

STEP 3 Identify A and B. $A = x$ and $B = 3x$

STEP 4 Substitute $A = x$ and $B = 3x$ into the sum identity. $y = 3(\underbrace{\sin x \cos 3x + \cos x \sin 3x}_{\sin 4x})$

Answer: a. $\dfrac{\sqrt{6} + \sqrt{2}}{4}$ **b.** $\dfrac{\sqrt{6} - \sqrt{2}}{4}$

STEP 5 Graph $y = 3 \sin 4x$.

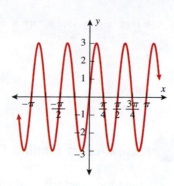

Sum and Difference Identities for Tangent

We now develop the sum and difference identities for tangent.

WORDS	**MATH**
Start with the quotient identity.	$\tan x = \dfrac{\sin x}{\cos x}$
Let $x = A + B$.	$\tan(A + B) = \dfrac{\sin(A + B)}{\cos(A + B)}$
Use the sum identities for sine and cosine.	$\tan(A + B) = \dfrac{\sin A \cos B + \cos A \sin B}{\cos A \cos B - \sin A \sin B}$

To be able to write the right-hand side in terms of tangents, we multiply the numerator and denominator by $\dfrac{1}{\cos A \cos B}$.

$$\tan(A + B) = \dfrac{\dfrac{\sin A \cos B + \cos A \sin B}{\cos A \cos B}}{\dfrac{\cos A \cos B - \sin A \sin B}{\cos A \cos B}} = \dfrac{\dfrac{\sin A \cancel{\cos B}}{\cos A \cancel{\cos B}} + \dfrac{\cancel{\cos A} \sin B}{\cancel{\cos A} \cos B}}{\dfrac{\cos A \cancel{\cos B}}{\cos A \cancel{\cos B}} - \dfrac{\sin A \sin B}{\cos A \cos B}}$$

Simplify.

$$\tan(A + B) = \dfrac{\left(\dfrac{\sin A}{\cos A}\right) + \left(\dfrac{\sin B}{\cos B}\right)}{1 - \left(\dfrac{\sin A}{\cos A}\right)\left(\dfrac{\sin B}{\cos B}\right)}$$

Write the expressions inside the parentheses in terms of tangent.

$$\tan(A + B) = \dfrac{\tan A + \tan B}{1 - \tan A \tan B}$$

Replace B with $-B$.

$$\tan(A - B) = \dfrac{\tan A + \tan(-B)}{1 - \tan A \tan(-B)}$$

Since tangent is an odd function, $\tan(-B) = -\tan B$.

$$\tan(A - B) = \dfrac{\tan A - \tan B}{1 + \tan A \tan B}$$

SUM AND DIFFERENCE IDENTITIES FOR TANGENT

Sum:
$$\tan(A + B) = \frac{\tan A + \tan B}{1 - \tan A \tan B}$$

Difference:
$$\tan(A - B) = \frac{\tan A - \tan B}{1 + \tan A \tan B}$$

EXAMPLE 5 Finding Exact Values for Tangent

Find the exact value of $\tan(\alpha + \beta)$ if $\sin\alpha = -\dfrac{1}{3}$ and $\cos\beta = -\dfrac{1}{4}$, given the terminal side of α lies in QIII and the terminal side of β lies in QII.

Solution:

STEP 1 Write the sum identity for tangent. $\tan(\alpha + \beta) = \dfrac{\tan\alpha + \tan\beta}{1 - \tan\alpha\tan\beta}$

STEP 2 Find $\tan\alpha$.

The terminal side of α lies in QIII.

$$\sin\alpha = \frac{y}{r} = -\frac{1}{3}$$

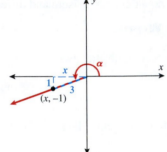

Solve for x.
$$x^2 + 1^2 = 3^2$$
$$x = \pm\sqrt{8}$$

Take the negative sign
since we are in QIII.
$$x = -2\sqrt{2}$$

Find $\tan\alpha$.
$$\tan\alpha = \frac{y}{x} = \frac{-1}{-2\sqrt{2}} = \frac{1}{2\sqrt{2}} \cdot \frac{\sqrt{2}}{\sqrt{2}} = \frac{\sqrt{2}}{4}$$

STEP 3 Find $\tan\beta$.

The terminal side of β lies in QII.

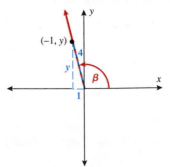

Solve for y.
$$(-1)^2 + y^2 = 4^2$$
$$y = \pm\sqrt{15}$$

Take the positive sign since
we are in QII.
$$y = \sqrt{15}$$

Find $\tan\beta$.
$$\tan\beta = \frac{y}{x} = \frac{\sqrt{15}}{-1} = -\sqrt{15}$$

STEP 4 Substitute $\tan \alpha = \dfrac{\sqrt{2}}{4}$ and $\tan \beta = -\sqrt{15}$ into the sum identity for tangent.

$$\tan(\alpha + \beta) = \frac{\dfrac{\sqrt{2}}{4} - \sqrt{15}}{1 - \left(\dfrac{\sqrt{2}}{4}\right)(-\sqrt{15})}$$

Multiply numerator and denominator by 4.

$$\tan(\alpha + \beta) = \frac{4\left(\dfrac{\sqrt{2}}{4} - \sqrt{15}\right)}{4\left(1 + \dfrac{\sqrt{30}}{4}\right)} = \frac{\sqrt{2} - 4\sqrt{15}}{4 + \sqrt{30}}$$

The expression $\tan(\alpha + \beta) = \dfrac{\sqrt{2} - 4\sqrt{15}}{4 + \sqrt{30}}$ can be simplified further if we rationalize the denominator.

It is important to note in Example 5 that triangles have been superimposed in the Cartesian plane. The coordinate pair (x, y) can have positive or negative values, and the radius, r, is always positive. When triangles are superimposed, one must understand that triangles have positive side lengths.

SECTION 5.2 SUMMARY

In this section we derived the sum and difference identities for cosine using the distance formula. We then used these identities to derive the cofunction identities. The cofunction identities and sum and difference identities for cosine were used to derive the sum and difference identities for sine. The sine and cosine sum and difference identities were combined to determine the tangent identities. The sum and difference identities were used to evaluate trigonometric expressions exactly.

$$\cos(A + B) = \cos A \cos B - \sin A \sin B \qquad \cos(A - B) = \cos A \cos B + \sin A \sin B$$

$$\sin(A + B) = \sin A \cos B + \cos A \sin B \qquad \sin(A - B) = \sin A \cos B - \cos A \sin B$$

$$\tan(A + B) = \frac{\tan A + \tan B}{1 - \tan A \tan B} \qquad \tan(A - B) = \frac{\tan A - \tan B}{1 + \tan A \tan B}$$

SECTION 5.2 EXERCISES

 **SKILLS**

In Exercises 1–14, find exact values for each trigonometric expression.

1. $\sin\left(\dfrac{\pi}{12}\right)$　　**2.** $\cos\left(\dfrac{\pi}{12}\right)$　　**3.** $\cos\left(-\dfrac{5\pi}{12}\right)$　　**4.** $\sin\left(-\dfrac{5\pi}{12}\right)$

5. $\tan\left(-\dfrac{\pi}{12}\right)$　　**6.** $\tan\left(\dfrac{13\pi}{12}\right)$　　**7.** $\sin(105°)$　　**8.** $\cos(195°)$

9. $\tan(-105°)$ **10.** $\tan(165°)$ **11.** $\cot\left(\dfrac{\pi}{12}\right)$ **12.** $\cot\left(-\dfrac{5\pi}{12}\right)$

13. $\sec\left(-\dfrac{11\pi}{12}\right)$ **14.** $\sec\left(-\dfrac{13\pi}{12}\right)$

In Exercises 15–24, write each expression as a single trigonometric function.

15. $\sin 2x \sin 3x + \cos 2x \cos 3x$

16. $\sin x \sin 2x - \cos x \cos 2x$

17. $\sin x \cos 2x - \cos x \sin 2x$

18. $\sin 2x \cos 3x + \cos 2x \sin 3x$

19. $(\sin A - \sin B)^2 + (\cos A - \cos B)^2 - 2$

20. $(\sin A + \sin B)^2 + (\cos A + \cos B)^2 - 2$

21. $2 - (\sin A + \cos B)^2 - (\cos A + \sin B)^2$

22. $2 - (\sin A - \cos B)^2 - (\cos A + \sin B)^2$

23. $\dfrac{\tan 49° - \tan 23°}{1 + \tan 49° \tan 23°}$

24. $\dfrac{\tan 49° + \tan 23°}{1 - \tan 49° \tan 23°}$

In Exercises 25–30, find the exact value of the indicated expression using the given information and identities.

25. Find the exact value of $\cos(\alpha + \beta)$ if $\cos\alpha = -\dfrac{1}{3}$ and $\cos\beta = -\dfrac{1}{4}$, if the terminal side of α lies in QIII and the terminal side of β lies in QII.

26. Find the exact value of $\cos(\alpha - \beta)$ if $\cos\alpha = \dfrac{1}{3}$ and $\cos\beta = -\dfrac{1}{4}$, if the terminal side of α lies in QIV and the terminal side of β lies in QII.

27. Find the exact value of $\sin(\alpha - \beta)$ if $\sin\alpha = -\dfrac{3}{5}$ and $\sin\beta = \dfrac{1}{5}$, if the terminal side of α lies in QIII and the terminal side of β lies in QI.

28. Find the exact value of $\sin(\alpha + \beta)$ if $\sin\alpha = -\dfrac{3}{5}$ and $\sin\beta = \dfrac{1}{5}$, if the terminal side of α lies in QIII and the terminal side of β lies in QII.

29. Find the exact value of $\tan(\alpha + \beta)$ if $\sin\alpha = -\dfrac{3}{5}$ and $\cos\beta = -\dfrac{1}{4}$, if the terminal side of α lies in QIII and the terminal side of β lies in QII.

30. Find the exact value of $\tan(\alpha - \beta)$ if $\sin\alpha = -\dfrac{3}{5}$ and $\cos\beta = -\dfrac{1}{4}$, if the terminal side of α lies in QIII and the terminal side of β lies in QII.

In Exercises 31–40, determine if each equation is conditional or an identity.

31. $\sin(A + B) + \sin(A - B) = 2 \sin A \cos B$

32. $\cos(A + B) + \cos(A - B) = 2 \cos A \cos B$

33. $\sin\left(x - \dfrac{\pi}{2}\right) = \cos\left(x + \dfrac{\pi}{2}\right)$

34. $\sin\left(x + \dfrac{\pi}{2}\right) = \cos\left(x + \dfrac{\pi}{2}\right)$

35. $\sin 2x = 2 \sin x \cos x$

36. $\cos 2x = \cos^2 x - \sin^2 x$

37. $\sin(A + B) = \sin A + \sin B$

38. $\cos(A + B) = \cos A + \cos B$

39. $\tan(\pi + B) = \tan B$

40. $\tan(A - \pi) = \tan A$

In Exercises 41–46, graph the following functions by first rewriting as a sine or cosine of a difference or sum.

41. $y = \cos\dfrac{\pi}{3} \sin x + \cos x \sin \dfrac{\pi}{3}$

42. $y = \cos\dfrac{\pi}{3} \sin x - \cos x \sin\dfrac{\pi}{3}$

43. $y = \sin x \sin\dfrac{\pi}{4} + \cos x \cos\dfrac{\pi}{4}$

44. $y = \sin x \sin\dfrac{\pi}{4} - \cos x \cos\dfrac{\pi}{4}$

45. $y = -\sin x \cos 3x - \cos x \sin 3x$

46. $y = \sin x \sin 3x + \cos x \cos 3x$

■ **APPLICATIONS**

The difference quotient, $f(x) = \dfrac{f(x + h) - f(x)}{h}$, is used to approximate the rate of change of the function f and will be used frequently in calculus.

47. Difference Quotient. Show that the difference quotient for $f(x) = \sin x$ is $\cos x \left(\dfrac{\sin h}{h} \right) - \sin x \left(\dfrac{1 - \cos h}{h} \right)$.

48. Difference Quotient. Show that the difference quotient for $f(x) = \cos x$ is $-\sin x \left(\dfrac{\sin h}{h} \right) - \cos x \left(\dfrac{1 - \cos h}{h} \right)$.

For Exercises 49 and 50, refer to the following:

A nonvertical line makes an angle with the x-axis. In the figure we see that the line L_1 makes an angle θ_1 with the x-axis. Similarly, the line L_2 makes an angle θ_2 with the x-axis. In Exercises 49 and 50 use the following:

$\tan \theta_1 = $ slope of $L_1 = m_1$

$\tan \theta_2 = $ slope of $L_2 = m_2$

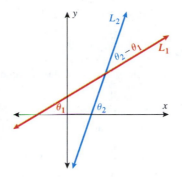

49. Angle Between Two Lines. Show that
$$\tan(\theta_2 - \theta_1) = \frac{m_2 - m_1}{1 + m_1 m_2}.$$

50. Angle Between Two Lines. Show that
$$\tan(\theta_1 - \theta_2) = \frac{m_1 - m_2}{1 + m_1 m_2}.$$

For Exercises 51 and 52, refer to the following:

An electric field, E, of a wave with constant amplitude A propagating a distance z is given by

$$E = A\cos(kz - ct)$$

where k is the propagation wave number, which is related to the wavelength, λ, by $k = \dfrac{2\pi}{\lambda}$. $c = 3.0 \times 10^8$ m/s is the speed of light in a vacuum and t is time in seconds.

51. Electromagnetic Wave Propagation. Use the cosine difference identity to express the electric field in terms of both sine and cosine functions. When the quotient of the propagation distance, t, and wavelength, λ, are equal to an integer, what do you notice?

52. Electromagnetic Wave Propagation. Use the cosine difference identity to express the electric field in terms of both sine and cosine functions. When $t = 0$ what do you notice?

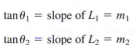

■ **CATCH THE MISTAKE**

In Exercises 53 and 54, explain the mistake that is made.

53. Find the exact value of $\tan\left(\dfrac{5\pi}{12}\right)$.

Solution:

Write $\dfrac{5\pi}{12}$ as a sum. $\tan\left(\dfrac{5\pi}{12}\right) = \tan\left(\dfrac{\pi}{4} + \dfrac{\pi}{6}\right)$

Distribute. $\tan\left(\dfrac{5\pi}{12}\right) = \tan\left(\dfrac{\pi}{4}\right) + \tan\left(\dfrac{\pi}{6}\right)$

Evaluate the tangent functions on the right. $\tan\left(\dfrac{5\pi}{12}\right) = 1 + \dfrac{\sqrt{3}}{3}$

This is incorrect. What mistake was made?

54. Find the exact value of $\tan\left(-\dfrac{7\pi}{6}\right)$.

Solution:

Tangent is an even function. $\tan\left(-\dfrac{7\pi}{6}\right) = \tan\left(\dfrac{7\pi}{6}\right)$

Write $\dfrac{7\pi}{6}$ as a sum. $= \tan\left(\pi + \dfrac{\pi}{6}\right)$

Use the tangent sum identity,
$$\tan(A + B) = \frac{\tan A + \tan B}{1 - \tan A \tan B}.$$ $= \dfrac{\tan \pi + \tan\left(\dfrac{\pi}{6}\right)}{1 - \tan \pi \tan\left(\dfrac{\pi}{6}\right)}$

Evaluate the tangent functions on the right. $= \dfrac{0 + \dfrac{1}{\sqrt{3}}}{1 - 0}$

Simplify. $= \dfrac{\sqrt{3}}{3}$

This is incorrect. What mistake was made?

■ CHALLENGE

55. T or F: $\cos(15°) = \cos(45°) - \cos(30°)$

56. T or F: $\sin\left(\dfrac{\pi}{2}\right) = \sin\left(\dfrac{\pi}{3}\right) + \sin\left(\dfrac{\pi}{6}\right)$

57. Verify that $\sin(A + B + C) = \sin A \cos B \cos C + \cos A \sin B \cos C + \cos A \cos B \sin C - \sin A \sin B \sin C$

58. Verify that $\cos(A + B + C) = \cos A \cos B \cos C - \sin A \sin B \cos C - \sin A \cos B \sin C - \cos A \sin B \sin C$

59. Although in general the statement $\sin(A - B) = \sin A - \sin B$ is not true, it is true for some values. Find infinitely many values of A and B that make this statement true.

60. Although in general the statement $\sin(A + B) = \sin A + \sin B$ is not true, it is true for some values. Find infinitely many values of A and B that make this statement true.

■ TECHNOLOGY

61. In Exercise 47, you showed that the difference quotient for $f(x) = \sin x$ is $\cos x\left(\dfrac{\sin h}{h}\right) - \sin x\left(\dfrac{1 - \cos h}{h}\right)$.

Plot $Y_1 = \cos x\left(\dfrac{\sin h}{h}\right) - \sin x\left(\dfrac{1 - \cos h}{h}\right)$ for

a. $h = 1$ **b.** $h = 0.1$ **c.** $h = 0.01$

What function does the difference quotient for $f(x) = \sin x$ resemble when h approaches zero?

62. Show that the difference quotient for $f(x) = \cos x$ is $-\sin x\left(\dfrac{\sin h}{h}\right) - \cos x\left(\dfrac{1 - \cos h}{h}\right)$. Plot

$Y_1 = -\sin x\left(\dfrac{\sin h}{h}\right) - \cos x\left(\dfrac{1 - \cos h}{h}\right)$ for

a. $h = 1$ **b.** $h = 0.1$ **c.** $h = 0.01$

What function does the difference quotient for $f(x) = \cos x$ resemble when h approaches zero?

SECTION 5.3 Double-Angle Identities

Skills Objectives

■ Use the double-angle identities to find exact values of trigonometric functions.
■ Use the double-angle identities to simplify verifying identities.

Conceptual Objective

■ Derive the double-angle identities from the sum identities.

Throughout this text much attention has been given to distinguishing between evaluating trigonometric functions exactly (for the special angles) or approximating with a calculator. In previous chapters we could only evaluate trigonometric functions exactly for reference angles of $30°$, $45°$, and $60°$ or $\dfrac{\pi}{6}$, $\dfrac{\pi}{4}$, and $\dfrac{\pi}{3}$. Now we can use *double-angle identities* to evaluate other angles that are even multiples of the special angles or to varify other trigonometric identities.

Derivation of Double-Angle Identities

To derive the *double-angle identities*, we let $A = B$ in the sum identities:

WORDS	MATH
Write the sine of a sum identity.	$\sin(A + B) = \sin A \cos B + \cos A \sin B$
Let $B = A$.	$\sin(A + A) = \sin A \cos A + \cos A \sin A$
Simplify.	$\sin(2A) = 2 \sin A \cos A$
Write the cosine of a sum identity.	$\cos(A + B) = \cos A \cos B - \sin A \sin B$
Let $B = A$.	$\cos(A + A) = \cos A \cos A - \sin A \sin A$
Simplify.	$\cos(2A) = \cos^2 A - \sin^2 A$

The *double-angle identity* for cosine can be written two other ways if we use the Pythagorean identity:

Write the cosine of a double angle.	$\cos(2A) = \cos^2 A - \sin^2 A$
Use the Pythagorean identity for cosine.	$\cos(2A) = \underbrace{\cos^2 A}_{1-\sin^2 A} - \sin^2 A$
Simplify.	$\cos(2A) = 1 - 2\sin^2 A$
Write the cosine of a double angle.	$\cos(2A) = \cos^2 A - \sin^2 A$
Use the Pythagorean identity for sine.	$\cos(2A) = \cos^2 A - \underbrace{\sin^2 A}_{1-\cos^2 A}$
Simplify.	$\cos(2A) = 2\cos^2 A - 1$
Write the tangent of a sum identity.	$\tan(A + B) = \dfrac{\tan A + \tan B}{1 - \tan A \tan B}$
Let $B = A$.	$\tan(A + A) = \dfrac{\tan A + \tan A}{1 - \tan A \tan A}$
Simplify.	$\tan(2A) = \dfrac{2\tan A}{1 - \tan^2 A}$

DOUBLE-ANGLE IDENTITIES

Sine	Cosine	Tangent
$\sin(2A) = 2 \sin A \cos A$	$\cos(2A) = \cos^2 A - \sin^2 A$	$\tan(2A) = \dfrac{2\tan A}{1 - \tan^2 A}$
	$\cos(2A) = 1 - 2\sin^2 A$	
	$\cos(2A) = 2\cos^2 A - 1$	

Applying Double-Angle Identities

EXAMPLE 1 Finding Exact Values Using Double-Angle Identities

If $\cos x = \dfrac{2}{3}$, find $\sin 2x$ given $\sin x < 0$.

Solution:

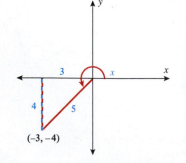

STEP 1 Find $\sin x$.

Use the Pythagorean identity.	$\sin^2 x + \cos^2 x = 1$
Substitute $\cos x = \dfrac{2}{3}$.	$\sin^2 x + \left(\dfrac{2}{3}\right)^2 = 1$
Solve for $\sin x$, which is negative.	$\sin x = -\sqrt{1 - \dfrac{4}{9}}$
Simplify.	$\sin x = -\dfrac{\sqrt{5}}{3}$

STEP 2 Find $\sin 2x$.

Use the double-angle formula for sine.	$\sin 2x = 2\sin x \cos x$
Substitute $\sin x = -\dfrac{\sqrt{5}}{3}$ and $\cos x = \dfrac{2}{3}$.	$\sin 2x = 2\left(-\dfrac{\sqrt{5}}{3}\right)\left(\dfrac{2}{3}\right)$
Simplify.	$\sin 2x = -\dfrac{4\sqrt{5}}{9}$

■ **YOUR TURN** If $\cos x = -\dfrac{1}{3}$, find $\sin 2x$ given $\sin x < 0$.

EXAMPLE 2 Finding Exact Values Using Double-Angle Identities

If $\sin x = -\dfrac{4}{5}$ and $\cos x < 0$, find $\sin 2x$, $\cos 2x$, and $\tan 2x$.

Solution:

STEP 1 Solve for $\cos x$.

Use the Pythagorean identity.	$\sin^2 x + \cos^2 x = 1$
Substitute $\sin x = -\dfrac{4}{5}$.	$\left(-\dfrac{4}{5}\right)^2 + \cos^2 x = 1$
Simplify.	$\cos^2 x = \dfrac{9}{25}$
Solve for $\cos x$, which is negative.	$\cos x = -\sqrt{\dfrac{9}{25}} = -\dfrac{3}{5}$

■ **Answer:** $\sin 2x = \dfrac{4\sqrt{2}}{9}$

STEP 2 Find $\sin 2x$.

Use the double-angle identity for sine. $\sin(2x) = 2\sin x \cos x$

Substitute $\sin x = -\dfrac{4}{5}$ and $\cos x = -\dfrac{3}{5}$. $\sin(2x) = 2\left(-\dfrac{4}{5}\right)\left(-\dfrac{3}{5}\right)$

Simplify. $\sin(2x) = \dfrac{24}{25}$

STEP 3 Find $\cos 2x$.

Use the double-angle identity for cosine. $\cos(2x) = \cos^2 x - \sin^2 x$

Substitute $\sin x = -\dfrac{4}{5}$ and $\cos x = -\dfrac{3}{5}$. $\cos 2x = \left(-\dfrac{3}{5}\right)^2 - \left(-\dfrac{4}{5}\right)^2$

Simplify. $\cos(2x) = -\dfrac{7}{25}$

STEP 4 Find $\tan 2x$.

Use the quotient identity. $\tan\theta = \dfrac{\sin\theta}{\cos\theta}$

Let $\theta = 2x$. $\tan 2x = \dfrac{\sin 2x}{\cos 2x}$

Substitute $\sin(2x) = \dfrac{24}{25}$ and $\cos(2x) = -\dfrac{7}{25}$. $\tan 2x = \dfrac{\dfrac{24}{25}}{-\dfrac{7}{25}}$

Simplify. $\tan 2x = -\dfrac{24}{7}$

Note: $\tan 2x$ could also have been found, first by finding $\tan x = \dfrac{\sin x}{\cos x}$ and then using the value for $\tan x$ in the double-angle identity, $\tan(2A) = \dfrac{2\tan A}{1 - \tan^2 A}$.

 ■ **YOUR TURN**   If $\cos x = \dfrac{3}{5}$ and $\sin x < 0$, find $\sin 2x$, $\cos 2x$, and $\tan 2x$.

EXAMPLE 3 Verifying Trigonometric Identities Using Double-Angle Identities

Verify the identity: $(\sin x - \cos x)^2 = 1 - \sin 2x$.

Solution:

Start with the left side of the equation. $(\sin x - \cos x)^2$

Square the binomial. $\sin^2 x - 2\sin x \cos x + \cos^2 x$

■ **Answer:** $\sin 2x = -\dfrac{24}{25}$, $\cos 2x = -\dfrac{7}{25}$, $\tan 2x = \dfrac{24}{7}$

TECHNOLOGY TIP

Graphs of
$y = (\sin x - \cos x)^2$ and
$y = 1 - \sin 2x$ follow.

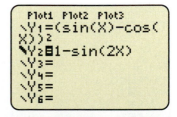

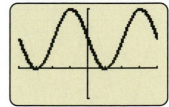

Group terms.

$$\sin^2 x + \cos^2 x - 2\sin x \cos x$$

Apply the Pythagorean identity.

$$\underbrace{\sin^2 x + \cos^2 x}_{1} - 2\sin x \cos x$$

Apply the sine double-angle identity.

$$1 - \underbrace{2\sin x \cos x}_{\sin 2x}$$

Simplify.

$$1 - \sin 2x$$

EXAMPLE 4 Verifying Multiple-Angle Identities

Verify the identity: $\cos 3x = [1 - 4\sin^2 x]\cos x$.

Solution:

Write $\cos 3x$ using the cosine sum identity.

$$\cos(A + B) = \cos A \cos B - \sin A \sin B$$

Let $A = 2x$ and $B = x$.

$$\cos(2x + x) = \cos 2x \cos x - \sin 2x \sin x$$

Apply the double-angle identities.

$$\cos(3x) = \underbrace{\cos 2x}_{1 - 2\sin^2 x} \cos x - \underbrace{\sin 2x}_{2\sin x \cos x} \sin x$$

Simplify.

$$\cos 3x = \cos x - 2\sin^2 x \cos x - 2\sin^2 x \cos x$$

$$\cos 3x = \cos x - 4\sin^2 x \cos x$$

Factor out the common cosine term.

$$\cos 3x = [1 - 4\sin^2 x]\cos x$$

TECHNOLOGY TIP

Graphs of $y = \dfrac{\cot x - \tan x}{\cot x + \tan x}$ and $y = \cos 2x$ follow.

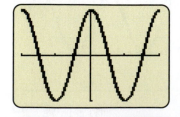

EXAMPLE 5 Simplifying Trigonometric Expressions Using Double-Angle Identities

Graph $y = \dfrac{\cot x - \tan x}{\cot x + \tan x}$.

Solution:

STEP 1 Simplify $y = \dfrac{\cot x - \tan x}{\cot x + \tan x}$.

Write cotangent and tangent in terms of sine and cosine.

$$y = \frac{\dfrac{\cos x}{\sin x} - \dfrac{\sin x}{\cos x}}{\dfrac{\cos x}{\sin x} + \dfrac{\sin x}{\cos x}}$$

Multiply numerator and denominator by $\sin x \cos x$.

$$y = \left(\frac{\dfrac{\cos x}{\sin x} - \dfrac{\sin x}{\cos x}}{\dfrac{\cos x}{\sin x} + \dfrac{\sin x}{\cos x}}\right)\left(\frac{\sin x \cos x}{\sin x \cos x}\right)$$

Simplify.

$$y = \frac{\cos^2 x - \sin^2 x}{\cos^2 x + \sin^2 x}$$

Use the double-angle and Pythagorean identities.

$$y = \frac{\overbrace{\cos^2 x - \sin^2 x}^{\cos 2x}}{\underbrace{\cos^2 x + \sin^2 x}_{1}}$$

$$y = \cos 2x$$

STEP 2 Graph $y = \cos 2x$.

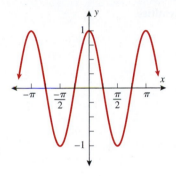

SECTION 5.3 SUMMARY

In this section we used the sum identities to derive the double-angle identities. We then used the double-angle identities to find exact values of trigonometric functions, verify other trigonometric identities, and simplify trigonometric expressions.

$$\sin(2A) = 2\sin A \cos A \qquad \cos(2A) = \cos^2 A - \sin^2 A \qquad \tan(2A) = \frac{2\tan A}{1 - \tan^2 A}$$

$$= 1 - 2\sin^2 A$$

$$= 2\cos^2 A - 1$$

There is no need to memorize the other forms of the cosine double-angle identity, since they can be derived from the first using the Pythagorean identity.

SECTION 5.3 EXERCISES

■ SKILLS

In Exercises 1–12, use the double-angle identities to answer the questions.

1. If $\sin x = \dfrac{1}{\sqrt{5}}$ and $\cos x < 0$, find $\sin 2x$.

2. If $\sin x = \dfrac{1}{\sqrt{5}}$ and $\cos x < 0$, find $\cos 2x$.

3. If $\cos x = \dfrac{5}{13}$ and $\sin x < 0$, find $\tan 2x$.

4. If $\cos x = -\dfrac{5}{13}$ and $\sin x < 0$, find $\tan 2x$.

5. If $\tan x = \dfrac{12}{5}$ and $\pi < x < \dfrac{3\pi}{2}$, find $\sin 2x$.

6. If $\tan x = \dfrac{12}{5}$ and $\pi < x < \dfrac{3\pi}{2}$, find $\cos 2x$.

7. If $\sec x = \sqrt{5}$ and $\sin x > 0$, find $\tan 2x$.

8. If $\sec x = \sqrt{3}$ and $\sin x < 0$, find $\tan 2x$.

9. If $\csc x = -2\sqrt{5}$ and $\cos x < 0$, find $\sin 2x$.

10. If $\csc x = -\sqrt{13}$ and $\cos x > 0$, find $\sin 2x$.

11. If $\cos x = -\dfrac{12}{13}$ and $\csc x < 0$, find $\cot 2x$.

12. If $\sin x = \dfrac{12}{13}$ and $\cot x < 0$, find $\csc 2x$.

In Exercises 13–18, simplify each expression. Evaluate exactly, if possible.

13. $\dfrac{2\tan 15°}{1 - \tan^2 15°}$

14. $\dfrac{2\tan \dfrac{\pi}{8}}{1 - \tan^2 \dfrac{\pi}{8}}$

15. $\sin \dfrac{\pi}{8} \cos \dfrac{\pi}{8}$

16. $\sin 15° \cos 15°$

17. $\cos^2 2x - \sin^2 2x$

18. $\cos^2(x + 2) - \sin^2(x + 2)$

In Exercises 19–34, verify the identities.

19. $\csc 2A = \dfrac{1}{2}\csc A \sec A$

20. $\cot 2A = \dfrac{1}{2}\left[\cot A - \tan A\right]$

21. $(\sin x - \cos x)(\cos x + \sin x) = -\cos 2x$

22. $(\sin x + \cos x)^2 = 1 + \sin 2x$

23. $\cos^2 x = \dfrac{1 + \cos 2x}{2}$

24. $\sin^2 x = \dfrac{1 - \cos 2x}{2}$

25. $\cos^4 x - \sin^4 x = \cos 2x$

26. $\cos^4 x + \sin^4 x = 1 - \dfrac{1}{2}\sin^2 2x$

27. $8\sin^2 x \cos^2 x = 1 - \cos 4x$

28. $(\cos 2x - \sin 2x)(\sin 2x + \cos 2x) = \cos 4x$

29. $-\dfrac{1}{2}\sec^2 x = -2\sin^2 x \csc^2 2x$

30. $4\csc 4x = \dfrac{\sec x \csc x}{\cos 2x}$

31. $\sin 3x = \sin x(4\cos^2 x - 1)$

32. $\tan 3x = \dfrac{\tan x(3 - \tan^2 x)}{(1 - 3\tan^2 x)}$

33. $\dfrac{1}{2}\sin 4x = 2\sin x \cos x - 4\sin^3 x \cos x$

34. $\cos 4x = (\cos 2x - \sin 2x)(\cos 2x + \sin 2x)$

In Exercises 35–38, graph the functions.

35. $y = \dfrac{\sin 2x}{1 - \cos 2x}$

36. $y = \dfrac{2\tan x}{2 - \sec^2 x}$

37. $y = \dfrac{\cot x + \tan x}{\cot x - \tan x}$

38. $y = \dfrac{1}{2}\tan x \cot x \sec x \csc x$

■ APPLICATIONS

For Exercises 39 and 40, refer to the following:

An ore crusher wheel consists of a heavy disk spinning on its axle. Its normal crushing force in pounds between the wheel and the inclined track is determined by

$$F = W\sin\theta + \dfrac{1}{2}\psi^2\left[\dfrac{C}{R}(1 - \cos 2\theta) + \dfrac{A}{l}\sin 2\theta\right]$$

where W is the weight of the wheel, θ is the angle of the axis, C and A are moments of inertia, R is the radius of the wheel, l is the distance from the wheel to the pin where the axle is attached, and ψ is the speed in rpm that the wheel is spinning.

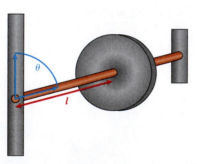

The optimum crushing force occurs when the angle is between 45° and 90°.

39. **Ore-Crusher Wheel.** Find F if the angle is 60°, W is 500 lb, and ψ is 200 rpm, $\dfrac{C}{R} = 750$, and $\dfrac{A}{l} = 3.75$.

40. **Ore-Crusher Wheel.** Find F if the angle is 75°, W is 500 lb, and ψ is 200 rpm, $\dfrac{C}{R} = 750$, and $\dfrac{A}{l} = 3.75$.

■ CATCH THE MISTAKE

In Exercises 41 and 42, explain the mistake that is made.

41. If $\cos x = \dfrac{1}{3}$, find $\sin 2x$ given $\sin x < 0$.

Solution:

Write the double-angle identity for sine. $\qquad \sin 2x = 2\sin x \cos x$

Solve for $\sin x$ using the Pythagorean identity. $\qquad \sin^2 x + \left(\dfrac{1}{3}\right)^2 = 1$

$$\sin x = \dfrac{2\sqrt{2}}{3}$$

Substitute $\cos x = \dfrac{1}{3}$

and $\sin x = \dfrac{2\sqrt{2}}{3}$. $\qquad \sin 2x = 2\left(\dfrac{2\sqrt{2}}{3}\right)\left(\dfrac{1}{3}\right)$

Simplify. $\qquad \sin 2x = \dfrac{4\sqrt{2}}{9}$

This is incorrect. What mistake was made?

42. If $\sin x = \dfrac{1}{3}$, find $\tan 2x$ given $\cos x < 0$.

Solution:

Use the quotient identity. $\qquad \tan 2x = \dfrac{\sin 2x}{\cos x}$

Use the double-angle formula for sine. $\qquad \tan 2x = \dfrac{2\sin x \cos x}{\cos x}$

Cancel the common cosine terms. $\qquad \tan 2x = 2\sin x$

Substitute $\sin x = \dfrac{1}{3}$. $\qquad \tan 2x = \dfrac{2}{3}$

This is incorrect. What mistake was made?

■ CHALLENGE

43. T or F: $\sin 2A + \sin 2A = \sin 4A$

44. T or F: $\cos 4A - \cos 2A = \cos 2A$

45. T or F: If $\tan x > 0$, then $\tan 2x > 0$.

46. T or F: If $\sin x > 0$, then $\sin 2x > 0$.

47. Express $\tan 4x$ in terms of functions of $\tan x$.

48. Express $\tan(-4x)$ in terms of functions of $\tan x$.

49. Is the identity $2\csc 2x = \dfrac{1 + \tan^2 x}{\tan x}$ true for $x = \dfrac{\pi}{2}$?

Explain.

50. Is the identity $\tan 2x = \dfrac{2\tan x}{1 - \tan^2 x}$ true for $x = \dfrac{\pi}{4}$?

Explain.

■ TECHNOLOGY

One cannot *prove* that an equation is an identity using technology, but can use technology as a first step to see whether or not the equation *seems* to be an identity.

51. Using a graphing calculator, plot

$Y_1 = (2x) - \dfrac{(2x)^3}{3!} + \dfrac{(2x)^5}{5!}$ and $Y_2 = \sin 2x$ for x

ranging $[-1, 1]$. Is Y_1 a good approximation to Y_2?

52. Using a graphing calculator, plot

$Y_1 = 1 - \dfrac{(2x)^2}{2!} + \dfrac{(2x)^4}{4!}$ and $Y_2 = \cos 2x$ for x ranging

$[-1, 1]$. Is Y_1 a good approximation to Y_2?

53. Using a graphing calculator, determine if

$\dfrac{\tan 4x - \tan 3x}{\tan x} = \dfrac{\csc 2x}{1 - \sec 2x}$ by plotting each side of the

equation and seeing if the graphs coincide.

54. Using a graphing calculator, determine if

$\csc 2x \sec 2x (\cos 2x - \sin 2x) = \dfrac{1 - 2\sin^2 x - 2\sin x \cos x}{2\sin x \cos x (\cos^2 x - \sin^2 x)}$

by plotting each side of the equation and seeing if the graphs coincide.

SECTION 5.4 Half-Angle Identities

Skills Objectives
- Use the half-angle identities to find exact values of trigonometric functions.
- Use the half-angle identities to verify other trigonometric identities.

Conceptual Objective
- Derive the half-angle identities from the double-angle identities.

We now use the *double-angle identities* from Section 5.3 to develop the *half-angle identities*. Like the double-angle identities, the half-angle *identities* will allow us to find exact values of trigonometric functions and to verify other trigonometric identities.

Half-Angle Identities

The *half-angle identities* come directly from the double-angle identities. We first start by rewriting the second and third forms of the cosine double-angle identity.

WORDS	MATH
Write the second form of the cosine double-angle identity.	$\cos(2A) = 1 - 2\sin^2 A$
Isolate the $2\sin^2 A$ term.	$2\sin^2 A = 1 - \cos 2A$
Divide by 2.	$\sin^2 A = \dfrac{1 - \cos 2A}{2}$
Write the third form of the cosine double-angle identity.	$\cos(2A) = 2\cos^2 A - 1$
Isolate the $2\cos^2 A$ term.	
Divide by 2.	$\cos^2 A = \dfrac{1 + \cos 2A}{2}$
Taking the quotient of these leads us to another identity.	$\tan^2 A = \dfrac{\sin^2 A}{\cos^2 A} = \dfrac{\frac{1 - \cos(2A)}{2}}{\frac{1 + \cos(2A)}{2}} = \dfrac{1 - \cos(2A)}{1 + \cos(2A)}$

These three identities are used in calculus as power reduction formulas, which reduces the power of the trigonometric function from 2 to 1.

$$\sin^2 A = \frac{1 - \cos 2A}{2} \qquad \cos^2 A = \frac{1 + \cos 2A}{2} \qquad \tan^2 A = \frac{1 - \cos 2A}{1 + \cos 2A}$$

We can now use these forms of the double-angle identities to derive the *half-angle identities*.

WORDS	MATH
Start with the double-angle formula involving both sine and cosine:	$\sin^2 x = \dfrac{1 - \cos 2x}{2}$
Solve for $\sin x$.	$\sin x = \pm\sqrt{\dfrac{1 - \cos 2x}{2}}$

Let $x = \dfrac{A}{2}$.

$$\sin\frac{A}{2} = \pm\sqrt{\frac{1 - \cos 2\left(\dfrac{A}{2}\right)}{2}}$$

Simplify.

$$\sin\frac{A}{2} = \pm\sqrt{\frac{1 - \cos A}{2}}$$

Start with the double-angle formula involving only cosine:

$$\cos^2 x = \frac{1 + \cos 2x}{2}$$

Solve for $\cos x$.

$$\cos x = \pm\sqrt{\frac{1 + \cos 2x}{2}}$$

Let $x = \dfrac{A}{2}$.

$$\cos\frac{A}{2} = \pm\sqrt{\frac{1 + \cos 2\left(\dfrac{A}{2}\right)}{2}}$$

Simplify.

$$\cos\frac{A}{2} = \pm\sqrt{\frac{1 + \cos A}{2}}$$

Start with the quotient identity.

$$\tan\frac{A}{2} = \frac{\sin\dfrac{A}{2}}{\cos\dfrac{A}{2}}$$

Substitute half-angle identities for sine and cosine.

$$\tan\frac{A}{2} = \frac{\pm\sqrt{\dfrac{1 - \cos A}{2}}}{\pm\sqrt{\dfrac{1 + \cos A}{2}}}$$

Simplify.

$$\tan\frac{A}{2} = \pm\sqrt{\frac{1 - \cos A}{1 + \cos A}}$$

Note: $\tan\dfrac{A}{2}$ can also be found by starting with the identity, $\tan^2 x = \dfrac{1 - \cos 2x}{1 + \cos 2x}$, solving for $\tan x$, and letting $x = \dfrac{A}{2}$. Tangent also has two other similar forms (see Exercises 49 and 50).

HALF-ANGLE IDENTITIES

Sine	Cosine	Tangent
$\sin\dfrac{A}{2} = \pm\sqrt{\dfrac{1 - \cos A}{2}}$	$\cos\dfrac{A}{2} = \pm\sqrt{\dfrac{1 + \cos A}{2}}$	$\tan\dfrac{A}{2} = \pm\sqrt{\dfrac{1 - \cos A}{1 + \cos A}}$
		$\tan\dfrac{A}{2} = \dfrac{\sin A}{1 + \cos A}$
		$\tan\dfrac{A}{2} = \dfrac{1 - \cos A}{\sin A}$

It is important to note that these identities hold for any real number, A, or any angle with either degree measure or radian measure, A, as long as both sides of the equation are defined. The $+$ or $-$ sign is determined by the sign of the trigonometric function in the quadrant that contains $\dfrac{A}{2}$.

EXAMPLE 1 Finding Exact Values Using Half-Angle Identities

Use a half-angle identity to find $\cos 15°$.

Solution:

Write $\cos 15°$ in terms of a half angle. $\qquad \cos 15° = \cos\left(\dfrac{30°}{2}\right)$

Write the half-angle identity for cosine. $\qquad \cos\dfrac{A}{2} = \pm\sqrt{\dfrac{1 + \cos A}{2}}$

Substitute $A = 30°$. $\qquad \cos\dfrac{30°}{2} = \pm\sqrt{\dfrac{1 + \cos 30°}{2}}$

Simplify. $\qquad \cos 15° = \pm\sqrt{\dfrac{1 + \dfrac{\sqrt{3}}{2}}{2}}$

$15°$ is in QI where cosine is positive. $\qquad \cos 15° = \sqrt{\dfrac{2 + \sqrt{3}}{4}}$

■ **YOUR TURN** Use a half-angle identity to find $\sin 22.5°$.

EXAMPLE 2 Finding Exact Values Using Half-Angle Identities

Use a half-angle identity to find $\tan\dfrac{11\pi}{12}$.

Solution:

Write $\tan\dfrac{11\pi}{12}$ in terms of a half angle. $\qquad \tan\dfrac{11\pi}{12} = \tan\left(\dfrac{\dfrac{11\pi}{6}}{2}\right)$

Write the half-angle identity for tangent.* $\qquad \tan\dfrac{A}{2} = \dfrac{1 - \cos A}{\sin A}$

*This form of the tangent half-angle identity was selected because of mathematical simplicity. If we had selected either of the other forms, we would have obtained the expressions that had square roots of square roots, or radicals in the denominator (requiring rationalization).

■ **Answer:** $\sin 22.5° = \dfrac{\sqrt{2 - \sqrt{2}}}{4}$

Substitute $A = \dfrac{11\pi}{6}$.

$$\tan\dfrac{\dfrac{11\pi}{6}}{2} = \dfrac{1 - \cos\dfrac{11\pi}{6}}{\sin\dfrac{11\pi}{6}}$$

Simplify.

$$\tan\dfrac{11\pi}{12} = \dfrac{1 - \dfrac{\sqrt{3}}{2}}{-\dfrac{1}{2}}$$

$$\boxed{\tan\dfrac{11\pi}{12} = \sqrt{3} - 2}$$

$\dfrac{11\pi}{12}$ is in QII where tangent is negative. Notice that if we approximate $\tan\dfrac{11\pi}{12}$ with a calculator, we find that $\tan\dfrac{11\pi}{12} \approx -0.2679$.

■ **YOUR TURN**   Use a half-angle identity to find $\tan\dfrac{\pi}{8}$.

EXAMPLE 3 Finding Exact Values Using Half-Angle Identities

If $\cos x = \dfrac{3}{5}$ and $\dfrac{3\pi}{2} < x < 2\pi$, find $\sin\dfrac{x}{2}$, $\cos\dfrac{x}{2}$, and $\tan\dfrac{x}{2}$.

Solution:

STEP 1   Determine what quadrant $\dfrac{x}{2}$ lies in, since $\dfrac{3\pi}{2} < x < 2\pi$.

Divide by 2.

$$\dfrac{3\pi}{4} < \dfrac{x}{2} < \pi$$

$\dfrac{x}{2}$ lies in QII; therefore sine is positive, and cosine and tangent are negative.

STEP 2   Use the half-angle identity for sine.

$$\sin\dfrac{x}{2} = \pm\sqrt{\dfrac{1 - \cos x}{2}}$$

Substitute $\cos x = \dfrac{3}{5}$.

$$\sin\dfrac{x}{2} = \pm\sqrt{\dfrac{1 - \dfrac{3}{5}}{2}}$$

Simplify.

$$\sin\dfrac{x}{2} = \pm\sqrt{\dfrac{1}{5}} = \pm\dfrac{\sqrt{5}}{5}$$

Since $\dfrac{x}{2}$ lies in QII, sine is positive.

$$\boxed{\sin\dfrac{x}{2} = \dfrac{\sqrt{5}}{5}}$$

■ **Answer:** $\dfrac{\sqrt{2}}{2 + \sqrt{2}}$ or $\sqrt{2} - 1$

STEP 3   Use the half-angle identity for cosine. $\cos\dfrac{x}{2} = \pm\sqrt{\dfrac{1+\cos x}{2}}$

Substitute $\cos x = \dfrac{3}{5}$. $\cos\dfrac{x}{2} = \pm\sqrt{\dfrac{1+\dfrac{3}{5}}{2}}$

Simplify. $\cos\dfrac{x}{2} = \pm\sqrt{\dfrac{4}{5}} = \pm\dfrac{2\sqrt{5}}{5}$

Since $\dfrac{x}{2}$ lies in QII, cosine is negative. $\boxed{\cos\dfrac{x}{2} = -\dfrac{2\sqrt{5}}{5}}$

STEP 4   Use the quotient identity for tangent. $\tan\dfrac{x}{2} = \dfrac{\sin\dfrac{x}{2}}{\cos\dfrac{x}{2}}$

Substitute $\sin\dfrac{x}{2} = \dfrac{\sqrt{5}}{5}$ and $\cos\dfrac{x}{2} = -\dfrac{2\sqrt{5}}{5}$. $\tan\dfrac{x}{2} = \dfrac{\dfrac{\sqrt{5}}{5}}{-\dfrac{2\sqrt{5}}{5}}$

Simplify. $\boxed{\tan\dfrac{x}{2} = -\dfrac{1}{2}}$

■ **YOUR TURN**   If $\cos x = -\dfrac{3}{5}$ and $\pi < x < \dfrac{3\pi}{2}$, find $\sin\dfrac{x}{2}$, $\cos\dfrac{x}{2}$, and $\tan\dfrac{x}{2}$.

TECHNOLOGY TIP

Graphs of $y = \cos^2\dfrac{x}{2}$ and

$y = \dfrac{\tan x + \sin x}{2\tan x}$ follow.

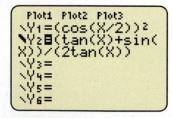

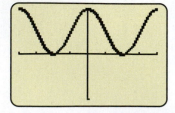

EXAMPLE 4 Verifying Identities Using Half-Angle Identities

Verify the identity: $\cos^2\dfrac{x}{2} = \dfrac{\tan x + \sin x}{2\tan x}$.

Solution:

Write the cosine half-angle identity. $\cos\dfrac{x}{2} = \pm\sqrt{\dfrac{1+\cos x}{2}}$

Square both sides of the equation. $\cos^2\dfrac{x}{2} = \dfrac{1+\cos x}{2}$

Multiply numerator and denominator on the right side by $\tan x$. $\cos^2\dfrac{x}{2} = \left(\dfrac{1+\cos x}{2}\right)\left(\dfrac{\tan x}{\tan x}\right)$

Simplify. $\cos^2\dfrac{x}{2} = \dfrac{\tan x + \cos x\tan x}{2\tan x}$

■ **Answer:** $\sin\dfrac{x}{2} = \dfrac{2\sqrt{5}}{5}$, $\cos\dfrac{x}{2} = -\dfrac{\sqrt{5}}{5}$, $\tan\dfrac{x}{2} = -2$

Note that $\cos x \tan x = \sin x$.

$$\cos^2 \frac{x}{2} = \frac{\tan x + \sin x}{2\tan x}$$

An alternative solution is to start with the right-hand side.

Solution (alternate):

Start with the right-hand side.

$$\frac{\tan x + \sin x}{2\tan x}$$

Write as the sum of two expressions.

$$= \frac{\tan x}{2\tan x} + \frac{\sin x}{2\tan x}$$

Simplify.

$$= \frac{1}{2} + \frac{1}{2}\cdot\frac{\sin x}{\tan x}$$

Write $\tan x = \frac{\sin x}{\cos x}$.

$$= \frac{1}{2} + \frac{1}{2}\cdot\frac{\sin x}{\frac{\sin x}{\cos x}}$$

$$= \frac{1}{2}(1 + \cos x)$$

$$= \cos^2\frac{x}{2}$$

EXAMPLE 5 Verifying Trigonometric Identities Using Half-Angle Identities

Verify the identity: $\tan x = \csc 2x - \cot 2x$.

Solution:

Write a half-angle formula for tangent.

$$\tan\frac{A}{2} = \frac{1 - \cos A}{\sin A}$$

Write the right side as a difference of two quotients.

$$\tan\frac{A}{2} = \frac{1}{\sin A} - \frac{\cos A}{\sin A}$$

Use reciprocal and quotient identities on the right.

$$\tan\frac{A}{2} = \csc A - \cot A$$

Let $A = 2x$.

$$\tan x = \csc 2x - \cot 2x$$

EXAMPLE 6 Simplifying Trigonometric Expressions Using Half-Angle Identities

Graph $y = \frac{\sin 2\pi x}{1 + \cos 2\pi x}$.

Solution:

STEP 1 Simplify the trigonometric expression using half-angle identities.

Write the tangent half-angle identity.

$$\tan\frac{A}{2} = \frac{\sin A}{1 + \cos A}$$

TECHNOLOGY TIP

Graphs of $y = \dfrac{\sin 2\pi x}{1 + \cos 2\pi x}$ and $y = \tan \pi x$ follow.

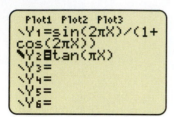

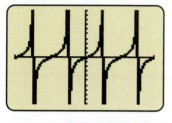

Let $A = 2\pi x$.

STEP 2 Graph $y = \tan \pi x$.

$$\tan \pi x = \frac{\sin 2\pi x}{1 + \cos 2\pi x}$$

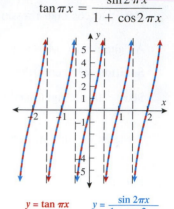

$y = \tan \pi x$ $y = \dfrac{\sin 2\pi x}{1 + \cos 2\pi x}$

■ **YOUR TURN** Graph $y = \dfrac{1 - \cos \pi x}{\sin \pi x}$.

After discussing double-angle identities in Section 5.3 and half-angle identities in this section, let us now address when to use each. For example, to evaluate $\sin 120°$ or $\sin 22.5°$ exactly, which would you use? Notice that $120°$ is twice (double) $60°$, which is a special angle. Therefore, we use the double-angle identity to find $\sin 120°$. Notice that $22.5°$ is half of $45°$, which is also a special angle. Therefore, we use the half-angle identity to evaluate $\sin 22.5°$.

SECTION 5.4   SUMMARY

In this section we used the double-angle identities to derive the half-angle identities. We then used the half-angle identities to find exact values of trigonometric functions, verify other trigonometric identities, and simplify trigonometric expressions.

$$\sin \frac{A}{2} = \pm\sqrt{\frac{1 - \cos A}{2}} \qquad \cos \frac{A}{2} = \pm\sqrt{\frac{1 + \cos A}{2}} \qquad \tan \frac{A}{2} = \pm\sqrt{\frac{1 - \cos A}{1 + \cos A}}$$

The sign, $+$ or $-$, is determined by first determining what quadrant contains $\dfrac{A}{2}$ and then determining the sign of the indicated trigonometric function in that quadrant.

There is no need to memorize the other forms of the tangent half-angle identity, since they can be derived from the first using the Pythagorean identity.

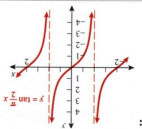

$y = \tan \dfrac{\pi}{2} x$

■ **Answer:**

SECTION 5.4 EXERCISES

 ■ **SKILLS**

In Exercises 1–12, use the half-angle identities to find the exact values of the trigonometric expressions.

1. $\sin 15°$

2. $\cos 22.5°$

3. $\cos\left(\dfrac{11\pi}{12}\right)$

4. $\sin\left(\dfrac{\pi}{8}\right)$

5. $\cos 75°$

6. $\sin 75°$

7. $\tan 67.5°$

8. $\tan 202.5°$

9. $\sec\left(-\dfrac{9\pi}{8}\right)$

10. $\csc\left(\dfrac{9\pi}{8}\right)$

11. $\cot\left(\dfrac{13\pi}{8}\right)$

12. $\cot\left(\dfrac{7\pi}{8}\right)$

In Exercises 13–22, use the half-angle identities to answer the questions.

13. If $\cos x = \dfrac{5}{13}$ and $\sin x < 0$, find $\sin\dfrac{x}{2}$.

14. If $\cos x = -\dfrac{5}{13}$ and $\sin x < 0$, find $\cos\dfrac{x}{2}$.

15. If $\tan x = \dfrac{12}{5}$ and $\pi < x < \dfrac{3\pi}{2}$, find $\sin\dfrac{x}{2}$.

16. If $\tan x = \dfrac{12}{5}$ and $\pi < x < \dfrac{3\pi}{2}$, find $\cos\dfrac{x}{2}$.

17. If $\sec x = \sqrt{5}$ and $\sin x > 0$, find $\tan\dfrac{x}{2}$.

18. If $\sec x = \sqrt{3}$ and $\sin x < 0$, find $\tan\dfrac{x}{2}$.

19. If $\csc x = 3$ and $\cos x < 0$, find $\sin\dfrac{x}{2}$.

20. If $\csc x = -3$ and $\cos x > 0$, find $\cos\dfrac{x}{2}$.

21. If $\cos x = -\dfrac{1}{4}$ and $\csc x < 0$, find $\cot\dfrac{x}{2}$.

22. If $\cos x = \dfrac{1}{4}$ and $\cot x < 0$, find $\csc\dfrac{x}{2}$.

In Exercises 23–26, simplify each expression using half-angle identities. Do not evaluate.

23. $\sqrt{\dfrac{1 + \cos\dfrac{5\pi}{6}}{2}}$

24. $\sqrt{\dfrac{1 - \cos\dfrac{\pi}{4}}{2}}$

25. $\dfrac{\sin 150°}{1 + \cos 150°}$

26. $\dfrac{1 - \cos 150°}{\sin 150°}$

In Exercises 27–36, verify the following identities.

27. $\sin^2\dfrac{x}{2} + \cos^2\dfrac{x}{2} = 1$

28. $\cos^2\dfrac{x}{2} - \sin^2\dfrac{x}{2} = \cos x$

29. $\tan^2\dfrac{x}{2} = \dfrac{1 - \cos x}{1 + \cos x}$

30. $\tan^2\dfrac{x}{2} = (\csc x - \cot x)^2$

31. $\tan\dfrac{A}{2} + \cot\dfrac{A}{2} = 2\csc A$

32. $\cot\dfrac{A}{2} - \tan\dfrac{A}{2} = 2\cot A$

33. $\csc^2\dfrac{A}{2} = \dfrac{2(1 + \cos A)}{\sin^2 A}$

34. $\sec^2\dfrac{A}{2} = \dfrac{2(1 - \cos A)}{\sin^2 A}$

35. $\csc\dfrac{A}{2} = \pm|\csc A|\sqrt{2 + 2\cos A}$

36. $\sec\dfrac{A}{2} = \pm|\csc A|\sqrt{2 - 2\cos A}$

In Exercises 37–40, graph the functions.

37. $y = 4\cos^2\dfrac{x}{2}$

38. $y = -6\sin^2\dfrac{x}{2}$

39. $y = \dfrac{1 - \tan^2\dfrac{x}{2}}{1 + \tan^2\dfrac{x}{2}}$

40. $y = 1 - \left(\sin\dfrac{x}{2} + \cos\dfrac{x}{2}\right)^2$

■ APPLICATIONS

41. Area of an Isosceles Triangle. A formula for the area of an isosceles triangle in terms of the common side and the subtended angle can be determined using half-angle identities. Consider the triangle below, where the vertex angle measures θ, the sides measure a, the height is h, and half the base is b. (In an isosceles triangle, the perpendicular dropped from the vertex angle divides the triangle into two congruent triangles.) The two triangles formed are right triangles.

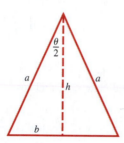

In the right triangles, $\sin\dfrac{\theta}{2} = \dfrac{b}{a}$ and $\cos\dfrac{\theta}{2} = \dfrac{h}{a}$. Multiply each side of each equation by a to get $b = a\sin\dfrac{\theta}{2}$, $h = a\cos\dfrac{\theta}{2}$.

The area of the entire isosceles triangle is $A = \dfrac{1}{2}(2b)h = bh$.

Substitute the values for b and h into the area formula. Show that the area is equivalent to $\dfrac{a^2}{2}\sin\theta$.

42. Area of an Isosceles Triangle. Use the results from Exercise 41 to find the area of an isosceles triangle whose sides measure 7 inches and whose base angles each measure 75°.

■ CATCH THE MISTAKE

In Exercises 43 and 44, explain the mistake that is made.

43. If $\cos x = \dfrac{1}{3}$, find $\sin\dfrac{x}{2}$ given $\pi < x < \dfrac{3\pi}{2}$.

Solution:

Write the half-angle identity for sine.
$$\sin\dfrac{x}{2} = \pm\sqrt{\dfrac{1 + \cos x}{2}}$$

Substitute $\cos x = \dfrac{1}{3}$.
$$\sin\dfrac{x}{2} = \pm\sqrt{\dfrac{1 + \dfrac{1}{3}}{2}}$$

Simplify.
$$\sin\dfrac{x}{2} = \pm\sqrt{\dfrac{2}{3}}$$

Sine is negative.
$$\sin\dfrac{x}{2} = -\sqrt{\dfrac{2}{3}}$$

This is incorrect. What mistake was made?

44. If $\cos x = \dfrac{1}{3}$ find $\tan^2\dfrac{x}{2}$ given $\cos x < 0$.

Solution:

Use the quotient identity.
$$\tan^2\left(\dfrac{1}{2}x\right) = \dfrac{\sin^2\left(\dfrac{1}{2}x\right)}{\cos^2 x}$$

Use the half-angle identity for sine.
$$\tan^2\dfrac{1}{2}x = \dfrac{\dfrac{1 - \cos x}{2}}{\cos^2 x}$$

Simplify.
$$\tan^2\dfrac{1}{2}x = \dfrac{1}{2}\left(\dfrac{1}{\cos^2 x} - \dfrac{\cos x}{\cos^2 x}\right)$$
$$\tan^2\dfrac{1}{2}x = \dfrac{1}{2}\left(\dfrac{1}{\cos^2 x} - \dfrac{1}{\cos x}\right)$$

Substitute $\cos x = \dfrac{1}{3}$.
$$\tan^2\dfrac{1}{2}x = \dfrac{1}{2}\left(\dfrac{1}{\dfrac{1}{9}} - \dfrac{1}{\dfrac{1}{3}}\right)$$
$$\tan^2\dfrac{1}{2}x = 3$$

This is incorrect. What mistake was made?

■ CHALLENGE

45. T or F: $\sin\dfrac{A}{2} + \sin\dfrac{A}{2} = \sin A$

46. T or F: $\cos\dfrac{A}{2} + \cos\dfrac{A}{2} = \cos A$

47. T or F: If $\tan x > 0$, then $\tan\dfrac{x}{2} > 0$.

48. T or F: If $\sin x > 0$, then $\sin\dfrac{x}{2} > 0$.

49. Given $\tan\dfrac{A}{2} = \pm\sqrt{\dfrac{1 - \cos A}{1 + \cos A}}$, verify $\tan\dfrac{A}{2} = \dfrac{\sin A}{1 + \cos A}$.
Substitute $A = \pi$ into the identity and explain your results.

50. Given $\tan\dfrac{A}{2} = \pm\sqrt{\dfrac{1 - \cos A}{1 + \cos A}}$, verify $\tan\dfrac{A}{2} = \dfrac{1 - \cos A}{\sin A}$.
Substitute $A = \pi$ into the identity and explain your results.

■ TECHNOLOGY

One cannot *prove* that an equation is an identity using technology, but can use it as a first step to see whether or not the equation *seems* to be an identity.

51. Using a graphing calculator, plot $Y_1 = \left(\dfrac{x}{2}\right) - \dfrac{\left(\dfrac{x}{2}\right)^3}{3!} + \dfrac{\left(\dfrac{x}{2}\right)^5}{5!}$ and $Y_2 = \sin\dfrac{x}{2}$ for x ranging $[-1, 1]$. Is Y_1 a good approximation to Y_2?

52. Using a graphing calculator, plot $Y_1 = 1 - \dfrac{\left(\dfrac{x}{2}\right)^2}{2!} + \dfrac{\left(\dfrac{x}{2}\right)^4}{4!}$ and $Y_2 = \cos\dfrac{x}{2}$ for x ranging $[-1, 1]$. Is Y_1 a good approximation to Y_2?

53. Using a graphing calculator, determine if $\csc^2\dfrac{x}{2} + \sec^2\dfrac{x}{2} = 4\csc^2 x$ is an identity by plotting each side of the equation and seeing if the graphs coincide.

54. Using a graphing calculator, determine if $\tan^2\dfrac{x}{2} + \cot^2\dfrac{x}{2} = -2\cot^2 x \sec x$ is an identity by plotting each side of the equation and seeing if the graphs coincide.

SECTION 5.5 Product-to-Sum and Sum-to-Product Identities

Skills Objectives

- Express products of trigonometric functions as sums of trigonometric functions.
- Express sums of trigonometric functions as products of trigonometric functions.

Conceptual Objectives

- Use the sum and difference identities to derive product-to-sum identities.
- Use the product-to-sum identities to derive the sum-to-product identities.

Often in calculus it will be helpful to write products of trigonometric functions as sums of other trigonometric functions and vice versa. In this section, we discuss the *product-to-sum identities,* which convert products to sums, and *sum-to-product identities,* which convert sums to products.

Product-to-Sum Identities

The *product-to-sum identities* are derived from the sum and difference identities.

WORDS	MATH
Write the identity for the cosine of a sum.	$\cos A \cos B - \sin A \sin B = \cos (A + B)$
Write the identity for the cosine of a difference.	$\cos A \cos B + \sin A \sin B = \cos (A - B)$
Add the two identities.	$2\cos A \cos B = \cos (A + B) + \cos (A - B)$

Divide both sides by 2.

$$\cos A \cos B = \frac{1}{2}[\cos (A + B) + \cos (A - B)]$$

Subtract the sum identity from the difference identity.

$$\cos A \cos B + \sin A \sin B = \cos (A - B)$$
$$-\cos A \cos B + \sin A \sin B = -\cos (A + B)$$
$$2\sin A \sin B = \cos (A - B) - \cos (A + B)$$

Divide both sides by 2.

$$\sin A \sin B = \frac{1}{2}[\cos (A - B) - \cos (A + B)]$$

Write the identity for the sine of a sum.	$\sin A \cos B + \cos A \sin B = \sin (A + B)$
Write the identity for the sine of a difference.	$\sin A \cos B - \cos A \sin B = \sin (A - B)$
Add the two identities.	$2\sin A \cos B = \sin (A + B) + \sin (A - B)$

Divide both sides by 2.

$$\sin A \cos B = \frac{1}{2}[\sin (A + B) + \sin (A - B)]$$

PRODUCT-TO-SUM IDENTITIES

1. $\cos A \cos B = \dfrac{1}{2}[\cos (A + B) + \cos (A - B)]$

2. $\sin A \sin B = \dfrac{1}{2}[\cos (A - B) - \cos (A + B)]$

3. $\sin A \cos B = \dfrac{1}{2}[\sin (A + B) + \sin (A - B)]$

EXAMPLE 1 Illustrating a Product-to-Sum Identity for Specific Values

Show that product-to-sum identity (3) is true when $A = 30°$ and $B = 90°$.

Solution:

Write product-to-sum identity (3).

$$\sin A \cos B = \frac{1}{2}[\sin (A + B) + \sin (A - B)]$$

Let $A = 30°$ and $B = 90°$.

$$\sin 30° \cos 90° = \frac{1}{2}[\sin (30° + 90°) + \sin (30° - 90°)]$$

Simplify.

$$\sin 30° \cos 90° = \frac{1}{2}[\sin (120°) + \sin (-60°)]$$

Evaluate the trigonometric functions.

$$\frac{1}{2} \cdot 0 = \frac{1}{2}\left[\frac{\sqrt{3}}{2} - \frac{\sqrt{3}}{2}\right]$$

Simplify.

$$0 = 0$$

EXAMPLE 2 Convert a Product to a Sum

Convert the product $\cos(4x)\cos(3x)$ to a sum.

Solution:

Write the product-to-sum identity (1).

$$\cos A \cos B = \frac{1}{2}\left[\cos(A+B)+\cos(A-B)\right]$$

Let $A = 4x$ and $B = 3x$.

$$\cos 4x \cos 3x = \frac{1}{2}\left[\cos(4x+3x)+\cos(4x-3x)\right]$$

Simplify.

$$\cos 4x \cos 3x = \frac{1}{2}\left[\cos(7x)+\cos(x)\right]$$

TECHNOLOGY TIP

Graphs of $y = \cos 4x \cos 3x$
and $y = \frac{1}{2}\left[\cos(7x)+\cos(x)\right]$
follow.

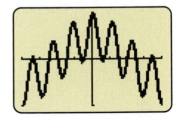

■ **YOUR TURN** Convert the product $\cos(2x)\cos(5x)$ to a sum.

EXAMPLE 3 Converting Products to Sums

Express $\sin(2x)\sin(3x)$ in terms of cosine.

COMMON MISTAKE

A common mistake that is often made is calling this expression the square of sine.

✓ CORRECT

Write product-to-sum identity (2).

$\sin A \sin B$

$= \frac{1}{2}\left[\cos(A-B)-\cos(A+B)\right]$

Let $A = 2x$ and $B = 3x$.

$\sin 2x \sin 3x$

$= \frac{1}{2}\left[\cos(2x-3x)-\cos(2x-3x)\right]$

Simplify.

$\sin 2x \sin 3x$

$= \frac{1}{2}\left[\cos(-x)-\cos(5x)\right]$

Cosine is an even function.

$\sin 2x \sin 3x$

$= \frac{1}{2}\left[\cos(x)-\cos(5x)\right]$

✗ INCORRECT

Multiply the two sine functions.

$\sin(2x)\sin(3x) = \sin^2 6x^2$ **ERROR**

CAUTION

1. $\sin A \sin B \neq \sin^2 AB$
2. The argument must be the same in order to use the identity, $\sin A \sin A = (\sin A)^2 = \sin^2 A$.

■ **Answer:** $\frac{1}{2}\left[\cos(7x)+\cos(3x)\right]$

■ **YOUR TURN** Express $\sin(x)\sin(2x)$ in terms of cosine.

Sum-to-Product Identities

The *sum-to-product identities* can be obtained from the product-to-sum identities.

WORDS	MATH
Write the product of the sine and cosine identity.	$\frac{1}{2}[\sin(x+y) + \sin(x-y)] = \sin x \cos y$
Let $x + y = A$ and $x - y = B$, then $x = \dfrac{A+B}{2}$ and $y = \dfrac{A-B}{2}$.	
Substitute these values into the identity.	$\frac{1}{2}[\sin A + \sin B] = \sin\left(\dfrac{A+B}{2}\right)\cos\left(\dfrac{A-B}{2}\right)$
Multiply by 2.	$\sin A + \sin B = 2\sin\left(\dfrac{A+B}{2}\right)\cos\left(\dfrac{A-B}{2}\right)$

The other three *sum-to-product* identities can be found similarly and are summarized in the following box.

SUM-TO-PRODUCT IDENTITIES

4. $\sin A + \sin B = 2\sin\left(\dfrac{A+B}{2}\right)\cos\left(\dfrac{A-B}{2}\right)$

5. $\sin A - \sin B = 2\sin\left(\dfrac{A-B}{2}\right)\cos\left(\dfrac{A+B}{2}\right)$

6. $\cos A + \cos B = 2\cos\left(\dfrac{A+B}{2}\right)\cos\left(\dfrac{A-B}{2}\right)$

7. $\cos A - \cos B = -2\sin\left(\dfrac{A+B}{2}\right)\sin\left(\dfrac{A-B}{2}\right)$

EXAMPLE 4 **Illustrating a Sum-to-Product Identity for Specific Values**

Show that sum-to-product identity (4) is true when $A = 30°$ and $B = 90°$.

Solution:

Write product-to-sum identity (7).
$$\cos A - \cos B = -2\sin\left(\dfrac{A+B}{2}\right)\sin\left(\dfrac{A-B}{2}\right)$$

■ **Answer:** $\frac{1}{2}[\cos x - \cos 3x]$

Let $A = 30°$ and $B = 90°$.

$$\cos 30° - \cos 90° = -2\sin\left(\frac{30° + 90°}{2}\right)\sin\left(\frac{30° - 90°}{2}\right)$$

Simplify.

$$\cos 30° - \cos 90° = -2\sin 60° \sin(-30°)$$

Sine is an odd function.

$$\cos 30° - \cos 90° = 2\sin 60° \sin 30°$$

Evaluate the trigonometric functions.

$$\frac{\sqrt{3}}{2} - 0 = 2\frac{\sqrt{3}}{2}\frac{1}{2}$$

Simplify.

$$\frac{\sqrt{3}}{2} = \frac{\sqrt{3}}{2}$$

EXAMPLE 5 Convert a Sum to a Product

Convert the sum $-9[\sin 2x - \sin 10x]$, to a product.

Solution:

The expression inside the brackets is in the form of identity (5).

$$\sin A - \sin B = 2\sin\left(\frac{A - B}{2}\right)\cos\left(\frac{A + B}{2}\right)$$

Let $A = 2x$ and $B = 10x$.

$$\sin 2x - \sin 10x = 2\sin\left(\frac{2x - 10x}{2}\right)\cos\left(\frac{2x + 10x}{2}\right)$$

Simplify.

$$\sin 2x - \sin 10x = 2\sin(-4x)\cos(6x)$$

Sine is an odd function.

$$\sin 2x - \sin 10x = -2\sin 4x \cos 6x$$

Multiply both sides by -9.

$$-9[\sin 2x - \sin 10x] = 18\sin 4x \cos 6x$$

TECHNOLOGY TIP

Graphs of
$y = -9[\sin 2x - \sin 10x]$ and
$y = 18\sin 4x \cos 6x$ follow.

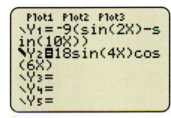

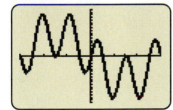

EXAMPLE 6 Simplifying a Trigonometric Expression

Simplify the expression $\sin\left(\frac{x + y}{2}\right)\cos\left(\frac{x - y}{2}\right) + \sin\left(\frac{x - y}{2}\right)\cos\left(\frac{x + y}{2}\right)$.

Solution:

Use identities (4) and (5).

$$\underbrace{\sin\left(\frac{x + y}{2}\right)\cos\left(\frac{x - y}{2}\right)}_{\frac{1}{2}[\sin x + \sin y]} + \underbrace{\sin\left(\frac{x - y}{2}\right)\cos\left(\frac{x + y}{2}\right)}_{\frac{1}{2}[\sin x - \sin y]}$$

$$= \frac{1}{2}\sin x + \frac{1}{2}\sin y + \frac{1}{2}\sin x - \frac{1}{2}\sin y$$

Simplify.

$$= \sin x$$

Applications

In music, a note is a fixed pitch (frequency) that is given a name. If two notes are sounded simultaneously, then they combine to produce another note often called a *beat*. The more rapid the beat, the further apart the two frequencies of the notes are. When musicians "tune" their instruments, they use a tuning fork to sound a note and then tune the instrument until the beat is eliminated; hence the fork and instrument are in tune. Mathematically, a note or tone is represented as $A\cos(2\pi f t)$, where A is the amplitude (loudness), f is the frequency in Hz, and t is time in seconds. The following table summarizes common notes and frequencies.

C	D	E	F	G	A	B
262 Hz	294 Hz	330 Hz	349 Hz	392 Hz	440 Hz	494 Hz

EXAMPLE 7 Music

Express the musical tone when a C and G are simultaneously struck (assume with the same loudness).

Find the beat frequency, $f_2 - f_1$. Assume uniform loudness, $A = 1$.

Solution:

Write the mathematical description of a C note. $\cos(2\pi f_1 t)$, $f_1 = 262$ Hz

Write the mathematical description of a G note. $\cos(2\pi f_2 t)$, $f_2 = 392$ Hz

Add the two notes. $\cos(524\pi t) + \cos(784\pi t)$

Use sum-product identities.

$$\cos(524\pi t) + \cos(784\pi t) = 2\cos\left(\frac{524\pi t + 784\pi t}{2}\right)\cos\left(\frac{524\pi t - 784\pi t}{2}\right)$$

Simplify. $= 2\cos(654\pi t)\cos(-130\pi t)$
                                                    $= 2\cos(654\pi t)\cos(130\pi t)$

Identify the beat frequency. $f_2 - f_1 = 392 - 262 = 130$ Hz

Therefore the tone of average frequency, 327 Hz, has a beat of 130 Hz (beats/sec).

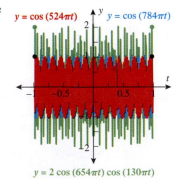

$y = \cos(524\pi t)$ $y = \cos(784\pi t)$

$y = 2\cos(654\pi t)\cos(130\pi t)$

SECTION 5.5   SUMMARY

In this section we used the sum and difference identities (Section 5.2) to derive the product-to-sum identities. The product-to-sum identities allow us to express products as sums.

$$\cos A \cos B = \frac{1}{2}\left[\cos(A+B) + \cos(A-B)\right]$$

$$\sin A \sin B = \frac{1}{2}\left[\cos(A-B) - \cos(A+B)\right]$$

$$\sin A \cos B = \frac{1}{2}\left[\sin(A+B) + \sin(A-B)\right]$$

We then used the product-to-sum identities to derive the sum-to-product identities. The sum-to-product identities allow us to express sums as products.

$$\sin A + \sin B = 2\sin\left(\frac{A+B}{2}\right)\cos\left(\frac{A-B}{2}\right)$$

$$\sin A - \sin B = 2\sin\left(\frac{A-B}{2}\right)\cos\left(\frac{A+B}{2}\right)$$

$$\cos A + \cos B = 2\cos\left(\frac{A+B}{2}\right)\cos\left(\frac{A-B}{2}\right)$$

$$\cos A - \cos B = -2\sin\left(\frac{A+B}{2}\right)\sin\left(\frac{A-B}{2}\right)$$

SECTION 5.5   EXERCISES

■ SKILLS

In Exercises 1–10, write each product as a sum or difference of sines and/or cosines.

1. $\sin 2x \cos x$

2. $\cos 10x \sin 5x$

3. $5\sin 4x \sin 6x$

4. $-3\sin 2x \sin 4x$

5. $4\cos(-x)\cos(2x)$

6. $-8\cos 3x \cos 5x$

7. $\sin\dfrac{3x}{2}\sin\dfrac{5x}{2}$

8. $\sin\dfrac{\pi x}{2}\sin\dfrac{5\pi x}{2}$

9. $\cos\dfrac{2}{3}x\cos\dfrac{4}{3}x$

10. $\sin\left(-\dfrac{\pi}{4}x\right)\cos\left(-\dfrac{\pi}{2}x\right)$

In Exercises 11–20, write each expression as a product of sines and/or cosines.

11. $\cos 5x + \cos 3x$

12. $\cos 2x - \cos 4x$

13. $\sin 3x - \sin x$

14. $\sin 10x + \sin 5x$

15. $\sin\dfrac{x}{2} - \sin\dfrac{5x}{2}$

16. $\cos\dfrac{x}{2} - \cos\dfrac{5x}{2}$

17. $\cos\dfrac{2}{3}x + \cos\dfrac{7}{3}x$

18. $\sin\dfrac{2}{3}x + \sin\dfrac{7}{3}x$

19. $\sin 0.4x + \sin 0.6x$

20. $\cos 0.3x - \cos 0.5x$

In Exercises 21–24, simplify the trigonometric expressions.

21. $\dfrac{\cos 3x - \cos x}{\sin 3x + \sin x}$

22. $\dfrac{\sin 4x + \sin 2x}{\cos 4x - \cos 2x}$

23. $\dfrac{\cos x - \cos 3x}{\sin 3x - \sin x}$

24. $\dfrac{\sin 4x + \sin 2x}{\cos 4x + \cos 2x}$

In Exercises 25–30, verify the identities.

25. $\dfrac{\sin A + \sin B}{\cos A + \cos B} = \tan \dfrac{A + B}{2}$

26. $\dfrac{\sin A - \sin B}{\cos A + \cos B} = \tan \dfrac{A - B}{2}$

27. $\dfrac{\cos A - \cos B}{\sin A + \sin B} = -\tan \dfrac{A - B}{2}$

28. $\dfrac{\cos A - \cos B}{\sin A - \sin B} = -\tan \dfrac{A + B}{2}$

29. $\dfrac{\sin A + \sin B}{\sin A - \sin B} = \tan\left(\dfrac{A + B}{2}\right)\cot\left(\dfrac{A - B}{2}\right)$

30. $\dfrac{\cos A - \cos B}{\cos A + \cos B} = -\tan\left(\dfrac{A + B}{2}\right)\tan\left(\dfrac{A - B}{2}\right)$

 ■ **APPLICATIONS**

31. Music. Write a mathematical description of a tone that results from simultaneously playing a G and a B. What is the beat frequency? What is the average frequency?

32. Music. Write a mathematical description of a tone that results from simultaneously playing an F and an A. What is the beat frequency? What is the average frequency?

33. Optics. Two optical signals with uniform $(A = 1)$ wavelengths of 1.55 μm and 0.63 μm are "beat" together. What is the resulting signal if their individual signals are given by $\sin\left[\dfrac{2\pi tc}{1.55\mu m}\right]$ and $\sin\left[\dfrac{2\pi tc}{0.63\mu m}\right]$ where $c = 3.0 \times 10^8$ m/s?

34. Optics. The two optical signals in Exercise 33 are beat together. What is the average frequency and the beat frequency?

Touch-Tone Dialing Touch-tone key pads have the following simultaneous low and high frequencies:

Frequency	1209 Hz	1336 Hz	1477 Hz
697 Hz	1	2	3
770 Hz	4	5	6
852 Hz	7	8	9
941 Hz	*	0	#

The signal made when a key is pressed is $\sin(2\pi f_1 t) + \sin(2\pi f_2 t)$, where f_1 is the low frequency and f_2 is the high frequency.

35. Touch-Tone Dialing. What is the mathematical function that models the sound of dialing 4?

36. Touch-Tone Dialing. What is the mathematical function that models the sound of dialing 3?

37. Area of a Triangle. A formula for finding the area of a triangle when given the measures of the angles and one side is area $= \dfrac{a^2 \sin B \sin C}{2 \sin A}$, where a is the side opposite angle A. If the measures of angles B and C are 52.5° and 7.5°, respectively, and if $a = 10$ feet, use the appropriate product-to-sum identity to change the formula so that you can solve for the area of the triangle exactly.

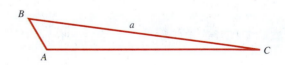

38. Area of a Triangle. A formula for finding the area of a triangle when given the measures of the angles and one side is area $= \dfrac{a^2 \sin B \sin C}{2 \sin A}$, where a is the side opposite angle A. If the measures of angles B and C are 75° and 45°, respectively, and if $a = 12$ inches, use the appropriate product-to-sum identity to change the formula so that you can solve for the area of the triangle exactly.

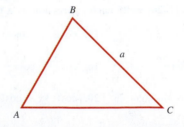

■ **CATCH THE MISTAKE**

In Exercises 39 and 40, explain the mistake that is made.

39. Simplify the expression:
$(\cos A - \cos B)^2 + (\sin A - \sin B)^2$.

Square binomials. $\cos^2 A - 2\cos A\cos B + \cos^2 B$
$+ \sin^2 A - 2\sin A\sin B + \sin^2 B$

Group terms. $\cos^2 A + \sin^2 A - 2\cos A\cos B$
$- 2\sin A\sin B + \cos^2 B + \sin^2 B$

Pythagorean identity. $\underbrace{\cos^2 A + \sin^2 A}_{1} - 2\cos A\cos B$

$- 2\sin A\sin B + \underbrace{\cos^2 B + \sin^2 B}_{1}$

Factor common 2. $2[1 - \cos A\cos B - \sin A\sin B]$

Simplify. $2[1 - \cos AB - \sin AB]$

This is not correct. What mistakes were made?

40. Simplify the expression: $(\sin A - \sin B)(\cos A + \cos B)$.

Multiply binomials.
$\sin A\cos A + \sin A\cos B - \sin B\cos A - \sin B\cos B$

Cancel the second and third terms.
$\sin A\cos A - \sin B\cos B$

Use the product-to-sum identity.
$$\underbrace{\sin A\cos A}_{\frac{1}{2}[\sin(A+A)+\sin(A-A)]} - \underbrace{\sin B\cos B}_{\frac{1}{2}[\sin(B+B)+\sin(B-B)]}$$

Simplify. $\dfrac{1}{2}\sin 2A - \dfrac{1}{2}\sin 2B$

This is incorrect. What mistake was made?

■ **CHALLENGE**

41. T or F: $\cos A\cos B = \cos AB$.

42. T or F: $\sin A\sin B = \sin AB$.

43. T or F: The product of two cosine functions is a sum of two other cosine functions.

44. T or F: The product of two sine functions is a difference of two cosine functions.

45. Write $\sin A\sin B\sin C$ as a sum or difference of sines and cosines.

46. Write $\cos A\cos B\cos C$ as a sum or difference of sines and cosines.

■ **TECHNOLOGY**

47. Suggest an identity $4\sin x\cos x\cos 2x = $ _____ by graphing $Y_1 = 4\sin x\cos x\cos 2x$ and determining the function based on the graph.

48. Suggest an identity $1 + \tan x\tan 2x = $ _____ by graphing $Y_1 = 1 + \tan x\tan 2x$ and determining the function based on the graph.

Q: Phil the physicist is calculating the strength of the electric field in an electromagnetic wave using

$$E = A\sin(x + ct)$$

where A is the amplitude of the wave, c is the speed of light (about 300,000 m/s), x is the distance from the source at which the measurement is taken, and t is time. He calculates E when $t = 10$ seconds, $x = 0$ meters, and $A = 10$ volts per meter and gets

$$E = 10\sin(300,000(10)) = 10\sin(3,000,000) \approx -8.8$$

volts per meter. Later, he needs to calculate E when $x = 1000$ meters and $t = 10$ seconds. He figures that he can use his previous result by calculating

$$E = A\sin(x + ct) = A\sin x + A\sin ct = 10\sin 100 + 10\sin(300,000(10)) \approx -0.5$$

volts per meter. However, this last calculation is not correct. What is his error?

A: $A\sin(x + ct) \neq A\sin x + A\sin ct$
Since the sum identity for sine is $\sin(A + B) = \sin A \cos B + \cos A \sin B$, then
$A\sin(x + ct) = A(\sin x \cos ct + \cos x \sin ct)$ and $10\sin(1000 + 300,000(10)) =$
$10[(\sin 1000)(\cos 3,000,000) + (\cos 1000)(\sin 3,000,000)] \approx -1.0$ volts per meter.
In applying the identity correctly, Phil can still use his previous calculation; however, he will have a simpler form to work with if he leaves his calculation as $10\sin(1000 + 3,000,000) = 10\sin(3,001,000)$. Use the form $E = A\sin(x + ct)$ and the form $E = A(\sin x \cos ct + \cos x \sin ct)$ to calculate E when $x = 100$ meters and $t = 5$ seconds (A is still 10 volts per meter) and verify that you get the same result.

TYING IT ALL TOGETHER

In calculus, trigonometric substitutions can be made in order to rewrite functions into a simpler form in which the operation of integration is easier to perform. Let a be a constant. If the function contains an expression of the form $\sqrt{a^2 - x^2}$, then we let $x = a\sin\theta$, where $-\dfrac{\pi}{2} \le \theta \le \dfrac{\pi}{2}$. If the function contains an expression of the form $\sqrt{a^2 + x^2}$, then we let $x = a\tan\theta$, where $-\dfrac{\pi}{2} < \theta < \dfrac{\pi}{2}$. Finally, if the function contains an expression of the form $\sqrt{x^2 - a^2}$, then we let $x = a\sec\theta$, where $0 \le \theta < \dfrac{\pi}{2}$ or $\pi \le \theta < \dfrac{3\pi}{2}$. Use an appropriate trigonometric substitution to rewrite the function $y = \dfrac{x}{\sqrt{9 - x^2}}$ in terms of θ.

Your final answer should contain no radicals and be simplified.

SECTION	TOPIC	PAGES	REVIEW EXERCISES	KEY CONCEPTS
5.1	Verifying trigonometric identities	246–252	1–16	Identities must hold for *all* values of x (not just some values of x) for which both sides of the equation are defined.
	Reciprocal identities	246	1–16	

Reciprocal Identities	Equivalent Forms
$\csc x = \dfrac{1}{\sin x}$	$\sin x = \dfrac{1}{\csc x}$
$\sec x = \dfrac{1}{\cos x}$	$\cos x = \dfrac{1}{\sec x}$
$\cot x = \dfrac{1}{\tan x}$	$\tan x = \dfrac{1}{\cot x}$

SECTION	TOPIC	PAGES	REVIEW EXERCISES	KEY CONCEPTS
	Quotient identities	247	1–16	$\tan x = \dfrac{\sin x}{\cos x} \qquad \cot x = \dfrac{\cos x}{\sin x}$
	Pythagorean identities	247	1–16	$\sin^2 x + \cos^2 x = 1$ $\tan^2 x + 1 = \sec^2 x \qquad 1 + \cot^2 x = \csc^2 x$
	Even-odd identities	247	1–16	Even: $f(-x) = f(x)$ symmetry at y-axis Odd: $f(-x) = -f(x)$ symmetry at origin Odd: $\qquad$ Even: $\sin(-x) = -\sin x \qquad \cos(-x) = \cos x$ $\tan(-x) = -\tan x \qquad \sec(-x) = \sec x$ $\csc(-x) = -\csc x$ $\cot(-x) = -\cot x$
5.2	Sum and difference identities	255–265	17–32	$f(A \pm B) \neq f(A) \pm f(B)$ For trigonometric functions we have the sum and difference identities.
	Cosine sum and difference	256–259	17–32	$\cos(A + B) = \cos A \cos B - \sin A \sin B$ $\cos(A - B) = \cos A \cos B + \sin A \sin B$
	Sine sum and difference	260–263	17–32	$\sin(A + B) = \sin A \cos B + \cos A \sin B$ $\sin(A - B) = \sin A \cos B - \cos A \sin B$
	Tangent sum and difference	263–265	17–32	$\tan(A + B) = \dfrac{\tan A + \tan B}{1 - \tan A \tan B}$ $\tan(A - B) = \dfrac{\tan A - \tan B}{1 + \tan A \tan B}$
5.3	Double-angle identities	268–273	33–48	$\sin(2A) = 2\sin A \cos A$ $\cos(2A) = \cos^2 A - \sin^2 A$ $\qquad = 1 - 2\sin^2 A = 2\cos^2 A - 1$ $\tan(2A) = \dfrac{2\tan A}{1 - \tan^2 A}$

SECTION	TOPIC	PAGES	REVIEW EXERCISES	KEY CONCEPTS
5.4	Half-angle identities	276–282	49–64	$\sin\dfrac{A}{2} = \pm\sqrt{\dfrac{1-\cos A}{2}}$ $\cos\dfrac{A}{2} = \pm\sqrt{\dfrac{1+\cos A}{2}}$ $\tan\dfrac{A}{2} = \pm\sqrt{\dfrac{1-\cos A}{1+\cos A}}$
5.5	Product-to-sum and sum-to-product identities	285–291	65–74	
	Product-to-sum identities	286–288	65–66	$\cos A \cos B = \dfrac{1}{2}\big[\cos(A+B) + \cos(A-B)\big]$ $\sin A \sin B = \dfrac{1}{2}\big[\cos(A-B) - \cos(A+B)\big]$ $\sin A \cos B = \dfrac{1}{2}\big[\sin(A+B) + \sin(A-B)\big]$
	Sum-to-product identities	288–289	67–74	$\sin A + \sin B = 2\sin\left(\dfrac{A+B}{2}\right)\cos\left(\dfrac{A-B}{2}\right)$ $\sin A - \sin B = 2\sin\left(\dfrac{A-B}{2}\right)\cos\left(\dfrac{A+B}{2}\right)$ $\cos A + \cos B = 2\cos\left(\dfrac{A+B}{2}\right)\cos\left(\dfrac{A-B}{2}\right)$ $\cos A - \cos B = -2\sin\left(\dfrac{A+B}{2}\right)\sin\left(\dfrac{A-B}{2}\right)$

5.1 Verifying Trigonometric Identities

Simplify the following trigonometric expressions.

1. $\tan x(\cot x + \tan x)$

2. $(\sec x + 1)(\sec x - 1)$

3. $\dfrac{\tan^4 x - 1}{\tan^2 x - 1}$

4. $\sec^2 x(\cot^2 x - \cos^2 x)$

5. $\cos x(\cos(-x) - \tan(-x)) - \sin x$

6. $\dfrac{\tan^2 x + 1}{2\sec^2 x}$

Verify the trigonometric identities.

7. $(\tan x + \cot x)^2 - 2 = \tan^2 x + \cot^2 x$

8. $\csc^2 x - \cot^2 x = 1$

9. $\dfrac{1}{\sin^2 x} - \dfrac{1}{\tan^2 x} = 1$

10. $\dfrac{1}{\csc x + 1} + \dfrac{1}{\csc x - 1} = \dfrac{2\tan x}{\cos x}$

11. $\dfrac{\tan^2 x - 1}{\sec^2 x + 3\tan x + 1} = \dfrac{\tan x - 1}{\tan x + 2}$

12. $\cot x(\sec x - \cos x) = \sin x$

Determine if the following equations are conditional or identities.

13. $2\tan^2 x + 1 = \dfrac{1 + \sin^2 x}{\cos^2 x}$

14. $\sin x - \cos x = 0$

15. $\cot^2 x - 1 = \tan^2 x$

16. $\cos^2 x(1 + \cot^2 x) = \cot^2 x$

5.2 Sum and Difference Identities

Find exact values for each trigonometric expression.

17. $\cos\dfrac{7\pi}{12}$

18. $\sin\dfrac{\pi}{12}$

19. $\tan(-15°)$

20. $\cot 105°$

Write each expression as a single trigonometric function.

21. $\sin 4x\cos 3x - \cos 4x\sin 3x$

22. $\sin(-x)\sin(-2x) + \cos(-x)\cos(-2x)$

23. $\dfrac{\tan 5x - \tan 4x}{1 + \tan 5x\tan 4x}$

24. $\dfrac{\tan\dfrac{\pi}{4} + \tan\dfrac{\pi}{3}}{1 - \tan\dfrac{\pi}{4}\tan\dfrac{\pi}{3}}$

Find the exact value of the indicated expression using the given information and identities.

25. Find the exact value of $\tan(\alpha - \beta)$ if $\sin\alpha = -\dfrac{3}{5}$ and $\sin\beta = -\dfrac{24}{25}$, if the terminal side of α lies in QIV and the terminal side of β lies in QIII.

26. Find the exact value of $\cos(\alpha + \beta)$ if $\cos\alpha = -\dfrac{5}{13}$ and $\sin\beta = \dfrac{7}{25}$, if the terminal side of α lies in QII and the terminal side of β also lies in QII.

27. Find the exact value of $\cos(\alpha - \beta)$ if $\cos\alpha = \dfrac{9}{41}$ and $\cos\beta = \dfrac{7}{25}$, if the terminal side of α lies in QIV and the terminal side of β lies in QI.

28. Find the exact value of $\sin(\alpha - \beta)$ if $\sin\alpha = -\dfrac{5}{13}$ and $\cos\beta = -\dfrac{4}{5}$, if the terminal side of α lies in QIII and the terminal side of β lies in QII.

Determine if each equation is conditional or an identity.

29. $2\cos A\cos B = \cos(A + B) + \cos(A - B)$

30. $2\sin A\sin B = \cos(A - B) - \cos(A + B)$

Graph the following functions.

31. $y = \cos\dfrac{\pi}{2}\cos x - \sin\dfrac{\pi}{2}\sin x$

32. $y = \sin\dfrac{2\pi}{3}\cos x + \cos\dfrac{2\pi}{3}\sin x$

5.3 Double-Angle Identities

Use double-angle identities to answer the following questions.

33. If $\sin x = \dfrac{3}{5}$ and $\dfrac{\pi}{2} < x < \pi$, find $\cos 2x$.

34. If $\cos x = \dfrac{7}{25}$ and $\dfrac{3\pi}{2} < x < 2\pi$, find $\sin 2x$.

35. If $\cot x = -\dfrac{11}{61}$ and $\dfrac{3\pi}{2} < x < 2\pi$, find $\tan 2x$.

36. If $\tan x = -\dfrac{12}{5}$ and $\dfrac{\pi}{2} < x < \pi$, find $\cos 2x$.

37. If $\sec x = \dfrac{25}{24}$ and $0 < x < \dfrac{\pi}{2}$, find $\sin 2x$.

38. If $\csc x = \dfrac{5}{4}$ and $\dfrac{\pi}{2} < x < \pi$, find $\tan 2x$.

Simplify each of the following. Evaluate exactly, if possible.

39. $\cos^2 15° - \sin^2 15°$

40. $\dfrac{2\tan\left(-\dfrac{\pi}{12}\right)}{1 - \tan^2\left(-\dfrac{\pi}{12}\right)}$

41. $6\sin\dfrac{\pi}{12}\cos\dfrac{\pi}{12}$

42. $1 - 2\sin^2\dfrac{\pi}{8}$

Verify the following identities.

43. $\sin^3 A - \cos^3 A = (\sin A - \cos A)\left(1 + \dfrac{1}{2}\sin 2A\right)$

44. $2\sin A \cos^3 A - 2\sin^3 A \cos A = \cos 2A \sin 2A$

45. $\tan A = \dfrac{\sin 2A}{1 + \cos 2A}$

46. $\tan A = \dfrac{1 - \cos 2A}{\sin 2A}$

Applications

47. Launching a Missile. When launching a missile for a given range, the minimum velocity that's needed is related to the angle of the launch, and the velocity is determined by $V = \dfrac{2\cos 2\theta}{1 + \cos 2\theta}$. Show that V is equivalent to $1 - \tan^2\theta$.

48. Launching a Missile. When launching a missile for a given range, the minimum velocity that's needed is related to the angle of the launch, and the velocity is determined by $V = \dfrac{2\cos 2\theta}{1 + \cos 2\theta}$. Find the value of V when $\theta = \dfrac{\pi}{6}$.

5.4 Half-Angle Identities

Use half-angle identities to find the exact values of the trigonometric expressions.

49. $\sin(-22.5°)$

50. $\cos 67.5°$

51. $\cot\dfrac{3\pi}{8}$

52. $\csc\left(-\dfrac{7\pi}{8}\right)$

Use half-angle identities to answer the following questions.

53. If $\sin x = -\dfrac{7}{25}$ and $\pi < x < \dfrac{3\pi}{2}$, find $\sin\dfrac{x}{2}$.

54. If $\cos x = -\dfrac{4}{5}$ and $\dfrac{\pi}{2} < x < \pi$, find $\cos\dfrac{x}{2}$.

55. If $\tan x = \dfrac{40}{9}$ and $\pi < x < \dfrac{3\pi}{2}$, find $\tan\dfrac{x}{2}$.

56. If $\sec x = \dfrac{17}{15}$ and $\dfrac{3\pi}{2} < x < 2\pi$, find $\sin\dfrac{x}{2}$.

Simplify each expression using half-angle identities. Do not evaluate.

57. $\sqrt{\dfrac{1 - \cos\dfrac{\pi}{6}}{2}}$

58. $\sqrt{\dfrac{1 - \cos\dfrac{11\pi}{6}}{1 + \cos\dfrac{11\pi}{6}}}$

Verify the following identities.

59. $\left(\sin\dfrac{A}{2} + \cos\dfrac{A}{2}\right)^2 = 1 + \sin A$

60. $\sec^2\dfrac{A}{2} + \tan^2\dfrac{A}{2} = \dfrac{3 - \cos A}{1 + \cos A}$

61. $\csc^2\dfrac{A}{2} + \cot^2\dfrac{A}{2} = \dfrac{3 + \cos A}{1 - \cos A}$

62. $\tan^2\dfrac{A}{2} + 1 = \sec^2\dfrac{A}{2}$

Graph the functions.

63. $y = \sqrt{\dfrac{1 - \cos\dfrac{\pi}{12}x}{2}}$

64. $y = \cos^2\dfrac{x}{2} - \sin^2\dfrac{x}{2}$

5.5 Product-to-Sum and Sum-to-Product Identities

Write each product as a sum or difference of sines and/or cosines.

65. $6\sin 5x \cos 2x$

66. $3\sin 4x \sin 2x$

Write each expression as a product of sines and/or cosines.

67. $\cos 5x - \cos 3x$

68. $\sin\dfrac{5x}{2} + \sin\dfrac{3x}{2}$

69. $\sin\dfrac{4x}{3} - \sin\dfrac{2x}{3}$

70. $\cos 7x + \cos x$

Simplify the trigonometric expressions.

71. $\dfrac{\cos 8x + \cos 2x}{\sin 8x - \sin 2x}$

72. $\dfrac{\sin 5x + \sin 3x}{\cos 5x + \cos 3x}$

Verify the identities.

73. $\dfrac{\sin A + \sin B}{\cos A - \cos B} = -\cot\left(\dfrac{A - B}{2}\right)$

74. $\dfrac{\sin A - \sin B}{\cos A - \cos B} = -\cot\left(\dfrac{A + B}{2}\right)$

1. Verify the identity: $\sec x - \tan x = \dfrac{1}{\sec x + \tan x}$.

2. For what values of x does the quotient identity, $\tan x = \dfrac{\sin x}{\cos x}$, not hold?

3. Is the equation $\sqrt{\sin^2 x + \cos^2 x} = \sin x + \cos x$ conditional or an identity?

4. Evaluate $\sin\left(-\dfrac{\pi}{8}\right)$ exactly.

5. Evaluate $\tan\left(\dfrac{7\pi}{12}\right)$ exactly.

6. If $\cos x = \dfrac{2}{5}$ and $\dfrac{3\pi}{2} < x < 2\pi$, find $\sin\dfrac{x}{2}$.

7. If $\sin x = -\dfrac{1}{5}$ and $\pi < x < \dfrac{3\pi}{2}$, find $\cos 2x$.

8. Write $\cos 7x \cos 3x - \sin 3x \sin 7x$ as a cosine or sine of a sum or difference.

9. Write $-\dfrac{2\tan x}{1 - \tan^2 x}$ as a single tangent function.

10. Write $\sqrt{\dfrac{1 + \cos(a + b)}{2}}$ as a single cosine function if $a + b$ is an angle in QII.

11. Write $2\sin\left(\dfrac{x + 3}{2}\right)\cos\left(\dfrac{x - 3}{2}\right)$ as a sum of two sine functions.

12. Write $10\cos(3 - x) + 10\cos(x + 3)$ as a product of two cosine functions.

13. In the expression $\sqrt{9 - u^2}$, let $u = 3\sin x$. What is the resulting expression?

Solving Trigonometric Equations

Photographer's Choice/Getty Images, Inc.

In the summer, St. Petersburg, Russia, has many more daylight hours than St. Petersburg, Florida (U.S.), because of their locations with respect to the equator. From late May to early July, nights are bright in St. Petersburg, Russia. The nature of the *white nights* can be explained by the geographical location of the St. Petersburg in Russia. It is one of the world's most northern cities: located at 59 degrees 57′ north (approximately the same latitude as Seward, Alaska). Due to such a high latitude, the Sun does not go under the horizon deep enough for the sky to get dark.

Trigonometric equations are used to represent the relationship between the number of daylight hours and the latitude of the city. *Inverse trigonometric functions* allow us to find the latitude given the hours of sunlight.

In this chapter we first develop the inverse trigonometric functions and then use these to assist us in solving trigonometric equations. As with all equations, the goal is to find the value(s) of the variable that make the equation true. You will find that trigonometric equations (like algebraic equations) can have no solution, one or more solutions, or an infinite number of solutions.

Solving Trigonometric Equations

Inverse Trigonometric Functions

$\sin^{-1}x$ or $\arcsin x$
$\cos^{-1}x$ or $\arccos x$
$\tan^{-1}x$ or $\arctan x$
$\cot^{-1}x$ or $\text{arccot}\, x$
$\sec^{-1}x$ or $\text{arcsec}\, x$
$\csc^{-1}x$ or $\text{arccsc}\, x$

Trigonometric Equations

• **Equations Involving Only One Trigonometric Function**
  • Solve by inspection
  • Solve trigonometric equations using inverse functions
  • Solve using algebraic techniques

• **Equations Involving Multiple Trigonometric Functions**
  • Use identities and algebraic techniques

CHAPTER OBJECTIVES

■ Develop inverse trigonometric functions.
■ Understand inverse trigonometric function notation.
■ Find the domain and range of inverse trigonometric functions.
■ Solve trigonometric equations by inspection.
■ Use algebraic techniques to solve trigonometric equations.
■ Solve trigonometric equations using inverse trigonometric functions.

NAVIGATION THROUGH SUPPLEMENTS

DIGITAL VIDEO SERIES #6

STUDENT SOLUTIONS MANUAL CHAPTER 6

BOOK COMPANION SITE
www.wiley.com/college/young

- Review inverse functions.
- Develop inverse trigonometric functions.
- Find values of inverse trigonometric functions.
- Graph inverse trigonometric functions.

- Understand the different notations for inverse trigonometric functions.
- Understand why domain restrictions on trigonometric functions are needed for inverse trigonometric functions to exist.
- Extend properties of inverse functions to develop inverse trigonometric identities.

Review of Inverse Functions

In Appendix A.6, one-to-one functions and inverse functions are discussed. Here we present a summary of that section. A function is one-to-one if it passes the horizontal line test. In other words, no two x values map to the same y value. Notice that the sine function does not pass the horizontal line test.

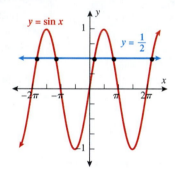

However, if we restrict the domain to $-\dfrac{\pi}{2} \le x \le \dfrac{\pi}{2}$, then the restricted function is one-to-one.

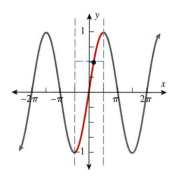

The notation is as follows: $y = f(x)$ corresponds to $x = f^{-1}(y)$.

The following are properties of inverse functions:

1. If f is a one-to-one function, then the inverse function, f^{-1}, exists.
2. The domain of f^{-1} = the range of f.
   The range of f^{-1} = the domain of f.

3. $f^{-1}(f(x)) = x$ for all x in the domain of f.

$f(f^{-1}(x)) = x$ for all x in the domain of f^{-1}.

4. The graphs of f and f^{-1} are symmetric about the line $y = x$. If the point (a, b) lies on the graph of a function, then the point (b, a) lies on the graph of its inverse.

Inverse Sine Function

Let us start with the sine function with a restricted domain:

$y = \sin x$ Domain: $\left[-\dfrac{\pi}{2}, \dfrac{\pi}{2}\right]$ Range: $[-1, 1]$

x	y
$-\dfrac{\pi}{2}$	-1
$-\dfrac{\pi}{4}$	$-\dfrac{\sqrt{2}}{2}$
0	0
$\dfrac{\pi}{4}$	$\dfrac{\sqrt{2}}{2}$
$\dfrac{\pi}{2}$	1

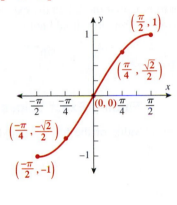

By the properties of inverses, the inverse sine function will have a domain of $[-1, 1]$ and a range of $\left[-\dfrac{\pi}{2}, \dfrac{\pi}{2}\right]$. To find the inverse sine function, we interchange the x and y values of $\sin x$.

$y = \sin^{-1} x$ Domain: $[-1, 1]$ Range: $\left[-\dfrac{\pi}{2}, \dfrac{\pi}{2}\right]$

x	y
-1	$-\dfrac{\pi}{2}$
$-\dfrac{\sqrt{2}}{2}$	$-\dfrac{\pi}{4}$
0	0
$\dfrac{\sqrt{2}}{2}$	$\dfrac{\pi}{4}$
1	$\dfrac{\pi}{2}$

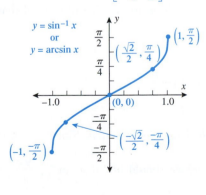

Notice that the inverse sine function, like the sine function, is an odd function (symmetric about the origin).

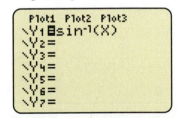
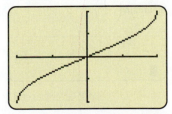

STUDY TIP

Trigonometric functions take angles and return real numbers. Inverse trigonometric functions take real numbers and return angles.

STUDY TIP

$\sin^{-1}x \neq \dfrac{1}{\sin x}$

If the sine of an angle is known, what is the angle? The inverse sine function determines the angle. Another notation for the inverse sine function is $\arcsin x$.

INVERSE SINE FUNCTION

$$y = \sin^{-1}x \text{ or } y = \arcsin x \qquad \text{means} \qquad x = \sin y$$

"y is the inverse sine of x"   "y is the angle or real number whose sine equals x"

$$\text{where } -1 \leq x \leq 1 \text{ and } -\dfrac{\pi}{2} \leq y \leq \dfrac{\pi}{2}$$

It is important to note that the -1 in the exponent indicates an inverse function. Therefore, the inverse sine function should not be interpreted as a reciprocal:

$$\sin^{-1}x \neq \dfrac{1}{\sin x}$$

TECHNOLOGY TIP

Set the TI/scientific calculator to degree mode by typing MODE.

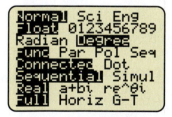

a. Using the TI calculator to find $\sin^{-1}\left(\dfrac{\sqrt{3}}{2}\right)$, type 2nd SIN for SIN^{-1} and 2nd x^2 for $\sqrt{\ }$.

```
sin-1(√(3)/2)
                    60
█
```

b. $\arcsin\left(-\dfrac{1}{2}\right)$

```
sin-1(-1/2)
                  -30
█
```

EXAMPLE 1 Finding Exact Values of an Inverse Sine Function

Find the exact value of the following expressions:

a. $\sin^{-1}\left(\dfrac{\sqrt{3}}{2}\right)$ **b.** $\arcsin\left(-\dfrac{1}{2}\right)$

Solution (a):

Let $\theta = \sin^{-1}\left(\dfrac{\sqrt{3}}{2}\right)$. $\sin\theta = \dfrac{\sqrt{3}}{2}$ when $-\dfrac{\pi}{2} \leq \theta \leq \dfrac{\pi}{2}$

What value of θ in the range $-\dfrac{\pi}{2} \leq \theta \leq \dfrac{\pi}{2}$ corresponds to a sine value of $\dfrac{\sqrt{3}}{2}$?

- The range $-\dfrac{\pi}{2} \leq \theta \leq \dfrac{\pi}{2}$ corresponds to Quadrants I and IV.
- Sine is positive in Quadrant I.
- We look for a value of θ in QI that has a sine value of $\dfrac{\sqrt{3}}{2}$. $\theta = \dfrac{\pi}{3}$

$\sin\dfrac{\pi}{3} = \dfrac{\sqrt{3}}{2}$ and $\dfrac{\pi}{3}$ is in the

interval $\left[-\dfrac{\pi}{2}, \dfrac{\pi}{2}\right]$.

$$\sin^{-1}\left(\dfrac{\sqrt{3}}{2}\right) = \dfrac{\pi}{3}$$

Calculator Confirmation: Since $\dfrac{\pi}{3} = 60°$ if our calculator is set in degrees mode, we should find that $\sin^{-1}\left(\dfrac{\sqrt{3}}{2}\right)$ is equal to $60°$.

Solution (b):

Let $\theta = \arcsin\left(-\dfrac{1}{2}\right)$. $\sin\theta = -\dfrac{1}{2}$ when $-\dfrac{\pi}{2} \leq \theta \leq \dfrac{\pi}{2}$

What value of θ in the range $-\dfrac{\pi}{2} \leq \theta \leq \dfrac{\pi}{2}$ corresponds to a sine value of $-\dfrac{1}{2}$?

- The range $-\dfrac{\pi}{2} \le \theta \le \dfrac{\pi}{2}$ corresponds to Quadrants I and IV.
- Sine is negative in Quadrant IV.
- We look for a value of θ in QIV that has a sine value of $-\dfrac{1}{2}$. $\theta = -\dfrac{\pi}{6}$

$\sin\left(-\dfrac{\pi}{6}\right) = -\dfrac{1}{2}$ and $-\dfrac{\pi}{6}$ is

in the interval $\left[-\dfrac{\pi}{2}, \dfrac{\pi}{2}\right]$.

$$\arcsin\left(-\dfrac{1}{2}\right) = -\dfrac{\pi}{6}$$

Calculator Confirmation: Since $-\dfrac{\pi}{6} = -30°$ if our calculator is set in degrees

mode, we should find that $\sin^{-1}\left(-\dfrac{1}{2}\right)$ is equal to $-30°$.

It is important to note that in part (a), both $60°$ and $120°$ correspond to the

sine function being equal to $\dfrac{\sqrt{3}}{2}$, which is why the domain restrictions are

necessary for inverse functions. Except for quadrantal angles, there are always
two angles (values) from 0 to $360°$ (or 0 to 2π) that correspond to the sine
function being equal to a particular value.

■ **YOUR TURN**   Find the exact value of the following expressions:

a. $\sin^{-1}\left(-\dfrac{\sqrt{3}}{2}\right)$ **b.** $\arcsin\left(\dfrac{1}{2}\right)$

It is important to note that the inverse sine function has a domain $[-1, 1]$. For example, $\sin^{-1}3$ does not exist because 3 is not in the domain. Notice that calculator evaluation of $\sin^{-1}3$ says *error*. Calculators can be used to evaluate inverse sine functions
when an exact evaluation is not possible, as we have seen in previous sections. For example, $\sin^{-1}0.3 = 17.46°$ or 0.305 radians.

We now state the properties relating the sine function and the inverse sine function
that follow directly from properties of inverses.

TECHNOLOGY TIP

Use a TI calculator to find
$\sin^{-1}3$ and $\sin^{-1}0.3$. Be sure to
set the calculator in radian mode.

```
sin-1(3)
■
```

```
ERR:DOMAIN
1■Quit
2:Goto
```

```
sin-1(3)
sin-1(0.3)
           .304692654
■
```

SINE-INVERSE SINE IDENTITIES

$$\sin^{-1}(\sin x) = x \qquad \text{for} \qquad -\dfrac{\pi}{2} \le x \le \dfrac{\pi}{2}$$
$$\sin(\sin^{-1}x) = x \qquad \text{for} \qquad -1 \le x \le 1$$

For example, $\sin^{-1}\left(\sin\dfrac{\pi}{12}\right) = \dfrac{\pi}{12}$, since $\dfrac{\pi}{12}$ is in the interval $\left[-\dfrac{\pi}{2}, \dfrac{\pi}{2}\right]$. However, you
must be careful not to overlook the domain restriction for which these identities hold, as
illustrated in the next example.

**EXAMPLE 2 Using Inverse Identities to Evaluate Expressions
Involving Inverse Sine Functions**

Find exact values of the trigonometric expressions.

a. $\sin\left(\sin^{-1}\left(\dfrac{\sqrt{2}}{2}\right)\right)$ **b.** $\sin^{-1}\left(\sin\dfrac{3\pi}{4}\right)$

■ **Answer: a.** $-\dfrac{\pi}{3}$ **b.** $\dfrac{\pi}{6}$

TECHNOLOGY TIP

a. Check the answer of $\sin\left(\sin^{-1}\left(\dfrac{\sqrt{2}}{2}\right)\right)$ using a calculator,

```
sin(sin⁻¹(√(2)/2)
)
         .7071067812
√(2)/2
         .7071067812
■
```

b. Check the answer of $\sin^{-1}\left(\sin\dfrac{3\pi}{4}\right)$

```
sin⁻¹(sin(3π/4))
         .7853981634
π/4
         .7853981634
■
```

Solution (a):

Write the appropriate identity. $\qquad$ $\sin(\sin^{-1}x) = x$ for $-1 \leq x \leq 1$

Let $x = \dfrac{\sqrt{2}}{2}$, which is in the interval $[-1, 1]$.

Since the domain restriction is met, the identity can be used. $\qquad$ $\sin\left(\sin^{-1}\left(\dfrac{\sqrt{2}}{2}\right)\right) = \dfrac{\sqrt{2}}{2}$

Solution (b):

COMMON MISTAKE

Ignoring the domain restrictions on inverse identities.

CORRECT

Write the appropriate identity.

$\sin^{-1}(\sin x) = x$ for $-\dfrac{\pi}{2} \leq x \leq \dfrac{\pi}{2}$

Let $x = \dfrac{3\pi}{4}$, which is *not* in the interval $\left[-\dfrac{\pi}{2}, \dfrac{\pi}{2}\right]$.

Since the domain restriction is not met, the identity cannot be used. Instead, we look for a value in the domain that corresponds to the same value of sine.

$$\sin\left(\dfrac{3\pi}{4}\right) = \sin\left(\dfrac{\pi}{4}\right)$$

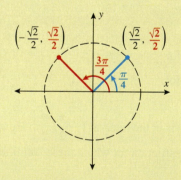

Substitute $\sin\dfrac{3\pi}{4} = \sin\dfrac{\pi}{4}$ into the expression.

$$\sin^{-1}\left(\sin\dfrac{3\pi}{4}\right) = \sin^{-1}\left(\sin\dfrac{\pi}{4}\right)$$

INCORRECT

Write the appropriate identity.

$\sin^{-1}(\sin x) = x$ $\qquad$ **ERROR**

Let $x = \dfrac{3\pi}{4}$.

$\sin^{-1}\left(\sin\dfrac{3\pi}{4}\right) = \dfrac{3\pi}{4}$

$\qquad\qquad\qquad$ **INCORRECT**

There are times when the expression actually does not exist. For example, $\sin(\sin^{-1}2)$ does not exist because the inverse function has a domain $[-1, 1]$. Notice that if you try to compute $\sin^{-1}2$, your calculator says *error*.

Since $\dfrac{\pi}{4}$ is in the interval $\left[-\dfrac{\pi}{2}, \dfrac{\pi}{2}\right]$, we can use the identity.

$$\sin^{-1}\left(\sin\dfrac{3\pi}{4}\right) = \sin^{-1}\left(\sin\dfrac{\pi}{4}\right) = \boxed{\dfrac{\pi}{4}}$$

■ **YOUR TURN** Find exact values of the trigonometric expressions.

a. $\sin\left(\sin^{-1}\left(-\dfrac{1}{2}\right)\right)$ **b.** $\sin^{-1}\left(\sin\dfrac{5\pi}{6}\right)$

Inverse Cosine Function

The cosine function is not a one-to-one function, so we must restrict the domain in order to develop the inverse cosine function.

$y = \cos x$ Domain: $[0, \pi]$ Range: $[-1, 1]$

x	y
0	1
$\dfrac{\pi}{4}$	$\dfrac{\sqrt{2}}{2}$
$\dfrac{\pi}{2}$	0
$\dfrac{3\pi}{4}$	$-\dfrac{\sqrt{2}}{2}$
π	-1

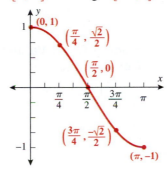

By properties of inverses, the inverse cosine function will have a domain of $[-1, 1]$ and a range of $[0, \pi]$. To find the inverse cosine function, we interchange the x and y values of $y = \cos x$.

$y = \cos^{-1}x$ Domain: $[-1, 1]$ Range: $[0, \pi]$

x	y
-1	π
$-\dfrac{\sqrt{2}}{2}$	$\dfrac{3\pi}{4}$
0	$\dfrac{\pi}{2}$
$\dfrac{\sqrt{2}}{2}$	$\dfrac{\pi}{4}$
1	0

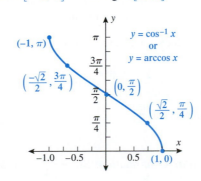

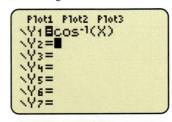
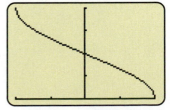

■ **Answer: a.** $-\dfrac{1}{2}$ **b.** $\dfrac{\pi}{6}$

Notice that the inverse cosine function, unlike the cosine function, is not symmetric about the y-axis or the origin. Although the inverse sine and inverse cosine functions have the same domain, they have different behavior. The inverse sine function increases on its domain (from left to right), whereas the inverse cosine function decreases on its domain (from left to right).

If the cosine of an angle is known, what is the angle? The inverse cosine function determines the angle. Another notation for the inverse cosine function is arccos x.

INVERSE COSINE FUNCTION

$$y = \cos^{-1}x \text{ or } y = \arccos x \qquad \text{means} \qquad x = \cos y$$

"y is the inverse cosine of x"   "y is the angle or real number whose cosine equals x"

where $-1 \leq x \leq 1$ and $0 \leq y \leq \pi$

STUDY TIP

$\cos^{-1}x \neq \dfrac{1}{\cos x}$

It is important to note that the inverse cosine function should not be interpreted as a reciprocal: $\cos^{-1}x \neq \dfrac{1}{\cos x}$.

EXAMPLE 3 **Finding Exact Values of an Inverse Cosine Function**

Find the exact value of the following expressions:

a. $\cos^{-1}\left(-\dfrac{\sqrt{2}}{2}\right)$ **b.** $\arccos(0)$

TECHNOLOGY TIP

a. Check the answer of $\cos^{-1}\left(-\dfrac{\sqrt{2}}{2}\right)$ using a calculator,

cos⁻¹(-√(2)/2)
                    135

Solution (a):

Let $\theta = \cos^{-1}\left(-\dfrac{\sqrt{2}}{2}\right)$.

$\cos \theta = -\dfrac{\sqrt{2}}{2}$ when $0 \leq \theta \leq \pi$

What value of θ in the range $0 \leq \theta \leq \pi$ corresponds to a cosine value of $-\dfrac{\sqrt{2}}{2}$?

■ The range $0 \leq \theta \leq \pi$ corresponds to Quadrants I and II.

■ Cosine is negative in Quadrant II.

■ We look for a value of θ in QII that has a cosine value of $-\dfrac{\sqrt{2}}{2}$. $\theta = \dfrac{3\pi}{4}$

$\cos\dfrac{3\pi}{4} = -\dfrac{\sqrt{2}}{2}$ and $\dfrac{3\pi}{4}$ is in the interval $[0, \pi]$.

$\boxed{\cos^{-1}\left(-\dfrac{\sqrt{2}}{2}\right) = \dfrac{3\pi}{4}}$

Calculator Confirmation: Since $\dfrac{3\pi}{4} = 135°$ if our calculator is set in degrees mode, we should find that $\cos^{-1}\left(-\dfrac{\sqrt{2}}{2}\right)$ is equal to 135°.

Solution (b):

Let $\theta = \arccos(0)$. $\cos\theta = 0$ when $0 \le \theta \le \pi$

What value of θ in the range $0 \le \theta \le \pi$
corresponds to a cosine value of 0? $\theta = \dfrac{\pi}{2}$

$\cos\left(\dfrac{\pi}{2}\right) = 0$ and $\dfrac{\pi}{2}$ is

in the interval $[0, \pi]$. $\arccos(0) = \dfrac{\pi}{2}$.

b. Check the answer of
arc cos(0), be sure to set the
calculator to degrees mode.

cos⁻¹(0)
                90

Calculator Confirmation: Since $\dfrac{\pi}{2} = 90°$ if our calculator is set in degrees

mode, we should find that $\cos^{-1}(0)$ is equal to 90°.

■ **YOUR TURN**   Find the exact value of the following expressions:

a. $\cos^{-1}\left(\dfrac{\sqrt{2}}{2}\right)$ **b.** $\arccos(1)$

We now state the properties relating the cosine function and the inverse cosine function that follow directly from properties of inverses.

COSINE-INVERSE COSINE IDENTITIES

$$\cos^{-1}(\cos x) = x \quad \text{for} \quad 0 \le x \le \pi$$
$$\cos(\cos^{-1}x) = x \quad \text{for} \quad -1 \le x \le 1$$

As was the case with inverse identities for sine, you must be careful not to overlook the domain restriction for which these identities hold.

EXAMPLE 4 Using Inverse Identities to Evaluate Expressions Involving Inverse Cosine Functions

Find exact values of the trigonometric expressions.

a. $\cos\left(\cos^{-1}\left(-\dfrac{1}{2}\right)\right)$ **b.** $\cos^{-1}\left(\cos\dfrac{7\pi}{4}\right)$

Solution (a):

Write the appropriate identity. $\cos(\cos^{-1}x) = x$ for $-1 \le x \le 1$

Let $x = -\dfrac{1}{2}$, which is in the interval $[-1, 1]$.

Since the domain restriction is met,
the identity can be used. $\cos\left(\cos^{-1}\left(-\dfrac{1}{2}\right)\right) = -\dfrac{1}{2}$

TECHNOLOGY TIP

a. Check the answer of
$\cos\left(\cos^{-1}\left(-\dfrac{1}{2}\right)\right)$ using a
calculator,

cos(cos⁻¹(-1/2))
                -.5
■

■ **Answers: a.** $\dfrac{\pi}{4}$ **b.** 0

Solution (b):

Write the appropriate identity. $\cos^{-1}(\cos x) = x$ for $0 \le x \le \pi$

Let $x = \dfrac{7\pi}{4}$, which is *not* in the interval $[0, \pi]$.

Since the domain restriction is not met, the identity cannot be used.

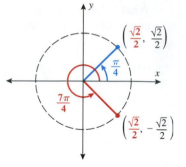

Instead, we find another angle in the interval that has the same value for cosine.

$$\cos\frac{7\pi}{4} = \cos\frac{\pi}{4}$$

Substitute $\cos\dfrac{7\pi}{4} = \cos\dfrac{\pi}{4}$ into the expression.

$$\cos^{-1}\!\left(\cos\frac{7\pi}{4}\right) = \cos^{-1}\!\left(\cos\frac{\pi}{4}\right)$$

Since $\dfrac{\pi}{4}$ is in the interval $[0, \pi]$,

we can use the identity.

$$= \frac{\pi}{4}$$

■ YOUR TURN   Find exact values of the trigonometric expressions.

a. $\cos\left(\cos^{-1}\left(\dfrac{1}{2}\right)\right)$ **b.** $\cos^{-1}\left(\cos\left(-\dfrac{\pi}{6}\right)\right)$

Inverse Tangent Function

The tangent function is not a one-to-one function (it fails the horizontal line test). Let us start with the tangent function with a restricted domain:

$y = \tan x$ Domain: $\left(-\dfrac{\pi}{2}, \dfrac{\pi}{2}\right)$ Range: $(-\infty, \infty)$

x	y
$-\dfrac{\pi}{2}$	$-\infty$
$-\dfrac{\pi}{4}$	-1
0	0
$\dfrac{\pi}{4}$	1
$\dfrac{\pi}{2}$	∞

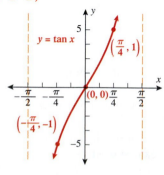

By properties of inverses, the inverse tangent function will have a domain of $(-\infty, \infty)$ and a range of $\left(-\dfrac{\pi}{2}, \dfrac{\pi}{2}\right)$. To find the inverse tangent function, interchange the x and y values.

$$y = \tan^{-1}x \qquad \text{Domain: } (-\infty, \infty) \qquad \text{Range: } \left(-\frac{\pi}{2}, \frac{\pi}{2}\right)$$

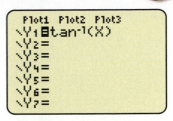

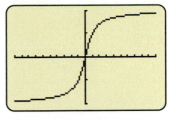

x	y
$-\infty$	$-\dfrac{\pi}{2}$
-1	$-\dfrac{\pi}{4}$
0	0
1	$\dfrac{\pi}{4}$
∞	$\dfrac{\pi}{2}$

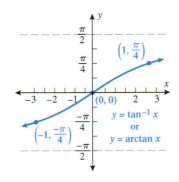

Notice that the inverse tangent function, like the tangent function, is an odd function (symmetric about the origin).

The inverse tangent function allows us to answer the question, if the tangent of an angle is known, what is the angle? Another notation for the inverse tangent function is arctan x.

INVERSE TANGENT FUNCTION

$$y = \tan^{-1}x \text{ or } y = \arctan x \qquad \text{means} \qquad x = \tan y$$

"y is the inverse tangent of x"   "y is the angle or real number whose tangent equals x"

$$\text{where } -\frac{\pi}{2} \leq y \leq \frac{\pi}{2}$$

EXAMPLE 5 Finding Exact Values of an Inverse Tangent Function

Find the exact value of the following expressions:

a. $\tan^{-1}(\sqrt{3})$ **b.** $\arctan(0)$

Solution (a):

Let $\theta = \tan^{-1}(\sqrt{3})$. $\tan\theta = \sqrt{3}$ when $-\dfrac{\pi}{2} < \theta < \dfrac{\pi}{2}$

What value of θ in the range $-\dfrac{\pi}{2} < \theta < \dfrac{\pi}{2}$ corresponds to a tangent value of $\sqrt{3}$? $\theta = \dfrac{\pi}{3}$

$\tan \dfrac{\pi}{3} = \sqrt{3}$ and $\dfrac{\pi}{3}$ is in

the interval $\left(-\dfrac{\pi}{2}, \dfrac{\pi}{2}\right)$. $\qquad \tan^{-1}(\sqrt{3}) = \dfrac{\pi}{3}$

Calculator Confirmation: Since $\dfrac{\pi}{3} = 60°$ if our calculator is set in degrees

mode, we should find that $\tan^{-1}(\sqrt{3})$ is equal to 60°.

Solution (b):

Let $\theta = \arctan(0)$. $\qquad\qquad \tan\theta = 0$ when $-\dfrac{\pi}{2} < \theta < \dfrac{\pi}{2}$.

What value of θ in the range

$-\dfrac{\pi}{2} < \theta < \dfrac{\pi}{2}$ corresponds to a

tangent value of 0? $\qquad\qquad\qquad \theta = 0$

$\tan 0 = 0$ and 0 is in

the interval $\left(-\dfrac{\pi}{2}, \dfrac{\pi}{2}\right)$. $\qquad \arctan(0) = 0$

Calculator Confirmation: $\tan^{-1} 0$ is equal to 0.

We now state the properties relating the tangent function and the inverse tangent function that follow directly from properties of inverses.

TANGENT-INVERSE TANGENT IDENTITIES

$$\tan^{-1}(\tan x) = x \qquad \text{for} \qquad -\dfrac{\pi}{2} < x < \dfrac{\pi}{2}$$

$$\tan\left(\tan^{-1} x\right) = x \qquad \text{for} \qquad -\infty < x < \infty$$

TECHNOLOGY TIP
a. Use a calculator to check the answer for $\tan(\tan^{-1} 17)$.

```
tan(tan-1(17))
                17
█
```

b. Use a calculator to check the answer for $\tan^{-1}\left(\tan\dfrac{2\pi}{3}\right)$.

```
tan-1(tan(2π/3)
       -1.047197551
-π/3
       -1.047197551
█
```

EXAMPLE 6 **Using Inverse Identities to Evaluate Expressions Involving Inverse Tangent Functions**

Find exact values of the trigonometric expressions.

a. $\tan(\tan^{-1} 17)$ $\qquad\qquad$ **b.** $\tan^{-1}\left(\tan\dfrac{2\pi}{3}\right)$

Solution (a):

Write the appropriate identity. $\qquad \tan\left(\tan^{-1} x\right) = x$ for $-\infty < x < \infty$

Let $x = 17$, which is in the interval $(-\infty, \infty)$.

Since the domain restriction is met, the identity can be used. $\qquad \tan(\tan^{-1} 17) = 17$

Solution (b):

Write the appropriate identity.

$$\tan^{-1}(\tan x) = x \text{ for } -\frac{\pi}{2} < x < \frac{\pi}{2}$$

Let $x = \dfrac{2\pi}{3}$, which is *not* in the interval $\left(-\dfrac{\pi}{2}, \dfrac{\pi}{2}\right)$.

Since the domain restriction is not met, the identity cannot be used.

Instead, we find another angle in the interval that has the same value for tangent.

$$\tan\frac{2\pi}{3} = \tan\left(-\frac{\pi}{3}\right)$$

Substitute $\tan\dfrac{2\pi}{3} = \tan\left(-\dfrac{\pi}{3}\right)$ into the expression.

$$\tan^{-1}\left(\tan\frac{2\pi}{3}\right) = \tan^{-1}\left(\tan\left(-\frac{\pi}{3}\right)\right)$$

Since $-\dfrac{\pi}{3}$ is in the interval $\left(-\dfrac{\pi}{2}, \dfrac{\pi}{2}\right)$, we can use the identity.

$$= -\frac{\pi}{3}$$

CONCEPT CHECK   What value in the interval $\left(-\dfrac{\pi}{2}, \dfrac{\pi}{2}\right)$ has the same tangent value as $\dfrac{7\pi}{6}$?

YOUR TURN   Find the exact value of $\tan^{-1}\left(\tan\dfrac{7\pi}{6}\right)$.

Remaining Inverse Trigonometric Functions

The remaining three inverse trigonometric functions are defined similarly.

- Inverse cotangent function: $\cot^{-1}x$ or $\operatorname{arccot} x$
- Inverse secant function: $\sec^{-1}x$ or $\operatorname{arcsec} x$
- Inverse cosecant function: $\csc^{-1}x$ or $\operatorname{arccsc} x$

■ **Answers:** $\dfrac{\pi}{6}$

A table summarizing all six of the inverse trigonometric functions is given below.

Inverse Function	Domain	Range	Graph
$y = \sin^{-1}x$	$[-1, 1]$	$\left[-\dfrac{\pi}{2}, \dfrac{\pi}{2}\right]$	
$y = \cos^{-1}x$	$[-1, 1]$	$[0, \pi]$	
$y = \tan^{-1}x$	$(-\infty, \infty)$	$\left(-\dfrac{\pi}{2}, \dfrac{\pi}{2}\right)$	
$y = \cot^{-1}x$	$(-\infty, \infty)$	$(0, \pi)$	
$y = \sec^{-1}x$	$(-\infty, -1]\cup[1, \infty)$	$\left[0, \dfrac{\pi}{2}\right)\cup\left(\dfrac{\pi}{2}, \pi\right]$	
$y = \csc^{-1}x$	$(-\infty, -1]\cup[1, \infty)$	$\left[-\dfrac{\pi}{2}, 0\right)\cup\left(0, \dfrac{\pi}{2}\right]$	

EXAMPLE 7 Finding the Exact Value of Inverse Trigonometric Functions

Find the exact value of the following expressions:

a. $\cot^{-1}\sqrt{3}$ **b.** $\csc^{-1}\sqrt{2}$ **c.** $\sec^{-1}\left(-\sqrt{2}\right)$

Solution (a):

Let $\theta = \cot^{-1}\sqrt{3}$. $\cot\theta = \sqrt{3}$ when $0 < \theta < \pi$

What value of θ in the range $0 < \theta < \pi$
corresponds to a cotangent value of $\sqrt{3}$? $\theta = \dfrac{\pi}{6}$

$\cot \dfrac{\pi}{6} = \sqrt{3}$ and $\dfrac{\pi}{6}$ is in

the interval $\left(-\dfrac{\pi}{2}, \dfrac{\pi}{2}\right)$. $\qquad\qquad\qquad \boxed{\cot^{-1}\sqrt{3} = \dfrac{\pi}{6}}$

Solution (b):

Let $\theta = \csc^{-1}\sqrt{2}$. $\qquad\qquad\qquad\qquad\qquad \csc \theta = \sqrt{2}$

What value of θ in the range $\left[-\dfrac{\pi}{2}, 0\right) \cup \left(0, \dfrac{\pi}{2}\right]$

corresponds to a cosecant value of $\sqrt{2}$? $\qquad\qquad\quad \theta = \dfrac{\pi}{4}$

$\csc \dfrac{\pi}{4} = \sqrt{2}$ and $\dfrac{\pi}{4}$ is in

the interval $\left[-\dfrac{\pi}{2}, 0\right) \cup \left(0, \dfrac{\pi}{2}\right]$. $\qquad\qquad \boxed{\csc^{-1}\sqrt{2} = \dfrac{\pi}{4}}$

Solution (c):

Let $\theta = \sec^{-1}(-\sqrt{2})$. $\qquad\qquad\qquad\qquad\quad \sec \theta = -\sqrt{2}$

What value of θ in the range $\left[0, \dfrac{\pi}{2}\right) \cup \left(\dfrac{\pi}{2}, \pi\right]$

corresponds to a secant value of $-\sqrt{2}$? $\qquad\qquad\quad \theta = \dfrac{3\pi}{4}$

$\sec \dfrac{3\pi}{4} = -\sqrt{2}$ and $\dfrac{3\pi}{4}$ is in

the interval $\left[0, \dfrac{\pi}{2}\right) \cup \left(\dfrac{\pi}{2}, \pi\right]$. $\qquad\quad \boxed{\sec^{-1}(-\sqrt{2}) = \dfrac{3\pi}{4}}$

How do we approximate inverse secant, inverse cosecant, and inverse cotangent functions with a calculator? Scientific calculators have keys ($\sin^{-1}$, $\cos^{-1}$, and $\tan^{-1}$) for three of the inverse trigonometric functions but not for the other three inverse trigonometric functions. We find the secant, cosecant, and cotangent function values by taking cosine, sine, or tangent, and finding the reciprocal.

$$\sec x = \frac{1}{\cos x} \qquad \csc x = \frac{1}{\sin x} \qquad \cot x = \frac{1}{\tan x}$$

That approach cannot be used for inverse functions. It is important to note that the other three inverse trigonometric functions cannot be found by finding the reciprocal of $\sin^{-1}x$, $\cos^{-1}x$, or $\tan^{-1}x$:

$$\sec^{-1}x \neq \frac{1}{\cos^{-1}x} \qquad \csc^{-1} \neq \frac{1}{\sin^{-1}x} \qquad \cot^{-1}x \neq \frac{1}{\tan^{-1}x}$$

Instead, we convert these other three inverse trigonometric functions into either $\sin^{-1}x$, $\cos^{-1}x$, or $\tan^{-1}x$, whichever has the same range.

STUDY TIP

$\sec^{-1}x \neq \dfrac{1}{\cos^{-1}x}$

$\csc^{-1}x \neq \dfrac{1}{\sin^{-1}x}$

$\cot^{-1}x \neq \dfrac{1}{\tan^{-1}x}$

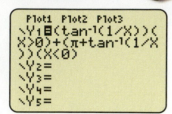

Start with the inverse secant function. $y = \sec^{-1} x$ for $x \le -1$ or $x \ge 1$

Write the equivalent expression. $\sec y = x$ for $0 \le y < \dfrac{\pi}{2}$ or $\dfrac{\pi}{2} < y \le \pi$

Use the reciprocal identity. $\dfrac{1}{\cos y} = x$

Use algebraic techniques to rewrite. $\cos y = \dfrac{1}{x}$

Use the definition of the inverse cosine function. $y = \cos^{-1}\left(\dfrac{1}{x}\right)$

Therefore we have this relationship: $\sec^{-1} x = \cos^{-1}\left(\dfrac{1}{x}\right)$ for $x \le -1$ or $x \ge 1$

The other relationships will be found in the Exercises. They are summarized below.

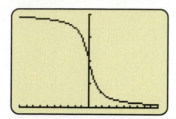

> **INVERSE SECANT, INVERSE COSECANT, AND INVERSE COTANGENT IDENTITIES**
>
> $$\sec^{-1} x = \cos^{-1}\left(\frac{1}{x}\right) \qquad \text{for } x \le -1 \text{ or } x \ge 1$$
>
> $$\csc^{-1} x = \sin^{-1}\left(\frac{1}{x}\right) \qquad \text{for } x \le -1 \text{ or } x \ge 1$$
>
> $$\cot^{-1} x = \begin{cases} \tan^{-1}\left(\dfrac{1}{x}\right) & x > 0 \\[2mm] \pi + \tan^{-1}\left(\dfrac{1}{x}\right) & x < 0 \end{cases}$$

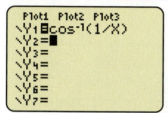

EXAMPLE 8 Using Inverse Identities

a. Find the exact value of $\sec^{-1} 2$.
b. Use a calculator to find the value of $\cot^{-1} 7$.

Solution (a):

Let $\theta = \sec^{-1} 2$, which means: $\sec \theta = 2$ on $\left[0, \dfrac{\pi}{2}\right) \cup \left(\dfrac{\pi}{2}, \pi\right]$

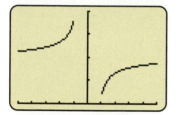

Use the reciprocal identity. $\dfrac{1}{\cos \theta} = 2$

Solve for $\cos \theta$. $\cos \theta = \dfrac{1}{2}$

The restriction interval $\left[0, \dfrac{\pi}{2}\right) \cup \left(\dfrac{\pi}{2}, \pi\right]$ corresponds to QI and QII.

Cosine is positive in QI. $\theta = \dfrac{\pi}{3}$

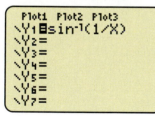

$$\boxed{\sec^{-1} 2 = \dfrac{\pi}{3}}$$

Solution (b):

Since we do not know exact values that correspond to the cotangent function being equal to 7, we proceed using identities and a calculator.

Select the correct identity given that $x = 7 > 0$.

$$\cot^{-1} x = \tan^{-1}\left(\frac{1}{x}\right)$$

Let $x = 7$.

$$\cot^{-1} 7 = \tan^{-1}\left(\frac{1}{7}\right)$$

Use a calculator to evaluate the right side.

$$\cot^{-1} 7 = 8.13°$$

Finding Exact Values for Expressions Involving Inverse Trigonometric Functions

Now we will find exact values of trigonometric expressions that involve inverse trigonometric functions.

EXAMPLE 9 Finding Exact Values of Trigonometric Expressions Involving Inverse Trigonometric Functions

Find the exact value of $\cos\left[\sin^{-1}\left(\frac{2}{3}\right)\right]$.

Solution: Let $\theta = \sin^{-1}\left(\frac{2}{3}\right)$ and find $\cos \theta$.

STEP 1 Let $\theta = \sin^{-1}\left(\frac{2}{3}\right)$.

$\sin \theta = \frac{2}{3}$ when $-\frac{\pi}{2} \leq \theta \leq \frac{\pi}{2}$

The range $-\frac{\pi}{2} \leq \theta \leq \frac{\pi}{2}$ corresponds to Quadrants I and IV.

Sine is positive in Quadrant I.

STEP 2 Draw angle θ in Quadrant I.

Label the sides known

from $\sin \theta = \frac{2}{3} = \frac{\text{opposite}}{\text{hypotenuse}}$.

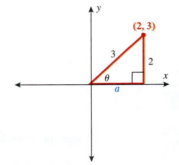

STEP 3 Find the unknown side length, a.

$$a^2 + 2^2 = 3^2$$

Solve for a.

$$a = \pm\sqrt{5}$$

Since θ is in Quadrant I, a is positive.

$$a = \sqrt{5}$$

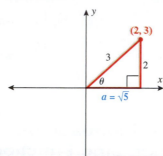
STEP 4 Find $\cos\left[\sin^{-1}\left(\dfrac{2}{3}\right)\right]$.

Substitute $\theta = \sin^{-1}\left(\dfrac{2}{3}\right)$. $\cos\left[\sin^{-1}\left(\dfrac{2}{3}\right)\right] = \cos\theta$

Find $\cos\theta$.

$$\cos\theta = \frac{\text{adjacent}}{\text{hypotenuse}} = \frac{\sqrt{5}}{3}$$

$$\cos\left[\sin^{-1}\left(\frac{2}{3}\right)\right] = \frac{\sqrt{5}}{3}$$

■ **YOUR TURN** Find the exact value of $\sin\left[\cos^{-1}\left(\dfrac{1}{3}\right)\right]$.

EXAMPLE 10 **Finding Exact Values of Trigonometric Expressions Involving Inverse Trigonometric Functions**

Find the exact value of $\tan\left[\cos^{-1}\left(-\dfrac{7}{12}\right)\right]$.

Solution: Let $\theta = \cos^{-1}\left(-\dfrac{7}{12}\right)$ then find $\tan\theta$.

STEP 1 Let $\theta = \cos^{-1}\left(-\dfrac{7}{12}\right)$ which means $\cos\theta = -\dfrac{7}{12}$ when $0 \le \theta \le \pi$

The range $0 \le \theta \le \pi$ corresponds to Quadrants I and II.

Cosine is negative in Quadrant II.

STEP 2 Draw angle θ in Quadrant II.

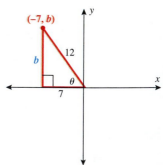

Label the sides known from $\cos\theta = -\dfrac{7}{12} = \dfrac{\text{adjacent}}{\text{hypotenuse}}$.

STEP 3 Find the unknown side length, b. $b^2 + (-7)^2 = 12^2$

Solve for b. $b = \pm\sqrt{95}$

Since θ is in Quadrant II, b is positive. $b = \sqrt{95}$

■ **Answer:** $\dfrac{2\sqrt{2}}{3}$

STEP 4 Find $\tan\left[\cos^{-1}\left(-\dfrac{7}{12}\right)\right]$.

TECHNOLOGY TIP
Use a TI calculator to check
the answer for
$$\tan\left(\cos^{-1}\left(-\dfrac{7}{12}\right)\right).$$

Substitute $\theta = \cos^{-1}\left(-\dfrac{7}{12}\right)$.

$$\tan\left[\cos^{-1}\left(-\dfrac{7}{12}\right)\right] = \tan\theta$$

```
tan(cos⁻¹(-7/12))
                 -1.392399192
-√(95)/7
                 -1.392399192
```

Find $\tan\theta$.

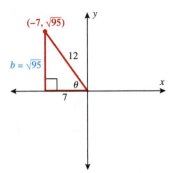

(−7, $\sqrt{95}$)

$b = \sqrt{95}$

12

7

θ

$$\tan\theta = \frac{\text{opposite}}{\text{adjacent}} = \frac{\sqrt{95}}{-7}$$

$$\tan\left[\cos^{-1}\left(-\frac{7}{12}\right)\right] = -\frac{\sqrt{95}}{7}$$

■ **YOUR TURN**   Find the exact value of $\tan\left[\sin^{-1}\left(-\dfrac{3}{7}\right)\right]$.

SECTION 6.1 SUMMARY

If a trigonometric function value of an angle is known, then what is the angle?
Inverse trigonometric functions determine the angle. To define the inverse trigonometric relations as functions, we first restrict the trigonometric functions to the
domains in which they are one-to-one functions. Techniques for finding exact values for the inverse trigonometric functions were demonstrated, as well as techniques for evaluating expressions involving inverse trigonometric functions. It is
important to note that the -1 in the exponent's spot indicates inverse functions,
not reciprocals.

SECTION 6.1 EXERCISES

■ **SKILLS**

In Exercises 1–12, find the exact value of each expression. Give the answer in radians.

1. $\arccos\dfrac{\sqrt{2}}{2}$

2. $\arccos\left(-\dfrac{\sqrt{2}}{2}\right)$

3. $\arcsin\left(-\dfrac{\sqrt{3}}{2}\right)$

4. $\arcsin\left(\dfrac{1}{2}\right)$

5. $\cot^{-1}(-1)$

6. $\tan^{-1}\left(\dfrac{\sqrt{3}}{3}\right)$

7. $\arcsec\left(\dfrac{2\sqrt{3}}{3}\right)$

8. $\arccsc(-1)$

9. $\csc^{-1}2$

10. $\sec^{-1}(-2)$

11. $\arctan(-\sqrt{3})$

12. $\operatorname{arccot}\sqrt{3}$

■ **Answer:** $-\dfrac{3\sqrt{10}}{20}$

In Exercises 13–24, find the exact value of each expression. Give the answer in degrees.

13. $\cos^{-1}\left(\dfrac{1}{2}\right)$ **14.** $\cos^{-1}\left(-\dfrac{\sqrt{3}}{2}\right)$ **15.** $\sin^{-1}\left(\dfrac{\sqrt{2}}{2}\right)$ **16.** $\sin^{-1}(0)$

17. $\cot^{-1}\left(-\dfrac{\sqrt{3}}{3}\right)$ **18.** $\tan^{-1}(\sqrt{3})$ **19.** $\arctan\left(\dfrac{\sqrt{3}}{3}\right)$ **20.** $\operatorname{arccot}(1)$

21. $\operatorname{arccsc}(-2)$ **22.** $\csc^{-1}\left(-\dfrac{2\sqrt{3}}{3}\right)$ **23.** $\operatorname{arcsec}(-\sqrt{2})$ **24.** $\operatorname{arccsc}(-\sqrt{2})$

In Exercises 25–34, use a calculator to evaluate each expression. Give the answer in degrees and round to two decimal places.

25. $\cos^{-1}0.5432$ **26.** $\sin^{-1}0.7821$ **27.** $\tan^{-1}1.895$ **28.** $\tan^{-1}3.2678$

29. $\sec^{-1}1.4973$ **30.** $\sec^{-1}2.7864$ **31.** $\csc^{-1}(-3.7893)$ **32.** $\csc^{-1}(-6.1324)$

33. $\cot^{-1}(-4.2319)$ **34.** $\cot^{-1}(-0.8977)$

In Exercises 35–44, use a calculator to evaluate each expression. Give the answer in radians and round to two decimal places.

35. $\sin^{-1}(-0.5878)$ **36.** $\sin^{-1}0.8660$ **37.** $\cos^{-1}0.1423$ **38.** $\tan^{-1}(-0.9279)$

39. $\tan^{-1}1.3242$ **40.** $\cot^{-1}2.4142$ **41.** $\cot^{-1}(-0.5774)$ **42.** $\sec^{-1}(-1.0422)$

43. $\csc^{-1}3.2361$ **44.** $\csc^{-1}(-2.9238)$

In Exercises 45–62, evaluate each expression exactly, if possible. If not possible, state why.

45. $\sin^{-1}\left(\sin\dfrac{5\pi}{12}\right)$ **46.** $\sin^{-1}\left(\sin\left(-\dfrac{5\pi}{12}\right)\right)$ **47.** $\sin(\sin^{-1}1.03)$ **48.** $\sin(\sin^{-1}1.1)$

49. $\sin^{-1}\left(\sin\left(-\dfrac{7\pi}{6}\right)\right)$ **50.** $\sin^{-1}\left(\sin\dfrac{7\pi}{6}\right)$ **51.** $\cos^{-1}\left(\cos\dfrac{4\pi}{3}\right)$ **52.** $\cos^{-1}\left(\cos\left(-\dfrac{5\pi}{3}\right)\right)$

53. $\cot(\cot^{-1}\sqrt{3})$ **54.** $\cot^{-1}\left(\cot\dfrac{5\pi}{4}\right)$ **55.** $\sec^{-1}\left(\sec\left(-\dfrac{\pi}{3}\right)\right)$ **56.** $\sec\left(\sec^{-1}\dfrac{1}{2}\right)$

57. $\csc\left(\csc^{-1}\dfrac{1}{2}\right)$ **58.** $\csc^{-1}\left(\csc\dfrac{7\pi}{6}\right)$ **59.** $\cot(\cot^{-1}0)$ **60.** $\cot^{-1}\left(\cot\left(-\dfrac{\pi}{4}\right)\right)$

61. $\tan^{-1}\left(\tan\left(-\dfrac{\pi}{4}\right)\right)$ **62.** $\tan^{-1}\left(\tan\dfrac{\pi}{4}\right)$

In Exercises 63–74, evaluate each expression exactly.

63. $\cos\left[\sin^{-1}\dfrac{3}{4}\right]$ **64.** $\sin\left[\cos^{-1}\dfrac{2}{3}\right]$ **65.** $\sin\left[\tan^{-1}\dfrac{12}{5}\right]$ **66.** $\cos\left[\tan^{-1}\dfrac{7}{24}\right]$

67. $\tan\left[\sin^{-1}\dfrac{3}{5}\right]$ **68.** $\tan\left[\cos^{-1}\dfrac{2}{5}\right]$ **69.** $\sec\left[\sin^{-1}\dfrac{\sqrt{2}}{5}\right]$ **70.** $\sec\left[\cos^{-1}\dfrac{\sqrt{7}}{4}\right]$

71. $\csc\left[\cos^{-1}\dfrac{1}{4}\right]$ **72.** $\csc\left[\sin^{-1}\dfrac{1}{4}\right]$ **73.** $\cot\left[\sin^{-1}\dfrac{60}{61}\right]$ **74.** $\cot\left[\sec^{-1}\dfrac{41}{9}\right]$

APPLICATIONS

75. Alternating Current. Alternating electrical current (amperes) is modeled by the equation $i = I\sin(2\pi ft)$, where i is the induced current, I is the maximum current, t is time in seconds, and f is frequency (Hz is number of cycles per second). If the frequency is 5 Hz, maximum current is 115 amps, what time, t, corresponds to a current of 85 amps? Find the smallest positive value of t.

76. Alternating Current. If the frequency is 100 Hz and maximum current is 240 amps, what time, t, corresponds to a current of 100 amps? Find the smallest positive value of t.

77. Hours of Daylight. The number of hours of daylight in San Diego, California, can be modeled with $H(t) = 12 + 2.4\sin(0.017t - 1.377)$, where t is the day of the year (January 1, $t = 1$, etc.). For what value of t is the number of hours equal to 14.4? May 31 is the 151st day of that year; what month and day corresponds to the value of t?

78. Hours of Daylight. Repeat Exercise 77. For what value of t is the number of hours equal to 9.6? What month and day corresponds to the value of t? (You may have to count backwards.)

79. Money. A young couple gets married and immediately starts saving money. They renovate a house and have less and less saved money. They have children after 10 years and are in debt until their children are in college. They then save until retirement. A formula that represents the percentage of their annual income that they either save (positive) or are in debt (negative) is given by $P(t) = 12.5\cos(0.157t) + 2.5$, where $t = 0$ corresponds to the year they get married. How many years into their marriage do they first accrue debt?

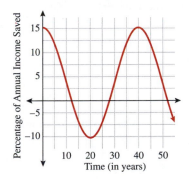

80. Money. How many years into their marriage are the couple in Exercise 79 back to saving 15% of their annual income?

81. Viewing Angle of Picture. A museum patron whose eye level is 5 feet above the floor is studying a painting that is 8 feet in height and mounted on the wall 4 feet above the floor. If the patron is x feet from the wall,

use $\tan(\alpha + \beta)$ to express $\tan(\theta)$, where θ is the angle that the patron's eye sweeps from the top to the bottom of the painting.

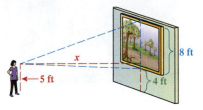

82. Viewing Angle of Picture. Using the equation for $\tan(\theta)$ in Exercise 81, solve for θ using the inverse tangent. Then find the measure of the angles θ for $x = 10$ and $x = 20$ (to the nearer degree).

83. Earthquake Movement. The horizontal movement of a point that is k kilometers away from an earthquake's fault line can be estimated with

$$M = \frac{f}{2}\left(1 - \frac{2\tan^{-1}\frac{k}{d}}{\pi}\right)$$

where M is the movement of the point in meters, f is the total horizontal displacement occurring along the fault line, k is the distance of the point from the fault line, and d is the depth in kilometers of the focal point of the earthquake. If an earthquake produced a displacement, f, of 2 meters and the depth of the focal point was 4 kilometers, then what is the movement, M, of a point that is 2 kilometers from the fault line? 10 kilometers from the fault line?

84. Earthquake Movement. Repeat Exercise 83. If an earthquake produced a displacement, f, of 3 meters and the depth of the focal point was 2.5 kilometers, then what is the movement, M, of a point that is 5 kilometers from the fault line? 10 kilometers from the fault line?

85. Laser Communication. A laser communication system is narrow-beam, and direct line of sight is necessary for communication links. If a transmitter/receiver for a laser system is placed in between two buildings (see the figure) and the other end of the system is located on a low Earth orbit satellite, then the link is only operational when the satellite and the ground system have line of sight. Find the angle, θ, that corresponds to the system being operational. Express θ in terms of inverse tangent functions and the distance from the shorter building, x.

Corbis Digital Stock

300 ft

150 ft

θ

200 ft

86. **Laser Communication.** Repeat Exercise 85, assuming the ground system is on top of a 20 foot tower.

■ CATCH THE MISTAKE

In Exercises 87–90, explain the mistake that is made.

87. Evaluate the expression exactly: $\sin^{-1}\left[\sin\dfrac{3\pi}{5}\right]$.

Solution:

Use the identity $\sin^{-1}(\sin x) = x$ on $0 \le x \le \pi$.

Since $\dfrac{3\pi}{5}$ is in the interval $[0,\pi]$,

the identity can be used. $\quad \sin^{-1}\left[\sin\dfrac{3\pi}{5}\right] = \dfrac{3\pi}{5}$

This is incorrect. What mistake was made?

88. Evaluate the expression exactly: $\cos^{-1}\left[\cos\left(-\dfrac{\pi}{5}\right)\right]$.

Solution:

Use the identity $\cos^{-1}(\cos x) = x$ on $-\dfrac{\pi}{2} \le x \le \dfrac{\pi}{2}$.

Since $-\dfrac{\pi}{5}$ is in the interval $\left[-\dfrac{\pi}{2}, \dfrac{\pi}{2}\right]$,

the identity can be used. $\quad \cos^{-1}\left[\cos\left(-\dfrac{\pi}{5}\right)\right] = -\dfrac{\pi}{5}$

This is incorrect. What mistake was made?

89. Evaluate the expression exactly: $\cot^{-1}(2.5)$.

Solution:

Use the reciprocal identity. $\quad \cot^{-1}(2.5) = \dfrac{1}{\tan^{-1}(2.5)}$

Evaluate $\tan^{-1}(2.5) = 1.19$. $\quad \cot^{-1}(2.5) = \dfrac{1}{1.19}$

Simplify. $\quad \cot^{-1}(2.5) = 0.8403$

This is incorrect. What mistake was made?

90. Evaluate the expression exactly: $\csc^{-1}\dfrac{1}{4}$.

Solution:

Use the reciprocal identity. $\quad \csc^{-1}\dfrac{1}{4} = \dfrac{1}{\sin^{-1}\dfrac{1}{4}}$

Evaluate $\sin^{-1}\dfrac{1}{4} = 14.478$. $\quad \csc^{-1}\dfrac{1}{4} = \dfrac{1}{14.478}$

Simplify. $\quad \csc^{-1}\dfrac{1}{4} = 0.0691$

This is incorrect. What mistake was made?

■ CHALLENGE

91. T or F: The inverse secant function is an even function.

92. T or F: The inverse cosecant function is an odd function.

93. Explain why $\sec^{-1}\dfrac{1}{2}$ does not exist.

94. Explain why $\csc^{-1}\dfrac{1}{2}$ does not exist.

95. Evaluate exactly: $\sin\left[\cos^{-1}\dfrac{\sqrt{2}}{2} + \sin^{-1}\left(-\dfrac{1}{2}\right)\right]$.

96. Evaluate exactly: $\sin\left[2\sin^{-1}1\right]$.

97. Let $f(x) = 2 - 4\sin\left(x - \dfrac{\pi}{2}\right)$.

    a. State the domain of $f(x)$ so that $f(x)$ is a one-to-one function.

    b. Find $f^{-1}(x)$ and state its domain.

98. Let $f(x) = 3 + \cos\left(x - \dfrac{\pi}{4}\right)$.

    a. State the domain of $f(x)$ so that $f(x)$ is a one-to-one function.

    b. Find $f^{-1}(x)$ and state its domain.

■ TECHNOLOGY

99. Use a graphing calculator to plot $Y_1 = \sin\left[\sin^{-1}x\right]$ and $Y_2 = x$ for the domain $-1 \le x \le 1$. If you then increase the domain to $-3 \le x \le 3$, you get a different result. Explain the result.

100. Use a graphing calculator to plot $Y_1 = \cos\left[\cos^{-1}x\right]$ and $Y_2 = x$ for the domain $-1 \le x \le 1$. If you then increase the domain to $-3 \le x \le 3$, you get a different result. Explain the result.

101. Use a graphing calculator to plot $Y_1 = \csc^{-1}\left[\csc x\right]$ and $Y_2 = x$. For what domain is the following statement true: $\csc^{-1}\left[\csc x\right] = x$. Give the domain in terms of π.

102. Use a graphing calculator to plot $Y_1 = \sec^{-1}\left[\sec x\right]$ and $Y_2 = x$. For what domain is the following statement true: $\sec^{-1}\left[\sec x\right] = x$? Give the domain in terms of π.

SECTION 6.2 Solving Trigonometric Equations that Involve Only One Trigonometric Function

Skills Objectives

■ Solve trigonometric equations by inspection.
■ Solve trigonometric equations using algebraic techniques.
■ Solve trigonometric equations using inverse functions.

Conceptual Objectives

■ Understand that solving trigonometric equations is similar to solving algebraic equations.
■ Realize that the goal in solving trigonometric equations is to find the value(s) for the independent variable that make the equation true.

In this section we will develop a strategy for solving trigonometric equations that involve only one trigonometric function. In the following section we will solve more complicated equations that involve multiple trigonometric functions and require the use of trigonometric identities.

Recall in solving algebraic equations that the goal is to find the value for the variable that makes the equation true. For example, the linear equation $2x - 5 = 7$ has only one value, $x = 6$, that makes the statement true. A quadratic equation, however, can have two solutions. The equation $x^2 = 9$ has two values, $x = \pm 3$, that make the statement true. In trigonometric equations, the goal is the same: find the value or values that make the equation true.

We will start with simple trigonometric equations that can be solved by inspection. Then we will use inverse functions and algebraic techniques to solve trigonometric equations involving a single trigonometric function.

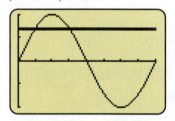
Solving Trigonometric Equations by Inspection

The goal in solving equations in one variable is to find the values for that variable that make the equation true. For example, $9x = 72$ can be solved by inspection by asking the question, "9 times what is 72?" The answer is $x = 8$. We approach simple trigonometric equations the same way we approach algebraic equations: we inspect the equation and determine the solution.

EXAMPLE 1 Solving a Trigonometric Equation by Inspection

Solve the equation $\sin x = \dfrac{\sqrt{2}}{2}$.

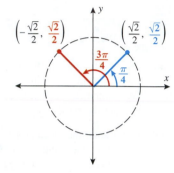

Solution:

STEP 1 Solve over one period, $[0, 2\pi)$.

Ask the question, "sine of what angle yields $\dfrac{\sqrt{2}}{2}$?"

Degrees	$x = 45°$ or $x = 135°$
Radians	$x = \dfrac{\pi}{4}$ or $x = \dfrac{3\pi}{4}$

Sine is positive in Quadrants I and II.

STEP 2 Solve over all x.

Since sine has period $360°$ or 2π, adding integer multiples of $360°$ or 2π will give the other solutions.

Degrees	$x = 45° + 360°n$ or $x = 135° + 360°n$
Radians	$x = \dfrac{\pi}{4} + 2n\pi$ or $x = \dfrac{3\pi}{4} + 2n\pi$
	where n is any integer

STUDY TIP

Find *all* solutions unless the domain is restricted.

■ **YOUR TURN** Solve the equation $\cos x = \dfrac{1}{2}$.

Notice that Example 1 and the Your Turn have an infinite number of solutions. Unless the domain is restricted, you must find *all* solutions.

		■ **Answer:**
Degrees	$x = 60° + 360°n$ or $x = 300° + 360°n$	
Radians	$x = \dfrac{\pi}{3} + 2n\pi$ or $x = \dfrac{5\pi}{3} + 2n\pi$	
	where n is any integer	

EXAMPLE 2 Solving a Trigonometric Equation by Inspection

Solve the equation $\tan 2x = -\sqrt{3}$.

Solution:

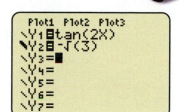

STEP 1 Solve over one period, $[0, \pi)$.

Ask the question, "tangent of what angle yields $-\sqrt{3}$?" Note that the angle in this case is $2x$.

Degrees	$2x = 120°$
Radians	$2x = \dfrac{2\pi}{3}$

Tangent is negative in Quadrants II and IV. Since $[0, \pi)$ includes QI and QII, we find only the angle in QII. (The solution corresponding to QIV will be found when we extend the solution over all real numbers.)

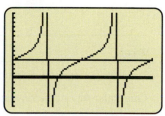

STEP 2 Solve over all x.

Since tangent has period 180° or π, Adding integer multiples of 180° or π will give the other solutions.

Degrees	$2x = 120° + 180°n$
Radians	$2x = \dfrac{2\pi}{3} + n\pi$
	where n is any integer

Solve for x by dividing by 2.

Degrees	$x = 60° + 90°n$
Radians	$x = \dfrac{\pi}{3} + \dfrac{n}{2}\pi$
	where n is any integer

To find the point of intersection, use [2nd] [TRACE] for [CALC], move the down arrow to [5: Intersect], type [ENTER] for the first curve, [ENTER] for the second curve, [0.8] for guess, and [ENTER].

Note:

■ There are infinitely many solutions. If we graph $y = \tan 2x$ and $y = -\sqrt{3}$, we see that there are infinitely many points of intersection.

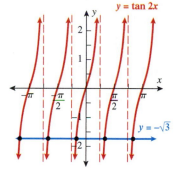

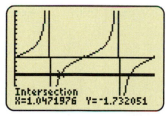

For the second answer, you need a number close to the answer for the guess. Type [2.5] for guess, and [ENTER].

■ Had we restricted the domain to $0 \le x < 2\pi$, the solutions (in radians) would be

n	$x = \dfrac{\pi}{3} + \dfrac{n}{2}\pi$
0	$x = \dfrac{\pi}{3}$
1	$x = \dfrac{5\pi}{6}$
2	$x = \dfrac{4\pi}{3}$
3	$x = \dfrac{11\pi}{6}$

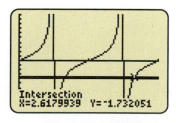

since only $n = 0, 1, 2, 3$ yield x values in the domain $0 \le x < 2\pi$.

Notice in Step 2 of Example 1, $2n\pi$ was added to get all of the solutions, whereas in Step 2 of Example 2, we added $n\pi$ to the argument of tangent. The difference is because the sine function has period 2π, whereas the tangent function has period π.

Solving Trigonometric Equations Using Algebraic Techniques

We now will use algebraic techniques to solve trigonometric equations. Let us first start with linear and quadratic equations. For linear equations, we solve for the variable by isolating it. For quadratic equations we often employ factoring or the quadratic formula. If we can let x represent the trigonometric function and the resulting equation is either linear or quadratic, then we use techniques learned in solving algebraic problems.

TYPE	EQUATION	SUBSTITUTION	ALGEBRAIC EQUATION
Linear trigonometric equation	$4\sin\theta - 2 = -4$	$x = \sin\theta$	$4x - 2 = -4$
Quadratic trigonometric equation	$2\cos^2\theta + \cos\theta - 1 = 0$	$x = \cos\theta$	$2x^2 + x - 1 = 0$

It is not necessary to make the substitution. Often, one can see how to factor a quadratic trigonometric equation without first converting it to an algebraic equation. In Example 3 we will not use the substitution. However, in Example 4, we will illustrate the use of a substitution.

EXAMPLE 3 Solving a Linear Trigonometric Equation

Solve $4\sin\theta - 2 = -4$ on $0 \le \theta < 2\pi$.

Solution:

STEP 1   Solve for the trigonometric function. $4\sin\theta - 2 = -4$

Add 2. $4\sin\theta = -2$

Divide by 4. $\sin\theta = -\dfrac{1}{2}$

STEP 2   Find the values of θ on $0 \le \theta < 2\pi$ that satisfy the equation $\sin\theta = -\dfrac{1}{2}$.

Sine is negative in Quadrants III and IV.

$\sin\dfrac{7\pi}{6} = -\dfrac{1}{2}$ and $\sin\dfrac{11\pi}{6} = -\dfrac{1}{2}$ $\boxed{\theta = \dfrac{7\pi}{6}}$ or $\boxed{\theta = \dfrac{11\pi}{6}}$

■ **YOUR TURN**   Solve $2\cos\theta + 1 = 2$ on $0 \le \theta < 2\pi$.

■ **Answer:** $\theta = \dfrac{\pi}{3}$ or $\theta = \dfrac{5\pi}{3}$

EXAMPLE 4 Solving a Quadratic Trigonometric Equation

TECHNOLOGY TIP

Solve $2\cos^2\theta + \cos\theta - 1 = 0$ on $0 \le \theta < 2\pi$.

Solution:

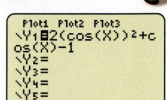

STEP 1	Solve for the trigonometric function.	$2\cos^2\theta + \cos\theta - 1 = 0$
	Let $x = \cos\theta$.	$2x^2 + x - 1 = 0$
	Factor the quadratic equation.	$(2x - 1)(x + 1) = 0$
	Set factors equal to 0.	$2x - 1 = 0$ or $x + 1 = 0$
	Solve for x.	$x = \dfrac{1}{2}$ or $x = -1$
	Substitute $x = \cos\theta$.	$\cos\theta = \dfrac{1}{2}$ or $\cos\theta = -1$

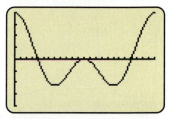

STEP 2 Find the values of θ on $0 \le \theta < 2\pi$ that satisfy the equation $\cos\theta = \dfrac{1}{2}$.

Cosine is positive in Quadrants I and IV.

$\cos\dfrac{\pi}{3} = \dfrac{1}{2}$ and $\cos\dfrac{5\pi}{3} = \dfrac{1}{2}$ $\boxed{\theta = \dfrac{\pi}{3} \quad \text{or} \quad \theta = \dfrac{5\pi}{3}}$

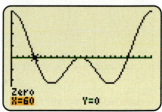

STEP 3 Find the values of θ on $0 \le \theta < 2\pi$ that satisfy the equation $\cos\theta = -1$.

$\cos\pi = -1$ $\boxed{\theta = \pi}$

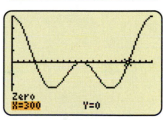

The solution to $2\cos^2\theta + \cos\theta - 1 = 0$ on $0 \le \theta < 2\pi$ is $\theta = \dfrac{\pi}{3}$, $\theta = \dfrac{5\pi}{3}$, or $\theta = \pi$.

✓**CONCEPT CHECK**   Factor the quadratic trigonometric expression $2\sin^2\theta - \sin\theta - 1$.

■ **YOUR TURN**   Solve $2\sin^2\theta - \sin\theta - 1 = 0$ on $0 \le \theta < 2\pi$.

Solving Trigonometric Equations that Require the Use of Inverse Functions

Thus far, we have been able to solve the trigonometric equations exactly. Now we turn our attention to those cases for which a calculator and inverse functions are required to approximate a solution to a trigonometric equation.

■ **Answer:** $\theta = \dfrac{\pi}{2}, \theta = \dfrac{7\pi}{6},$ or $\theta = \dfrac{11\pi}{6}$

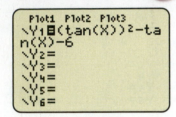

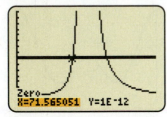

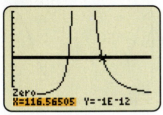

EXAMPLE 5 Solving a Trigonometric Equation that Requires the Use of Inverse Functions

Solve $\tan^2 \theta - \tan \theta = 6$ on $0° \le \theta < 180°$

Solution:

STEP 1 Solve for the trigonometric function.

Subtract 6.	$\tan^2 \theta - \tan \theta - 6 = 0$
Factor the quadratic trigonometric expression on the left.	$(\tan \theta - 3)(\tan \theta + 2) = 0$
Set the factors equal to 0.	$\tan \theta - 3 = 0$ or $\tan \theta + 2 = 0$
Solve for $\tan \theta$.	$\tan \theta = 3$ or $\tan \theta = -2$

STEP 2 Solve $\tan \theta = 3$ on $0° \le \theta < 180°$.

Tangent is positive on $0° \le \theta < 180°$ only in Quadrant I.

Write the equivalent inverse notation to $\tan \theta = 3$. $\theta = \tan^{-1} 3$

Use a calculator to evaluate (approximate) θ. $\theta = 71.6°$

STEP 3 Solve $\tan \theta = -2$ on $0° \le \theta < 180°$.

Tangent is negative on $0° \le \theta < 180°$ only in Quadrant II.

A calculator gives values of inverse tangent in Quadrants I and IV.

We will call the reference angle in Quadrant IV, α.

Write the equivalent inverse notation
to $\tan \alpha = -2$. $\alpha = \tan^{-1}(-2)$

Use a calculator to evaluate (approximate) α. $\alpha = -63.4°$

To find the value of θ in QII, add 180°. $\theta = \alpha + 180°$

$\theta = 116.6°$

The solution to $\tan^2 \theta - \tan \theta = 6$ on $0° \le \theta < 180°$ is $\theta = 71.6°$ or $\theta = 116.6°$.

YOUR TURN Solve $\tan^2 \theta + \tan \theta = 6$ on $0° \le \theta < 180°$.

Recall in solving algebraic quadratic equations that one method (when factoring is not obvious or possible) is to use the quadratic formula.

$$ax^2 + bx + c = 0 \text{ has solutions } x = \frac{-b \pm \sqrt{b^2 - 4ac}}{2a}$$

Answer: $\theta = 63.4°$ or $\theta = 108.4°$

EXAMPLE 6 Solving a Quadratic Trigonometric Equation that Requires the Use of the Quadratic Formula and Inverse Functions

Solve $2\cos^2\theta + 5\cos\theta - 6 = 0$ on $0° \le \theta < 360°$.

Solution:

STEP 1 Solve for the trigonometric function.

$$2\cos^2\theta + 5\cos\theta - 6 = 0$$

Let $x = \cos\theta$.

$$2x^2 + 5x - 6 = 0$$

Use the quadratic formula, $a = 2, b = 5, c = -6$.

$$x = \frac{-5 \pm \sqrt{5^2 - 4(2)(-6)}}{2(2)}$$

Simplify.

$$x = \frac{-5 \pm \sqrt{73}}{4}$$

Use a calculator to approximate. $x = -3.3860$ or $x = 0.8860$

Let $x = \cos\theta$. $\cos\theta = -3.3860$ or $\cos\theta = 0.8860$

STEP 2 Solve $\cos\theta = -3.3860$ on $0° \le \theta < 360°$.

Recall that the range of cosine is $[-1, 1]$; therefore cosine can never equal a number outside that range (-3.3860).

No solution from this equation.

STEP 3 Solve $\cos\theta = 0.8860$ on $0° \le \theta < 360°$.

Cosine is positive in QI and QIV. Since a calculator gives inverse cosine values only in QI and QII, we will have to use a reference angle to get the QIV solution.

Write the equivalent inverse notation for $\cos\theta = 0.8860$. $\theta = \cos^{-1} 0.8860$

Use a calculator to evaluate (approximate). $\theta = 27.6°$

To find the second solution (in QIV), subtract the reference angle from $360°$. $\theta = 360° - 27.6°$

$$\theta = 332.4°$$

The solution to $2\cos^2\theta + 5\cos\theta - 6 = 0$ on $0° \le \theta < 360°$ is $\theta = 27.6°$ or $\theta = 332.4°$.

TECHNOLOGY TIP

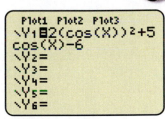

```
Plot1  Plot2  Plot3
\Y1∎2(cos(X))²+5
cos(X)-6
\Y2=
\Y3=
\Y4=
\Y5=
\Y6=
```

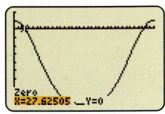

```
Zero
X=27.62505 _Y=0
```

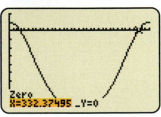

```
Zero
X=332.37495 _Y=0
```

■ **YOUR TURN**   Solve $2\sin^2\theta - 5\sin\theta - 6 = 0$ on $0° \le \theta < 360°$.

■ **Answer:** $\theta = 242.4°$ or $\theta = 297.6°$

Applications

 EXAMPLE 7 Applications Involving Trigonometric Equations

Light bends according to Snell's law, which states:

$$n_i \sin(\theta_i) = n_r \sin(\theta_r)$$

- n_i is the refractive index of the medium the light is leaving.

- θ_i is the incident angle between the light ray and the normal (perpendicular) to the interface between mediums.

- n_r is the refractive index of the medium the light is entering.

- θ_r is the refractive angle between the light ray and the normal (perpendicular) to the interface between mediums.

Janis Christie/Getty Images, Inc.

Assume that light is going from air into a diamond. Calculate the refractive angle, θ_r, if the incidence angle is $\theta_i = 32°$ and the index of refraction values for air and diamond are $n_i = 1.00$ and $n_r = 2.417$, respectively.

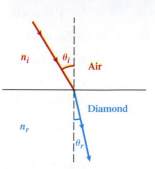

Solution:

Write Snell's law.

$$n_i \sin(\theta_i) = n_r \sin(\theta_r)$$

Substitute $\theta_i = 32°$, $n_i = 1.00$ and $n_r = 2.417$.

$$\sin 32° = 2.417 \sin\theta_r$$

Isolate $\sin\theta_r$ and simplify.

$$\sin\theta_r = \frac{\sin 32°}{2.417} = 0.21925$$

Solve for θ_r using the inverse sine function.

$$\theta_r = \sin^{-1}(0.21925) = 12.665°$$

Round to the nearest degree.

$$\theta_r \approx 13°$$

SECTION 6.2 ## SUMMARY

In this section we solve basic trigonometric equations that contained only one trigonometric function. We solved such equations exactly by inspection, and then later using algebraic techniques similar to linear and quadratic equations. Calculators and inverse functions are needed when exact values are not known. It is important to note that calculators only give the inverse function in half of the quadrants. The other quadrant solutions must be found using reference angles.

SECTION 6.2 EXERCISES

▪ SKILLS

In Exercises 1–14, solve the given trigonometric equation exactly over the indicated interval.

1. $\cos \theta = -\dfrac{\sqrt{2}}{2}, 0 \le \theta < 2\pi$

2. $\sin \theta = -\dfrac{\sqrt{2}}{2}, 0 \le \theta < 2\pi$

3. $\csc \theta = -2, 0 \le \theta < 4\pi$

4. $\sec \theta = -2, 0 \le \theta < 4\pi$

5. $\tan \theta = 0$, all real numbers

6. $\cot \theta = 0$, all real numbers

7. $\sin 2\theta = -\dfrac{1}{2}, 0 \le \theta < 2\pi$

8. $\cos 2\theta = \dfrac{\sqrt{3}}{2}, 0 \le \theta < 2\pi$

9. $\sin \dfrac{\theta}{2} = -\dfrac{1}{2}$, all real numbers

10. $\cos \dfrac{\theta}{2} = -1$, all real numbers

11. $\tan 2\theta = \sqrt{3}, -2\pi \le \theta < 2\pi$

12. $\tan 2\theta = -\sqrt{3}$, all real numbers

13. $\sec \theta = -2, -2\pi \le \theta < 0$

14. $\csc \theta = \dfrac{2\sqrt{3}}{3}, -\pi \le \theta < \pi$

In Exercises 15–30, solve the given trigonometric equation exactly on $0 \le \theta < 2\pi$.

15. $2 \sin 2\theta = \sqrt{3}$

16. $2 \cos \dfrac{\theta}{2} = -\sqrt{2}$

17. $3 \tan 2\theta - \sqrt{3} = 0$

18. $4 \tan \dfrac{\theta}{2} - 4 = 0$

19. $2 \cos 2\theta + 1 = 0$

20. $4 \csc 2\theta + 8 = 0$

21. $\sqrt{3} \cot \dfrac{\theta}{2} - 3 = 0$

22. $\sqrt{3} \sec 2\theta + 2 = 0$

23. $\tan^2 \theta - 1 = 0$

24. $\sin^2 \theta + 2 \sin \theta + 1 = 0$

25. $2 \cos^2 \theta - \cos \theta = 0$

26. $\tan^2 \theta - \sqrt{3} \tan \theta = 0$

27. $\csc^2 \theta + 3 \csc \theta + 2 = 0$

28. $\cot^2 \theta = 1$

29. $\sin^2 \theta + 2 \sin \theta - 3 = 0$

30. $2 \sec^2 \theta + \sec \theta - 1 = 0$

In Exercises 31–48, solve the given trigonometric equation on $0° \le \theta < 360°$ and express the answer in degrees to two decimal places.

31. $\sin 2\theta = -0.7843$

32. $\cos 2\theta = 0.5136$

33. $\tan \dfrac{\theta}{2} = -0.2343$

34. $\sec \dfrac{\theta}{2} = 1.4275$

35. $5 \cot \theta - 9 = 0$

36. $5 \sec \theta + 6 = 0$

37. $4 \sin \theta + \sqrt{2} = 0$

38. $3 \cos \theta - \sqrt{5} = 0$

39. $4 \cos^2 \theta + 5 \cos \theta - 6 = 0$

40. $6 \sin^2 \theta - 13 \sin \theta - 5 = 0$

41. $6 \tan^2 \theta - \tan \theta - 12 = 0$

42. $6 \sec^2 \theta - 7 \sec \theta - 20 = 0$

43. $15 \sin^2 2\theta + \sin 2\theta - 2 = 0$

44. $12 \cos^2 \dfrac{\theta}{2} - 13 \cos \dfrac{\theta}{2} + 3 = 0$

45. $\cos^2 \theta - 6 \cos \theta + 1 = 0$

46. $\sin^2 \theta + 3 \sin \theta - 3 = 0$

47. $2 \tan^2 \theta - \tan \theta - 7 = 0$

48. $3 \cot^2 \theta + 2 \cot \theta - 4 = 0$

▪ APPLICATIONS

49. **Sales.** The monthly sales of soccer balls is approximated by $S = 400 \sin \dfrac{\pi}{6} x + 2000$, where x is the number of the month (January is $x = 1$, etc.). During which month do the sales reach 2400?

50. **Sales.** The monthly sales of soccer balls is approximated by $S = 400 \sin \dfrac{\pi}{6} x + 2000$, where x is the number of the month (January is $x = 1$, etc.). During which two months do the sales reach 1800?

51. **Home Improvement.** A rain gutter is constructed from a single strip of sheet metal by bending it, as shown below, so that the base and sides are the same length. Express the area of the cross section of the rain gutter as a function of the angle θ (note that the expression will also involve x).

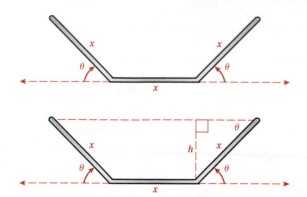

52. **Home Improvement**. A rain gutter is constructed from a single strip of sheet metal by bending it, as shown above, so that the base and sides are the same length. When the area of the cross section of the rain gutter is expressed as a function of the angle θ, you can then use calculus to determine the value of θ that produces the cross section with the greatest possible area. The angle is found by solving the equation $\cos^2\theta - \sin^2\theta + \cos\theta = 0$. What angle gives the maximum area?

53. **Deer Population.** The number of deer on an island is given by $D = 200 + 100\sin\frac{\pi}{2}x$, where x is the number of years since 2000. What is the first year after 2000 that the number of deer reaches 300?

54. **Deer Population.** The number of deer on an island is given by $D = 200 + 100\sin\frac{\pi}{6}x$, where x is the number of years since 2000. During what year is it the first time after 2000 that the number of deer reaches 150?

55. **Optics.** Assume that light is going from air into a diamond. Calculate the refractive angle, θ_r, if the incidence angle is $\theta_i = 75°$, and the index of refraction values for air and diamond are $n_i = 1.00$ and $n_r = 2.417$, respectively. Round to the nearest degree.

56. **Optics.** Assume that light is going from a diamond into air. Calculate the refractive angle, θ_r, if the incidence angle is $\theta_i = 15°$, and the index of refraction values for diamond and air are $n_i = 2.417$ and $n_r = 1.00$, respectively. Round to the nearest degree.

■ CATCH THE MISTAKE

In Exercises 57–60, explain the mistake that is made.

57. Solve $\cos\theta = 0.5899$ on $0° \le \theta < 360°$.

    Solution:

    Write in equivalent inverse notation. $\theta = \cos^{-1}0.5899$

    Use a calculator to approximate inverse cosine. $\theta = 53.85°$

    This is incorrect. What mistake was made?

58. Find all solutions to the equation $\sin\theta = \frac{3}{5}$ on $0° \le \theta \le 360°$.

    Solution:

    Write in equivalent inverse notation. $\theta = \sin^{-1}\frac{3}{5}$

    Use a calculator to approximate inverse sine. $\theta = 36.87°$

    This is incorrect. What mistake was made?

59. Solve $\sqrt{2 + \sin\theta} = \sin\theta$ on $0 \le \theta < 2\pi$.

    Solution:

Square both sides.	$2 + \sin\theta = \sin^2\theta$
Gather all terms to one side.	$\sin^2\theta - \sin\theta - 2 = 0$
Factor.	$(\sin\theta - 2)(\sin\theta + 1) = 0$
Set the factors equal to zero.	$\sin\theta - 2 = 0$ or $\sin\theta + 1 = 0$
Solve for $\sin\theta$.	$\sin\theta = 2$ or $\sin\theta = -1$
Solve $\sin\theta = 2$ for θ.	No solution
Solve $\sin\theta = -1$ for θ.	$\theta = \frac{3\pi}{2}$

    This is incorrect. What mistake was made?

60. Solve $\sqrt{3\sin\theta - 2} = -\sin\theta$ on $0 \le \theta < 2\pi$

Solution:

Square both sides.	$3\sin\theta - 2 = \sin^2\theta$
Gather all terms to one side.	$\sin^2\theta - 3\sin\theta + 2 = 0$
Factor.	$(\sin\theta - 2)(\sin\theta - 1) = 0$
Set the factors equal to zero.	$\sin\theta - 2 = 0$ or
	$\sin\theta - 1 = 0$

Solve for $\sin\theta$.	$\sin\theta = 2$ or $\sin\theta = 1$
Solve $\sin\theta = 2$ for θ.	No solution
Solve $\sin\theta = 1$ for θ.	$\theta = \dfrac{\pi}{2}$

This is incorrect. What mistake was made?

■ **CHALLENGE**

61. T or F: Linear trigonometric equations always have one solution on $[0, 2\pi]$.

62. T or F: Quadratic trigonometric equations always have two solutions on $[0, 2\pi]$.

63. Solve $16\sin^4\theta - 8\sin^2\theta = -1$ over $0 \le \theta \le 2\pi$.

64. Solve $\left|\cos\left(\theta + \dfrac{\pi}{4}\right)\right| = \dfrac{\sqrt{3}}{2}$ over all real numbers.

■ **TECHNOLOGY**

Graphing calculators can be used to find approximate solutions to trigonometric equations. For the equation $f(x) = g(x)$, let $Y_1 = f(x)$ and $Y_2 = g(x)$. The x values that correspond to points of intersections represent solutions.

65. Use a graphing utility to solve the equation $\sin\theta = \cos 2\theta$ on $0 \le \theta < \pi$.

66. Use a graphing utility to solve the equation $\csc\theta = \sec\theta$ on $0 \le \theta < \dfrac{\pi}{2}$.

67. Use a graphing utility to solve the equation $\sin\theta = \sec\theta$ on $0 \le \theta < \pi$.

68. Use a graphing utility to solve the equation $\cos\theta = \csc\theta$ on $0 \le \theta < \pi$.

69. Use a graphing utility to find all solutions to the equation $\sin\theta = e^\theta$ for $\theta \ge 0$.

70. Use a graphing utility to find all solutions to the equation $\cos\theta = e^\theta$ for $\theta \ge 0$.

Section **6.3** Solving Trigonometric Equations that Involve Multiple Trigonometric Functions

Skills Objective

■ Use trigonometric identities in solving trigonometric equations.

Conceptual Objective

■ Extend the strategies for solving trigonometric equations involving one trigonometric function to equations involving multiple trigonometric functions.

We now consider trigonometric equations that involve more than one trigonometric function. Trigonometric identities are an important part of solving these types of equations.

EXAMPLE 1 Using Trigonometric Identities in Solving Trigonometric Equations

Solve $\sin x + \cos x = 1$ on $0 \le x < 2\pi$.

Solution:

Square both sides.	$\sin^2 x + 2\sin x \cos x + \cos^2 x = 1$
Use the Pythagorean identity.	$\underbrace{\sin^2 x + \cos^2 x}_{1} + 2\sin x \cos x = 1$
Subtract 1.	$2\sin x \cos x = 0$
Use the zero product property.	$\sin x = 0 \text{ or } \cos x = 0$
Solve for x on $0 \le x < 2\pi$.	$x = 0 \text{ or } x = \pi \text{ or } x = \dfrac{\pi}{2} \text{ or } x = \dfrac{3\pi}{2}$

Because we squared the equation, we have to check for extraneous solutions.

Check $x = 0$: $\qquad\qquad \sin(0) + \cos(0) = 0 + 1 = 1 \quad \checkmark$

Check $x = \pi$: $\qquad\qquad \sin(\pi) + \cos(\pi) = 0 - 1 = -1 \quad$ X

Check $x = \dfrac{\pi}{2}$: $\qquad \sin\left(\dfrac{\pi}{2}\right) + \cos\left(\dfrac{\pi}{2}\right) = 1 + 0 = 1 \quad \checkmark$

Check $x = \dfrac{3\pi}{2}$: $\qquad \sin\left(\dfrac{3\pi}{2}\right) + \cos\left(\dfrac{3\pi}{2}\right) = -1 + 0 = -1 \quad$ X

The solution to $\sin x + \cos x = 1$ on $0 \le x < 2\pi$ is $\boxed{x = 0}$ or $\boxed{x = \dfrac{\pi}{2}}$.

TECHNOLOGY TIP

Find the points of intersection of $y = \sin x + \cos x$ and $y = 1$.

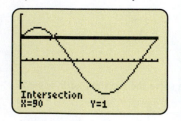

Intersection
X=90   Y=1

■ **YOUR TURN** Solve $\sin x - \cos x = 1$ on $0 \le x < 2\pi$.

EXAMPLE 2 Using Trigonometric Identities in Solving Trigonometric Equations

Solve $\sin 2x = \sin x$ on $0 \le x < 2\pi$.

COMMON MISTAKE

Dividing by a trigonometric function (which could be equal to zero).

✓ CORRECT

Use the double-angle formula for sine.

$$\underbrace{\sin 2x}_{2\sin x \cos x} = \sin x$$

Subtract $\sin x$.

$$2\sin x \cos x - \sin x = 0$$

✗ INCORRECT

Use the double-angle formula for sine.

$$\underbrace{\sin 2x}_{2\sin x \cos x} = \sin x$$

■ **Answer:** $x = \dfrac{\pi}{2}$ or $x = \pi$

Factor the common $\sin x$.

$$\sin x(2\cos x - 1) = 0$$

Set the factors equal to 0.

$$\sin x = 0 \text{ or } 2\cos x - 1 = 0$$

$$\sin x = 0 \text{ or } \cos x = \frac{1}{2}$$

Solve $\sin x = 0$ for x
on $0 \le x < 2\pi$.

$$\boxed{x = 0} \text{ or } \boxed{x = \pi}$$

Solve $\cos x = \frac{1}{2}$ for x
on $0 \le x < 2\pi$.

$$\boxed{x = \frac{\pi}{3}} \text{ or } \boxed{x = \frac{5\pi}{3}}$$

Simplify.

$$2\sin x \cos x = \sin x$$

Divide by $\sin x$.

$$2\cos x = 1 \quad \textbf{ERROR}$$

Solve for $\cos x$.

$$\cos x = \frac{1}{2}$$

Solve $\cos x = \frac{1}{2}$ for x on

$0 \le x < 2\pi$.

$$x = \frac{\pi}{3} \text{ or } x = \frac{5\pi}{3}$$

**Incorrect Solution: Two
additional solutions missing.**

CAUTION
Do not divide
equations by
trigonometric
functions as they
can sometimes equal
zero.

■ **YOUR TURN**   Solve $\sin 2x = \cos x$ on $0 \le x < 2\pi$.

EXAMPLE 3 Using Trigonometric Identities to Solve Trigonometric Equations

Solve $\sin x + \csc x = -2$.

Solution:

Use the reciprocal identity.

$$\sin x + \underbrace{\csc x}_{\frac{1}{\sin x}} = -2$$

Add 2.

$$\sin x + 2 + \frac{1}{\sin x} = 0$$

Multiply by $\sin x$. (Note $\sin x \ne 0$.)

$$\sin^2 x + 2\sin x + 1 = 0$$

Factor into a perfect square.

$$(\sin x + 1)^2 = 0$$

Solve for $\sin x$.

$$\sin x = -1$$

Solve for x on one period of sine, $[0, 2\pi)$.

$$x = \frac{3\pi}{2}$$

Add integer multiples of 2π to obtain all solutions.

$$\boxed{x = \frac{3\pi}{2} + 2n\pi}$$

■ **Answer** $x = \dfrac{\pi}{2}, x = \dfrac{3\pi}{2}, x = \dfrac{\pi}{6} \text{ or } x = \dfrac{5\pi}{6}$

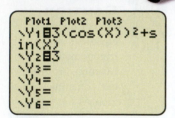

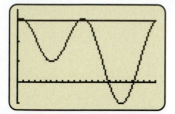

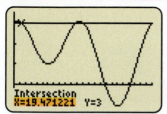

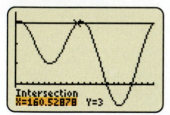

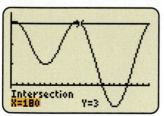

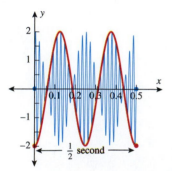

EXAMPLE 4 Using Trigonometric Identities and Inverse Functions to Solve Trigonometric Equations

Solve $3\cos^2\theta + \sin\theta = 3$ on $0° \le \theta < 360°$.

Solution:

Use the Pythagorean identity.
$$\underbrace{3\cos^2\theta}_{1-\sin^2\theta} + \sin\theta = 3$$

Subtract 3.
$$3(1 - \sin^2\theta) + \sin\theta - 3 = 0$$

Eliminate parentheses.
$$3 - 3\sin^2\theta + \sin\theta - 3 = 0$$

Simplify.
$$-3\sin^2\theta + \sin\theta = 0$$

Factor a common $\sin\theta$.
$$(\sin\theta)(1 - 3\sin\theta) = 0$$

Set the factors equal to 0.
$$\sin\theta = 0 \text{ or } 1 - 3\sin\theta = 0$$

Solve for $\sin\theta$.
$$\sin\theta = 0 \text{ or } \sin\theta = \frac{1}{3}$$

Solve $\sin\theta = 0$ for x on $0° \le \theta < 360°$.
$$\theta = 0° \quad \text{or} \quad \theta = 180°$$

Solve $\sin\theta = \dfrac{1}{3}$ for x on $0° \le \theta < 360°$.

Sine is positive in QI and QII.

A calculator gives inverse values only in QI.

Write the equivalent inverse notation for $\sin\theta = \dfrac{1}{3}$.
$$\theta = \sin^{-1}\frac{1}{3}$$

Use a calculator to approximate the QI solution.
$$\theta = 19.5°$$

To find the QII solution, subtract the reference angle from 180°.
$$\theta = 180° - 19.5°$$
$$\theta = 160.5°$$

Applications

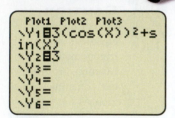

Musical tones can be represented mathematically with sinusoidal functions. Recall that a tone of 48 Hz (cycles per second) can be represented by the function $A\sin[2\pi(48)t]$, where A is the amplitude (loudness) and 48 Hz is the frequency. If we play two musical tones simultaneously, the combined tone can be found using the sum-product identities. For example, if a tone of 48 Hz is simultaneously played with a tone of 56 Hz, the result (assuming uniform amplitude, $A = 1$) is $\sin 96\pi t + \sin 112\pi t = 2\sin(104\pi t)\cos(8\pi t)$, whose graph is given on the left.

The $\sin(104\pi t)$ term represents a sound of average frequency 52 Hz. The $2\cos(8\pi t)$ represents a time-varying amplitude and a "beat" frequency that corresponds to 8 beats per second. Notice in the graph that there are 4 beats in a $\frac{1}{2}$ second.

EXAMPLE 5 Tuning Fork

A tuning fork is used to help musicians tune their instruments. They simultaneously play the fork and a note and tune the instrument until the beat frequency is eliminated.

If an E (659 Hz) tuning fork is used and the musician hears 4 beats per second, find the frequency of the note from the instrument if the loudest note is on the third beat. Assume each note has an amplitude of 1.

abzee/iStockphoto

Solution:

Express the E note mathematically.

$$\sin 2\pi(659)t$$

Express the instrument's note mathematically.

$$\sin 2\pi(f)t$$

Add the two notes (combined sound).

$$\sin 2\pi(659)t + \sin 2\pi ft$$

Use the sum-product identity.

$$2\sin\left[2\left(\frac{f+659}{2}\right)\pi t\right]\cos\left[2\left(\frac{f-659}{2}\right)\pi t\right]$$

The beat frequency is 4 Hz.

$$2\sin\left[2\left(\frac{f+659}{2}\right)\pi t\right]\cos[4\pi t]$$

Since cosine is an even function, $\cos(-x) = \cos x$, we equate the absolute value of the cosine arguments:

$$\left|2\left(\frac{f-659}{2}\right)\pi t\right| = |4\pi t|$$

Solve for f.

$$f - 659 = -4 \quad \text{or} \quad f - 659 = 4$$

f is either 654 Hz or 663 Hz.

$$f = 654 \quad \text{or} \quad f = 663$$

The loudest note is the peak amplitude.

$$|A| = \underset{2}{|2|}\;\underset{1}{\underbrace{|\sin(f+659)\pi t|}}\;\underset{1}{\underbrace{|\cos(4\pi t)|}}$$

The loudest note, $|A| = 2$, is heard on the third beat, $t = \dfrac{3}{4}$ seconds.

$$2 = 2\sin\left[(f+659)\frac{3}{4}\pi\right]\cos\left[4\pi\frac{3}{4}\right]$$

Simplify.

$$\sin\left[(f+659)\frac{3}{4}\pi\right] = -1$$

The inverse does not exist unless we restrict the domain of sine. Therefore we inspect the two choices, 655 Hz or 663 Hz, and see that both satisfy the equation. The instrument can be playing either frequency based on the information given.

SECTION 6.3　SUMMARY

In this section we solved trigonometric equations when more than one trigonometric function was involved. Trigonometric identities were used to transform equations into equations involving only one trigonometric function, and then algebraic techniques were used.

SECTION 6.3 EXERCISES

 ■ SKILLS

In Exercises 1–26, solve the trigonometric equations exactly on the indicated interval, $0 \leq x < 2\pi$.

1. $\sin x = \cos x$

2. $\sin x = -\cos x$

3. $\sec x + \cos x = -2$

4. $\sin x + \csc x = 2$

5. $\sec x - \tan x = \dfrac{\sqrt{3}}{3}$

6. $\sec x + \tan x = 1$

7. $\csc x + \cot x = \sqrt{3}$

8. $\csc x - \cot x = \dfrac{\sqrt{3}}{3}$

9. $2 \sin x - \csc x = 0$

10. $2 \sin x + \csc x = 3$

11. $\sin 2x = 4 \cos x$

12. $\sin 2x = \sqrt{3} \sin x$

13. $\sqrt{2} \sin x = \tan x$

14. $\cos 2x = \sin x$

15. $\tan 2x = \cot x$

16. $3 \cot 2x = \cot x$

17. $\sqrt{3} \sec x = 4 \sin x$

18. $\sqrt{3} \tan x = 2 \sin x$

19. $\sin^2 x - \cos 2x = -\dfrac{1}{4}$

20. $\sin^2 x - 2 \sin x = 0$

21. $\cos^2 x + 2 \sin x + 2 = 0$

22. $2 \cos^2 x = \sin x + 1$

23. $2 \sin^2 x + 3 \cos x = 0$

24. $4 \cos^2 x - 4 \sin x = 5$

25. $\cos 2x + \cos x = 0$

26. $2 \cot x = \csc x$

In Exercises 27–34, solve the trigonometric equations on $0° \leq \theta < 360°$. Give the answers in degrees and round to two decimal places.

27. $\cos 2x + \dfrac{1}{2} \sin x = 0$

28. $\sec^2 x = \tan x + 1$

29. $6 \cos^2 x + \sin x = 5$

30. $\sec^2 x = 2 \tan x + 4$

31. $\cot^2 x - 3 \csc x - 3 = 0$

32. $\csc^2 x + \cot x = 7$

33. $2 \sin^2 x + 2 \cos x - 1 = 0$

34. $\sec^2 x + \tan x - 2 = 0$

■ APPLICATIONS

35. Air in Lungs. If a person breathes in and out every 3 seconds, the volume of air in his lungs can be modeled by $A = 2 \sin\left(\dfrac{\pi}{3} x\right) \cos\left(\dfrac{\pi}{3} x\right) + 3$, where A is in liters of air and x is in seconds. How many seconds into the cycle is the volume of air equal to 4 liters?

36. Air in Lungs. If a person breathes in and out every 3 seconds, the volume of air in her lungs can be modeled by $A = 2 \sin\left(\dfrac{\pi}{3} x\right) \cos\left(\dfrac{\pi}{3} x\right) + 3$, where A is in liters of air and x is in seconds. How many seconds into the cycle is the volume of air equal to 2 liters?

37. Finding Turning Points. The figure on the right shows the graph of $y = 2 \cos x - \cos 2x$ between -2π and 2π. The maximum and minimum values of the curve occur at the *turning points* and are found in the solutions of the equation $-2 \sin x + 2 \sin 2x = 0$. Solve for the

coordinates of the turning points of the curve between 0 and 2π.

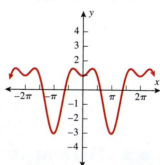

38. Finding Turning Points. The figure above shows the graph of $y = 2 \cos x - \cos 2x$ between -2π and 2π. The maximum and minimum values of the curve occur at the *turning points* and are found in the solutions of the equation $-2 \sin x + 2 \sin 2x = 0$. Solve for the coordinates of the turning points of the curve between -2π and 0.

■ CATCH THE MISTAKE

In Exercises 39 and 40, explain the mistake that is made.

39. Solve $3 \sin 2x = 2 \cos x$ on $0° \leq \theta < 180°$.

Solution:

Use the double-angle
identity for sine. $3 \underbrace{\sin 2x}_{2\sin x\, \cos x} = 2 \cos x$

Simplify. $6 \sin x \cos x = 2 \cos x$

Divide by $2 \cos x$. $3 \sin x = 1$

Divide by 3. $\sin x = \dfrac{1}{3}$

Write equivalent inverse
notation. $x = \sin^{-1}\left(\dfrac{1}{3}\right)$

Use a calculator to
approximate. $x = 19.47°$, QI solution

The QII solution is $x = 180° - 19.47° = 160.53°$

This is incorrect. What mistake was made?

40. Solve $\sqrt{1 + \sin x} = \cos x$ on $0 \leq x < 2\pi$.

Solution:

Square both sides. $1 + \sin x = \cos^2 x$

Use the Pythagorean
identity. $1 + \sin x = \underbrace{\cos^2 x}_{1 - \sin^2 x}$

Simplify. $\sin^2 x + \sin x = 0$

Factor. $\sin x (\sin x + 1) = 0$

Set factors equal to zero. $\sin x = 0$ or $\sin x + 1 = 0$

Solve for $\sin x$. $\sin x = 0$ or $\sin x = -1$

Solve for x. $x = 0, \pi, \dfrac{3\pi}{2}$

This is incorrect. What mistake was made?

■ CHALLENGE

41. T or F: If a trigonometric equation has all real numbers as its solution, then it is an identity.

42. T or F: If a trigonometric equation has an infinite number of solutions, then it is an identity.

43. Solve for the smallest positive x that makes this statement true:

$$\sin\left(x + \frac{\pi}{4}\right) + \sin\left(x - \frac{\pi}{4}\right) = \frac{\sqrt{2}}{2}$$

44. Solve for the smallest positive x that makes this statement true:

$$\cos x \cos 15° + \sin x \sin 15° = 0.7$$

■ TECHNOLOGY

In Exercises 45–48, find the smallest positive value of x that makes the statement true. Give the answer in degrees and round to two decimal places.

45. $\sec 3x + \csc 2x = 5$

46. $\cot 5x + \tan 2x = -3$

47. $e^x - \tan x = 0$

48. $e^x + 2 \sin x = 1$

Q: Kyle draws a ray from the point $(-1, \sqrt{3})$ to the origin and wants to find the angle that this ray makes with the positive x-axis (call this angle θ). He knows that if given a point (x, y), the relationship among x, y, and θ is $\tan\theta = \dfrac{y}{x}$. Thus, he calculates that

$$\tan\theta = \frac{-\sqrt{3}}{-1}$$

$$\tan\theta = \sqrt{3}$$

$$\theta = \tan^{-1}\sqrt{3} = \frac{\pi}{3}$$

However, this is not correct. Why not?

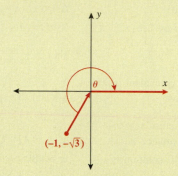

A: Kyle actually calculated the angle between the ray and the negative x-axis. The range of the inverse tangent function is $\left(-\dfrac{\pi}{2}, \dfrac{\pi}{2}\right)$. However, the angle that Kyle wants is in the third quadrant. Thus, he needs to add π to $\tan^{-1}\sqrt{3}$ to get the correct angle in the third quadrant. Thus, $\theta = \pi + \tan^{-1}\sqrt{3} = \dfrac{4\pi}{3}$. Now draw a ray from the point $(-4, 4)$ to the origin and find the angle that this ray makes with the positive x-axis.

TYING IT ALL TOGETHER

Phil the physicist is measuring the strength of the electric field in radio waves originating from a transmitter that has a power output of 50,000 watts. At a distance of 10,000 meters from the transmitter, he finds that the strength of the electric field is 10 volts per meter. The amplitude of the wave is 20 volts per meter and the wavelength is 50 meters. Phil also wants to know the corresponding value of time for his measurement, but it was too quick for him to measure on his stopwatch. Since he needs this information, he decides to calculate t using the fact that

$$E = A\sin(x - ct)$$

where A is the amplitude of the wave, c is the speed of light (about 300,000 m/s), x is the distance from the source at which the measurement is taken, and t is time. Calculate the time t at which Phil made his measurement. Take the smallest positive value of t as your answer.

SECTION	TOPIC	PAGES	REVIEW EXERCISES	KEY CONCEPTS
6.1	Inverse trigonometric functions	302–319	1–30	$\sin^{-1}x$ or $\arcsin x$ $\cos^{-1}x$ or $\arccos x$ $\tan^{-1}x$ or $\arctan x$ $\cot^{-1}x$ or $\text{arccot}\, x$ $\sec^{-1}x$ or $\text{arcsec}\, x$ $\csc^{-1}x$ or $\text{arccsc}\, x$
	Inverse sine function	303–307	1–30	**Definition:** $y = \sin^{-1}x$ means $x = \sin y$ $-1 \le x \le 1$ and $-\dfrac{\pi}{2} \le y \le \dfrac{\pi}{2}$

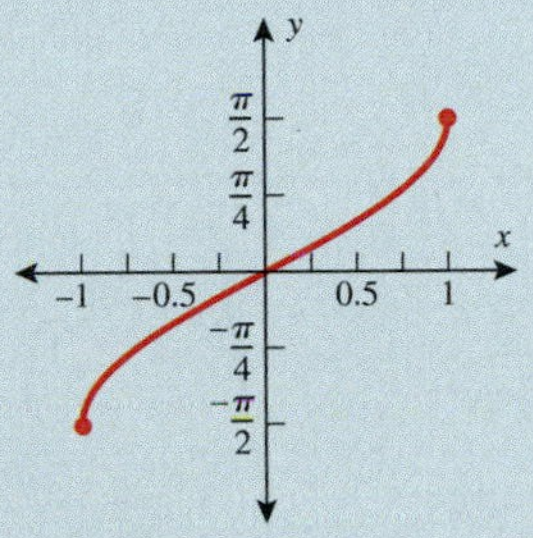

Identities:

$$\sin^{-1}(\sin x) = x \text{ for } -\frac{\pi}{2} \le x \le \frac{\pi}{2}$$

$$\sin(\sin^{-1}x) = x \text{ for } -1 \le x \le 1$$

| | Inverse cosine function | 307–310 | 1–30 | **Definition:**
 $y = \cos^{-1}x$ means $x = \cos y$
 $-1 \le x \le 1$ and $0 \le y \le \pi$ |

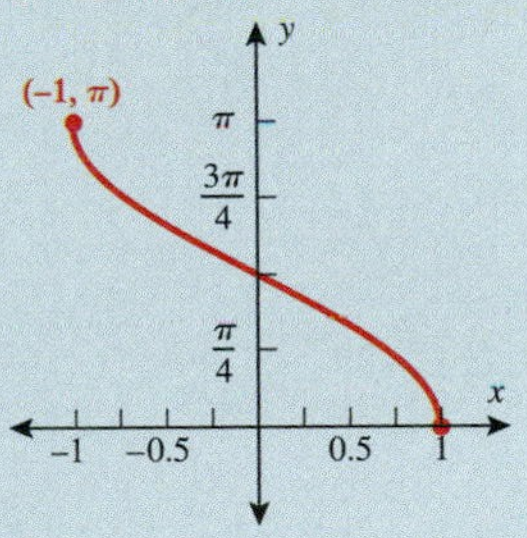

Identities:

$$\cos^{-1}(\cos x) = x \text{ for } 0 \le x \le \pi$$

$$\cos(\cos^{-1}x) = x \text{ for } -1 \le x \le 1$$

SECTION	TOPIC	PAGES	REVIEW EXERCISES	KEY CONCEPTS
	Inverse tangent function	310–313	1–30	

Definition:

$y = \tan^{-1} x$ means $x = \tan y$

$-\infty < x < \infty$ and $-\dfrac{\pi}{2} < y < \dfrac{\pi}{2}$

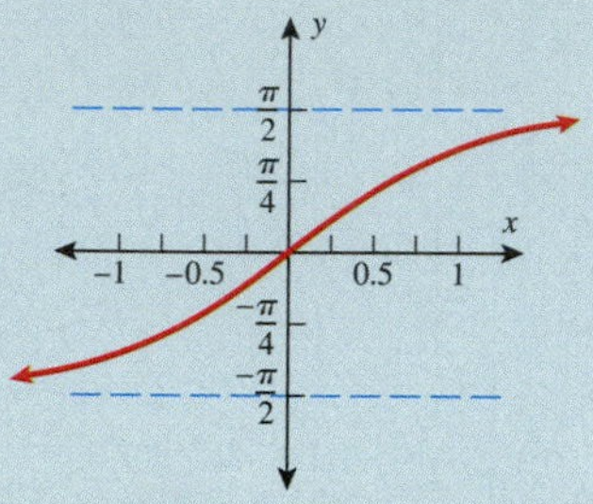

Identities:

$\tan^{-1}(\tan x) = x$ for $-\dfrac{\pi}{2} < x < \dfrac{\pi}{2}$

$\tan(\tan^{-1} x) = x$ for $-\infty < x < \infty$

SECTION	TOPIC	PAGES	REVIEW EXERCISES	KEY CONCEPTS
	Inverse cotangent function	313–317	1–30	

Definition:

$y = \cot^{-1} x$ means $x = \cot y$

$-\infty < x < \infty$ and $0 < y < \pi$

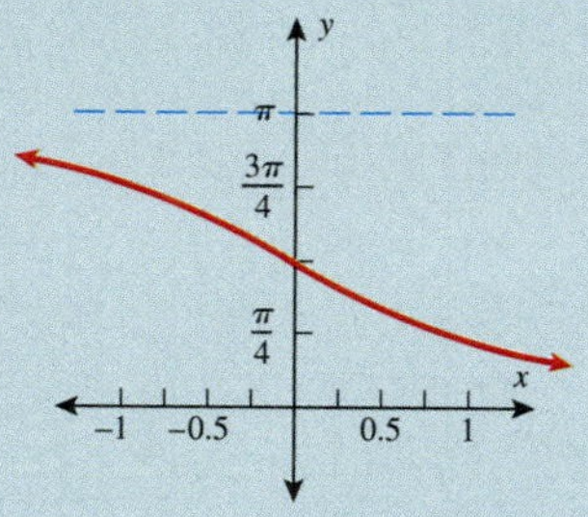

Identity:

$$\cot^{-1} x = \begin{cases} \tan^{-1}\left(\dfrac{1}{x}\right), & x > 0 \\[2ex] \pi + \tan^{-1}\left(\dfrac{1}{x}\right), & x < 0 \end{cases}$$

SECTION	TOPIC	PAGES	REVIEW EXERCISES	KEY CONCEPTS
	Inverse secant function	313–317	1–30	**Definition:**

Definition:

$y = \sec^{-1} x$ means $x = \sec y$

$x \leq -1$ or $x \geq 1$ and

$$0 \leq y < \frac{\pi}{2} \text{ or } \frac{\pi}{2} < y \leq \pi$$

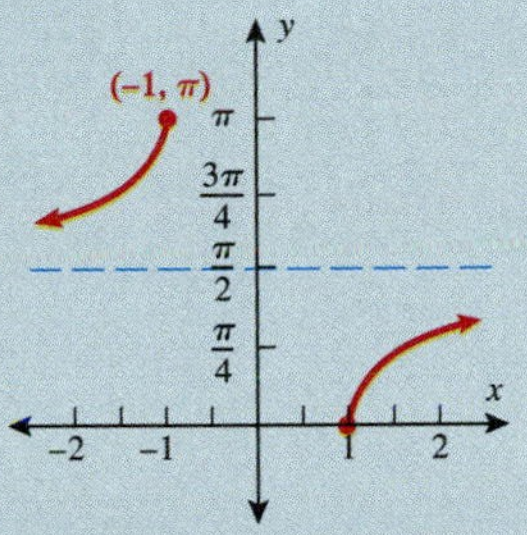

Identity:

$$\sec^{-1} x = \cos^{-1}\left(\frac{1}{x}\right) \text{ for } x \leq -1 \text{ or } x \geq 1$$

	Inverse cosecant function	313–317	1–30	**Definition:**

Definition:

$y = \csc^{-1} x$ means $x = \csc y$

$x \leq -1$ or $x \geq 1$ and

$$-\frac{\pi}{2} \leq y < 0 \text{ or } 0 < y \leq \frac{\pi}{2}$$

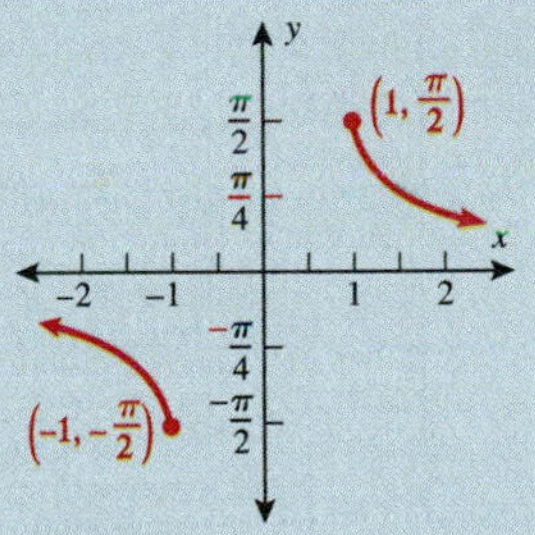

Identity:

$$\csc^{-1} x = \sin^{-1}\left(\frac{1}{x}\right) \text{ for } x \leq -1 \text{ or } x \geq 1$$

SECTION	TOPIC	PAGES	REVIEW EXERCISES	KEY CONCEPTS
6.2	Solving trigonometric equations that involve only one trigonometric function	323–330	31–46	**Goal:** Find the values of the variable that make the equation true.
	Solving *exactly* by inspection	324–326	31–34	Solve: $\sin\theta = \dfrac{\sqrt{2}}{2}$ on $0 \le \theta \le 2\pi$ Answer: $\theta = \dfrac{\pi}{4}$ or $\theta = \dfrac{3\pi}{4}$ Solve $\sin\theta = \dfrac{\sqrt{2}}{2}$ on all real numbers. Answer: $\theta = \begin{cases} \dfrac{\pi}{4} + 2n\pi \\ \dfrac{3\pi}{4} + 2n\pi \end{cases}$, where n is an integer
	Solving *exactly* using algebraic methods	326–327	35–40	Trigonometric equations can be transformed into linear or quadratic algebraic equations by making a substitution such as $x = \sin\theta$. Algebraic methods are then used for solving linear and quadratic equations. If an expression is ever squared, always check for extraneous solutions.
	Using a calculator and inverse functions to solve trigonometric equations	328–330	41–46	Follow the same procedures outlined by inspection or algebraic methods. Finding the solution requires the use of inverse functions and a calculator. Be careful: calculators only give one solution (the one in the domain of the inverse function).
6.3	Solving trigonometric equations that involve multiple trigonometric functions	333–337	47–62	Use trigonometric identities to transform an equation with multiple trigonometric functions into an equation with only one trigonometric function. Then use the methods outlined in Section 6.2.

6.1 Inverse Trigonometric Functions

Find the exact value of each expression. Give the answer in radians.

1. $\arctan 1$

2. $\operatorname{arccsc}(-2)$

3. $\cos^{-1} 0$

4. $\sin^{-1}(-1)$

Find the exact value of each expression. Give the answer in degrees.

5. $\csc^{-1}(-1)$

6. $\arctan(-1)$

7. $\operatorname{arccot}\left(\dfrac{\sqrt{3}}{3}\right)$

8. $\cos^{-1}\dfrac{\sqrt{2}}{2}$

Use a calculator to evaluate each expression. Give the answer in degrees and round to two decimal places.

9. $\sin^{-1}(-0.6088)$

10. $\tan^{-1} 1.1918$

11. $\sec^{-1} 1.0824$

12. $\cot^{-1}(-3.7321)$

Use a calculator to evaluate each expression. Give the answer in radians and round to two decimal places.

13. $\cos^{-1}(-0.1736)$

14. $\tan^{-1} 0.1584$

15. $\csc^{-1}(-10.0167)$

16. $\sec^{-1}(-1.1223)$

Evaluate each expression exactly if possible. If not possible, state why.

17. $\sin^{-1}\left[\sin\left(-\dfrac{\pi}{4}\right)\right]$

18. $\cos\left[\cos^{-1}\left(-\dfrac{\sqrt{2}}{2}\right)\right]$

19. $\tan[\tan^{-1}\sqrt{3}]$

20. $\cot^{-1}\left[\cot\dfrac{11\pi}{6}\right]$

21. $\csc^{-1}\left[\csc\dfrac{2\pi}{3}\right]$

22. $\sec\left[\sec^{-1}\left(-\dfrac{2\sqrt{3}}{3}\right)\right]$

Evaluate each expression exactly.

23. $\sin\left[\cos^{-1}\dfrac{11}{61}\right]$

24. $\cos\left[\tan^{-1}\dfrac{40}{9}\right]$

25. $\tan\left[\cot^{-1}\dfrac{6}{7}\right]$

26. $\cot\left[\sec^{-1}\dfrac{25}{7}\right]$

27. $\sec\left[\sin^{-1}\dfrac{1}{6}\right]$

28. $\csc\left[\cot^{-1}\dfrac{5}{12}\right]$

Applications.

29. Average Temperature. If the average temperature in Chicago, Illinois, can be modeled with the formula $T(m) = 26\sin(0.48m - 1.84) + 47$, where m is the month of the year (January corresponds to $m = 1$, etc.), then during which month is the average temperature 73 degrees?

30. Average Temperature. If the average temperature in Chicago, Illinois, can be modeled with the formula $T(m) = 26\sin(0.48m - 1.84) + 47$, where m is the month of the year (January corresponds to $m = 1$, etc.), then during which month is the average temperature 21 degrees?

6.2 Solving Trigonometric Equations that Involve Only One Trigonometric Function

Solve the given trigonometric equation over the indicated interval.

31. $\sin 2\theta = -\dfrac{\sqrt{3}}{2}, 0 \leq \theta < 2\pi$

32. $\sec\dfrac{\theta}{2} = 2, -2\pi \leq \theta < 2\pi$

33. $\sin\dfrac{\theta}{2} = -\dfrac{\sqrt{2}}{2}, -2\pi \leq \theta < 2\pi$

34. $\csc 2\theta = 2, 0 \leq \theta < 2\pi$

Solve the trigonometric equation exactly on $0 \leq \theta < 2\pi$.

35. $4\cos 2\theta + 2 = 0$

36. $\sqrt{3}\tan\dfrac{\theta}{2} - 1 = 0$

37. $2\tan 2\theta + 2 = 0$

38. $2\sin^2\theta + \sin\theta - 1 = 0$

39. $\tan^2\theta + \tan\theta = 0$

40. $\sec^2\theta - 3\sec\theta + 2 = 0$

Solve the given trigonometric equation on $0° \leq \theta < 360°$ and express the answer in degrees to two decimal places.

41. $\tan 2\theta = -0.3459$

42. $6\sin\theta - 5 = 0$

43. $4\cos^2\theta + 3\cos\theta = 0$

44. $12\cos^2\theta - 7\cos\theta + 1 = 0$

45. $\csc^2\theta - 3\csc\theta - 1 = 0$

46. $2\cot^2\theta + 5\cot\theta - 4 = 0$

6.3 Solving Trigonometric Equations that Involve Multiple Trigonometric Functions

Solve the trigonometric equations exactly on the interval $0 \leq \theta < 2\pi$.

47. $\sec x = 2\sin x$

48. $3\tan x + \cot x = 2\sqrt{3}$

49. $\sqrt{3}\tan x - \sec x = 1$

50. $2\sin 2x = \cot x$

51. $\sqrt{3}\tan x = 2\sin x$

52. $2\sin x = 3\cot x$

53. $\cos^2 x + \sin x + 1 = 0$

54. $2\cos^2 x - \sqrt{3}\cos x = 0$

55. $\cos 2x + 4\cos x + 3 = 0$

56. $\sin 2x + \sin x = 0$

Solve the trigonometric equations on $0° \leq \theta < 360°$. Give the answers in degrees and round to two decimal places.

57. $\csc^2 x + \cot x = 1$

58. $8\cos^2 x + 6\sin x = 9$

59. $\sin^2 x + 2 = 2\cos x$

60. $\cos 2x = 3\sin x - 1$

61. $\cos x - 1 = \cos 2x$

62. $12\cos^2 x + 4\sin x = 11$

CHAPTER 6 PRACTICE TEST

State the interval over x, for which the indicated identity is valid.

1. $\sin\left(\sin^{-1}x\right) = x$

2. $\tan^{-1}(\tan x) = x$

3. $\cos^{-1}(\cos x) = x$

Evaluate *exactly*.

4. $\csc^{-1}\sqrt{2}$

5. $\cot^{-1}(-\sqrt{3})$

Evaluate the expressions.

6. $\sin^{-1}\left(\sin\dfrac{3\pi}{4}\right)$

7. $\cos\left(\tan^{-1}2\right)$

8. $\sec\left(\sec^{-1}2\right)$

Solve the trigonometric equations exactly if possible; otherwise use a calculator to approximate solution(s).

9. $2\sin\theta = -\sqrt{3}$ on all real numbers.

10. $3\tan\theta = 1$ on $0 \le \theta < 180°$

11. $2\cos^2\theta + \cos\theta - 1 = 0$ on $0 \le \theta < 2\pi$

12. $\sin^2\theta + 3\sin\theta - 1 = 0$ on $0 \le \theta < 360°$

13. $\sin 2\theta = \dfrac{1}{2}\cos\theta$ over $0 \le \theta < 360°$

14. $\sin^2\theta - 3\cos\theta - 3 = 0$ over $0 \le \theta < 360°$

15. $\sqrt{\sin x} + \cos x = -1$ over $0 \le \theta < 2\pi$

Applications of Trigonometry: Triangles and Vectors

Florida/Bermuda/Puerto Rico

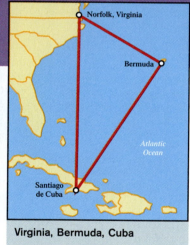

Virginia, Bermuda, Cuba

The "Bermuda Triangle" is an imaginary area located off the southeastern Atlantic coast of the United States, which is noted for a high incidence of losses of ships, small boats, and aircraft. The U.S. Board of Geographic Names does not recognize the Bermuda Triangle as an official name and does not maintain an official file on the area.

The vertices of the triangle are generally accepted to be Bermuda, Ft. Lauderdale (or Miami, Florida), and San Juan (Puerto Rico). However, others have argued that the vertices are instead Norfolk (Virginia), Bermuda, and Santiago (Cuba).

In this chapter you will develop a formula that determines the area of a triangle from the perimeter and side lengths. Which Bermuda Triangle has a larger area: the Miami-Bermuda-Puerto Rico one or the Norfolk-Bermuda-Cuba one? You will calculate the answer in this chapter.

In this chapter we discuss oblique (nonright) triangles. We use the law of sines and the law of cosines to solve oblique triangles. We then use the law of sines and the law of cosines (combined with trigonometric identities) to develop formulas for calculating the area of an oblique triangle. We also define vectors and use the laws of cosines and sines to determine resulting velocity and force vectors. Finally, we use dot products (product of two vectors) to assist in physical problems such as calculating work.

Applications of Trigonometry: Triangles and Vectors

Oblique (nonright) Triangles

- **Solving Oblique Triangles**
  - The Law of Sines
  - The Law of Cosines
- **Areas of Oblique Triangles**

Vectors

- **Magnitude of a Vector**
- **Unit Vector**
- **Scalar Multiplication**
- **Dot Product**

CHAPTER OBJECTIVES

- Solve oblique triangles using the law of sines or the law of cosines.
- Find areas of oblique triangles.
- Draw vectors.
- Find the direction and magnitude of a vector.
- Add and subtract vectors.
- Perform scalar multiplication and dot products.
- Determine if two vectors are perpendicular.
- Solve application problems involving triangles and vectors.

NAVIGATION THROUGH SUPPLEMENTS

DIGITAL VIDEO SERIES #7

STUDENT SOLUTIONS MANUAL CHAPTER 7

BOOK COMPANION SITE
www.wiley.com/college/young

Skills Objectives

- Solve AAS or ASA triangle cases.
- Solve ambiguous SSA triangle cases.
- Solve application problems involving oblique triangles.

Conceptual Objectives

- Classify an oblique triangle as one of four cases.
- Derive the law of sines.
- Understand that the ambiguous case can yield no triangle, one triangle, or two triangles.

Solving Oblique Triangles: Four Cases

Thus far we have only discussed *right* triangles. There are, however, two types of triangles, right and *oblique*. An **oblique triangle** is any triangle that does not have a right angle. An oblique triangle will be either an **acute triangle**, having three acute (less than 90°) angles; or an **obtuse triangle**, having one obtuse (between 90°and 180°) angle.

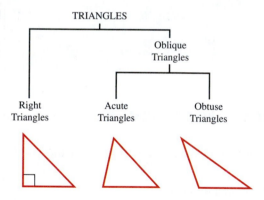

It is customary to label oblique triangles the following way:

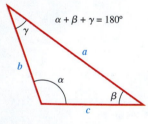

- angle α (alpha) opposite side a.
- angle β (beta) opposite side b.
- angle γ (gamma) opposite side c.

Remember that the sum of the three angles of any triangle must equal 180°. Recall in Section 1.5 that we solved right triangles. In this chapter we solve oblique triangles, which means we find the lengths of all three sides and the measures of all three angles. In order to solve an oblique triangle, *we need to know the length of one side* and one of the following three:

- two angles
- one angle and another side
- the other two sides

This leads to four possible cases to consider:

REQUIRED INFORMATION TO SOLVE OBLIQUE TRIANGLES

CASE	WHAT'S GIVEN	EXAMPLES/NAMES
Case 1	One side/two angles	**AAS: Angle-Angle-Side**
		ASA: Angle-Side-Angle
Case 2	Two sides and the angle opposite one of them	**SSA: Side-Side-Angle**
Case 3	Two sides and the angle between them	**SAS: Side-Angle-Side**
Case 4	Three sides	**SSS: Side-Side-Side**

Notice that there is no AAA case. This is because two similar triangles can have the same angle measures but different side lengths, so at least one side must be known.

STUDY TIP

To solve triangles, at least one side must be known.

In this section we will derive the law of sines, which will enable us to solve case 1 and case 2 problems. In the next section we will derive the law of cosines, which will enable us to solve case 3 and case 4 problems.

Derivation of the Law of Sines

Let us start with two oblique triangles (an acute triangle and an obtuse triangle):

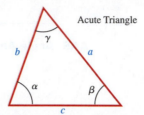
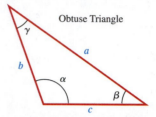

The following discussion applies to both triangles. First, construct an altitude (perpendicular), h, from the vertex at angle γ to the side (or its extension) opposite γ.

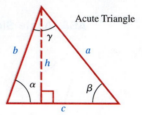
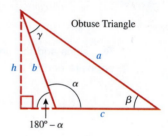

WORDS	MATH
Formulate sine ratios for an acute triangle.	$\sin\alpha = \dfrac{h}{b}$ and $\sin\beta = \dfrac{h}{a}$
Formulate sine ratios for an obtuse triangle.	$\sin(180° - \alpha) = \dfrac{h}{b}$ and $\sin\beta = \dfrac{h}{a}$
Apply the sine difference identity.	$\sin(180° - \alpha) = \sin 180° \cos\alpha - \cos 180° \sin\alpha$ $= 0 \cdot \cos\alpha - (-1)\sin\alpha$ $= \sin\alpha$
Therefore in both triangles we find the same equations.	$\sin\alpha = \dfrac{h}{b}$ and $\sin\beta = \dfrac{h}{a}$
Solve for h in both equations.	$h = b\sin\alpha$ and $h = a\sin\beta$
Since h is equal to itself, equate the expressions for h.	$b\sin\alpha = a\sin\beta$
Divide both sides by ab.	$\dfrac{b\sin\alpha}{ab} = \dfrac{a\sin\beta}{ab}$
Divide out common factors.	$\dfrac{\sin\alpha}{a} = \dfrac{\sin\beta}{b}$

In a similar manner, we can extend an altitude (perpendicular) from angle α, and we will find that $\dfrac{\sin \gamma}{c} = \dfrac{\sin \beta}{b}$. Equating these two expressions leads us to the third ratio

of the *law of sines*: $\dfrac{\sin \alpha}{a} = \dfrac{\sin \gamma}{c}$.

THE LAW OF SINES

For a triangle with sides a, b, and c and opposite angles α, β, and γ, the following is true:

$$\frac{\sin \alpha}{a} = \frac{\sin \beta}{b} = \frac{\sin \gamma}{c}$$

In other words, the ratio of the sine of an angle to its opposite side is equal to the ratios of the sine of the other two angles to their opposite sides.

STUDY TIP

The longest side is opposite the largest angle; the shortest side is opposite the smallest angle.

A few things to note before we begin solving oblique triangles:

- The angles and sides share the same progression of magnitude:
  - The longest side of a triangle is opposite the largest angle.
  - The shortest side of a triangle is opposite the smallest angle.
- Draw the triangle and label the angles and sides.
- If two angles are known, start by determining the third angle.
- Always use given values rather than calculated (approximated) values for better accuracy.

Keeping these in mind will help you determine if your answers seem reasonable. Also, recall (Section 1.5) the accuracy relationship between side lengths and angle measures.

STUDY TIP

Always use given values rather than calculated (approximated) values for better accuracy.

ANGLE TO NEAREST	SIGNIFICANT DIGITS FOR SIDE MEASURE
1°	2
0.1°	3
0.01°	4

Case 1: Two Angles and One Side (AAS or ASA)

EXAMPLE 1 Using the Law of Sines to Solve a Triangle (AAS)

Solve the triangle:

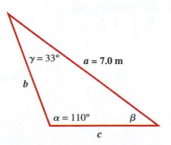

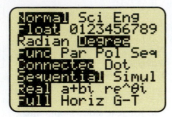
Solution:

This is an AAS (angle-angle-side) case because two angles and a side are given and the side is opposite one of the angles.

STEP 1 Find β.

The sum of the measures of the angles in a triangle is 180°.	$\alpha + \beta + \gamma = 180°$
Let $\alpha = 110°$ and $\gamma = 33°$.	$110° + \beta + 33° = 180°$
Solve for β.	$\beta = 37°$

STEP 2 Find b.

Use the law of sines with the known side, a.	$\dfrac{\sin \alpha}{a} = \dfrac{\sin \beta}{b}$
Isolate b.	$b = \dfrac{a \sin \beta}{\sin \alpha}$
Let $\alpha = 110°$, $\beta = 37°$, and $a = 7$ m.	$b = \dfrac{7 \sin 37°}{\sin 110°}$
Use a calculator to approximate b.	$b = 4.483067$ m
Since angles are given to the nearest degrees, round b to two significant digits.	$b = 4.5$ m

STEP 3 Find c.

Use the law of sines with the known side, a.	$\dfrac{\sin \alpha}{a} = \dfrac{\sin \gamma}{c}$
Isolate c.	$c = \dfrac{a \sin \gamma}{\sin \alpha}$
Let $\alpha = 110°$, $\gamma = 33°$, and $a = 7$ m.	$c = \dfrac{7 \sin 33°}{\sin 110°}$
Use a calculator to approximate c.	$c = 4.057149$ m
Since angles are given to the nearest degrees, round c to two significant digits.	$c = 4.1$ m

STEP 4 Draw and label the triangle.

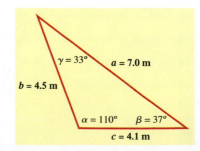

■ **YOUR TURN** Solve the triangle:

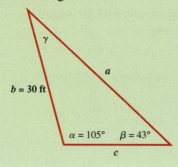

γ

a

$b = 30$ ft

$\alpha = 105°$ $\beta = 43°$

c

EXAMPLE 2 Using the Law of Sines to Solve a Triangle (ASA)

Solve the triangle:

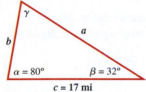

γ

a

b

$\alpha = 80°$ $\beta = 32°$

$c = 17$ mi

Solution:

This is an ASA (angle-side-angle) case, because two angles and a side are given and the side is not opposite one of the angles.

STEP 1 Find γ.

The sum of the measures of the
angles in a triangle is 180°. $\alpha + \beta + \gamma = 180°$

Let $\alpha = 80°$ and $\beta = 32°$. $80° + 32° + \gamma = 180°$

Solve for γ. $\gamma = 68°$

STEP 2 Find b.

Use the law of sines with the
known side, c. $\dfrac{\sin \beta}{b} = \dfrac{\sin \gamma}{c}$

Isolate b. $b = \dfrac{c \sin \beta}{\sin \gamma}$

Let $\beta = 32°$, $\gamma = 68°$, and
$c = 17$ miles. $b = \dfrac{17 \sin 32°}{\sin 68°}$

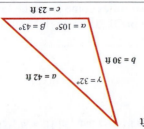

$c = 23$ ft

$\alpha = 105°$ $\beta = 43°$

$b = 30$ ft

$a = 42$ ft

$\gamma = 32°$

■ **Answer:** $\gamma = 32°$, $a = 42$ ft, and $c = 23$ ft ■

Step 2: Use the calculator to find $b = \dfrac{17\sin(32°)}{\sin(68°)}$.

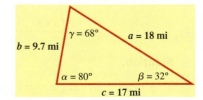

17sin(32)/sin(68
)
                9.716117734

Step 3: Use the calculator to find $a = \dfrac{17\sin(80°)}{\sin(68°)}$.

17sin(80)/sin(68
)
                18.0565394
▪

Use a calculator to approximate b.	$b = 9.7161177$
Since angles are given to the nearest degrees, round b to two significant digits.	$b = 9.7$ miles

STEP 3 Find a.

Use the law of sines with the known side, c.	$\dfrac{\sin\alpha}{a} = \dfrac{\sin\gamma}{c}$
Isolate a.	$a = \dfrac{c\sin\alpha}{\sin\gamma}$
Let $\alpha = 80°$, $\gamma = 68°$, and $c = 17$ miles.	$a = \dfrac{17\sin 80°}{\sin 68°}$
Use a calculator to approximate a.	$a = 18.056539$
Since angles are given to the nearest degrees, round a to two significant digits.	$a = 18$ miles

STEP 4 Draw and label the triangle.

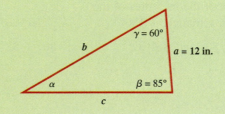

▪ **YOUR TURN** Solve the triangle:

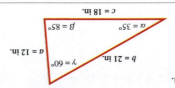

Case 2 (Ambiguous Case): Two Sides and One Angle (SSA)

If we are given two sides and an angle opposite one of the sides, we call that case 2, SSA (side-side-angle). This case is called the ambiguous case because one triangle, two triangles, or no triangle at all are possible. If the angle given is acute, then the possibilities are zero, one, or two triangles. If the angle given is obtuse, then the possibilities are zero or one triangle. The possibilities come from the fact that $\sin\alpha = c$, where $0 < c < 1$, has two solutions. One in QI (acute angle) and one in QII (obtuse

▪ **Answer:** $\alpha = 35°$, $b = 21$ in., and $c = 18$ in.

angle). In the figure below, note that $h = b \sin \alpha$, and a may be smaller than, equal to, or larger than h. Since $0 < \sin \alpha < 1$, then $h < b$.

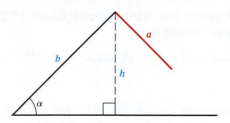

When the Given Angle (α) Is Acute

CONDITION	PICTURE	NUMBER OF TRIANGLES
$0 < a < h$ $\sin \beta > 1$	No Triangle	0
$a = h$ $\sin \beta = 1$	Right Triangle	1
$h < a < b$ $0 < \sin \beta < 1$	Acute Triangle Obtuse Triangle	2
$a \geq b$ $0 < \sin \beta < 1$	Acute Triangle	1

When the Given Angle (α) Is Obtuse

CONDITION	PICTURE	NUMBER OF TRIANGLES
$a \leq b$ $\sin \beta \geq 1$		0
$a > b$ $0 < \sin \beta < 1$		1

Before working the next example, make sure your calculator is in degree mode.

EXAMPLE 3 **Solving the Ambiguous Case (SSA)—One Triangle**

Solve the triangle: $a = 23$ ft, $b = 11$ ft, and $\alpha = 122°$.

Solution:

This is an ambiguous case because two sides and an angle opposite one of those sides is given. Since the given angle, α, is obtuse and $a > b$, we expect one triangle.

STEP 1 Find β.

Use the law of sines. $\qquad\qquad \dfrac{\sin \alpha}{a} = \dfrac{\sin \beta}{b}$

Isolate $\sin \beta$. $\qquad\qquad \sin \beta = \dfrac{b \sin \alpha}{a}$

Let $a = 23$ ft, $b = 11$ ft, and $\alpha = 122°$. $\quad \sin \beta = \dfrac{(11 \text{ ft}) \sin 122°}{23 \text{ ft}}$

Use a calculator to evaluate $\sin \beta$. $\qquad \sin \beta = 0.40558822$

Solve for β using the inverse sine function. $\qquad\qquad \beta = \sin^{-1}(0.40558822)$

Round to the nearest degree. $\qquad\qquad \boxed{\beta = 24°}$

STEP 2 Find γ.

The measures of angles in a triangle sum to 180°. $\qquad\qquad \alpha + \beta + \gamma = 180°$

Substitute $\alpha = 122°$ and $\beta = 24°$. $\qquad 122° + 24° + \gamma = 180°$

Solve for γ. $\qquad\qquad \boxed{\gamma = 34°}$

STEP 3 Find c.

Use the law of sines. $\qquad\qquad \dfrac{\sin \alpha}{a} = \dfrac{\sin \gamma}{c}$

Isolate c. $\qquad\qquad c = \dfrac{a \sin \gamma}{\sin \alpha}$

Substitute $a = 23$ ft, $\alpha = 122°$, and $\gamma = 34°$. $\qquad\qquad c = \dfrac{(23 \text{ ft}) \sin 34°}{\sin 122°}$

Use a calculator to evaluate c. $\qquad\qquad \boxed{c = 15 \text{ ft}}$

STEP 4 Draw and label the triangle.

$\gamma = 34°$ $a = 23$ ft
$b = 11$ ft
$\alpha = 122°$ $\beta = 24°$
$c = 15$ ft

CONCEPT CHECK If $\alpha = 133°$, can you determine if the triangle is acute or obtuse with just this knowledge?

YOUR TURN Solve the triangle: $\alpha = 133°$, $a = 48$ mm, $c = 17$ mm.

EXAMPLE 4 Solving the Ambiguous Case (SSA)—Two Triangles

Solve the triangle: $a = 8.1$ m, $b = 8.3$ m, and $\alpha = 72°$.

Solution:

This is an ambiguous case because two sides and an angle opposite one of those sides is given. Since the given angle, α, is acute and $a < b$, we expect two triangles.

STEP 1 Find β.

Use the law of sines.

$$\frac{\sin \alpha}{a} = \frac{\sin \beta}{b}$$

Isolate $\sin \beta$.

$$\sin \beta = \frac{b \sin \alpha}{a}$$

Let $a = 8.1$ m, $b = 8.3$ m, and $\alpha = 72°$.

$$\sin \beta = \frac{(8.3 \text{ m})\sin 72°}{8.1 \text{ m}}$$

Use a calculator to evaluate $\sin \beta$.

$$\sin \beta = 0.974539393$$

β can be acute or obtuse.

Solve for β using the inverse sine function.

$$\beta = \sin^{-1}(0.974539393)$$

This is the quadrant I solution (β is acute).

$$\beta_1 = 77°$$

The quadrant II solution is $\beta_2 = 180 - \beta_1$.

$$\beta_2 = 103°$$

STEP 2 Find γ.

The measures of the angles in a triangle sum to 180°.

$$\alpha + \beta + \gamma = 180°$$

Substitute $\alpha = 72°$ and $\beta_1 = 77°$.

$$72° + 77° + \gamma_1 = 180°$$

Solve for γ_1.

$$\gamma_1 = 31°$$

Substitute $\alpha = 72°$ and $\beta_2 = 103°$.

$$72° + 103° + \gamma_2 = 180°$$

Solve for γ_2.

$$\gamma_2 = 5°$$

Answer: $\beta = 32°$, $\gamma = 15°$, and $b = 35$ mm

STEP 3 Find c.

Use the law of sines.

$$\frac{\sin \alpha}{a} = \frac{\sin \gamma}{c}$$

Isolate c.

$$c = \frac{a \sin \gamma}{\sin \alpha}$$

Substitute $a = 8.1$ m, $\alpha = 122°$, and $\gamma_1 = 31°$.

$$c_1 = \frac{(8.1 \text{ m}) \sin 31°}{\sin 122°}$$

Use a calculator to evaluate c_1.

$$c_1 = 4.9 \text{ m}$$

Substitute $a = 8.1$ m, $\alpha = 122°$, and $\gamma_2 = 5°$.

$$c_2 = \frac{(8.1 \text{ m}) \sin 5°}{\sin 122°}$$

Use a calculator to evaluate c_2.

$$c_2 = 0.83 \text{ m}$$

STEP 4 Draw and label the two triangles.

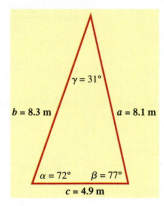

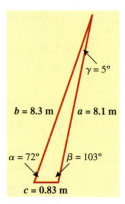

EXAMPLE 5 Solving the Ambiguous Case (SSA)—No Triangle

Solve the triangle: $\alpha = 107°$, $a = 6$, and $b = 8$.

Solution:

This is an ambiguous case because two sides and an angle opposite one of those sides is given. Since the given angle, α, is obtuse and $a < b$, we expect no triangle.

Use the law of sines.

$$\frac{\sin \alpha}{a} = \frac{\sin \beta}{b}$$

Isolate $\sin \beta$.

$$\sin \beta = \frac{b \sin \alpha}{a}$$

Let $\alpha = 107°$, $a = 6$, and $b = 8$.

$$\sin \beta = \frac{(8) \sin 107°}{6}$$

Use a calculator to evaluate $\sin \beta$.

$$\sin \beta \approx 1.28$$

Since the range of the sine function is $[-1, 1]$, there is no angle, β, such that $\sin \beta \approx 1.28$. Therefore, there is   no triangle   with the given measurements.

Applications

The solution of oblique triangles has applications in astronomy, surveying, aircraft design, piloting, and many other areas.

EXAMPLE 6 How Far Is the Tower of Pisa Leaning?

The Tower of Pisa was originally built 56 meters tall. Because of poor soil in the foundation, it started to lean. At a distance 44 meters from the base of the tower, the angle of inclination is 55°. How much is the Tower of Pisa leaning away from the vertical position? (What is the angle of inclination?)

Corbis Digital Stock

Solution:

$\beta = 55°$, $c = 44$ m, and $b = 56$ m is the given information: SSA.

STEP 1 Find γ.

Use the law of sines.	$\dfrac{\sin \beta}{b} = \dfrac{\sin \gamma}{c}$
Isolate $\sin \gamma$.	$\sin \gamma = \dfrac{c \sin \beta}{b}$
Substitute $\beta = 55°$, $c = 44$ m, and $b = 56$ m.	$\sin \gamma = \dfrac{(44 \text{ m}) \sin 55°}{56 \text{ m}}$
Evaluate the right side using a calculator.	$\sin \gamma = 0.643619463$
Solve for γ using the inverse sine function.	$\gamma = \sin^{-1}(0.643619463)$
Round to the nearest degree.	$\gamma = 40°$

STEP 2 Find α.

The measures of angles in a triangle
sum to 180°. $\alpha + \beta + \gamma = 180°$

Substitute $\beta = 55°$ and $\gamma = 40°$. $\alpha + 55° + 40° = 180°$

Solve for α. $\alpha = 85°$

The Tower of Pisa makes an angle of 85° with the ground (it is leaning at an
angle of 5°).

SECTION 7.1   SUMMARY

In this section, we solved for oblique triangles. When given three pieces of information about a triangle, we classify the triangle according to the data (sides and
angles). Four cases arise:

- one side/two angles (AAS or ASA)
- two sides/one angle opposite one of the sides (SSA)
- two sides/one angle between sides (SAS)
- three sides (SSS)

The law of sines can be used to solve the first two cases (AAS/ASA and SSA). It
is important to note that the SSA case is called the ambiguous case because one of
three things can result: no triangle, one triangle, or two triangles.

SECTION 7.1 EXERCISES

 **SKILLS**

In Exercises 1–6, classify each triangle as AAS, ASA, SAS, SSA, SAS, or SSS given the following information.

1. c, a, and α **2.** c, a, and γ

3. a, b, and c **4.** a, b, and γ

5. α, β, and c **6.** β, γ, and a

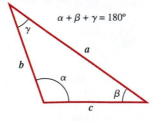

In Exercises 7–16, solve the given triangles.

7. $\alpha = 45°$, $\beta = 60°$, $a = 10$ m   **8.** $\beta = 75°$, $\gamma = 60°$, $b = 25$ in.

9. $\alpha = 46°$, $\gamma = 72°$, $b = 200$ cm   **10.** $\gamma = 100°$, $\beta = 40°$, $a = 16$ ft

11. $\alpha = 16.3°$, $\gamma = 47.6°$, $c = 211$ yd.   **12.** $\beta = 104.2°$, $\gamma = 33.6°$, $a = 26$ in.

13. $\alpha = 30°$, $\beta = 30°$, $c = 12$ m   **14.** $\alpha = 45°$, $\gamma = 75°$, $c = 9$ in.

15. $\beta = 26°$, $\gamma = 57°$, $c = 100$ yd.`   **16.** $\alpha = 80°$, $\gamma = 30°$, $b = 3$ ft.

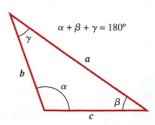

In Exercises 17–26, two sides and an angle are given. Determine if a triangle (or two) exist and if so, solve those triangles.

17. $a = 4, b = 5, \alpha = 16°$

18. $b = 30, c = 20, \beta = 70°$

19. $a = 12, c = 12, \gamma = 40°$

20. $b = 111, a = 80, \alpha = 25°$

21. $a = 21, b = 14, \beta = 100°$

22. $a = 13, b = 26, \alpha = 120°$

23. $b = 500, c = 330, \gamma = 40°$

24. $b = 16, a = 9, \beta = 137°$

25. $a = \sqrt{2}, b = \sqrt{7}, \beta = 106°$

26. $b = 15.3, c = 27.2, \gamma = 11.6°$

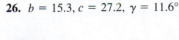

For Exercises 27 and 28, refer to the following:

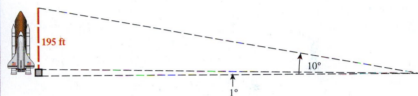

On the launch pad at Kennedy Space Center, the astronauts have an escape basket that can hold four astronauts. The basket slides down a wire that is attached 195 feet high, above the base of the launch pad. The angle measured from where the basket would touch the ground to the base of the launch pad is 1°, and the angle from that same point to where the wire is attached is 10°.

27. NASA. How long is the wire?

28. NASA. How far from the launch pad does the basket touch the ground?

29. Hot Air Balloon. A hot air balloon is sighted at the same time by two friends who are 1 mile apart on the same side of the balloon. The angles of elevation from the two friends are 20.5° and 25.5°. How high is the balloon?

30. Hot Air Balloon. A hot air balloon is sighted at the same time by two friends who are 2 miles apart on the same side of the balloon. The angles of elevation from the two friends are 10° and 15°. How high is the balloon?

31. Rocket Tracking. A tracking station has two telescopes that are 1 mile apart. The telescopes can lock onto a rocket after it is launched and record the angles of elevation to the rocket. If the angles of elevation from telescopes A and B are 30° and 80°, respectively, then how far is the rocket from telescope A?

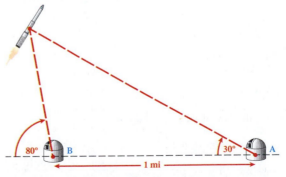

32. Rocket Tracking. A tracking station has two telescopes that are 1 mile apart. The telescopes can lock onto a rocket after it is launched and record the angles of elevation to the rocket. If the angles of elevation from telescopes A and B are 30° and 80°, respectively, then how far is the rocket from telescope B?

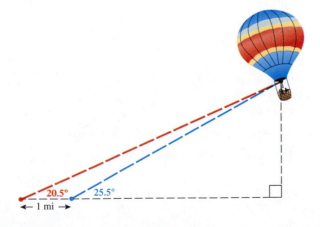

33. **Distance Across River.** An engineer wants to construct a bridge across a fast moving river. Using a straight line segment between two points that are 100 feet apart along his side of the river, he measures the angles formed when sighting the point on the other side where he wants to have the bridge end. If the angles formed at points A and B are 65° and 25°, respectively, how far is it from point A to the point on the other side of the river?

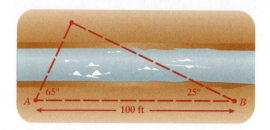

34. **Distance Across River.** An engineer wants to construct a bridge across a fast-moving river. Using a straight line segment between two points that are 100 feet apart along his side of the river, he measures the angles formed when sighting the point on the other side where he wants to have the bridge end. If the angles formed at points A and B are 65° and 25°, respectively, how far is it from point B to the point on the other side of the river?

■CATCH THE MISTAKE

In Exercises 35 and 36, explain the mistake that is made.

35. Solve the triangle: $\alpha = 120°$, $a = 7$, and $b = 9$.

    Solution:

    Use the law of sines to find β. $\quad \dfrac{\sin \alpha}{a} = \dfrac{\sin \beta}{b}$

    Let $\alpha = 120°$, $a = 7$, and $b = 9$. $\quad \dfrac{\sin 120°}{7} = \dfrac{\sin \beta}{9}$

    Solve for $\sin \beta$. $\quad \sin \beta = 1.113$

    Solve for β. $\quad \beta = 42°$

    Sum the angle measures to 180°. $\quad 120° + 42° + \gamma = 180°$

    Solve for γ. $\quad \gamma = 18°$

    Use the law of sines to find c. $\quad \dfrac{\sin \alpha}{a} = \dfrac{\sin \gamma}{c}$

    Let $\alpha = 120°$, $a = 7$,

    and $\gamma = 18°$ $\quad \dfrac{\sin 120°}{7} = \dfrac{\sin 18°}{c}$

    Solve for c. $\quad c = 2.5$

    $\alpha = 120°$, $\beta = 42°$, $\gamma = 18°$, $a = 7$, $b = 9$, and $c = 2.5$

    This is incorrect. The longest side is not opposite the longest angle. There is no triangle that makes the original measurements work. What mistake was made?

36. Solve the triangle: $\alpha = 40°$, $a = 7$, and $b = 9$.

    Solution:

    Use the law of sines to find β. $\quad \dfrac{\sin \alpha}{a} = \dfrac{\sin \beta}{b}$

    Let $\alpha = 40°$, $a = 7$, and $b = 9$ $\quad \dfrac{\sin 40°}{7} = \dfrac{\sin \beta}{9}$

    Solve for $\sin \beta$. $\quad \sin \beta = 0.826441212$

    Solve for β. $\quad \beta = 56°$

    Find γ. $\quad 40° + 56° + \gamma = 180°$

    $\quad \gamma = 84°$

    Use the law of sines to find c. $\quad \dfrac{\sin \alpha}{a} = \dfrac{\sin \gamma}{c}$

    Let $\alpha = 40°$, $a = 7$,
    and $\gamma = 84°$. $\quad \dfrac{\sin 40°}{7} = \dfrac{\sin 84°}{c}$

    Solve for c. $\quad c = 11$

    $\alpha = 40°$, $\beta = 56°$, $\gamma = 84°$, $a = 7$, $b = 9$ and $c = 11$

    This is incorrect. What mistake was made?

■ **CHALLENGE**

37. T or F: The law of sines applies to right triangles.

38. T or F: If you are given two sides and any angle, there is a unique solution for the triangle.

39. T or F: An acute triangle is an oblique triangle.

40. T or F: An obtuse triangle is an oblique triangle.

41. Mollweide's Identity. For any triangle, the following identity is true. It is often used to check the solution of a triangle since all six pieces of information (three sides and three angles) are involved. Derive the identity using the law of sines.

$$(a + b) \sin\left(\frac{1}{2}\gamma\right) = c \cos\left[\frac{1}{2}(\alpha - \beta)\right]$$

42. The Law of Tangents. Use the law of sines and trigonometric identities to show that for any triangle, the following is true.

$$\frac{a - b}{a + b} = \frac{\tan\left[\dfrac{1}{2}(\alpha - \beta)\right]}{\tan\left[\dfrac{1}{2}(\alpha + \beta)\right]}$$

■ **TECHNOLOGY**

For Exercises 43–46, let A, B, and C be the lengths of the three sides with X, Y, and Z as the corresponding angles. Write a program using a calculator to solve the given triangle. Go to the book website (www.wiley.com/college/young) for instructions or help on how to write a program in a graphing calculator.

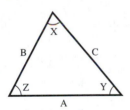

43. A = 10, Y = 40°, and Z = 72°

44. B = 42.8, X = 31.6°, and Y = 82.2°

45. A = 22, B = 17, and X = 105°

46. B = 16.5, C = 9.8, and Z = 79.2°

SECTION 7.2 The Law of Cosines

Skills Objectives

■ Solve SAS triangles.
■ Solve SSS triangles.
■ Solve application problems involving oblique triangles.

Conceptual Objectives

■ Derive the law of cosines.
■ Develop a strategy for which angles to select (longer or shorter) and which method (the law of sines or the law of cosines) to use to solve oblique triangles.

In the previous section (Section 7.1) we learned that to solve oblique triangles means to find all three side lengths and angle measures. At least one side length must be known. Two additional pieces of information are needed to solve a triangle (combinations of side lengths and/or angles). We found that there are four cases:

- Case 1: AAS or ASA (two angles and a side are given)
- Case 2: SSA (two sides and an angle opposite one of the sides are given)
- Case 3: SAS (two sides and the angle between them are given)
- Case 4: SSS (three sides are given)

We used the law of sines to solve case 1 and case 2 triangles. Now, we use the *law of cosines* to solve case 3 and case 4 triangles.

Derivation of the Law of Cosines

WORDS	MATH
Start with a triangle.	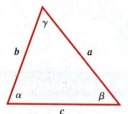
Drop a perpendicular line from γ with height h. The result is two triangles within the larger triangle.	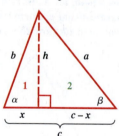
Use the Pythagorean theorem on both right triangles.	
Triangle 1:	$x^2 + h^2 = b^2$
Triangle 2:	$(c - x)^2 + h^2 = a^2$
Solve for h^2.	
Triangle 1:	$h^2 = b^2 - x^2$
Triangle 2:	$h^2 = a^2 - (c - x)^2$
Since the segment of length h is shared, set $h^2 = h^2$ for the two triangles.	$b^2 - x^2 = a^2 - (c - x)^2$
Multiply out the squared binomial on the right.	$b^2 - x^2 = a^2 - (c^2 - 2cx + x^2)$
Eliminate the parentheses.	$b^2 - x^2 = a^2 - c^2 + 2cx - x^2$
Add x^2 to both sides.	$b^2 = a^2 - c^2 + 2cx$
Isolate a^2.	$a^2 = b^2 + c^2 - 2cx$
Notice that $\cos\alpha = \dfrac{x}{b}$. Let $x = b\cos\alpha$.	$a^2 = b^2 + c^2 - 2bc\cos\alpha$

Note: If we instead drop the perpendicular line with length h from the angle α or the angle β, we can derive the other two parts of the law of cosines:

$$b^2 = a^2 + c^2 - 2ac\cos\beta \quad \text{and} \quad c^2 = a^2 + b^2 - 2ab\cos\gamma \,.$$

THE LAW OF COSINES

For a triangle with sides a, b, and c and opposite angles α, β, and γ, the following is true:

$$a^2 = b^2 + c^2 - 2bc\cos\alpha$$
$$b^2 = a^2 + c^2 - 2ac\cos\beta$$
$$c^2 = a^2 + b^2 - 2ab\cos\gamma$$

It is important to note that the law of cosines can be used to find side lengths or angles. As long as three of the four variables in the equation are known, the fourth can be calculated.

Notice that in the special case of a right triangle ($\alpha = 90°$),

$$a^2 = b^2 + c^2 - 2\,bc\,\underset{0}{\underline{\cos 90°}}$$

then the first law of cosines reduces to the Pythagorean theorem:

$$\underset{\text{hyp}}{\underline{a^2}} = \underset{\text{leg}}{\underline{b^2}} + \underset{\text{leg}}{\underline{c^2}}$$

The Pythagorean theorem is a special case of the law of cosines.

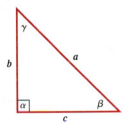

STUDY TIP

The Pythagorean theorem is a special case of the law of cosines.

Case 3: Solving Oblique Triangles (SAS)

We now solve SAS-type triangles, where the angle between two sides is given. We start by using the law of cosines to solve for the side opposite the given angle. We then can use either the law of sines or the law of cosines to find the second angle.

EXAMPLE 1 Using the Law of Cosines to Solve a Triangle (SAS)

Solve the triangle.

Solution:

Two sides and the angle between them are given (SAS).

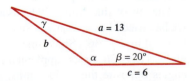

Notice that the law of sines cannot be used because it requires knowledge of at least one angle and the side opposite that angle.

TECHNOLOGY TIP

Step 1: Use a calculator
to find the value of b.

```
13²+6²-2*13*6*co
s(20)
        58.40795116
√(Ans)
        7.64250948
■
```

Step 2: Use a calculator to find γ.

```
6sin(20)/7.6
        .2700159026
sin⁻¹(Ans)
        15.66521315
sin⁻¹(6sin(20)/7.
6)
        15.66521315
■
```

STEP 1 Find b.

Use the law of cosines that involves β.	$b^2 = a^2 + c^2 - 2ac\cos\beta$
Let $a = 13$, $c = 6$, and $\beta = 20°$.	$b^2 = 13^2 + 6^2 - 2(13)(6)\cos 20°$
Evaluate the right side using a calculator.	$b^2 = 58.40795$
Solve for b.	$b = \pm 7.6425$
Round to two significant digits, and b can only be positive.	$b = 7.6$

STEP 2 Find γ.

Use the law of sines.	$\dfrac{\sin\gamma}{c} = \dfrac{\sin\beta}{b}$
Isolate $\sin\gamma$.	$\sin\gamma = \dfrac{c\sin\beta}{b}$
Let $b = 7.6$, $c = 6$, and $\beta = 20°$.	$\sin\gamma = \dfrac{(6)\sin 20°}{(7.6)}$
Use the inverse sine function.	$\gamma = \sin^{-1}\left(\dfrac{(6)\sin 20°}{(7.6)}\right)$
Evaluate the right side using a calculator.	$\gamma = 15.66521°$
Round to the nearest degree.	$\gamma = 16°$

STEP 3 Find α.

The angle measures must sum to 180°.	$\alpha + 20° + 16° = 180°$
Solve for α.	$\alpha = 144°$

Notice the steps we took in solving a SAS triangle:

1. Find the side opposite the given angle using the law of cosines.

2. Solve for the smaller angle using the law of sines.

3. Solve for the larger angle using properties of triangles.

You may be thinking, does it matter if we solved for α before solving for γ? Yes, it does matter—in this problem you cannot solve for α by the law of sines before finding γ. The law of sines can only be used on the smaller angle (opposite the shortest side). If we had tried to use the law of sines with the obtuse angle, α, the inverse sine would have resulted in $\alpha = 36°$. Since sine is positive in QI and QII, we would not know if that angle was $\alpha = 36°$ or its supplementary angle $\alpha = 144°$. Notice that $c < a$; therefore the angles opposite those sides must have the same relationship $\gamma < \alpha$. We choose the smaller angle first. Alternatively, if we wanted to solve for the obtuse angle first, we could have used the law of cosines to solve for α.

CONCEPT CHECK Given the triangle, $b = 4.2$, $c = 1.8$, and $\alpha = 35.0°$, after side a is calculated, which angle, β or γ, would you find next using the law of sines?

■ **YOUR TURN** Solve the triangle: $b = 4.2$, $c = 1.8$, and $\alpha = 35°$.

EXAMPLE 2 Using the Law of Cosines in an Application (SAS)

In an AKC (American Kennel Club)-sanctioned field trial, a judge sets up a mark (bird) that requires the dog to swim (the dogs are judged on how closely they adhere to the straight line to the bird, not the time it takes to retrieve the bird). The judge is trying to calculate how far the dogs would have to swim to this mark, so she walks off the two legs across the land and measures the angle as shown in the figure. How far will the dog swim from the starting line to the bird?

152 yd

117°

176 yd   Pond

Solution:

Label the triangle.

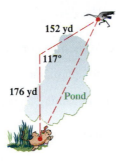

$b = 152$ yd

$\gamma = 117°$

c

$a = 176$ yd   Pond

TECHNOLOGY TIP

Use a calculator to find the value of c.

```
176²+152²-2*176*
152cos(117)
        78370.3077
√(Ans)
        279.946973
```

Use the law of cosines: $c^2 = a^2 + b^2 - 2ab\cos\gamma$

Let $a = 176$ yards, $b = 152$ yards, and $\gamma = 117°$. $c^2 = 176^2 + 152^2 - 2(176)(152)\cos 117°$

Simplify. $c^2 = 78370.03077$

Solve for c and round to the nearest yard. $c = 280$ yards

Case 4: Solving Oblique Triangles (SSS)

We now solve oblique triangles when all three side lengths are given (SSS). In this case, start by finding the largest angle (opposite the largest side) using the law of cosines. Then use the law of sines to find either of the remaining two angles. Last, find the third angle using the triangle angle sum identity.

EXAMPLE 3 Using the Law of Cosines to Solve a Triangle (SSS)

Solve the triangle: $a = 8$, $b = 6$, and $c = 7$.

Solution:

STEP 1 Identify the largest angle, which is α.

Use the law of cosines that involve α.

$$a^2 = b^2 + c^2 - 2bc\cos\alpha$$

Let $a = 8$, $b = 6$, and $c = 7$.

$$8^2 = 6^2 + 7^2 - 2(6)(7)\cos\alpha$$

Simplify and isolate $\cos\alpha$.

$$\cos\alpha = \frac{6^2 + 7^2 - 8^2}{2(6)(7)} = 0.25$$

Use the inverse cosine function.

$$\alpha = \cos^{-1}0.25$$

Use a calculator to approximate.

$$\boxed{\alpha = 75.5°}$$

STEP 2 Find either of the remaining angles. To solve for β:

Use the law of sines.

$$\frac{\sin\alpha}{a} = \frac{\sin\beta}{b}$$

Isolate $\sin\beta$.

$$\sin\beta = \frac{b\sin\alpha}{a}$$

Let $a = 8$, $b = 6$, and $\alpha = 75.5°$.

$$\sin\beta = \frac{6\sin75.5°}{8}$$

Use the inverse sine function.

$$\beta = \sin^{-1}\left(\frac{6\sin75.5°}{8}\right)$$

Use a calculator to approximate.

$$\boxed{\beta = 46.6°}$$

STEP 3 Find the third angle, γ.

The sum of the angle measures is 180°.

$$75.5° + 46.6° + \gamma = 180°$$

Solve for γ.

$$\boxed{\gamma = 57.9°}$$

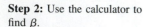
CONCEPT CHECK Given the triangle, $a = 5$, $b = 7$, and $c = 8$, which angle would you solve for first?

■ **YOUR TURN** Solve the triangle: $a = 5$, $b = 7$, and $c = 8$.

■ **Answer:** $\alpha = 38.2°$, $\beta = 60.0°$, and $\gamma = 81.8°$

In the next example, instead of immediately substituting values in for the law of cosines equation, we will solve for the angle in general, and then substitute in values.

EXAMPLE 4 Using the Law of Cosines in an Application (SSS)

In recent decades, many people have come to believe that an imaginary area called the Bermuda Triangle, located off the southeastern Atlantic coast of the United States, has been the site, over the centuries, of a high incidence of losses of ships, small boats, and aircraft. Assume for the moment, without judging the merits of the hypothesis, that this Bermuda Triangle has vertices in Norfolk (Virginia), Bermuda, and Santiago (Cuba), and find the angles of the Bermuda Triangle given the following distances:

LOCATION	LOCATION	DISTANCE (NAUTICAL MILES)
Norfolk	Bermuda	850
Bermuda	Santiago	810
Norfolk	Santiago	894

(Ignore the curvature of the Earth in your calculations.)

Solution:

Draw and label the triangle.

STEP 1 Find β (the largest angle).

Use the law of cosines.

$$b^2 = a^2 + c^2 - 2ac\cos\beta$$

Isolate $\cos\beta$.

$$\cos\beta = \frac{a^2 + c^2 - b^2}{2ac}$$

Use the inverse cosine function to solve for β.

$$\beta = \cos^{-1}\left(\frac{a^2 + c^2 - b^2}{2ac}\right)$$

Let $a = 810$, $b = 894$, and $c = 850$.

$$\beta = \cos^{-1}\left(\frac{810^2 + 850^2 - 894^2}{2(810)(850)}\right)$$

Use a calculator to approximate.

$$\beta \approx 65°$$

STEP 2 Find α.

Use the law of sines.

$$\frac{\sin\alpha}{a} = \frac{\sin\beta}{b}$$

Isolate $\sin\alpha$.

$$\sin\alpha = \frac{a\sin\beta}{b}$$

TECHNOLOGY TIP

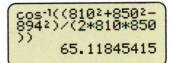

Step 1: Use a calculator to find the value of β.

```
cos⁻¹((810²+850²−
894²)/(2*810*850
))
          65.11845415
```

Step 2: Use the calculator to find α.

```
sin⁻¹(810sin(65)/
894)
          55.200215
```

Instead of using the approximated value of β to calculate α, you can use ANS to retrieve the actual value of β.

```
cos⁻¹((810²+850²−
894²)/(2*810*850
))
          65.11845415
sin⁻¹(810sin(Ans)
/894)
          55.27959284
```

Use the inverse sine function to solve for α.

$$\alpha = \sin^{-1}\left(\frac{a\sin\beta}{b}\right)$$

Let $a = 810$, $b = 894$, and $\beta = 65°$.

$$\alpha = \sin^{-1}\left(\frac{810\sin 65°}{894}\right)$$

Use a calculator to approximate, and round to nearest degree.

$\alpha \approx 55°$

STEP 3 Find γ.

The angle measures must sum to $180°$.

$55° + 65° + \gamma = 180°$

Solve for γ.

$\gamma = 60°$

SECTION 7.2 **SUMMARY**

We can solve any triangle given three pieces of information, as long as one of the pieces is a side length. Depending on the information given, we either use the law of sines and the angle sum identity or we use a combination of the law of cosines, then law of sines, and the angle sum identity. The table summarizes the strategies for solving oblique triangles.

OBLIQUE TRIANGLE	WHAT IS KNOWN	PROCEDURE FOR SOLVING
AAS or ASA	Two angles and a side	Step 1: Find the remaining angle using $\alpha + \beta + \gamma = 180°$. Step 2: Find the remaining sides using the law of sines.
SSA	Two sides and an angle opposite one of the sides	This is the ambiguous case, so there is either no triangle, one triangle, or two triangles. If the given angle is obtuse, then there is either one or no triangles. If the given angle is acute, then there is 0, 1, or 2 triangles. Step 1: Use the law of sines to find one of the angles. Step 2: Find the remaining angle using $\alpha + \beta + \gamma = 180°$. Step 3: Find the remaining side using the law of sines. If two triangles exist, then the angle found in step 1 can be either acute or obtuse, and steps 2 and 3 must be performed for each.
SAS	Two sides and an angle between the sides	Step 1: Find the third side using the law of cosines. Step 2: Find the smaller angle using the law of sines. Step 3: Find the remaining angle using $\alpha + \beta + \gamma = 180°$.
SSS	Three sides	Step 1: Find the largest angle using the law of cosines. Step 2: Find either angle using the law of sines. Step 3: Find the remaining angle using $\alpha + \beta + \gamma = 180°$.

SECTION 7.2 EXERCISES

■ SKILLS

For each of the given triangles in Exercises 1–8, the angle sum identity, $\alpha + \beta + \gamma = 180°$, will be used in solving the triangle. Label the problem as S if only the law of sines is needed to solve the triangle. Label the problem as C if the law of cosines is needed to solve the triangle.

1. a, b, and c are given.
2. a, b, and γ are given.
3. α, β, and c are given.
4. a, b, and α are given.

5. a, β, and γ are given.
6. α, β, and a are given.
7. b, c, and α are given.
8. b, c, and β are given.

In Exercises 9–36, solve each triangle.

9. $a = 4$, $c = 3$, $\beta = 100°$
10. $a = 6$, $b = 10$, $\gamma = 80°$
11. $b = 7$, $c = 2$, $\alpha = 16°$

12. $b = 5$, $a = 6$, $\gamma = 170°$
13. $b = 5$, $c = 5$, $\alpha = 20°$
14. $a = 4.2$, $b = 7.3$, $\gamma = 25°$

15. $a = 9$, $c = 12$, $\beta = 23°$
16. $b = 6$, $c = 13$, $\alpha = 16°$
17. $a = 4$, $c = 8$, $\beta = 60°$

18. $b = 3$, $c = \sqrt{18}$, $\alpha = 45°$
19. $a = 8$, $b = 5$, $c = 6$
20. $a = 6$, $b = 9$, $c = 12$

21. $a = 4$, $b = 4$, $c = 5$
22. $a = 17$, $b = 20$, $c = 33$
23. $a = 8.2$, $b = 7.1$, $c = 6.3$

24. $a = 1492$, $b = 2001$, $c = 1776$
25. $a = 4$, $b = 5$, $c = 10$
26. $a = 1.3$, $b = 2.7$, $c = 4.2$

27. $a = 12$, $b = 5$, $c = 13$
28. $a = 4$, $b = 5$, $c = \sqrt{41}$
29. $\alpha = 40°$, $\beta = 35°$, $a = 6$

30. $b = 11.2$, $a = 19.0$, $\gamma = 13.3°$
31. $\alpha = 31°$, $b = 5$, $a = 12$
32. $a = 11$, $c = 12$, $\gamma = 60°$

33. $a = \sqrt{7}$, $b = \sqrt{8}$, $c = \sqrt{3}$
34. $\beta = 106°$, $\gamma = 43°$, $a = 1$
35. $b = 11$, $c = 2$, $\beta = 10°$

36. $\alpha = 25°$, $a = 6$, $c = 9$

■ APPLICATIONS

37. **Airplane Speed.** A plane flew due north at 500 mph for 3 hours. A second plane, starting at the same point and at the same time, flew southeast at an angle 150° clockwise from due north at 435 mph for 3 hours. At the end of the 3 hours, how far apart were the two planes? Round to the nearest mile.

38. **Airplane Speed.** A plane flew due north at 400 mph for 4 hours. A second plane, starting at the same point and at the same time, flew southeast at an angle 120° clockwise from due north at 300 mph for 4 hours. At the end of the 4 hours, how far apart were the two planes? Round to the nearest mile.

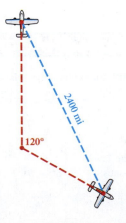

39. Baseball. A baseball diamond is actually a square that is 90 feet on each side. The pitcher's mound is located 60.5 feet from home plate. How far is it from third base to the pitcher's mound?

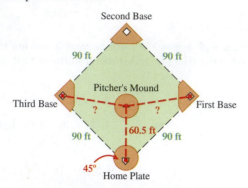

Second Base

90 ft   90 ft

Pitcher's Mound

Third Base   ?   ?   First Base

90 ft   60.5 ft   90 ft

45°

Home Plate

40. Aircraft Wing. Given the acute angle and two sides of the stealth bomber, what is the unknown length?

? ft
18 ft   13.2°
25 ft

Stone/Getty Images, Inc.

41. Sliding Board. A 40 foot slide leaning against the bottom of a building's window makes a 55° angle with the building. The angle formed with the building with the line of sight from the top of the window to the point on the ground where the slide ends is 40°. How tall is the window?

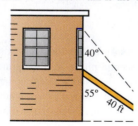

40°

55°
40 ft

42. Airplane Slide. An airplane door is 6 feet high. If a slide attached to the bottom of the open door is at an angle of 40° with the ground, and the angle formed by the line of sight from where the slide touches the ground to the top of the door is 45°, then how long is the slide?

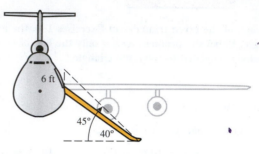

6 ft

45°
40°

For Exercises 43 and 44, refer to the following:

Peg and Meg live one mile apart. The school that they attend lies on a street that makes a 60° angle with the street connecting their houses when measured from Peg's house. The street connecting Meg's house and the school makes a 50° angle with the street connecting them.

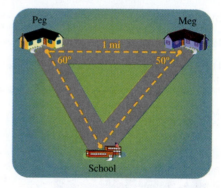

Peg   Meg

1 mi

60°   50°

School

43. Distance to School. How far is it from Peg's house to the school?

44. Distance to School. How far is it from Meg's house to the school?

■ **CATCH THE MISTAKE**

In Exercises 45 and 46, explain the mistake that is made.

45. Solve the triangle: $b = 3$, $c = 4$, and $\alpha = 30°$.

Solution:

Step 1: Find a.

Use the law of cosines. $a^2 = b^2 + c^2 - 2bc\cos\alpha$

$b = 3$, $c = 4$, and $\alpha = 30°$. $a^2 = 3^2 + 4^2 - 2(3)(4)\cos 30°$

Solve for a. $a = 2.1$

Step 2: Find γ.

Use the law of sines. $\dfrac{\sin\alpha}{a} = \dfrac{\sin\gamma}{c}$

Solve for $\sin\gamma$.

$$\sin\gamma = \frac{c\sin\alpha}{a}$$

Solve for γ.

$$\gamma = \sin^{-1}\left(\frac{c\sin\alpha}{a}\right)$$

Let $a = 2.1$, $c = 4$,
and $\alpha = 30°$.

$$\gamma = 72°$$

Step 3: Find β.

$\alpha + \beta + \gamma = 180°$ $\qquad$ $30° + \beta + 72° = 180°$

Solve for β. $\qquad\qquad$ $\beta = 78°$

$a = 2.1$, $b = 3$, $c = 4$, $\alpha = 30°$, $\beta = 78°$, and $\gamma = 72°$

This is incorrect. The longest side is not opposite the largest angle. What mistake was made?

46. Solve the triangle: $a = 6$, $b = 2$, $c = 5$.

Solution:

Step 1: Find β.

Use the law
of cosines. $\qquad$ $b^2 = a^2 + c^2 - 2ac\cos\beta$

Solve for β.

$$\beta = \cos^{-1}\left(\frac{a^2 + c^2 - b^2}{2ac}\right)$$

Let $a = 6$, $b = 2$,
$c = 5$. $\qquad\qquad$ $\beta \approx 18°$

Step 2: Find α.

Use the law of sines. $\qquad$ $\dfrac{\sin\alpha}{a} = \dfrac{\sin\beta}{b}$

Solve for α. $\qquad\qquad$ $\alpha = \sin^{-1}\left(\dfrac{a\sin\beta}{b}\right)$

Let $a = 6$, $b = 2$,
and $\beta = 18°$. $\qquad\qquad$ $\alpha \approx 68°$

Step 3: Find γ.

$\alpha + \beta + \gamma = 180°$ $\qquad$ $68° + 18° + \gamma = 180°$

$$\gamma = 94°$$

$a = 6$, $b = 2$, $c = 5$, $\alpha = 68°$, $\beta = 18°$, and $\gamma = 94°$

This is incorrect. The longest side is not opposite the largest angle. What mistake was made?

 ■ **CHALLENGE**

47. T or F: Given three sides of a triangle, there is insufficient information to solve the triangle.

48. T or F: Given three angles of a triangle, there is insufficient information to solve the triangle.

49. T or F: The Pythagorean theorem is a special case of the law of cosines.

50. T or F: The law of cosines is a special case of the Pythagorean theorem.

51. Show that $\dfrac{\cos\alpha}{a} + \dfrac{\cos\beta}{b} + \dfrac{\cos\gamma}{c} = \dfrac{a^2 + b^2 + c^2}{2abc}$. *Hint:* Use the law of cosines.

52. Show that $a = c\cos\beta + b\cos\gamma$. *Hint:* Use the law of cosines.

■ **TECHNOLOGY**

For Exercises 53–56, let A, B, and C be the lengths of the three sides with X, Y, and Z as the corresponding angles. Write a program using TI to solve the given triangle.

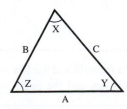

53. $B = 45$, $C = 57$, and $X = 43°$

54. $B = 24.5$, $C = 31.6$, and $X = 81.5°$

55. $A = 29.8$, $B = 37.6$, and $C = 53.2$

56. $A = 100$, $B = 170$, and $C = 250$

Skills Objectives

- Find the area of triangles in the SAS case.
- Find the area of triangles in the SSS case.

Conceptual Objectives

- Use the law of sines to derive a formula for an area of a triangle (SAS case).
- Use the law of cosines to derive a formula for an area of a triangle (SSS case).

In Sections 7.1 and 7.2 we used the law of sines and the law of cosines to solve oblique triangles, which means to find all of the side lengths and angle measures. Now, we use these laws to derive formulas for *areas* of triangles (SAS and SSS).

Our starting point for both cases is the standard formula for the area of a triangle:

$$A = \frac{1}{2}bh$$

WORDS	MATH
Start with a triangle with base, b, and height, h.	
Construct a rectangle with base, b, and height, h, around the triangle.	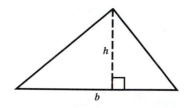
The area of the rectangle is base times height.	$A_{\text{rectangle}} = bh$
Triangles 1 and 2 have the same area. Triangles 3 and 4 have the same area. Therefore, the area of the triangle is half the area of the rectangle.	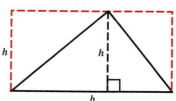
Area of a triangle with base b and height h.	$A_{\text{triangle}} = \frac{1}{2}bh$

Area of a Triangle (SAS)

We now can use the general formula for area of a triangle and the law of sines to develop a formula for the area of a triangle when two sides and the angle between them are given.

WORDS	MATH

WORDS

MATH

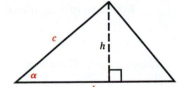

Start with a triangle (SAS) (given b, c, and α).

Form the sine ratio for angle α. $\sin \alpha = \dfrac{h}{c}$

Solve for h. $h = c \sin \alpha$

Write the formula for area of a triangle. $A_{\text{triangle}} = \dfrac{1}{2} bh$

Substitute $h = c \sin \alpha$ for the SAS triangle. $A_{\text{SAS triangle}} = \dfrac{1}{2} bc \sin \alpha$

Now, the area of this triangle can be calculated with the given information (two sides and the angle between them: b, c, and α). Similarly, it can be shown that the other formulas for SAS triangles are

$$A_{\text{SAS triangle}} = \dfrac{1}{2} ab \sin \gamma \quad \text{and} \quad A_{\text{SAS triangle}} = \dfrac{1}{2} ac \sin \beta$$

AREA OF A TRIANGLE (SAS)

For any triangle where two sides and the angle between them are known, the area for that triangle is given by one of the following formulas (depending on which angle and sides are given):

$A_{\text{SAS triangle}} = \dfrac{1}{2} bc \sin \alpha$ when b, c, and α are known.

$A_{\text{SAS triangle}} = \dfrac{1}{2} ab \sin \gamma$ when a, b, and γ are known.

$A_{\text{SAS triangle}} = \dfrac{1}{2} ac \sin \beta$ when a, c, and β are known.

In other words, the area of a triangle equals one-half the product of two of its sides and the sine of the angle between them.

EXAMPLE 1 Finding the Area of a Triangle (SAS)

Find the area of the triangle: $a = 7.0$ ft, $b = 9.3$ ft, and $\gamma = 86°$.

Solution:

Use the area formula where a, b, and γ are given. $A = \dfrac{1}{2} ab \sin \gamma$

Substitute $a = 7.0$ ft, $b = 9.3$ ft, and $\gamma = 86°$. $A = \dfrac{1}{2}(7.0 \text{ ft})(9.3 \text{ ft}) \sin 86°$

TECHNOLOGY TIP

Use a calculator to find A.

```
1/2*7*9.3*sin(86
)
          32.47070984
```

Use a calculator to approximate. $\qquad$ $A = 32.47071$ sq ft

Round to two significant digits. $\qquad$ $A = 32$ sq ft

CONCEPT CHECK Which formula for area would you use given a, c, and β?

■ **YOUR TURN** Find the area of the triangle: $a = 3.2$ m, $c = 5.1$ m, and $\beta = 49°$.

Area of a Triangle (SSS)

We used the law of sines to develop a formula for area of an SAS triangle. That formula, several trigonometric identities, and the law of cosines can be used to develop a formula for the area of an SSS triangle, called **Heron's formula**.

WORDS	MATH
Start with any of the formulas for SAS triangles.	$A = \dfrac{1}{2}\, ab \sin \gamma$
Square both sides.	$A^2 = \dfrac{1}{4}\, a^2 b^2 \sin^2 \gamma$
Isolate $\sin^2 \gamma$.	$\dfrac{4A^2}{a^2 b^2} = \sin^2 \gamma$
Use the Pythagorean identity.	$\dfrac{4A^2}{a^2 b^2} = 1 - \cos^2 \gamma$
Factor the difference of the two squares on the right.	$\dfrac{4A^2}{a^2 b^2} = (1 - \cos \gamma)(1 + \cos \gamma)$
Solve the law of cosines, $c^2 = a^2 + b^2 - 2\, ab \cos \gamma$, for $\cos \gamma$.	$\cos \gamma = \dfrac{a^2 + b^2 - c^2}{2ab}$
Substitute $\cos \gamma = \dfrac{a^2 + b^2 - c^2}{2ab}$ into $\dfrac{4A^2}{a^2 b^2} = (1 - \cos \gamma)(1 + \cos \gamma)$.	$\dfrac{4A^2}{a^2 b^2} = \left[1 - \dfrac{a^2 + b^2 - c^2}{2ab}\right]\left[1 + \dfrac{a^2 + b^2 - c^2}{2ab}\right]$
Combine the expressions in brackets.	$\dfrac{4A^2}{a^2 b^2} = \left[\dfrac{2ab - a^2 - b^2 + c^2}{2ab}\right]\left[\dfrac{2ab + a^2 + b^2 - c^2}{2ab}\right]$
Group the terms in the numerators.	$\dfrac{4A^2}{a^2 b^2} = \left[\dfrac{-(a^2 - 2ab + b^2) + c^2}{2ab}\right]\left[\dfrac{(a^2 + 2ab + b^2) - c^2}{2ab}\right]$
Write the numerators as the difference of two squares.	$\dfrac{4A^2}{a^2 b^2} = \left[\dfrac{c^2 - (a - b)^2}{2ab}\right]\left[\dfrac{(a + b)^2 - c^2}{2ab}\right]$

■ **Answer:** 6.2 m^2

Factor the numerators:
$x^2 - y^2 = (x - y)(x + y)$.

$$\frac{4A^2}{a^2b^2} = \left[\frac{(c - [a - b])(c + [a - b])}{2ab}\right]\left[\frac{([a + b] - c)([a + b] + c)}{2ab}\right]$$

Simplify.

$$\frac{4A^2}{a^2b^2} = \left[\frac{(c - a + b)(c + a - b)}{2ab}\right]\left[\frac{(a + b - c)(a + b + c)}{2ab}\right]$$

$$\frac{4A^2}{a^2b^2} = \frac{(c - a + b)(c + a - b)(a + b - c)(a + b + c)}{4a^2b^2}$$

Multiply by $\dfrac{a^2b^2}{4}$.

$$A^2 = \frac{1}{16}(c - a + b)(c + a - b)(a + b - c)(a + b + c)$$

The semiperimeter, s, is half the perimeter of the triangle.

$$2s = a + b + c$$

Manipulate each of the four factors:

$c - a + b = a + b + c - 2a = 2s - 2a = 2(s - a)$

$c + a - b = a + b + c - 2b = 2s - 2b = 2(s - b)$

$a + b - c = a + b + c - 2c = 2s - 2c = 2(s - c)$

$a + b + c = 2s$

Substitute these values in for the four factors.

$$A^2 = \frac{1}{16} \cdot 2(s - a) \cdot 2(s - b) \cdot 2(s - c) \cdot 2s$$

Simplify.

$$A^2 = s(s - a)(s - b)(s - c)$$

Solve for A (area is always positive).

$$A = \sqrt{s(s - a)(s - b)(s - c)}$$

AREA OF A TRIANGLE (SSS CASE, HERON'S FORMULA)

For any triangle where the lengths of the three sides are known, the area for that triangle is given by the following formula:

$$A_{\text{SSS triangle}} = \sqrt{s(s - a)(s - b)(s - c)}$$

where a, b, and c are the lengths of the sides of the triangle and s is half the perimeter of the triangle, called the semiperimeter:

$$s = \frac{a + b + c}{2}$$

EXAMPLE 2 Finding the Area of a Triangle (SSS)

Find the area of the triangle: $a = 5$, $b = 6$, and $c = 9$.

Solution:

Find the semiperimeter, s.

$$s = \frac{a + b + c}{2}$$

Substitute : $a = 5$, $b = 6$, and $c = 9$.

$$s = \frac{5 + 6 + 9}{2}$$

Simplify.

$$s = 10$$

TECHNOLOGY TIP

Use a calculator to find A.

$$\sqrt{(10(10-5)(10-6)} \\ (10-9)) \\ \qquad 14.14213562$$

Write the formula for the area of an SSS triangle.	$A = \sqrt{s(s-a)(s-b)(s-c)}$
Substitute $a = 5$, $b = 6$, $c = 9$, and $s = 10$.	$A = \sqrt{10(10-5)(10-6)(10-9)}$
Simplify the radicand.	$A = \sqrt{10 \cdot 5 \cdot 4 \cdot 1}$
Evaluate the radical.	$A = 10\sqrt{2} \approx 14$

CONCEPT CHECK What is the semiperimeter of the triangle: $a = 3$, $b = 5$, and $c = 6$?

■ **YOUR TURN** Find the area of the triangle: $a = 3$, $b = 5$, and $c = 6$.

SECTION 7.3 **SUMMARY**

In this section we derived formulas for calculating the areas of triangles (SAS and SSS). The law of sines leads to three area formulas for SAS triangles depending on which angles and sides are given. The law of cosines was instrumental in developing a formula for the area of a triangle (SSS) when all three sides are given.

SECTION 7.3 EXERCISES

■ **SKILLS**

In Exercises 1–24, find the area of each triangle described.

1. $a = 8$, $c = 16$, $\beta = 60°$

2. $b = 6$, $c = 4\sqrt{3}$, $\alpha = 30°$

3. $a = 1$, $b = \sqrt{2}$, $\alpha = 45°$

4. $b = 2\sqrt{2}$, $c = 4$, $\beta = 45°$

5. $a = 6$, $b = 8$, $\gamma = 80°$

6. $b = 9$, $c = 10$, $\alpha = 100°$

7. $a = 4$, $c = 7$, $\beta = 27°$

8. $a = 6.3$, $b = 4.8$, $\gamma = 17°$

9. $b = 100$, $c = 150$, $\alpha = 36°$

10. $c = 0.3$, $a = 0.7$, $\beta = 145°$

11. $a = \sqrt{5}$, $b = 5\sqrt{5}$, $\gamma = 50°$

12. $b = \sqrt{11}$, $c = \sqrt{11}$, $\alpha = 21°$

13. $a = 15$, $b = 15$, $c = 15$

14. $a = 1$, $b = 1$, $c = 1$

15. $a = 7$, $b = \sqrt{51}$, $c = 10$

16. $a = 9$, $b = 40$, $c = 41$

17. $a = 6$, $b = 10$, $c = 9$

18. $a = 40$, $b = 50$, $c = 60$

19. $a = 14.3$, $b = 15.7$, $c = 20.1$

20. $a = 146.5$, $b = 146.5$, $c = 100$

21. $a = 14{,}000$, $b = 16{,}500$, $c = 18{,}700$

22. $a = \sqrt{2}$, $b = \sqrt{3}$, $c = \sqrt{5}$

23. $a = 80$, $b = 75$, $c = 160$

24. $a = 19$, $b = 23$, $c = 3$

■ APPLICATIONS

25. Bermuda Triangle. Calculate the area of the Bermuda Triangle, described in Example 4 of Section 7.2, if, as some people define it, its vertices are located in Norfolk, Bermuda, and Santiago:

Location	Location	Distance (nautical miles)
Norfolk	Bermuda	850
Bermuda	Santiago	810
Norfolk	Santiago	894

Again, ignore the curvature of the Earth in your calculations.

26. Bermuda Triangle. Calculate the area of the Bermuda Triangle if, as other people define it, its vertices are located in Miami, Bermuda, and San Juan:

Location	Location	Distance (nautical miles)
Miami	Bermuda	898
Bermuda	San Juan	831
Miami	San Juan	890

27. Triangular Tarp. A large triangular tarp is needed to cover a playground when it rains. If the sides of the tarp measure 160 feet by 140 feet by 175 feet, then what is the area of the tarp (to the nearest square foot)?

28. Flower Seed. A triangular garden measures 41 feet by 16 feet by 28 feet. You are going to plant wildflower seed that costs $4 per bag. Each bag of flower seed covers an area of 20 square feet. How much will the seed cost? (Assume you have to buy a whole bag—you can't split one.)

29. Mural. Some students are painting a mural on the side of a building. They have enough paint for a 1000 square foot area triangle. If two sides of the triangle measure 60 feet and 120 feet, then what angle (to the nearest degree) should the two sides form in order to create a triangle that uses up all the paint?

30. Mural. Some students are painting a mural on the side of a building. They have enough paint for a 500 square foot area triangle. If two sides of the triangle measure 40 feet and 60 feet, then what angle (to the nearest degree) should the two sides form in order to create a triangle that uses up all the paint?

31. Insect Infestation. Some very destructive beetles have made their way into a forest preserve. The rangers are trying to keep track of their spread and how well preventive

measures are working. In a triangular area that is 22.5 miles on one side, 28.1 miles on the second, and 38.6 miles on the third, what is the total area the rangers are covering?

32. Real Estate. A real estate agent needs to determine the area of a triangular lot. Two sides of the lot are 150 feet and 60 feet. The angle between the two measured sides is 43°. What is the area of the lot?

33. Pizza. Big Augie's Pizza now offers equilateral triangle-shaped pizzas. How long must the sides of the pizza be to the nearest inch to equal the size of a round pizza with a 9 inch diameter?

34. Pizza. Big Augie's Pizza now offers equilateral triangle-shaped pizzas. How long must the sides of the pizza be to the nearest inch to equal the size of a round pizza with a 15 inch diameter?

35. Parking Lot. A parking lot is to have the shape of a parallelogram that has adjacent sides measuring 200 feet and 260 feet. The angle between the two sides is 65°. What is the area of the parking lot?

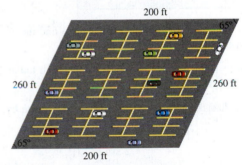

36. Parking Lot. A parking lot is to have the shape of a parallelogram that has adjacent sides measuring 250 feet and 300 feet. The angle between the two sides is 55°. What is the area of the parking lot?

37. Regular Hexagon. A regular hexagon has sides measuring 3 feet. What is its area? Recall that the measure of an angle of a regular n-gon is given by the formula angle $= \dfrac{180(n-2)}{n}$.

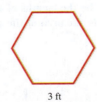
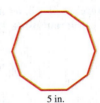

3 ft   5 in.

38. Regular Decagon. A regular decagon has sides measuring 5 inches. What is its area?

■ CATCH THE MISTAKE

In Exercises 39 and 40, explain the mistake that is made.

39. Calculate the area of the triangle: $a = 2, b = 6, c = 7$.

Solution:

Find the
semiperimeter. $s = a + b + c = 2 + 6 + 7 = 15$

Write the formula
for the area of the
triangle. $A = \sqrt{s(s - a)(s - b)(s - c)}$

Let $a = 2, b = 6,$
$c = 7$, and $s = 15$. $A = \sqrt{15(15 - 2)(15 - 6)(15 - 7)}$

Simplify. $A = \sqrt{14040} \approx 118$

This is not correct. What mistake was made?

40. Calculate the area of the triangle: $a = 2, b = 6, c = 5$.

Solution:

Find the
semiperimeter. $s = \dfrac{a + b + c}{2} = \dfrac{2 + 6 + 5}{2} = 6.5$

Write the formula
for area of a triangle. $A = s(s - a)(s - b)(s - c)$

Let $a = 2, b = 6,$
$c = 5$, and $s = 6.5$. $A = 6.5(4.5)(0.5)(1.5)$

Simplify. $A \approx 22$

This is incorrect. What mistake was made?

■ CHALLENGE

41. T or F: Heron's formula can be used to find the area of right triangles.

42. T or F: Heron's formula can be used to find the area of isosceles triangles.

43. Show that the area for an SAA triangle is given by:

$$A = \frac{a^2 \sin \beta \sin \gamma}{2 \sin \alpha}$$

Assume that $\alpha, \beta,$ and a are given.

44. Show that the area of an isosceles triangle with equal sides of length s is given by

$$A_{\text{isosceles}} = \frac{1}{2} s^2 \sin \theta$$

θ is the angle between the two equal sides.

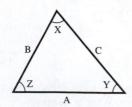

■ TECHNOLOGY

For Exercises 45–48, let A, B, and C be the lengths of the three sides with X, Y, and Z as the corresponding angles. Write a program using TI to calculate the area of the given triangle.

45. A = 35, B = 47, and Z = 68°

46. A = 1241 B = 1472, and Z = 56°

47. A = 85, B = 92, and C = 123

48. A = $\sqrt{167}$, B = $\sqrt{113}$, C = $\sqrt{203}$

- Represent vectors geometrically and algebraically.
- Find the magnitude and direction of a vector.
- Add and subtract vectors.
- Perform scalar multiplication of a vector.
- Find unit vectors.
- Express a vector in terms of its horizontal and vertical components.

- Understand the difference between scalars and vectors.
- Relate the geometric and algebraic representations of vectors.

Introduction to Vectors

What is the difference between velocity and speed? Speed only has *magnitude*, whereas velocity has *magnitude* and *direction*. We use **scalars**, which are real numbers, to denote magnitudes such as speed and weight. We use **vectors**, which have magnitude *and* direction, to denote quantities such as velocity (speed in a certain direction) and force (weight in a certain direction).

A vector quantity is geometrically denoted by a **directed line segment**, which is a line segment with an arrow representing direction. There are many ways to denote a vector. For example, the following vector can be denoted as **u**, **AB**, $\vec{u}$ or $\overrightarrow{AB}$, where A is the **initial point** and B is the **terminal point**.

It is customary in books to use the bold letter to denote a vector and when handwritten (as in your class notes and homework) to use the arrow on top to represent a vector.

In this book we will limit our discussion to vectors in a plane (two-dimensional). It is important to note that geometric representation can be extended to three dimensions, and algebraic representation can be extended to any higher dimension, as you will see in the exercises.

Geometric Interpretation of Vectors

The *magnitude* of a vector can be denoted one of two ways: $|\mathbf{u}|$ or $\|\mathbf{u}\|$. We will use the former notation.

MAGNITUDE: $|\mathbf{u}|$

The **magnitude** of a vector **u**, denoted $|\mathbf{u}|$, is the length of the directed line segment.

Two vectors have the **same direction** if they are parallel and point in the same direction. Two vectors have **opposite direction** if they are parallel and point in opposite directions.

> **EQUAL VECTORS: u = v**
>
> Two vectors are **equal**, **u = v**, if they have the same magnitude, |**u**| = |**v**|, and have the same direction.

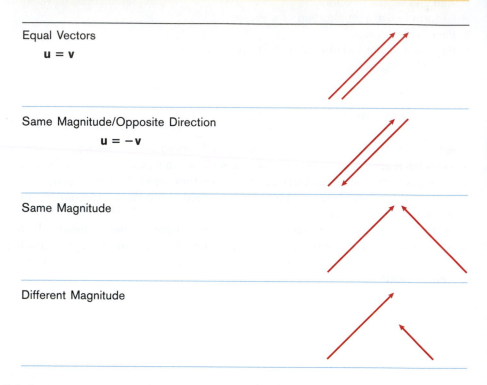

Equal Vectors

    u = v

Same Magnitude/Opposite Direction

      u = −v

Same Magnitude

Different Magnitude

It is important to note that vectors do not have to coincide to be equal.

> **VECTOR ADDITION: u + v**
>
> Two vectors, **u** and **v**, can be added together using the **tail-to-tip** rule.
>
>
>
> Translate **v** so that its tail end (the initial point) is located at the tip end (the terminal point) of **u**.
>
>
>
> The **sum**, **u + v**, is the **resultant** vector from the tail end of **u** to the tip end of **v**.

The difference, **u** − **v**, is the resultant vector from the tip of **v** to the tip of **u**.

Algebraic Interpretation of Vectors

Since equal vectors have the same direction and magnitude, any vector can be translated to an equal vector whose initial point is located at the origin in the Cartesian plane. Therefore, we now consider vectors in a rectangular coordinate system.

A vector with its initial point at the origin is called a **position vector**, or in **standard position**. A position vector, **u**, with its terminal point at (a, b) is denoted:

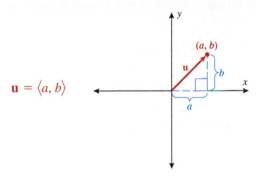

$$\mathbf{u} = \langle a, b \rangle$$

where the real numbers a and b are called the **components** of vector **u**.

Notice the subtle difference between coordinate notation and vector notation. The point is denoted with parentheses, (a, b), whereas the vector is denoted with angled brackets, $\langle a, b \rangle$. The notation $\langle a, b \rangle$ denotes a vector whose tail point is $(0, 0)$ and terminal point is (a, b).

The vector with initial point $(3, 4)$ and terminal point $(8, 9)$ is equal to the vector $\langle 5, 5 \rangle$, which has initial point $(0, 0)$ and terminal point $(5, 5)$.

Recall that the geometric definition of the *magnitude* of a vector is the *length* of the vector.

MAGNITUDE: $|\mathbf{u}|$

The **magnitude** (or norm) of a vector, $\mathbf{u} = \langle a, b \rangle$, is

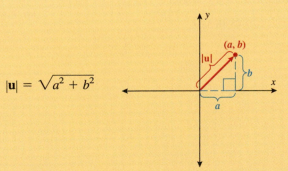

$$|\mathbf{u}| = \sqrt{a^2 + b^2}$$

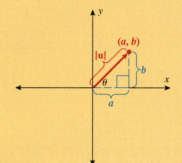

$|\mathbf{u}| = 5$

$(3, -4)$

EXAMPLE 1 Finding the Magnitude of a Vector

Find the magnitude of the vector: $\mathbf{u} = \langle 3, -4 \rangle$.

Solution:

Write the formula for magnitude of a vector. $|\mathbf{u}| = \sqrt{a^2 + b^2}$

Let $a = 3$ and $b = -4$. $|\mathbf{u}| = \sqrt{3^2 + (-4)^2}$

Simplify. $|\mathbf{u}| = \sqrt{25} = 5$

Note: If we graph the vector, $\mathbf{u} = \langle 3, -4 \rangle$, we see that the distance from the origin to the point $(3, -4)$ is 5 units.

■ **YOUR TURN**   Find the magnitude of the vector: $\mathbf{v} = \langle -1, 5 \rangle$.

DIRECTION ANGLE OF A VECTOR

The positive angle between the x-axis and a position vector is called the **direction angle**, denoted θ.

$$\tan \theta = \frac{b}{a} \text{ where } a \neq 0$$

y

(a, b)

$|\mathbf{u}|$

b

θ

a

x

TECHNOLOGY TIP

Use a calculator to find θ.

```
tan⁻¹(-5)
        -78.69006753
```

EXAMPLE 2 Finding the Direction Angle of a Vector

Find the direction angle of the vector: $\mathbf{v} = \langle -1, 5 \rangle$.

Solution:

Start with $\tan \theta = \frac{b}{a}$ and let $a = -1$ and $b = 5$. $\tan \theta = \frac{5}{-1}$

Use a calculator to find $\tan^{-1}(-5)$. $\tan^{-1}(-5) = -78.7°$

The calculator gave a QIV angle.

The point $(-1, 5)$ lies in QII.

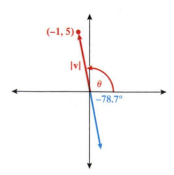

$(-1, 5)$

$|\mathbf{v}|$

θ

$-78.7°$

■ **Answer:** $\sqrt{26}$

Add 180°.
$$\theta = -78.7° + 180° = 101.3°$$
$$\theta = 101.3°$$

■ **YOUR TURN**   Find the direction angle of the vector: $\mathbf{u} = \langle 3, -4 \rangle$.

Recall that two vectors are equal if they have the same magnitude and direction. Algebraically, this corresponds to their components (a and b) being equal.

EQUAL VECTORS: u = v

The vectors $\mathbf{u} = \langle a, b \rangle$ and $\mathbf{v} = \langle c, d \rangle$ are **equal**, $\mathbf{u} = \mathbf{v}$, if $a = c$ and $b = d$.

Vector addition geometrically is done with the tail-to-tip rule. Algebraically, vector addition is performed component by component.

VECTOR ADDITION: u + v

If $\mathbf{u} = \langle a, b \rangle$ and $\mathbf{v} = \langle c, d \rangle$, then $\mathbf{u} + \mathbf{v} = \langle a + c, b + d \rangle$.

EXAMPLE 3 Adding Vectors

Let $\mathbf{u} = \langle 2, -7 \rangle$ and $\mathbf{v} = \langle -3, 4 \rangle$ and find $\mathbf{u} + \mathbf{v}$.

Solution:

Let $\mathbf{u} = \langle 2, -7 \rangle$ and
$\mathbf{v} = \langle -3, 4 \rangle$ in the
addition formula.
$$\mathbf{u} + \mathbf{v} = \langle 2 + (-3), -7 + 4 \rangle$$

Simplify.
$$\mathbf{u} + \mathbf{v} = \langle -1, -3 \rangle$$

■ **YOUR TURN**   Let $\mathbf{u} = \langle 1, 2 \rangle$ and $\mathbf{v} = \langle -5, -4 \rangle$ and find $\mathbf{u} + \mathbf{v}$.

Vector Operations

We now summarize vector operations. Addition and subtraction are defined component by component. Multiplication, however, is not as straightforward. Scalar multiplication of a vector (a real number times a vector) is performed component by component. A form of multiplication for two vectors, however, is not always defined, and when it is, the procedure is called the dot product (Section 7.5).

■ **Answer:** $\mathbf{u} + \mathbf{v} = \langle -4, -2 \rangle$ ■ **Answer:** $-53.1°$

If k is a scalar (real number) and $\mathbf{u} = \langle a, b \rangle$, then

$$k\mathbf{u} = k\langle a, b \rangle = \langle ka, kb \rangle$$

Scalar multiplication corresponds to

- Increasing the length of the vector: $|k| > 1$
- Decreasing the length of the vector: $|k| < 1$
- Changing the direction of the vector: $k < 0$

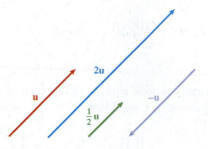

The following box is a summary of vector operations.

VECTOR OPERATIONS

If $\mathbf{u} = \langle a, b \rangle$ and $\mathbf{v} = \langle c, d \rangle$ and k is a scalar, then:

$$\mathbf{u} + \mathbf{v} = \langle a + c, b + d \rangle$$

$$\mathbf{u} - \mathbf{v} = \langle a - c, b - d \rangle$$

$$k\mathbf{u} = k\langle a, b \rangle = \langle ka, kb \rangle$$

The zero vector, $\mathbf{0} = \langle 0, 0 \rangle$, is a vector in any direction with a magnitude equal to zero. We now can state the algebraic properties (associative, commutative, and distributive):

ALGEBRAIC PROPERTIES OF VECTORS

$$\mathbf{u} + \mathbf{v} = \mathbf{v} + \mathbf{u}$$

$$(\mathbf{u} + \mathbf{v}) + \mathbf{w} = \mathbf{u} + (\mathbf{v} + \mathbf{w})$$

$$(k_1 k_2)\mathbf{u} = k_1(k_2\mathbf{u})$$

$$k(\mathbf{u} + \mathbf{v}) = k\mathbf{u} + k\mathbf{v}$$

$$(k_1 + k_2)\mathbf{u} = k_1\mathbf{u} + k_2\mathbf{u}$$

$$0\mathbf{u} = \mathbf{0} \qquad 1\mathbf{u} = \mathbf{u} \qquad -1\mathbf{u} = -\mathbf{u}$$

$$\mathbf{u} + (-\mathbf{u}) = \mathbf{0}$$

Horizontal and Vertical Components

The **horizontal component**, a, and **vertical component**, b, of a vector **u** are related to the magnitude of the vector, $|\mathbf{u}|$, through the sine and cosine functions.

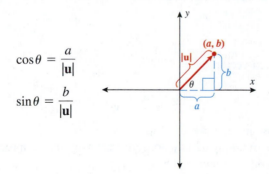

$$\cos\theta = \frac{a}{|\mathbf{u}|}$$

$$\sin\theta = \frac{b}{|\mathbf{u}|}$$

HORIZONTAL AND VERTICAL COMPONENTS

The horizontal and vertical components of a vector **u**, with magnitude $|\mathbf{u}|$ and direction angle θ, are given by

horizontal component: $a = |\mathbf{u}| \cos\theta$

vertical component: $b = |\mathbf{u}| \sin\theta$

The vector, **u**, can then be written as $\mathbf{u} = \langle a, b \rangle = \langle |\mathbf{u}| \cos\theta, |\mathbf{u}| \sin\theta \rangle$.

EXAMPLE 4 Finding the Horizontal and Vertical Components of a Vector

Find the vector that has a magnitude of 6 and a direction angle of 15°.

Solution:

Write the horizontal and vertical components of a vector **u**.	$a = |\mathbf{u}| \cos\theta$ and $b = |\mathbf{u}| \sin\theta$
Let $|\mathbf{u}| = 6$ and $\theta = 15°$.	$a = 6 \cos 15°$ and $b = 6 \sin 15°$
Evaluate sine and cosine for 15°.	$a = 5.8$ and $b = 1.6$
Let $\mathbf{u} = \langle a, b \rangle$.	$\mathbf{u} = \langle 5.8, 1.6 \rangle$

■ **YOUR TURN** Find the vector that has a magnitude of 3 and direction angle of 75°.

Unit Vectors

A **unit vector** is any vector with magnitude equal to one, $|\mathbf{u}| = 1$. It is often useful to be able to find a unit vector in the same direction of some vector $\mathbf{v}$. A unit vector can be formed from any nonzero vector as follows:

FINDING A UNIT VECTOR

If $\mathbf{v}$ is a nonzero vector, then

$$\mathbf{u} = \frac{\mathbf{v}}{|\mathbf{v}|}$$

is a **unit vector** in the same direction as $\mathbf{v}$.
In other words, multiplying any nonzero vector by the reciprocal of its magnitude results in a unit vector.

STUDY TIP

Multiplying a nonzero vector by the reciprocal of its magnitude results in a unit vector.

It is important to notice that since magnitude is always a scalar, then the reciprocal of magnitude is always a scalar. A scalar times a vector is a vector.

EXAMPLE 5 Finding a Unit Vector

Find a unit vector in the same direction as $\mathbf{v} = \langle -3, -4 \rangle$.

Solution:

Find the magnitude of the vector $\mathbf{v} = \langle -3, -4 \rangle$.

$$|\mathbf{v}| = \sqrt{(-3)^2 + (-4)^2}$$

Simplify.

$$|\mathbf{v}| = 5$$

Multiply $\mathbf{v}$ by the reciprocal of its magnitude.

$$\frac{1}{|\mathbf{v}|}\mathbf{v}$$

Let $|\mathbf{v}| = 5$ and $\mathbf{v} = \langle -3, -4 \rangle$.

$$\frac{1}{5}\langle -3, -4 \rangle$$

Simplify.

$$\left\langle -\frac{3}{5}, -\frac{4}{5} \right\rangle$$

Check: The unit vector, $\left\langle -\frac{3}{5}, -\frac{4}{5} \right\rangle$, should have a magnitude of 1.

$$\sqrt{\left(-\frac{3}{5}\right)^2 + \left(-\frac{4}{5}\right)^2} = \sqrt{\frac{25}{25}} = 1$$

■ **YOUR TURN** Find a unit vector in the same direction as $\mathbf{v} = \langle 5, -12 \rangle$.

■ **Answer:** $\left\langle \frac{5}{13}, -\frac{12}{13} \right\rangle$

Two important unit vectors are the horizontal and vertical unit vectors **i** and **j**. The unit vector **i** has an initial point at the origin and terminal point at $(1, 0)$. The unit vector **j** has an initial point at the origin and terminal point at $(0, 1)$. These unit vectors can be used to represent vectors algebraically: $\langle 3, -4 \rangle = 3i - 4j$.

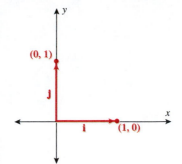

Applications

There are many applications in which vectors arise. **Velocity vectors** and **force vectors** are two that we will discuss. For example, you might be at the beach and you "think" you are swimming straight out at a certain speed (magnitude and direction). This is your **apparent velocity** with respect to the water. But after a few minutes you turn around to look at shore, and you are farther out than you thought and you also appear to have drifted down the beach. This is because of the current of the water. When the **current velocity** and the apparent velocity are added together, the result is the **actual** or **resultant velocity**.

EXAMPLE 6 Resultant Velocities

A boat's speedometer reads 25 mph (which is relative to the water) and sets a course due east (90° from due north). If the river is moving 10 mph due north, what is the resultant (actual) velocity of the boat?

Solution:

Draw a picture.

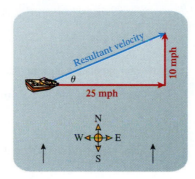

Label the horizontal and vertical components of the resultant vector. $\langle 25, 10 \rangle$

Determine the magnitude of the resultant vector. $\sqrt{25^2 + 10^2} = 5\sqrt{29} \approx 27$ mph

Determine the direction angle. $\tan \theta = \dfrac{10}{25}$

Solve for θ. $\theta = 21.8°$

The actual velocity of the boat has magnitude 27 mph and is headed 21.8° north of east .

In Example 6, the three vectors formed a right triangle. In Example 7, the three vectors form an oblique triangle.

EXAMPLE 7 Resultant Velocities

A speedboat traveling 30 mph has a compass heading of 100° (from due north). The current velocity has a magnitude of 15 mph and its heading is 22° (from due north). Find the resultant (actual) velocity of the boat.

Solution:

Draw a picture.

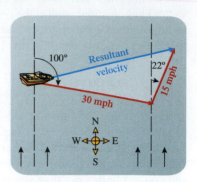

Label the supplementary angles to 100°.

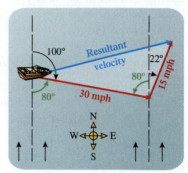

Draw and label the oblique triangle.

The magnitude of the actual (resultant) velocity is b.

The heading of the actual (resultant) velocity is $100° - \alpha$.

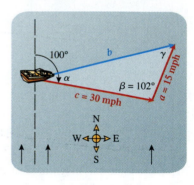

Use the law of sines and the law of cosines to solve for α and b.

Find b: Use the law of cosines.

$$b^2 = a^2 + c^2 - 2\,ac\,\cos \beta$$

Let $a = 15$, $c = 30$, and $\beta = 102°$.

$$b^2 = 15^2 + 30^2 - 2(15)(30)\cos 102°$$

Solve for b.

$b \approx 36$ mph

Find α: Use the law of sines.

$$\frac{\sin \alpha}{a} = \frac{\sin \beta}{b}$$

Isolate $\sin \alpha$.

$$\sin \alpha = \frac{a}{b}\sin \beta$$

Let $a = 15$, $b = 36$, and $\beta = 102°$.

$$\sin \alpha = \frac{15}{36}\sin 102°$$

Use the inverse sine function to solve for α.

$$\alpha = \sin^{-1}\left(\frac{15}{36}\sin 102°\right)$$

Use a calculator to approximate α.

$$\alpha \approx 24°$$

Actual heading: $100° - \alpha = 100° - 24° = 76°$

The actual velocity vector of the boat has magnitude 36 mph and is headed $76°$ east of north .

Two vectors combine to yield a resultant vector. The opposite vector to the resultant vector is called the **equilibrant**.

EXAMPLE 8 Finding an Equilibrant

A skier is being pulled up a handle lift. Let $\mathbf{F}_1$ represent the vertical force due to gravity and $\mathbf{F}_2$ represent the force of the skier pushing against the side of the mountain, at an angle of $35°$ to the horizontal. If the weight of the skier is 145 pounds, $|\mathbf{F}_1| = 145$, find the magnitude of the equilibrant (required force), $\mathbf{F}_3$, to hold the skier in place (not let the skier slide down the mountain). Assume that the side of the mountain is a frictionless surface.

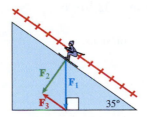

Solution:

The angle between vectors $\mathbf{F}_1$ and $\mathbf{F}_2$ is $35°$.

The magnitude of vector $\mathbf{F}_3$ is the force required to hold the skier in place.

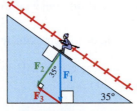

Relate the magnitudes (side lengths) to an angle using the sine ratio.

$$\sin 35° = \frac{|\mathbf{F}_3|}{|\mathbf{F}_1|}$$

Solve for $|\mathbf{F}_3|$.

$$|\mathbf{F}_3| = |\mathbf{F}_1|\sin 35°$$

Let $|\mathbf{F}_1| = 145$.

$$|\mathbf{F}_3| = 145\sin 35°$$

$$|\mathbf{F}_3| = 83.16858$$

A force of approximately 83 pounds is required to keep the skier from sliding down the hill.

EXAMPLE 9 Resultant Forces

A barge runs aground outside the channel. A single tugboat cannot generate enough force to pull the barge off the sandbar. A second tugboat comes to assist. The following diagram illustrates the force vectors, $\mathbf{F}_1$ and $\mathbf{F}_2$, from the tugboats. What is the resultant force vector of the two tugboats?

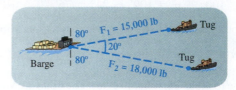

Solution:

Using the tail-to-tip rule, we can add these two vectors and form a triangle:

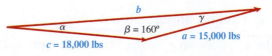

TECHNOLOGY TIP

Use a calculator to find b.

```
15000²+18000²−2*
15000*18000cos(1
60)
          1056434015
√(Ans)
          32502.83088
```

Use a calculator to find α.

```
15000/32503sin(1
60
          .1578408808
sin⁻¹(Ans)
          9.08159538
```

Solve for b: Use the law of cosines. $\quad b^2 = a^2 + c^2 - 2\,ac\cos\beta$

Let $a = 15,000$,
$c = 18,000$, and
$\beta = 160°$. $\qquad b^2 = 15,000^2 + 18,000^2 - 2(15,000)(18,000)\cos 160°$

Solve for b. $\qquad b = 32,503 \text{ lb}$

Solve for α: Use the law of sines. $\qquad \dfrac{\sin\alpha}{a} = \dfrac{\sin\beta}{b}$

Isolate $\sin\alpha$. $\qquad \sin\alpha = \dfrac{a}{b}\sin\beta$

Let $a = 15,000$, $b = 32,503$, $\beta = 160°$. $\qquad \sin\alpha = \dfrac{15,000}{32,503}\sin 160°$

Use the inverse sine function to solve for α. $\quad \alpha = \sin^{-1}\left(\dfrac{15,000}{32,503}\sin 160°\right)$

Use a calculator to approximate α. $\qquad \alpha \approx 9.08°$

The resulting force is 32,503 lb at an angle of 9° from the tug pulling with a force of 18,000 pounds.

SECTION 7.4 SUMMARY

In this section we discussed scalars (real numbers) and vectors. Scalars only have magnitude, whereas vectors have both magnitude and direction. We defined vectors both algebraically and geometrically and gave interpretations of magnitude and vector addition in both ways. Vector addition is performed algebraically component by component. The trigonometric functions are used to express the horizontal and vertical components of a vector. Velocity and force vectors illustrate examples of the law of sines and the law of cosines.

SECTION 7.4 EXERCISES

■ **SKILLS**

In Exercises 1–6, find the magnitude of the vector AB.

1. $A = (2, 7)$ and $B = (5, 9)$

2. $A = (-2, 3)$ and $B = (3, -4)$

3. $A = (4, 1)$ and $B = (-3, 0)$

4. $A = (-1, -1)$ and $B = (2, -5)$

5. $A = (0, 7)$ and $B = (-24, 0)$

6. $A = (-2, 1)$ and $B = (4, 9)$

In Exercises 7–16, find the magnitude and direction angle of the given vector.

7. $\mathbf{u} = \langle 3, 8 \rangle$

8. $\mathbf{u} = \langle 4, 7 \rangle$

9. $\mathbf{u} = \langle 5, -1 \rangle$

10. $\mathbf{u} = \langle -6, -2 \rangle$

11. $\mathbf{u} = \langle -4, 1 \rangle$

12. $\mathbf{u} = \langle -6, 3 \rangle$

13. $\mathbf{u} = \langle -8, 0 \rangle$

14. $\mathbf{u} = \langle 0, 7 \rangle$

15. $\mathbf{u} = \langle \sqrt{3}, 3 \rangle$

16. $\mathbf{u} = \langle -5, -5 \rangle$

In Exercises 17–24, perform the indicated vector operation, given $\mathbf{u} = \langle -4, 3 \rangle$ and $\mathbf{v} = \langle 2, -5 \rangle$.

17. $\mathbf{u} + \mathbf{v}$

18. $\mathbf{u} - \mathbf{v}$

19. $3\mathbf{u}$

20. $-2\mathbf{u}$

21. $2\mathbf{u} + 4\mathbf{v}$

22. $5(\mathbf{u} + \mathbf{v})$

23. $6(\mathbf{u} - \mathbf{v})$

24. $2\mathbf{u} - 3\mathbf{v} + 4\mathbf{u}$

In Exercises 25–34, find the vector, given its magnitude and direction angle.

25. $|\mathbf{u}| = 7, \theta = 25°$

26. $|\mathbf{u}| = 5, \theta = 75°$

27. $|\mathbf{u}| = 16, \theta = 100°$

28. $|\mathbf{u}| = 8, \theta = 200°$

29. $|\mathbf{u}| = 4, \theta = 310°$

30. $|\mathbf{u}| = 8, \theta = 225°$

31. $|\mathbf{u}| = 9, \theta = 335°$

32. $|\mathbf{u}| = 3, \theta = 315°$

33. $|\mathbf{u}| = 2, \theta = 120°$

34. $|\mathbf{u}| = 6, \theta = 330°$

In Exercises 35–44, find a unit vector in the direction of the given vector.

35. $\mathbf{v} = \langle -5, -12 \rangle$

36. $\mathbf{v} = \langle 3, 4 \rangle$

37. $\mathbf{v} = \langle 60, 11 \rangle$

38. $\mathbf{v} = \langle -7, 24 \rangle$

39. $\mathbf{v} = \langle 24, -7 \rangle$

40. $\mathbf{v} = \langle -10, 24 \rangle$

41. $\mathbf{v} = \langle -9, -12 \rangle$

42. $\mathbf{v} = \langle 40, -9 \rangle$

43. $\mathbf{v} = \langle \sqrt{2}, 3\sqrt{2} \rangle$

44. $\mathbf{v} = \langle -4\sqrt{3}, -2\sqrt{3} \rangle$

In Exercises 45–50, express the vector in terms of unit vectors i and j.

45. $\langle 7, 3 \rangle$

46. $\langle -2, 4 \rangle$

47. $\langle 5, -3 \rangle$

48. $\langle -6, -2 \rangle$

49. $\langle -1, 0 \rangle$

50. $\langle 0, 2 \rangle$

In Exercises 51–56, perform the indicated vector operation.

51. $(5\mathbf{i} - 2\mathbf{j}) + (-3\mathbf{i} + 2\mathbf{j})$

52. $(4\mathbf{i} - 2\mathbf{j}) + (3\mathbf{i} - 5\mathbf{j})$

53. $(-3\mathbf{i} + 3\mathbf{j}) - (2\mathbf{i} - 2\mathbf{j})$

54. $(\mathbf{i} - 3\mathbf{j}) - (-2\mathbf{i} + \mathbf{j})$

55. $(5\mathbf{i} + 3\mathbf{j}) + (2\mathbf{i} - 3\mathbf{j})$

56. $(-2\mathbf{i} + \mathbf{j}) + (2\mathbf{i} - 4\mathbf{j})$

■ **APPLICATIONS**

57. Bullet Speed. A bullet is fired from ground level at a speed of 2200 ft/sec at an angle of 30° from the horizontal. Find the magnitude of the horizontal and vertical components of the velocity vector.

58. Weightlifting. A 50 pound weight lies on an inclined bench that makes an angle of 40° with the horizontal. Find the component of the weight directed perpendicular to the bench and also the component of the weight parallel to the inclined bench.

59. **Weight of a Boat.** A force of 630 pounds is needed to pull a speedboat and its trailer up a ramp that has an incline of 13°. What is the weight of the boat and its trailer?

60. **Weight of a Boat.** A force of 500 pounds is needed to pull a speedboat and its trailer up a ramp that has an incline of 16°. What is the weight of the boat and its trailer?

61. **Speed and Direction of a Ship.** A ship's captain sets a course due north at 10 mph. The water is moving at 6 mph due west. What is the actual velocity of the ship, and in what direction is it traveling?

62. **Speed and Direction of a Ship.** A ship's captain sets a course due west at 12 mph. The water is moving at 3 mph due north. What is the actual velocity of the ship, and in what direction is it traveling?

63. **Heading and Airspeed.** A plane has a compass heading of 60° (east of due north) and an airspeed of 300 mph. The wind is blowing at 40 mph with a heading of 30° (west of due north). What is the plane's actual heading and airspeed?

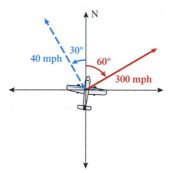

64. **Heading and Airspeed.** A plane has a compass heading of 30° (east of due north) and an airspeed of 400 mph. The wind is blowing at 30 mph with a heading of 60° (west of due north). What is the plane's actual heading and airspeed?

65. **Sliding Box.** A box weighing 500 pounds is held in place on an inclined plane that has an angle of 30°. What force is required to hold it in place?

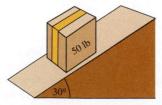

66. **Sliding Box.** A box weighing 500 pounds is held in place on an inclined plane that has an angle of 10°. What force is required to hold it in place?

67. **Baseball.** A baseball player throws a ball with an initial velocity of 80 feet per second at an angle of 40° with the horizontal. What are the vertical and horizontal components of the velocity?

68. **Baseball.** A baseball pitcher throws a ball with an initial velocity of 100 feet per second at an angle of 5° with the horizontal. What are the vertical and horizontal components of the velocity?

For Exercises 69 and 70, refer to the following:

A post pattern in football is when the receiver in motion runs past the quarterback parallel to the line of scrimmage (A), runs 12 yards perpendicular to the line of scrimmage (B), and then cuts toward the goal post (C).

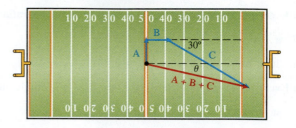

69. **Football.** A receiver runs the post pattern. If the magnitudes of the vectors are $|A| = 4$ yards, $|B| = 12$ yards, and $|C| = 20$ yards, find the magnitude of the resultant vector, **A + B + C**.

70. **Football.** A receiver runs the post pattern. If the magnitudes of the vectors are $|A| = 4$ yards, $|B| = 12$ yards, and $|C| = 20$ yards, find the direction angle, θ.

71. **Resultant Force.** A force with a magnitude of 100 pounds and another with a magnitude of 400 pounds are acting on an object. The two forces have an angle of 60° between them. What is the direction of the resultant force with respect to the force pulling 400 pounds?

72. **Resultant Force.** A force with a magnitude of 100 pounds and another with a magnitude of 400 pounds are acting on an object. The two forces have an angle of 60° between them. What is the magnitude of the resultant force?

73. **Resultant Force.** A force of 1000 pounds is acting on an object at an angle of 45° from the horizontal. Another force of 500 pounds is acting at an angle of −40° from the horizontal. What is the magnitude of the resultant force?

74. **Resultant Force.** A force of 1000 pounds is acting on an object at an angle of 45° from the horizontal. Another force of 500 pounds is acting at an angle of −40° from the horizontal. What is the angle of the resultant force?

■ CATCH THE MISTAKE

In Exercises 75 and 76, explain the mistake that is made.

75. Find the magnitude of the vector: $\langle -2, -8 \rangle$.

Solution:

Factor the -1. $\qquad -\langle 2, 8 \rangle$

Find the magnitude of $\langle 2, 8 \rangle$.

$$|\langle 2, 8 \rangle| = \sqrt{2^2 + 8^2}$$
$$= \sqrt{68} = 2\sqrt{17}$$

Write the magnitude of $\langle -2, -8 \rangle$.

$$|\langle -2, -8 \rangle| = -2\sqrt{17}$$

This is incorrect. What mistake was made?

76. Find the direction angle of the vector: $\langle -2, -8 \rangle$.

Solution:

Write the formula for the direction angle of $\langle a, b \rangle$.

$$\tan \theta = \frac{b}{a}$$

Let $a = -2, b = -8$.

$$\tan \theta = \frac{-8}{-2}$$

Use the inverse tangent function.

$$\theta = \tan^{-1} 4$$

Use a calculator to evaluate.

$$\theta = 76°$$

This is incorrect. What mistake was made?

■ CHALLENGE

77. T or F: The magnitude of the vector **i** is the imaginary number i.

78. T or F: The arrow components of equal vectors must coincide.

79. T or F: The magnitude of a vector is always greater than or equal to the magnitude of its horizontal component.

80. T or F: The magnitude of a vector is always greater than or equal to the magnitude of its vertical component.

81. Would a scalar or a vector represent the following: A car is driving 72 mph due east (90° with respect to north).

82. Would a scalar or vector represent the following: the granite weighs 290 lb.

83. Find the magnitude of the vector $\langle -a, b \rangle$ if $a > 0$ and $b > 0$.

84. Find the direction angle of the vector $\langle -a, b \rangle$ if $a > 0$ and $b > 0$.

■ TECHNOLOGY

Vectors can be represented as column matrices. For example, the vector $\mathbf{u} = \langle 3, -4 \rangle$ can be represented as a 2×1 column matrix $\begin{bmatrix} 3 \\ -4 \end{bmatrix}$. Using TI-83, vectors can be entered as matrices in two ways, directly or via MATRIX .

Directly:

```
[[3][-4]]
        [[3 ]
         [-4]]
■
```

Matrix:

```
[A]
        [[3 ]
         [-4]]
```

Use a calculator to perform the vector operation given $\mathbf{u} = \langle 8, -5 \rangle$ and $\mathbf{v} = \langle -7, 11 \rangle$.

85. $\mathbf{u} + 3\mathbf{v}$

86. $-9(\mathbf{u} - 2\mathbf{v})$

Use a calculator to find a unit vector in the direction of the given vector.

87. $\mathbf{u} = \langle 10, -24 \rangle$

88. $\mathbf{u} = \langle -9, -40 \rangle$

SECTION 7.5 Dot Product

Skills Objectives
- Find the dot product of two vectors.
- Use the dot product to find the angle between two vectors.
- Determine if two vectors are parallel or perpendicular.
- Use dot products to calculate the amount of work associated with a physical problem.

Conceptual Objectives
- Understand the dot product of parallel or perpendicular vectors as limiting cases of the dot product of two vectors.
- Project a vector onto another vector.

Multiplying Two Vectors

In this course we study two types of multiplication defined for vectors: scalar multiplication and the dot product. Scalar multiplication (Section 7.4) is multiplication of a scalar by a vector; the result is a vector. Now, we discuss the *dot product* of two vectors. In this case there are two important things to note: (1) the dot product of two vectors is only defined if the vectors have the same number of components, and (2) if the dot product does exist, then the result is a scalar.

DOT PRODUCT

The **dot product** of two vectors $\mathbf{u} = \langle a, b \rangle$ and $\mathbf{v} = \langle c, d \rangle$ is given by

$$\mathbf{u} \cdot \mathbf{v} = ac + bd$$

$\mathbf{u} \cdot \mathbf{v}$ is pronounced "u dot v."

EXAMPLE 1 Finding the Dot Product of Two Vectors

Find the dot product: $\langle -7, 3 \rangle \cdot \langle 2, 5 \rangle$.

Solution:

Sum the products of the first components and the products of the second components.

$$\langle -7, 3 \rangle \cdot \langle 2, 5 \rangle = (-7)(2) + (3)(5)$$

Simplify.

$$\langle -7, 3 \rangle \cdot \langle 2, 5 \rangle = 1$$

STUDY TIP

The dot product of two vectors is a scalar.

YOUR TURN   Find the dot product: $\langle 6, 1 \rangle \cdot \langle -2, 3 \rangle$.

The following box summarizes the properties of the dot product.

PROPERTIES OF THE DOT PRODUCT

1. $\mathbf{u} \cdot \mathbf{v} = \mathbf{v} \cdot \mathbf{u}$
2. $\mathbf{u} \cdot \mathbf{u} = |\mathbf{u}|^2$
3. $\mathbf{0} \cdot \mathbf{u} = 0$
4. $k(\mathbf{u} \cdot \mathbf{v}) = (k\mathbf{u}) \cdot \mathbf{v} = \mathbf{u} \cdot (k\mathbf{v})$
5. $(\mathbf{u} + \mathbf{v}) \cdot \mathbf{w} = \mathbf{u} \cdot \mathbf{w} + \mathbf{v} \cdot \mathbf{w}$
6. $\mathbf{u} \cdot (\mathbf{v} + \mathbf{w}) = \mathbf{u} \cdot \mathbf{v} + \mathbf{u} \cdot \mathbf{w}$

These properties are verified in the exercises.

Angle Between Two Vectors

We can use these properties to develop an equation that relates the angle between two vectors and the dot product of the vectors.

WORDS	**MATH**
Let $\mathbf{u}$ and $\mathbf{v}$ be two vectors with the same initial point and let θ be the angle between them.	
The vector $\mathbf{u} - \mathbf{v}$ is opposite angle θ.	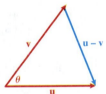
A triangle is formed with side lengths equal to the magnitudes of the three vectors.	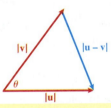

Use the law of cosines.

$$|\mathbf{u} - \mathbf{v}|^2 = |\mathbf{u}|^2 + |\mathbf{v}|^2 - 2|\mathbf{u}||\mathbf{v}|\cos\theta$$

Use properties of the dot product to rewrite the left side of equation:

Property (2): $\qquad |\mathbf{u} - \mathbf{v}|^2 = (\mathbf{u} - \mathbf{v}) \cdot (\mathbf{u} - \mathbf{v})$

Property (6): $\qquad\qquad\quad = \mathbf{u} \cdot (\mathbf{u} - \mathbf{v}) - \mathbf{v} \cdot (\mathbf{u} - \mathbf{v})$

Property (6): $\qquad\qquad\quad = \mathbf{u} \cdot \mathbf{u} - \mathbf{u} \cdot \mathbf{v} - \mathbf{v} \cdot \mathbf{u} + \mathbf{v} \cdot \mathbf{v}$

Property (2): $\qquad\qquad\quad = |\mathbf{u}|^2 - \mathbf{u} \cdot \mathbf{v} - \mathbf{v} \cdot \mathbf{u} + |\mathbf{v}|^2$

Property (1): $\qquad\qquad\quad = |\mathbf{u}|^2 - 2(\mathbf{u} \cdot \mathbf{v}) + |\mathbf{v}|^2$

Substitute this last expression for the left side of the original law of cosines equation.

$$|\mathbf{u}|^2 - 2(\mathbf{u} \cdot \mathbf{v}) + |\mathbf{v}|^2 = |\mathbf{u}|^2 + |\mathbf{v}|^2 - 2|\mathbf{u}||\mathbf{v}|\cos\theta$$

Simplify.

$$-2(\mathbf{u} \cdot \mathbf{v}) = -2|\mathbf{u}||\mathbf{v}|\cos\theta$$

Isolate $\cos\theta$.

$$\cos\theta = \frac{\mathbf{u} \cdot \mathbf{v}}{|\mathbf{u}||\mathbf{v}|}$$

Notice that $\mathbf{u}$ and $\mathbf{v}$ have to be nonzero vectors since we divided by them in the last step.

ANGLE BETWEEN TWO VECTORS

If θ is the angle between two nonzero vectors $\mathbf{u}$ and $\mathbf{v}$, where $0° \leq \theta \leq 180°$, then

$$\cos\theta = \frac{\mathbf{u} \cdot \mathbf{v}}{|\mathbf{u}||\mathbf{v}|}$$

In the cartesian plane, there are two angles between two vectors. We assume θ is the "smaller" angle.

EXAMPLE 2 Finding the Angle Between Two Vectors

Find the angle between $\langle 2, -3 \rangle$ and $\langle -4, 3 \rangle$.

Solution:

Let $\mathbf{u} = \langle 2, -3 \rangle$ and $\mathbf{v} = \langle -4, 3 \rangle$.

STEP 1 Find $\mathbf{u} \cdot \mathbf{v}$.

$$\begin{aligned} \mathbf{u} \cdot \mathbf{v} &= \langle 2, -3 \rangle \cdot \langle -4, 3 \rangle \\ &= (2)(-4) + (-3)(3) \\ &= -17 \end{aligned}$$

TECHNOLOGY TIP
Use a calculator to find θ,
$\cos\theta = \dfrac{-17}{5\sqrt{13}}$.

```
-17/(5*√(13))
      -.9429903336
cos⁻¹(Ans)
      160.5599652
```

STEP 2 Find $|\mathbf{u}|$.

$$|\mathbf{u}| = \sqrt{\mathbf{u} \cdot \mathbf{u}} = \sqrt{2^2 + (-3)^2} = \sqrt{13}$$

STEP 3 Find $|\mathbf{v}|$.

$$|\mathbf{v}| = \sqrt{\mathbf{v} \cdot \mathbf{v}} = \sqrt{(-4)^2 + 3^2} = \sqrt{25} = 5$$

STEP 4 Find θ.

$$\cos\theta = \frac{\mathbf{u} \cdot \mathbf{v}}{|\mathbf{u}||\mathbf{v}|} = \frac{-17}{5\sqrt{13}}$$

Use a calculator to approximate θ.

$$\theta = \cos^{-1}\left(-\frac{17}{5\sqrt{13}}\right) = 160.559965°$$

$$\theta = 161°$$

STEP 5 Draw a picture to confirm the answer.

Draw vectors $\langle 2, -3 \rangle$ and $\langle -4, 3 \rangle$. 161° appears to be correct.

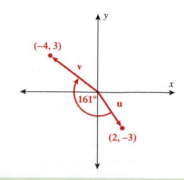

■ YOUR TURN Find the angle between $\langle 1, 5 \rangle$ and $\langle -2, 4 \rangle$.

Answer: $\theta = 38°$

When two vectors are **parallel**, the angle between them is 0° or 180°.

$\theta = 0°$ $\theta = 180°$

When two vectors are **perpendicular (orthogonal)**, the angle between them is 90°.

$\theta = 90°$

Note: We did not include 270° because the angle between two vectors, $0° \leq \theta \leq 180°$, is taken to be the smaller angle.

When two vectors **u** and **v** are perpendicular, $\theta = 90°$.

$$\cos 90° = \frac{\mathbf{u} \cdot \mathbf{v}}{|\mathbf{u}||\mathbf{v}|}$$

Substitute $\cos 90° = 0$.

$$0 = \frac{\mathbf{u} \cdot \mathbf{v}}{|\mathbf{u}||\mathbf{v}|}$$

Therefore the dot product of **u** and **v** must be zero.

$$\mathbf{u} \cdot \mathbf{v} = \mathbf{0}$$

ORTHOGONAL VECTORS

Two vectors, **u** and **v**, are **orthogonal** (perpendicular) if and only if their dot product is zero.

$$\mathbf{u} \cdot \mathbf{v} = \mathbf{0}$$

EXAMPLE 3 Determining If Vectors Are Orthogonal

Determine if each pair of vectors is orthogonal:

a. $\mathbf{u} = \langle 2, -3 \rangle$ and $\mathbf{v} = \langle 3, 2 \rangle$ **b.** $\mathbf{u} = \langle -7, -3 \rangle$ and $\mathbf{v} = \langle 7, 3 \rangle$

Solution (a):

Find the dot product, $\mathbf{u} \cdot \mathbf{v}$.

$$\mathbf{u} \cdot \mathbf{v} = (2)(3) + (-3)(2)$$

Simplify.

$$\mathbf{u} \cdot \mathbf{v} = 0$$

Vectors **u** and **v** are orthogonal since $\mathbf{u} \cdot \mathbf{v} = 0$.

Solution (b):

Find the dot product, $\mathbf{u} \cdot \mathbf{v}$.

$$\mathbf{u} \cdot \mathbf{v} = (-7)(7) + (-3)(3)$$

Simplify.

$$\mathbf{u} \cdot \mathbf{v} = -58$$

Vectors **u** and **v** are not orthogonal since $\mathbf{u} \cdot \mathbf{v} \neq 0$.

Work

If you had to carry either of the following items for 1 mile, which would you choose?

You probably would pick the pillows over the barbell and weights because the pillows are lighter. It requires less work to carry the pillows than it does to carry the weights. If asked to carry either of them 1 mile or 10 miles, you would probably pick 1 mile, because it's a shorter distance and requires less work. **Work** *is done when a force causes an object to move a certain distance.*

The simplest case is when the force is in the same direction as the displacement—for example, a stagecoach (the horses pull with a force in the same direction). In this case, the work is defined as the magnitude of the force times the magnitude of the displacement, distance *d*.

$$W = |\mathbf{F}|d$$

Notice that the magnitude of the force is a scalar, the distance *d* is a scalar, and hence the product is a scalar.

If the horses pull with a force of 1000 lb and they move the stagecoach 100 feet, then the work done by the force is

$$W = (1000 \text{ lb})(100 \text{ ft}) = 100,000 \text{ ft-lb}$$

In many physical applications, however, the force is not in the same direction as the displacement, and hence vectors (not just their magnitudes) are required.

We often want to know how much of a vector is applied in a certain direction. For example, when your car runs out of gasoline and you try to push it, the force vector, F_1, you generate from pushing has some of that force translate into the horizontal direction, F_2 (hence the car moves).

If we let θ be the angle between the vectors F_1 and F_2, then the horizontal component of F_1 is F_2 where $|F_2| = |F_1|\cos\theta$.

If the couple pushes at an angle of 25° with a force of 150 lb, then the horizontal component of the force vector, F_1, is

$$(150 \text{ lb})(\cos 25°) \approx 136 \text{ ft-lb}$$

To develop a generalized formula when the force exerted and the displacement are not in the same direction, we start with the formula: $\cos\theta = \dfrac{\mathbf{u}\cdot\mathbf{v}}{|\mathbf{u}||\mathbf{v}|}$

Isolate the dot product, $\mathbf{u}\cdot\mathbf{v}$. $\mathbf{u}\cdot\mathbf{v} = |\mathbf{u}||\mathbf{v}|\cos\theta$

Let $\mathbf{u} = \mathbf{F}$ and $\mathbf{v} = \mathbf{d}$. $W = \mathbf{F}\cdot\mathbf{d} = |\mathbf{F}||\mathbf{d}|\cos\theta = \underbrace{|\mathbf{F}|\cos\theta}_{\substack{\text{magnitude of force}\\\text{in direction of displacement}}} \cdot \underbrace{|\mathbf{d}|}_{\text{distance}}$

WORK

If an object is moved from point A to point B by a constant force, then the work associated with this displacement is

$$W = \mathbf{F}\cdot\mathbf{d}$$

where $\mathbf{d} = \mathbf{AB}$ is the displacement vector and $\mathbf{F}$ is the force vector.

Work is typically expressed in one of two units:

SYSTEM	FORCE	DISTANCE	WORK
American	pound	foot	ft-lb
SI	newton	meter	N-m

EXAMPLE 4 Calculating Work

How much work is done when a force (in pounds), $\mathbf{F} = \langle 2, 4\rangle$, moves an object from $(0, 0)$ to $(5, 9)$ (the distance is in feet)?

Solution:

Find the displacement vector, $\mathbf{d}$.	$\mathbf{d} = \langle 5, 9\rangle$
Use the work formula, $W = \mathbf{F}\cdot\mathbf{d}$.	$W = \langle 2, 4\rangle \cdot \langle 5, 9\rangle$
Calculate the dot product.	$W = (2)(5) + (4)(9)$
Simplify.	$W = 46$ ft-lb

■ **YOUR TURN** How much work is done when a force (in newtons), $\mathbf{F} = \langle 1, 3\rangle$, moves an object from $(0, 0)$ to $(4, 7)$ (the distance is in meters)?

SECTION 7.5   SUMMARY

In this section we defined a dot product as a form of multiplication of two vectors. A scalar times a vector results in a vector, whereas the dot product of two vectors is a scalar. We developed a formula that determines the angle between two vectors. Orthogonal (perpendicular) vectors have an angle of 90° between them, and consequently the dot product of two orthogonal vectors is equal to zero. Work is the result of a force displacing an object. When the force and displacement are in the same direction, the work is equal to the product of the magnitude of the force and the distance (magnitude of the displacement). When the force and displacement are not in the same direction, work is the dot product of the force vector and displacement vector.

■ **Answer: 25 N-m**

SECTION 7.5 EXERCISES

■ SKILLS

In Exercises 1–12, find the indicated dot product.

1. $\langle 4, -2 \rangle \cdot \langle 3, 5 \rangle$

2. $\langle 7, 8 \rangle \cdot \langle 2, -1 \rangle$

3. $\langle -5, 6 \rangle \cdot \langle 3, 2 \rangle$

4. $\langle 6, -3 \rangle \cdot \langle 2, 1 \rangle$

5. $\langle -7, -4 \rangle \cdot \langle -2, -7 \rangle$

6. $\langle 5, -2 \rangle \cdot \langle -1, -1 \rangle$

7. $\langle \sqrt{3}, -2 \rangle \cdot \langle 3\sqrt{3}, -1 \rangle$

8. $\langle 4\sqrt{2}, \sqrt{7} \rangle \cdot \langle -\sqrt{2}, -\sqrt{7} \rangle$

9. $\langle 5, a \rangle \cdot \langle -3a, 2 \rangle$

10. $\langle 4x, 3y \rangle \cdot \langle 2y, -5x \rangle$

11. $\langle 0.8, -0.5 \rangle \cdot \langle 2, 6 \rangle$

12. $\langle -18, 3 \rangle \cdot \langle 10, -300 \rangle$

In Exercises 13–24, find the angle (round to the nearest degree) between each pair of vectors.

13. $\langle -4, 3 \rangle$ and $\langle -5, -9 \rangle$

14. $\langle 2, -4 \rangle$ and $\langle 4, -1 \rangle$

15. $\langle -2, -3 \rangle$ and $\langle -3, 4 \rangle$

16. $\langle 6, 5 \rangle$ and $\langle 3, -2 \rangle$

17. $\langle -4, 6 \rangle$ and $\langle -6, 8 \rangle$

18. $\langle 1, 5 \rangle$ and $\langle -3, -2 \rangle$

19. $\langle -2, 2\sqrt{3} \rangle$ and $\langle -\sqrt{3}, 1 \rangle$

20. $\langle -3\sqrt{3}, -3 \rangle$ and $\langle -2\sqrt{3}, 2 \rangle$

21. $\langle -5\sqrt{3}, -5 \rangle$ and $\langle \sqrt{2}, -\sqrt{2} \rangle$

22. $\langle -5, -5\sqrt{3} \rangle$ and $\langle 2, -\sqrt{2} \rangle$

23. $\langle 4, 6 \rangle$ and $\langle -6, -9 \rangle$

24. $\langle 2, 8 \rangle$ and $\langle -12, 3 \rangle$

In Exercises 25–36, determine if each pair of vectors is orthogonal.

25. $\langle -6, 8 \rangle$ and $\langle -8, 6 \rangle$

26. $\langle 5, -2 \rangle$ and $\langle -5, 2 \rangle$

27. $\langle 6, -4 \rangle$ and $\langle -6, -9 \rangle$

28. $\langle 8, 3 \rangle$ and $\langle -6, 16 \rangle$

29. $\langle 0.8, 4 \rangle$ and $\langle 3, -6 \rangle$

30. $\langle -7, 3 \rangle$ and $\left\langle \dfrac{1}{7}, -\dfrac{1}{3} \right\rangle$

31. $\langle 5, -0.4 \rangle$ and $\langle 1.6, 20 \rangle$

32. $\langle 12, 9 \rangle$ and $\langle 3, -4 \rangle$

33. $\langle \sqrt{3}, \sqrt{6} \rangle$ and $\langle -\sqrt{2}, 1 \rangle$

34. $\langle \sqrt{7}, -\sqrt{3} \rangle$ and $\langle 3, 7 \rangle$

35. $\left\langle \dfrac{4}{3}, \dfrac{8}{15} \right\rangle$ and $\left\langle -\dfrac{1}{12}, \dfrac{5}{24} \right\rangle$

36. $\left\langle \dfrac{5}{6}, \dfrac{6}{7} \right\rangle$ and $\left\langle \dfrac{36}{25}, -\dfrac{49}{36} \right\rangle$

■ APPLICATIONS

37. **Lifting Weights.** How much work does it take to lift 100 pounds vertically 4 feet?

38. **Lifting Weights.** How much work does it take to lift 150 pounds vertically 3.5 feet?

39. **Raising Wrecks.** How much work is done by a crane to lift a 2 ton car to a level of 20 feet?

40. **Raising Wrecks.** How much work is done by a crane to lift a 2.5 ton car to a level of 25 feet?

41. **Work.** To slide a crate across the floor, a force of 50 pounds at a 30° angle is needed. How much work is done if the crate is dragged 30 feet?

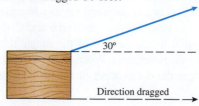

42. Work. To slide a crate across the floor, a force of 800 pounds at a 20° angle is needed. How much work is done if the crate is dragged 50 feet?

43. Close a Door. A sliding door is closed by pulling a cord with a constant force of 35 pounds at a constant angle of 45°. The door is moved 6 feet to close it. How much work is done?

44. Close a Door. A sliding door is closed by pulling a cord with a constant force of 45 pounds at a constant angle of 55°. The door is moved 6 feet to close it. How much work is done?

45. Braking Power. A car that weighs 2500 lb is parked on a hill in San Francisco with a slant of 40° from the horizontal. How much force will keep it from rolling down the hill?

46. Towing Power. A car that weighs 2500 lb is parked on a hill in San Francisco with a slant of 40° from the horizontal. A tow truck has to remove the car from its parking spot and move it 120 feet up the hill. How much work is required?

47. Towing Power. A semi-trailer truck that weighs 40,000 lb is parked on a hill in San Francisco with a slant of 10° from the horizontal. A tow truck has to remove the truck from its parking spot and move it 100 feet up the hill. How much work is required?

48. Braking Power. A car that weighs 40,000 lb is parked on a hill in San Francisco with a slant of 10° from the horizontal. How much force will keep it from rolling down the hill?

■ CATCH THE MISTAKE

In Exercises 49 and 50, explain the mistake that is made.

49. Find the dot product: $\langle -3, 2 \rangle \cdot \langle 2, 5 \rangle$.

Solution:

Multiply component by component. $\quad \langle -3, 2 \rangle \cdot \langle 2, 5 \rangle = \langle (-3)(2), (2)(5) \rangle$

Simplify. $\quad \langle -3, 2 \rangle \cdot \langle 2, 5 \rangle = \langle -6, 10 \rangle$

This is incorrect. What mistake was made?

50. Find the dot product: $\langle 11, 12 \rangle \cdot \langle -2, 3 \rangle$.

Solution:

Multiply the outer and inner components.
$$\langle 11, 12 \rangle \cdot \langle -2, 3 \rangle = (11)(3) + (12)(-2)$$

Simplify. $\quad \langle 11, 12 \rangle \cdot \langle -2, 3 \rangle = 9$

This is incorrect. What mistake was made?

■ CHALLENGE

51. T or F: A dot product of two vectors is a vector.

52. T or F: A dot product of two vectors is a scalar.

53. T or F: Orthogonal vectors have a dot product equal to zero.

54. T or F: If the dot product of two nonzero vectors is equal to zero, then the vectors must be perpendicular.

For Exercises 55 and 56, refer to the following to find the dot product.

The dot product of vectors with n component is

$$\langle a_1, a_2, \ldots, a_n \rangle \cdot \langle b_1, b_2, \ldots, b_n \rangle = a_1 b_1 + a_2 b_2 + \ldots + a_n b_n.$$

55. $\langle 3, 7, -5 \rangle \cdot \langle -2, 4, 1 \rangle$

56. $\langle 1, 0, -2, 3 \rangle \cdot \langle 5, 2, 3, 1 \rangle$

In Exercises 57–60, given $\mathbf{u} = \langle a, b \rangle$ and $\mathbf{v} = \langle c, d \rangle$, show that the following properties are true.

57. $\mathbf{u} \cdot \mathbf{v} = \mathbf{v} \cdot \mathbf{u}$

58. $\mathbf{u} \cdot \mathbf{u} = |\mathbf{u}|^2$

59. $\mathbf{0} \cdot \mathbf{u} = 0$

60. $k(\mathbf{u} \cdot \mathbf{v}) = (k\mathbf{u}) \cdot \mathbf{v} = \mathbf{u} \cdot (k\mathbf{v})$, $\quad k$ is a scalar

■ TECHNOLOGY

For Exercises 61 and 62, use a calculator to find the indicated dot product.

61. $\langle -11, 34 \rangle \cdot \langle 15, -27 \rangle$

62. $\langle 23, -350 \rangle \cdot \langle 45, 202 \rangle$

63. A rectangle has sides with lengths 18 units and 11 units. Find the angle, to one decimal place, between the diagonal and the side with length of 18 units. [Hint: Set up a rectangular coordinate system and use vectors $\langle 18, 0 \rangle$ to represent the side of length 18 units and $\langle 18, 11 \rangle$ to represent the diagonal.]

64. The definition of a dot product and the formula to find the angle between two vectors can be extended and applied to vectors with more than two components. A rectangular box has sides with lengths 12 feet, 7 feet, and 9 feet. Find the angle, to the nearest degree, between the diagonal and the side with length 7 feet.

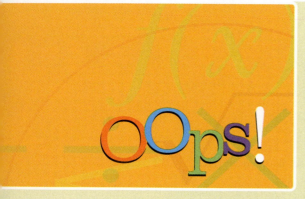

Q: Josie is moving to a new apartment. She places a box of books on a wheeled cart and pulls it by an attached rope at an angle of 45° with the horizontal with a constant force of 20 lb. She moves the cart a distance of 12 feet from the bookshelf to the door. Since Josie is waiting for her friend to help her and she has recently studied work in physics, she decides to calculate how much work she did in moving the box of books to the door. She defines a coordinate system with the bookshelf at the origin and considers the positive x-direction to be the direction from the bookshelf to the door. Thus, the direction vector is $\mathbf{d} = \langle 12, 0 \rangle$. The horizontal component of the force vector is $20\cos 45° = 20\left(\dfrac{\sqrt{2}}{2}\right) = 10\sqrt{2}$, and the vertical component of the force vector is $20\sin 45° = 10\sqrt{2}$. Thus, $\mathbf{F} = \langle 10\sqrt{2}, 10\sqrt{2} \rangle$. She then calculates the work as follows:

$$W = \mathbf{F} \cdot \mathbf{d}$$
$$= \langle 10\sqrt{2}, 10\sqrt{2} \rangle \cdot \langle 12, 0 \rangle$$
$$= \langle 120\sqrt{2}, 0 \rangle$$

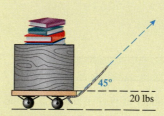

However, this last calculation is not correct. What is Josie's error?

A: Work is the dot product of the force and distance vectors. Thus, work is a scalar, not a vector. The work done is

$$W = \mathbf{F} \cdot \mathbf{d} = \langle 10\sqrt{2}, 10\sqrt{2} \rangle \cdot \langle 12, 0 \rangle = 12(10\sqrt{2}) + 0(10\sqrt{2}) = 120\sqrt{2} \text{ ft-lb}$$

Find the work done if Josie uses the same cart to move another box of books from the bookshelf to the door and she pulls the attached rope at an angle of 60° with the horizontal with a constant force of 35 lb.

TYING IT ALL TOGETHER

Jessica and Mike are at the airport waiting for their brother to arrive on a delayed flight. They are bored, and pass some time by riding the escalators. There are two escalators going up: escalator A has a speed of 90 feet per minute and escalator B has a speed of 96 feet per minute. Both escalators travel parallel and in the same direction as the vector $\mathbf{v} = \langle \sqrt{3}, 1 \rangle$, and both escalators are 40 feet in length. Jessica is 5 feet 4 inches tall and weighs 120 lb. She travels downward on escalator A at a rate of 440 feet per minute (this would be her speed if the escalator were still). Mike is 5 feet 11 inches tall and weighs 175 lb. He travels downward on escalator B at a rate of 432 feet per minute (this would be his speed if the escalator were still). Find vectors that represent the velocity of each escalator and the resultant velocities of Jessica and Mike's movement along the escalators. Who reaches the bottom of the escalators first?

SECTION	TOPIC	PAGES	REVIEW EXERCISES	KEY CONCEPTS
7.1	Oblique triangles and the law of sines	350–362	1-18	
	Oblique triangles: acute and obtuse	350–352	1-18	Oblique (Nonright) Triangles

Acute Triangle

Obtuse Triangle

	The law of sines	352–362	1-18	$\dfrac{\sin \alpha}{a} = \dfrac{\sin \beta}{b} = \dfrac{\sin \gamma}{c}$ ■ AAS (or ASA) triangles ■ SSA triangles (ambiguous case)
7.2	The law of cosines	365–372	19-40	$a^2 = b^2 + c^2 - 2bc \cos \alpha$ $b^2 = a^2 + c^2 - 2ac \cos \beta$ $c^2 = a^2 + b^2 - 2ab \cos \gamma$ ■ SAS triangles ■ SSS triangles
7.3	Area of a Triangle	376–380	41-50	Use the law of sines for SAS triangles:

$$A_{\text{SAS triangle}} = \frac{1}{2} bc \sin \alpha \text{ when } b, c, \text{ and } \alpha \text{ are known.}$$

$$A_{\text{SAS triangle}} = \frac{1}{2} ab \sin \gamma \text{ when } a, b, \text{ and } \gamma \text{ are known.}$$

$$A_{\text{SAS triangle}} = \frac{1}{2} ac \sin \beta \text{ when } a, c, \text{ and } \beta \text{ are known.}$$

Use Heron's formula for SSS triangles:

$$A_{\text{SSS triangle}} = \sqrt{s(s - a)(s - b)(s - c)}$$

where a, b, and c are the lengths of the sides of the triangle and s is half the perimeter of the triangle, called the semiperimeter:

$$s = \frac{a + b + c}{2}$$

Section	Topic	Pages	Review Exercises	Key Concepts
7.4	Vectors	383	51-76	Vector, $\mathbf{u}$ or $\mathbf{AB}$

Section	Topic	Pages	Review Exercises	Key Concepts
	Magnitude of a vector	383–384 385–386	51-58	$\mathbf{u} = \langle a, b \rangle$ Geometric: length of a vector Algebraic: $|\mathbf{u}| = \sqrt{a^2 + b^2}$
	Direction angle of a vector	386–387	55-58	$\tan \theta = \dfrac{b}{a}$
	Adding two vectors	384–385 387–388	59-62 69-70	$\mathbf{u} = \langle a, b \rangle$ and $\mathbf{v} = \langle c, d \rangle$ Geometric: tail-to-tip

Algebraic: $\mathbf{u} + \mathbf{v} = \langle a + c, b + d \rangle$

Section	Topic	Pages	Review Exercises	Key Concepts
	Scalar multiplication	387–388	59-62	$k\langle a, b \rangle = \langle ka, kb \rangle$
	Horizontal and vertical components	389	63-66	**Horizontal component**: $a = |\mathbf{u}| \cos \theta$ **Vertical component**: $b = |\mathbf{u}| \sin \theta$
	Unit vectors	390–391	67-68	$\mathbf{u} = \dfrac{\mathbf{v}}{|\mathbf{v}|}$
7.5	Dot product	398–403	71-90	▪ The product of a scalar and vector is a vector. ▪ The product of two vectors is a scalar.
	$\mathbf{u}$ dot $\mathbf{v}$	398–399	71-76	$\mathbf{u} = \langle a, b \rangle$ and $\mathbf{v} = \langle c, d \rangle$ $\mathbf{u} \cdot \mathbf{v} = ac + bd$
	Angle between two vectors	399–400	77-82	If θ is the angle between two nonzero vectors $\mathbf{u}$ and $\mathbf{v}$, where $0 \le \theta \le 180°$, then $\cos \theta = \dfrac{\mathbf{u} \cdot \mathbf{v}}{|\mathbf{u}||\mathbf{v}|}$
	Orthogonal (perpendicular) vectors	401	83-90	$\mathbf{u} \cdot \mathbf{v} = 0$
	Work	401–403		When force and displacement are in the same direction: $W = |\mathbf{F}||\mathbf{d}|$. When force and displacement are not in the same direction $W = \mathbf{F} \cdot \mathbf{d}$.

CHAPTER 7 REVIEW EXERCISES

7.1 Oblique Triangles and the Law of Sines

Solve the given triangles.

1. $\alpha = 10°, \beta = 20°, a = 4$

2. $\beta = 40°, \gamma = 60°, b = 10$

3. $\alpha = 5°, \beta = 45°, c = 10$

4. $\beta = 60°, \gamma = 70°, a = 20$

5. $\gamma = 11°, \alpha = 11°, c = 11$

6. $\beta = 20°, \gamma = 50°, b = 8$

7. $\alpha = 45°, \gamma = 45°, b = 2$

8. $\alpha = 60°, \beta = 20°, c = 17$

9. $\alpha = 12°, \gamma = 22°, a = 99$

10. $\beta = 102°, \gamma = 27°, a = 24$

Two sides and an angle are given. Determine if a triangle (or two) exist and, if so, solve those triangles.

11. $a = 7, b = 9, \alpha = 20°$

12. $b = 24, c = 30, \beta = 16°$

13. $a = 10, c = 12, \alpha = 24°$

14. $b = 100, c = 116, \beta = 12°$

15. $a = 40, b = 30, \beta = 150°$

16. $b = 2, c = 3, \gamma = 165°$

17. $a = 4, b = 6, \alpha = 10°$

18. $c = 25, a = 37, \gamma = 4°$

7.2 The Law of Cosines

19. **How Far from Home?** Gary walked 8 miles due north, made a 130° turn to the southeast, and walked for another 6 miles. How far was Gary from home?

20. **How Far from Home?** Mary walked 10 miles due north, made a 55° turn to the northeast, and walked for another 3 miles. How far was Mary from home?

Solve each triangle.

21. $a = 40, b = 60, \gamma = 50°$

22. $b = 15, c = 12, \alpha = 140°$

23. $a = 24, b = 25, c = 30$

24. $a = 6, b = 6, c = 8$

25. $a = \sqrt{11}, b = \sqrt{14}, c = 5$

26. $a = 22, b = 120, c = 122$

27. $b = 7, c = 10, \alpha = 14°$

28. $a = 6, b = 12, \gamma = 80°$

29. $b = 10, c = 4, \alpha = 90°$

30. $a = 4, b = 5, \gamma = 75°$

31. $a = 10, b = 11, c = 12$

32. $a = 22, b = 24, c = 25$

33. $b = 16, c = 18, \alpha = 100°$

34. $a = 25, c = 25, \beta = 9°$

35. $b = 12, c = 40, \alpha = 10°$

36. $a = 26, b = 20, c = 10$

37. $a = 26, b = 40, c = 13$

38. $a = 1, b = 2, c = 3$

39. $a = 6.3, b = 4.2, \alpha = 15°$

40. $b = 5, c = 6, \beta = 35°$

7.3 Area of a Triangle

Find the area of each triangle described.

41. $b = 16, c = 18, \alpha = 100°$

42. $a = 25, c = 25, \beta = 9°$

43. $a = 10, b = 11, c = 12$

44. $a = 22, b = 24, c = 25$

45. $a = 26, b = 20, c = 10$

46. $a = 24, b = 32, c = 40$

47. $b = 12, c = 40, \alpha = 10°$

48. $a = 21, c = 75, \beta = 60°$

49. **Area of Inscribed Triangle.** The area of a triangle inscribed in a circle can be found if you know the lengths of the sides of the triangle and the radius of the circle. The formula is $A = \dfrac{abc}{4r}$. Find the radius of a triangle circumscribed by a circle if all the sides of the triangle measure 9 inches and the area of the triangle is 35 square inches.

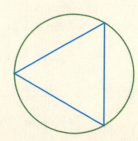

50. Area of Inscribed Triangle. The area of a triangle inscribed in a circle can be found if you know the lengths of the sides of the triangle and the radius of the circle. The formula is $A = \dfrac{abc}{4r}$. Find the radius of a triangle circumscribed by a circle if the sides of the triangle measure 9, 12, and 15 inches and the area of the triangle is 54 square inches.

7.4 Vectors

Find the magnitude of vector AB.

51. $A = (4, -3)$ and $B = (-8, 2)$

52. $A = (-2, 11)$ and $B = (2, 8)$

53. $A = (0, -3)$ and $B = (5, 9)$

54. $A = (3, -11)$ and $B = (9, -3)$

Find the magnitude and direction angle of the given vector.

55. $\mathbf{u} = \langle -10, 24 \rangle$ **56.** $\mathbf{u} = \langle -5, -12 \rangle$

57. $\mathbf{u} = \langle 16, -12 \rangle$ **58.** $\mathbf{u} = \langle 0, 3 \rangle$

Perform the given vector operation, given that $\mathbf{u} = \langle 7, -2 \rangle$ and $\mathbf{v} = \langle -4, 5 \rangle$.

59. $2\mathbf{u} + 3\mathbf{v}$ **60.** $\mathbf{u} - \mathbf{v}$

61. $6\mathbf{u} + \mathbf{v}$ **62.** $-3(\mathbf{u} + 2\mathbf{v})$

Find the vector, given its magnitude and direction angle.

63. $|\mathbf{u}| = 10, \theta = 75°$ **64.** $|\mathbf{u}| = 8, \theta = 225°$

65. $|\mathbf{u}| = 12, \theta = 105°$ **66.** $|\mathbf{u}| = 20, \theta = 15°$

Find a unit vector in the direction of the given vector.

67. $\mathbf{v} = \langle \sqrt{6}, -\sqrt{6} \rangle$ **68.** $\mathbf{v} = \langle -11, 60 \rangle$

Perform the indicated vector operation.

69. $(3\mathbf{i} - 4\mathbf{j}) + (2\mathbf{i} + 5\mathbf{j})$ **70.** $(-6\mathbf{i} + \mathbf{j}) - (9\mathbf{i} - \mathbf{j})$

7.5 Dot Product

Find the indicated dot product.

71. $\langle 6, -3 \rangle \cdot \langle 1, 4 \rangle$

72. $\langle -6, 5 \rangle \cdot \langle -4, 2 \rangle$

73. $\langle 3, 3 \rangle \cdot \langle 3, -6 \rangle$

74. $\langle -2, -8 \rangle \cdot \langle -1, 1 \rangle$

75. $\langle 0, 8 \rangle \cdot \langle 1, 2 \rangle$

76. $\langle 4, -3 \rangle \cdot \langle -1, 0 \rangle$

Find the angle (round to the nearest degree) between each pair of vectors.

77. $\langle 3, 4 \rangle$ and $\langle -5, 12 \rangle$

78. $\langle -4, 5 \rangle$ and $\langle 5, -4 \rangle$

79. $\langle 1, \sqrt{2} \rangle$ and $\langle -1, 3\sqrt{2} \rangle$

80. $\langle 7, -24 \rangle$ and $\langle -6, 8 \rangle$

81. $\langle 3, 5 \rangle$ and $\langle -4, -4 \rangle$

82. $\langle -1, 6 \rangle$ and $\langle 2, -2 \rangle$

Determine if each pair of vectors is orthogonal.

83. $\langle 8, 3 \rangle$ and $\langle -3, 12 \rangle$

84. $\langle -6, 2 \rangle$ and $\langle 4, 12 \rangle$

85. $\langle 5, -6 \rangle$ and $\langle -12, -10 \rangle$

86. $\langle 1, 1 \rangle$ and $\langle -4, 4 \rangle$

87. $\langle 0, 4 \rangle$ and $\langle 0, -4 \rangle$

88. $\langle -7, 2 \rangle$ and $\left\langle \dfrac{1}{7}, -\dfrac{1}{2} \right\rangle$

89. $\langle 6z, a - b \rangle$ and $\langle a + b, -6z \rangle$

90. $\langle a - b, -1 \rangle$ and $\langle a + b, a^2 - b^2 \rangle$

Solve the triangles (if possible).

1. $\alpha = 30°, \beta = 40°, b = 10$

2. $\alpha = 47°, \beta = 98°, \gamma = 35°$

3. $a = 7, b = 9, c = 12$

4. $\alpha = 45°, a = 8, b = 10$

Find the area of the given triangles.

5. $\gamma = 72°, a = 10, b = 12$

6. $a = 7, b = 10, c = 13$

7. Find the magnitude and direction angle of the vector:
$\mathbf{u} = \langle -5, 12 \rangle$.

8. Find a unit vector pointing in the same direction as
$\mathbf{v} = \langle -3, -4 \rangle$.

9. Perform the indicated operation: (a) $2\langle -1, 4 \rangle - 3\langle 4, 1 \rangle$
and (b) $\langle -7, -1 \rangle \cdot \langle 2, 2 \rangle$.

10. Are the vectors $\mathbf{u} = \langle 1, 5 \rangle$ and $\mathbf{v} = \langle -4, 1 \rangle$ perpendicular?

11. Find the smallest positive angle between the two vectors:
$\mathbf{u} = \langle 3, 4 \rangle$ and $\mathbf{v} = \langle -5, 12 \rangle$.

12. Calculate the work involved in pushing a lawnmower 100
feet if the constant force you exert equals 30 pounds and
makes an angle of 60° with the ground.

13. Explain why the magnitude of a vector can never be
negative.

14. A post pattern in football is when the receiver in motion
runs past the quarterback parallel to the line of
scrimmage (A), runs 12 yards perpendicular to the line of
scrimmage (B), and then cuts toward the goal post (C).

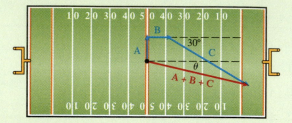

A receiver runs the post pattern. If the magnitudes of the
vectors are $|A| = 3$ yards, $|B| = 12$ yards, and
$|C| = 18$ yards, find the magnitude of the resultant vector,
$\mathbf{A} + \mathbf{B} + \mathbf{C}$ and the direction angle, θ.

15. The dot product of two vectors is a _____
as long as the number of components of each vector
is _____.

Complex Numbers, Polar Coordinates, and Parametric Equations

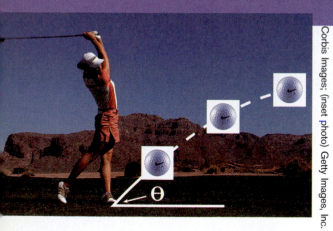

Corbis Images; (inset photo) Getty Images, Inc.

If Michelle Wie tees off with an initial velocity of v_0 ft/sec and initial angle of trajectory, θ, we can describe the position of the ball (x, y) with *parametric equations*. Parametric equations are a set of equations that express a set of quantities, such as x and y coordinates, as explicit functions of a number of independent variables, known as *parameters*. At some time t (sec), the horizontal distance, x (feet), from Michelle down the fairway and the height above the ground, y (feet), are given by the parametric equations:

$$x = (v_0 \cos \theta)t \quad \text{and} \quad y = (v_0 \sin \theta)t - 16t^2$$

where we have neglected air resistance, and t is the parameter. These *parametric equations* essentially map the path of the ball over time.

In this chapter we will review complex numbers. We will discuss the trigonometric (polar) form of complex numbers and operations on complex numbers. We will then introduce polar coordinates which is often a preferred coordinate system over the rectangular system. We will graph polar equations in the polar coordinate system and finally discuss parametric equations and their graphs.

Complex Numbers, Polar Coordinates, and Parametric Equations

Complex Numbers

- Rectangular form: $a + bi$
- Polar form: $r(\cos\theta + i\sin\theta)$
- Products and Quotients
- Powers and Roots

Polar Coordinates

- Rectangular and Polar Coordinates
- Polar Equations
- Graphs of Polar Equations

Parametric Equations

- Trace a Path
  - Spirals
  - Projectiles

CHAPTER OBJECTIVES

- Write complex numbers in polar form.
- Find products, quotients, powers, and roots of complex numbers using polar form.
- Convert between rectangular and polar coordinates.
- Solve polar equations.
- Graph polar equations.
- Use parametric equations to model paths: spirals and projectiles.

NAVIGATION THROUGH SUPPLEMENTS

DIGITAL VIDEO SERIES #8

STUDENT SOLUTIONS MANUAL CHAPTER 8

BOOK COMPANION SITE
www.wiley.com/college/young

SECTION 8.1 Complex Numbers

Skills Objectives

- Write radicals with negative arguments as imaginary numbers.
- Add and subtract complex numbers.
- Solve quadratic equations that lead to complex solutions.
- Multiply complex numbers.
- Express quotients of complex numbers in standard form.

Conceptual Objectives

- Understand that real numbers and imaginary numbers are subsets of complex numbers.
- Understand how to eliminate imaginary numbers in denominators.

Introduction

TECHNOLOGY TIP

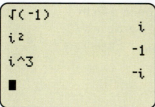

For some equations like $x^2 = 1$, the solutions are always real numbers, $x = \pm 1$. However, there are some equations like $x^2 = -1$ that do not have real solutions because the square of a real number cannot be negative. In order to solve such equations, mathematicians created a new set of numbers based on a number, called the *imaginary unit*, which when squared would give the negative quantity -1. This new set of numbers is called *imaginary numbers*.

> **DEFINITION   IMAGINARY UNIT, *i***
>
> The **imaginary unit** is denoted by the letter i and is defined as
> $$i = \sqrt{-1}$$
> where $i^2 = -1$.

$$i = \sqrt{-1}$$
$$i^2 = -1$$
$$i^3 = i^2 \cdot i = (-1)i = -i$$
$$i^4 = i^2 \cdot i^2 = (-1)(-1) = 1$$

Note that i raised to the fourth power is one. In simplifying imaginary numbers, we factor out i raised to the largest multiple of four.

EXAMPLE 1 Imaginary Unit Raised to Powers

Simplify.

a. i^7 **b.** i^{13} **c.** i^{100}

Solution:

a. $i^7 = i^4 \cdot i^3 = (1)(-i) = -i$

b. $i^{13} = i^{12} \cdot i = (i^4)^3 \cdot i = 1^3 \cdot i = i$

c. $i^{100} = (i^4)^{25} = 1^{25} = 1$

■ **YOUR TURN** Simplify i^{27}.

■ **Answer:** $-i$

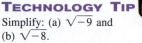

EXAMPLE 2 Using Imaginary Numbers to Simplify Radicals

Simplify using imaginary numbers.

a. $\sqrt{-9}$　　　　**b.** $\sqrt{-8}$

Solution:

a. $\sqrt{-9} = \sqrt{9} \cdot \sqrt{-1} = 3i$

b. $\sqrt{-8} = \sqrt{8} \cdot \sqrt{-1} = 2\sqrt{2} \cdot i = 2i\sqrt{2}$

■ **YOUR TURN** Simplify $\sqrt{-144}$.

TECHNOLOGY TIP

Simplify: (a) $\sqrt{-9}$ and
(b) $\sqrt{-8}$.

```
√(-9)
               3i
√(-8)
        2.828427125i
√(8)
          2.828427125
```

Complex Numbers

A *complex number* is a number that consists of a real and an imaginary term, as follows:

DEFINITION　COMPLEX NUMBER

A **complex number** in standard form is defined as

$$a + bi$$

where a and b are real numbers and i is the imaginary unit.

In the above definition, we denote a as the *real part* of the complex number and b as the *imaginary part* of the complex number.

$$2 - 3i \qquad -5 + i$$

are examples of complex numbers.

The set of real numbers and the set of imaginary numbers are both subsets of the set of complex numbers.

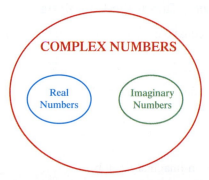

COMPLEX NUMBERS

Real Numbers

Imaginary Numbers

STUDY TIP

The set of real numbers and the set of imaginary numbers are subsets of the set of complex numbers.

DEFINITION　EQUALITY OF COMPLEX NUMBERS

Two complex numbers in **standard form**, $a + bi$ and $c + di$, are **equal** if and only if

$$a = c \quad \text{and} \quad b = d$$

In other words, *two complex numbers are equal if both real parts are equal and both imaginary parts are equal.*

■ **Answer:** $12i$

We can treat complex numbers, $a + bi$, similar to the way we treat binomials, $a + bx$. When adding or subtracting binomials, combine like terms. Similarly, when adding or subtracting complex numbers, real parts are combined with real parts and imaginary parts are combined with imaginary parts.

TECHNOLOGY TIP

Simplify:
(a) $(3 - 2i) + (-1 + i)$ and
(b) $(2 - i) - (3 - 4i)$.

```
(3-2i)+(-1+i)
                  2-i
(2-i)-(3-4i)
                -1+3i
■
```

EXAMPLE 3 Adding and Subtracting Complex Numbers

Perform the indicated operation and simplify.

a. $(3 - 2i) + (-1 + i)$ **b.** $(2 - i) - (3 - 4i)$

Solution (a):

Eliminate the parentheses. $3 - 2i - 1 + i$

Group real and imaginary numbers respectively. $(3 - 1) + (-2i + i)$

Simplify. $2 - i$

Solution (b):

Eliminate the parentheses (distribute negative). $2 - i - 3 + 4i$

Group real and imaginary numbers respectively. $(2 - 3) + (-i + 4i)$

Simplify. $-1 + 3i$

■ **YOUR TURN**   Perform the indicated operation and simplify:
$(4 + i) - (3 - 5i)$.

Complex numbers often arise as solutions to quadratic equations. In Example 4 they arise when using the quadratic formula, $x = \dfrac{-b \pm \sqrt{b^2 - 4ac}}{2a}$, when $b^2 - 4ac < 0$.

TECHNOLOGY TIP

Using the quadratic formula,
$a = 1, b = 2, c = 2$, simplify:
$x = \dfrac{-2 \pm \sqrt{2^2 - 4(1)(2)}}{2(1)} = \dfrac{-2 \pm \sqrt{-4}}{2}$.

```
(-2+√(2²-4*1*2))
/2
                -1+i
(-2-√(2²-4*1*2))
/2
                -1-i
■
```

EXAMPLE 4 Solving Quadratic Equations

Solve the quadratic equation: $x^2 + 2x + 2 = 0$.

Solution:

Use the quadratic formula, $a = 1, b = 2, c = 2$. $x = \dfrac{-2 \pm \sqrt{2^2 - 4(1)(2)}}{2(1)}$

Simplify. $x = \dfrac{-2 \pm \sqrt{-4}}{2}$

Write the radical as an imaginary number,
$\sqrt{-4} = i\sqrt{4} = 2i$. $x = \dfrac{-2 \pm 2i}{2}$

Simplify. $x = -1 \pm i$

■ **YOUR TURN**   Solve the quadratic equation: $x^2 + 3x + 9 = 0$.

■ **Answer:** $1 + 6i$ ■ **Answer:** $x = -\dfrac{3}{2} \pm \dfrac{5}{2}i$

Multiplying Complex Numbers

When multiplying complex numbers, you apply all of the methods for multiplying binomials. It is important to remember that $i^2 = -1$.

EXAMPLE 5 Multiplying Complex Numbers

Multiply the complex numbers and express the result in standard form, $a \pm bi$.

a. $(3 - i)(2 + i)$ **b.** $i(-3 + i)$

Solution (a):

Use the FOIL method to multiply.	$(3 - i)(2 + i) = 3(2) + 3(i) - i(2) - i(i)$
Eliminate parentheses.	$= 6 + 3i - 2i - i^2$
Substitute $i^2 = -1$.	$= 6 + 3i - 2i - (-1)$
Group like terms.	$= (6 + 1) + (3i - 2i)$
Simplify.	$= 7 + i$

Solution (b):

Use the distributive property to multiply.	$i(-3 + i) = -3i + i^2$
Substitute $i^2 = -1$.	$= -3i - 1$
Write in standard form.	$= -1 - 3i$

TECHNOLOGY TIP

Simplify: (a) $(3 - i)(2 + i)$
and (b) $i(-3 + i)$.

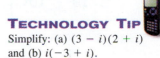

> ■ **YOUR TURN**   Multiply the complex numbers and express result in standard form, $a \pm bi$.
>
> $$(4 - 3i)(-1 + 2i)$$

Dividing Complex Numbers

Recall the special product that produces a difference of two squares, $(a + b)(a - b) = a^2 - b^2$. This special product has only first and last terms because the products of the outer and inner terms sum to zero. Similarly, if we multiply complex conjugates in the same manner, the result is a real number because the imaginary terms sum to zero.

> **PRODUCT OF COMPLEX CONJUGATES**
>
> The product of a complex number, $a + bi$, and its **complex conjugate**, $a - bi$, is a real number
>
> $$(a + bi)(a - bi) = a^2 + \cancel{abi} - \cancel{abi} - b^2 i^2 = a^2 - b^2 \underset{-1}{\underline{i^2}} = a^2 + b^2$$

■ **Answer:** $2 + 11i$

In order to write quotients of complex numbers in standard form, $a + bi$, multiply the numerator and denominator by the complex conjugate of the denominator.

EXAMPLE 6 Dividing Complex Numbers

Write the quotient of the complex numbers in standard form: $\dfrac{2 - i}{1 + 3i}$.

Solution:

Multiply numerator and denominator by the complex conjugate of the denominator, $1 - 3i$.

$$\left(\dfrac{2 - i}{1 + 3i}\right)\left(\dfrac{1 - 3i}{1 - 3i}\right)$$

Multiply numerators and denominators.

$$\dfrac{(2 - i)(1 - 3i)}{(1 + 3i)(1 - 3i)}$$

Use the FOIL method to multiply the binomials.

$$\dfrac{2 - 7i + 3i^2}{1 - 9i^2}$$

Substitute $i^2 = -1$.

$$\dfrac{2 - 7i - 3}{1 + 9}$$

Simplify numerator and denominator.

$$\dfrac{-1 - 7i}{10}$$

Write in standard form.

$$-\dfrac{1}{10} - \dfrac{7}{10}i$$

■ **YOUR TURN** Write the complex number in standard form: $\dfrac{3 + 2i}{4 - i}$.

Applications

Although the letter i is used to represent $\sqrt{-1}$, in some fields such as electrical engineering and physics, the letter j is used instead, because the letter i is typically a symbol for electrical current. Special attention should be paid to the notation, as some textbooks let $j = i$ and others let $j = -i$.

Electrical impedance, Z, is a measure of opposition to a sinusoidal electrical current and is a complex number.

$$Z = R + jX \qquad j = \sqrt{-1}$$

R is the real part and is referred to as the resistive part of impedance (resistors), whereas X is the imaginary part and is referred to as the reactive part of impedance (capacitor and inductors).

Two impedances, Z_1 and Z_2, are combined differently if they are in series or in parallel. If they are in series they are added, $Z_1 + Z_2$, whereas if they are in parallel the resulting impedance is given by $\dfrac{Z_1 Z_2}{Z_1 + Z_2}$.

■ **Answer:** $\dfrac{10}{17} + \dfrac{11}{17}i$

SECTION 8.1   SUMMARY

Imaginary numbers are a concept that makes it possible to represent solutions to equations that have no real solutions, such as $x^2 = -1$, because they require taking the square root of a negative number. The imaginary unit is $i = \sqrt{-1}$. Complex numbers are numbers that include both real and imaginary terms. To add or subtract complex numbers, simply add or subtract the real parts and imaginary parts respectively. Complex solutions arise in solving some quadratic equations. Complex numbers can be multiplied using the same FOIL method that is used for multiplying binomials. It is important to remember that $i^2 = -1$. In order to write a quotient of complex numbers in standard form, multiply the numerator and denominator by the complex conjugate of the denominator.

SECTION 8.1   EXERCISES

■ SKILLS

In Exercises 1–8, simplify.

1. i^{15} **2.** i^{99} **3.** i^{40} **4.** i^{18}

5. $\sqrt{-16}$ **6.** $\sqrt{-100}$ **7.** $\sqrt{-20}$ **8.** $\sqrt{-24}$

In Exercises 9–20, perform the indicated operation, simplify, and express in standard form.

9. $(3 - 7i) + (-1 - 2i)$ **10.** $(1 + i) + (9 - 3i)$ **11.** $(4 - 5i) - (2 - 3i)$ **12.** $(-2 + i) - (1 - i)$

13. $(1 - i)(3 + 2i)$ **14.** $(-3 + 2i)(1 - 3i)$ **15.** $(4 + 7i)(4 - 7i)$ **16.** $(-2 + 3i)(-2 - 3i)$

17. $\dfrac{1}{3 - i}$ **18.** $\dfrac{1}{3 + 2i}$ **19.** $\dfrac{4 - 5i}{7 + 2i}$ **20.** $\dfrac{1 - i}{1 + i}$

In Exercises 21–26, solve the quadratic equations. Write the answers in standard form.

21. $2x^2 - 3x + 2 = 0$ **22.** $3x^2 + 8x + 9 = 0$ **23.** $x^2 + 5x + 25 = 0$

24. $x^2 + x + 1 = 0$ **25.** $4x^2 - 2x + 5 = 0$ **26.** $5x^2 + 4x + 8 = 0$

In Exercises 27–30, write the negative radicals in terms of imaginary numbers and then perform the indicated operation and simplify.

27. $\sqrt{-16} + 2\sqrt{-25} - 3\sqrt{-9}$ **28.** $\sqrt{-8} + \sqrt{-18}$

29. $\left(\sqrt{-4} + 1\right)\left(3 - \sqrt{-25}\right)$ **30.** $\left(3 - \sqrt{-16}\right)\left(2 - \sqrt{-9}\right)$

■ APPLICATIONS

For Exercises 31–34, assume there are two impedances, $Z_1 = R_1 + jX_1$ and $Z_2 = R_2 + jX_2$ (where $j = \sqrt{-1}$), and the two impedances are either:

■ in series $Z_{\text{series}} = Z_1 + Z_2$ or

■ in parallel $Z_{\text{parallel}} = \dfrac{Z_1 Z_2}{Z_1 + Z_2}$

31. Calculate the real part of the impedance, R, in terms of R_1 and R_2 for two impedances in series.

32. Calculate the imaginary part of the impedance, X, in terms of X_1 and X_2 for two impedances in series.

33. Calculate the real part of the impedance, R, in terms of R_1, R_2, X_1, and X_2 for two impedances in parallel.

34. Calculate the imaginary part of the impedance, X, in terms of R_1, R_2, X_1, and X_2 for two impedances in parallel.

■ CATCH THE MISTAKE

In Exercises 35–38, explain the mistake that is made.

35. Simplify: $\sqrt{-4} \cdot \sqrt{-4}$.

Solution:

Multiply the two radicals. $\sqrt{-4} \cdot \sqrt{-4} = \sqrt{(-4)(-4)}$

Simplify. $= \sqrt{16}$

Evaluate the radical. $= 4$

This is incorrect. What mistake was made?

36. Simplify: i^{103}.

Solution: $i^{103} = (i^{100})(i^3) = (-1)(-i) = i$

This is incorrect. What mistake was made?

37. Write the quotient in standard form: $\dfrac{2}{4 - i}$.

Solution:

Multiply numerator and denominator by $4 - i$. $\dfrac{2}{4 - i} \cdot \dfrac{4 - i}{4 - i}$

Multiply the numerator using distributive property and the denominator using the FOIL method. $\dfrac{8 - 2i}{16 - 1}$

Simplify. $\dfrac{8 - 2i}{15}$

Write in standard form. $\dfrac{8}{15} - \dfrac{2}{15}i$

This is incorrect. What mistake was made?

38. Write the product in standard form: $(2 - 3i)(5 + 4i)$.

Solution:

Use the FOIL method to multiply the complex numbers. $10 - 7i - 12i^2$

Simplify. $-2 - 7i$

This is incorrect. What mistake was made?

■ CHALLENGE

39. T or F: The product $(a + bi)(a - bi)$ is a real number if a and b are real numbers.

40. T or F: Imaginary numbers are a subset of the complex numbers.

41. T or F: Real numbers are a subset of the complex numbers.

42. T or F: There is no complex number that equals its conjugate.

43. Simplify: $i^{26} - i^{15} + i^{17} - i^{1000}$.

44. Solve: $x^4 + 2x^2 + 1 = 0$.

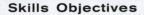

SECTION 8.2 Polar (Trigonometric) Form of Complex Numbers

Skills Objectives

- Graph a point in the complex plane.
- Convert complex numbers in rectangular form to polar form.
- Convert complex numbers in polar form to rectangular form.

Conceptual Objectives

- Understand that a complex number can be represented in either rectangular or polar form.
- Relate the horizontal axis in the complex plane to the real component of a complex number.
- Relate the vertical axis in the complex plane to the imaginary component of a complex number.

Complex Numbers in Rectangular Form

We are already familiar with the **rectangular coordinate system**, where the horizontal axis is called the x-axis and the vertical axis is called the y-axis. In our study of complex numbers, we refer to the **standard (rectangular) form** as $a + bi$, where a represents the real part and b represents the imaginary part. If we let the horizontal axis be the **real axis** and the vertical axis be the **imaginary axis**, the result is the **complex plane**. The point, $a + bi$, is located in the complex plane by finding the coordinates (a, b).

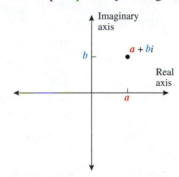

When $b = 0$ the result is a real number; and therefore any numbers along the horizontal axis are real numbers. When $a = 0$ the result is an imaginary number, so any numbers along the vertical axis are imaginary numbers.

The variable z is often used to represent a complex number: $z = x + iy$. Complex numbers are analogous to vectors. Suppose we have a vector $\mathbf{z} = \langle x, y \rangle$ whose initial point is the origin and whose terminal point is (x, y), then the magnitude of that vector would be $|\mathbf{z}| = \sqrt{x^2 + y^2}$. Similarly, the magnitude, or *modulus*, of a complex number is defined like the magnitude of a position vector in the x, y plane: as the distance from the origin $(0, 0)$ to the point (x, y) in the complex plane.

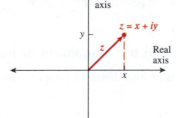

> **DEFINITION**   **MODULUS OF A COMPLEX NUMBER**
>
> The **modulus**, or magnitude, of a complex number $z = x + iy$ is the distance from the origin to the point (x, y) in the complex plane given by
>
> $$|z| = \sqrt{x^2 + y^2}$$

Recall from Section 8.1 that a complex number $z = x + iy$ has a complex conjugate $\bar{z} = x - iy$. The bar above a complex number denotes its conjugate. Notice that

$$z\bar{z} = (x + iy)(x - iy) = x^2 - i^2y^2 = x^2 + y^2$$

and therefore the modulus can also be written as

$$|z| = \sqrt{z\bar{z}}$$

EXAMPLE 1 Finding the Modulus of a Complex Number

Find the modulus of $z = -3 + 2i$.

> **COMMON MISTAKE**
>
> Including the i in the imaginary part.
>
>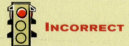
>
> **CORRECT**
>
> Let $x = -3$ and $y = 2$ in $|z| = \sqrt{x^2 + y^2}$.
>
> $$|-3 + 2i| = \sqrt{(-3)^2 + 2^2}$$
>
> Eliminate parentheses.
>
> $$|-3 + 2i| = \sqrt{9 + 4}$$
>
> Simplify.
>
> $$|z| = |-3 + 2i| = \sqrt{13}$$
>
> **INCORRECT**
>
> Let $x = -3$ and $y = 2i$ in $|z| = \sqrt{x^2 + y^2}$.
>
> $$|-3 + 2i| = \sqrt{(-3)^2 + (2i)^2}$$
>
> **ERROR**
>
> Eliminate parentheses.
>
> $$|-3 + 2i| = \sqrt{9 - 4}$$
>
> Simplify.
>
> $$|-3 + 2i| = \sqrt{5}$$
>
> The i is not included in the formula; only the imaginary part (coefficient of i) is used.

TECHNOLOGY TIP

Find the modulus of $z = -3 + 2i$.

| MATH | ▶ | CPX | ▼ | 5: abs (|

| ENTER | (−) | 3 | + | 2 | 2nd |

| · |) | ENTER |

```
abs( -3+2i)
        3.605551275
√(13)
        3.605551275
■
```

■ **YOUR TURN**   Find the modulus of $z = 2 - 5i$.

EXAMPLE 2 Calculating the Magnitude of Impedance

Calculate the magnitude of impedance in terms of the resistive and reactive parts.

■ **Answer:** $|z| = |2 - 5i| = \sqrt{29}$

Solution:

Write the impedance.

$$Z = R + jX$$

The magnitude of a complex number is the square root of the sum of the squares of the real and imaginary parts.

$$|Z| = \sqrt{R^2 + X^2}$$

Complex Numbers in Polar Form

We say that a complex number $z = x + iy$ is in *rectangular* form because it is located at the point (x, y), which is expressed in rectangular coordinates in the complex plane. Another convenient way of expressing complex numbers is in *polar* form. Recall in our study of vectors (Section 7.4) that vectors have both magnitude and a direction angle. The same is true of numbers in the complex plane. Let r represent the magnitude, or distance from the origin to the point (x, y), and θ represent the direction angle. Then we have the following relationships:

$$r = \sqrt{x^2 + y^2}$$

$$\sin\theta = \frac{y}{r} \quad \cos\theta = \frac{x}{r} \quad \tan\theta = \frac{y}{x} \quad (x \neq 0)$$

Isolating x and y in the sinusoidal functions, we find:

$$x = r\cos\theta \qquad y = r\sin\theta$$

Using these expressions for x and y, a complex number can be written in *polar* form:

$$z = x + yi = (r\cos\theta) + (r\sin\theta)i = r(\cos\theta + i\sin\theta)$$

POLAR (TRIGONOMETRIC) FORM OF COMPLEX NUMBERS

The following expression is the **polar form** of a complex number:

$$z = r(\cos\theta + i\sin\theta)$$

where r represents the **modulus** (magnitude) of the complex number and θ represents the **argument** of z.

The following is standard notation for modulus and argument:

$$r = \text{mod } z = |z|$$
$$\theta = \text{Arg } z \qquad 0 \leq \theta < 2\pi$$

Converting Complex Numbers Between Rectangular and Polar Forms

We can convert back and forth between rectangular and polar (trigonometric) forms of complex numbers using the modulus and trigonometric ratios:

$$r = \sqrt{x^2 + y^2}, \quad \sin\theta = \frac{y}{r}, \quad \cos\theta = \frac{x}{r}, \quad \text{and} \quad \tan\theta = \frac{y}{x} \quad (x \neq 0)$$

CONVERTING COMPLEX NUMBERS FROM RECTANGULAR FORM TO POLAR FORM

Step 1: Plot the point, $z = x + yi$, in the complex plane (note the quadrant).

Step 2: Find r. Use $r = \sqrt{x^2 + y^2}$.

Step 3: Find θ. Use $\tan \theta = \dfrac{y}{x}$, $x \neq 0$, where θ is in the quadrant found in step 1.

Step 4: Write the complex number in polar form: $z = r(\cos \theta + i \sin \theta)$.

TECHNOLOGY TIP

To convert complex numbers from rectangular to polar form, set the calculator to degree mode. For points in QII, QIII, and QIV, use the inverse tangent function to find the reference angle and then the actual angles.

Express the complex number $z = \sqrt{3} - i$ in polar form.

Method I: Use $\tan^{-1}\dfrac{1}{\sqrt{3}}$ to find the reference angle for θ, which is in QIV.

```
tan⁻¹(1/√(3))
                  30
360-Ans
                 330
■
```

Method II: Use the $\boxed{\text{angle (}}$ feature on the calculator to find θ. You still have to find the actual angle in QIV. Press

$\boxed{\text{MATH}}$ $\boxed{\blacktriangleright}$ $\boxed{\text{CPX}}$ $\boxed{\blacktriangledown}$ $\boxed{\text{4: angle (}}$
$\boxed{\text{ENTER}}$ $\boxed{\text{2nd}}$ $\boxed{x^2}$ $\boxed{3}$ $\boxed{)}$ $\boxed{-}$
$\boxed{\text{2nd}}$ $\boxed{\cdot}$ $\boxed{)}$ $\boxed{\text{ENTER}}$

```
angle(√(3)-i)
                 -30
360+Ans
                 330
■
```

EXAMPLE 3 Converting from Rectangular to Polar Form

Express the complex number $z = \sqrt{3} - i$ in polar form.

Solution:

STEP 1 Plot the point.
The point lies in **QIV**.

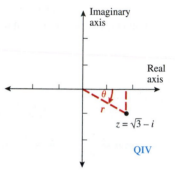

$z = \sqrt{3} - i$

QIV

STEP 2 Find r.

Let $x = \sqrt{3}$ and $y = -1$
in $r = \sqrt{x^2 + y^2}$. $\qquad r = \sqrt{(\sqrt{3})^2 + (-1)^2}$

Eliminate parentheses. $\qquad r = \sqrt{3 + 1}$

Simplify. $\qquad \boxed{r = 2}$

STEP 3 Find θ.

Let $x = \sqrt{3}$ and $y = -1$ in $\tan \theta = \dfrac{y}{x}$. $\qquad \tan \theta = -\dfrac{1}{\sqrt{3}}$

Find the reference angle. $\qquad$ reference angle $= \dfrac{\pi}{6}$

The complex number lies in QIV. $\qquad \boxed{\theta = \dfrac{11\pi}{6}}$

STEP 4 Write the complex number in polar form.

$z = r(\cos \theta + i \sin \theta)$ $\qquad \boxed{z = 2\left(\cos \dfrac{11\pi}{6} + i \sin \dfrac{11\pi}{6}\right)}$

Note: An alternative form is in degrees: $z = 2(\cos 330° + i \sin 330°)$.

■ **YOUR TURN** Express the complex number $z = 1 - i\sqrt{3}$ in polar form.

You must be very careful in converting from rectangular to polar form. Remember that the inverse tangent function is a one-to-one function and will yield values in QI and

■ **Answer:** $z = 2\left(\cos \dfrac{5\pi}{3} + i\sin \dfrac{5\pi}{3}\right)$ or $2(\cos 300° + i\sin 300°)$

QIV. If the point lies in QII or QIII, add 180° to the angle found through the inverse tangent function.

 EXAMPLE 4 Converting from Rectangular to Polar Form

COMMON MISTAKE

Forgetting to confirm the quadrant, which results in using the reference angle instead of the actual angle.

Express the complex number $z = -2 + i$ in polar form.

 CORRECT

STEP 1: Plot the point.

The point lies in **QII**.

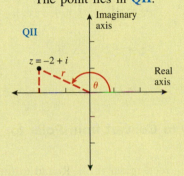

STEP 2: Find r.

Let $x = -2$ and $y = 1$ in $r = \sqrt{x^2 + y^2}$.

$$r = \sqrt{(-2)^2 + 1^2}$$

Simplify.

$$r = \sqrt{5}$$

STEP 3: Find θ.

Let $x = -2$ and $y = 1$ in $\tan \theta = \dfrac{y}{x}$.

$$\tan \theta = -\frac{1}{2}$$

$$\theta = \tan^{-1}\left(-\frac{1}{2}\right)$$

$$= -26.565°$$

The complex number lies in QII.

$$\theta = -26.6° + 180° = 153.4°$$

STEP 4: Write the complex number in polar form.

$$z = r(\cos \theta + i \sin \theta)$$

$$z = \sqrt{5}\,(\cos 153.4° + i \sin 153.4°)$$

INCORRECT

Find r.

Let $x = -2$ and $y = 1$ in $r = \sqrt{x^2 + y^2}$.

$$r = \sqrt{(-2)^2 + 1^2}$$

Simplify. $r = \sqrt{5}$

Find θ.

Let $x = -2$ and $y = 1$ in

$\tan \theta = \dfrac{y}{x}$. $\tan \theta = -\dfrac{1}{2}$

Use a calculator to evaluate the inverse function.

$$\theta = \tan^{-1}\left(-\frac{1}{2}\right) = -26.565°$$

Write the complex number in polar form.

$$z = r(\cos \theta + i \sin \theta)$$
$$z = \sqrt{5}(\cos(-26.6°) + i \sin(-26.6°))$$

$\theta = -26.565°$ lies in QIV, whereas the original point lies in QII. Therefore, we should have added 180° to θ in order to arrive at a point in QII.

TECHNOLOGY TIP

Express the complex number $z = -2 + i$ in polar form.

```
abs(-2+i)
            2.236067977
√(5)
            2.236067977
angle(-2+i)
            153.4349488
```

■ **YOUR TURN** Express the complex number $z = -1 + 2i$ in polar form.

To convert from polar to rectangular form, simply evaluate the trigonometric functions.

EXAMPLE 5 Converting from Polar to Rectangular Form

Express $z = 4(\cos 120° + i\sin 120°)$ in rectangular form.

Solution:

Evaluate the trigonometric functions exactly.

$$z = 4(\underbrace{\cos 120°}_{-\frac{1}{2}} + i\underbrace{\sin 120°}_{\frac{\sqrt{3}}{2}})$$

Distribute the 4.

$$z = 4\left(-\frac{1}{2}\right) + 4\left(\frac{\sqrt{3}}{2}\right)i$$

Simplify.

$$z = -2 + 2\sqrt{3}i$$

■ **YOUR TURN** Express $z = 2(\cos 210° + i\sin 210°)$ in rectangular form.

EXAMPLE 6 Using a Calculator to Convert from Polar to Rectangular Form

Express $z = 3(\cos 109° + i\sin 109°)$ in rectangular form. Round to four decimal places.

Solution:

Use a calculator to evaluate the trigonometric functions.

$$z = 3(\underbrace{\cos 109°}_{-0.3256} + i\underbrace{\sin 109°}_{0.9455})$$

Distribute the 3.

$$z = 3(-0.3256) + (3)(0.9455)i$$

Simplify.

$$z = -0.9768 + 2.8365i$$

■ **YOUR TURN** Express $z = 7(\cos 217° + i\sin 217°)$ in rectangular form. Round to four decimal places.

SECTION 8.2 **SUMMARY**

In the complex plane, the horizontal axis is the real axis and the vertical axis is the imaginary axis. Complex numbers can be expressed in either rectangular form, $z = x + iy$, or polar form, $z = r(\cos\theta + i\sin\theta)$. To convert from rectangular to polar form, we use the relationships $r = \sqrt{x^2 + y^2}$ and $\tan\theta = \frac{y}{x}$, $x \neq 0$ and $0 \leq \theta < 2\pi$. It is important to note in which quadrant the point lies. To convert from polar to rectangular form, simply evaluate the trigonometric functions:

$$x = r\cos\theta \quad \text{and} \quad y = r\sin\theta$$

■ **Answer:** $z = \sqrt{5}(\cos(116.6°) + i\sin(116.6°))$ ■ **Answer:** $z = -\sqrt{3} - i$

■ **Answer:** $z = -5.5904 - 4.2127i$

SECTION 8.2 EXERCISES

 SKILLS

In Exercises 1–8, graph each complex number.

1. $7 + 8i$

2. $3 + 5i$

3. $-2 - 4i$

4. $-3 - 2i$

5. 2

6. 7

7. $-3i$

8. $-5i$

In Exercises 9–18, express each complex number in polar form.

9. $1 - i$

10. $2 + 2i$

11. $1 + \sqrt{3}i$

12. $-3 - \sqrt{3}i$

13. $-4 + 4i$

14. $\sqrt{5} - \sqrt{5}i$

15. $\sqrt{3} - 3i$

16. $-\sqrt{3} + i$

17. $3 + 0i$

18. $-2 + 0i$

In Exercises 19–28, use a calculator to express each complex number in polar form.

19. $3 - 7i$

20. $2 + 3i$

21. $-6 + 5i$

22. $-4 - 3i$

23. $-5 + 12i$

24. $24 + 7i$

25. $8 - 6i$

26. $-3 + 4i$

27. $a - 2ai$, where $a > 0$

28. $-3a - 4ai$, where $a > 0$

In Exercises 29–38, express each complex number in rectangular form.

29. $5(\cos 180° + i \sin 180°)$

30. $2(\cos 135° + i \sin 135°)$

31. $2(\cos 315° + i \sin 315°)$

32. $3(\cos 270° + i \sin 270°)$

33. $-4(\cos 60° + i \sin 60°)$

34. $-4(\cos 210° + i \sin 210°)$

35. $\sqrt{3}(\cos 150° + i \sin 150°)$

36. $\sqrt{3}(\cos 330° + i \sin 330°)$

37. $\sqrt{2}\left(\cos \dfrac{\pi}{4} + i \sin \dfrac{\pi}{4} \right)$

38. $2\left(\cos \dfrac{5\pi}{6} + i \sin \dfrac{5\pi}{6} \right)$

In Exercises 39–48, use a calculator to express each complex number in rectangular form.

39. $5(\cos 295° + i \sin 295°)$

40. $4(\cos 35° + i \sin 35°)$

41. $3(\cos 100° + i \sin 100°)$

42. $6(\cos 250° + i \sin 250°)$

43. $-7(\cos 140° + i \sin 140°)$

44. $-5(\cos 320° + i \sin 320°)$

45. $3\left(\cos \dfrac{11\pi}{12} + i \sin \dfrac{11\pi}{12} \right)$

46. $2\left(\cos \dfrac{4\pi}{7} + i \sin \dfrac{4\pi}{7} \right)$

47. $-2\left(\cos \dfrac{3\pi}{5} + i \sin \dfrac{3\pi}{5} \right)$

48. $-4\left(\cos \dfrac{15\pi}{11} + i \sin \dfrac{15\pi}{11} \right)$

APPLICATIONS

49. Resultant Force. Force A, at 100 pounds, and force B, at 120 pounds, make an angle of 30° with each other. Represent their respective vectors as complex numbers written in trigonometric form, and solve for the resultant force.

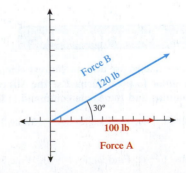

50. Resultant Force. Force A, at 40 pounds, and force B, at 50 pounds, make an angle of 45° with each other. Represent their respective vectors as complex numbers written in trigonometric form, and solve for the resultant force.

51. Resultant Force. Force A, at 80 pounds, and force B, at 150 pounds, make an angle of 30° with each other. Represent their respective vectors as complex numbers written in trigonometric form, and solve for the resultant angle.

52. Resultant Force. Force A, at 20 pounds, and force B, at 60 pounds, make an angle of 60° with each other. Represent their respective vectors as complex numbers written in trigonometric form, and solve for the resultant angle.

■ CATCH THE MISTAKE

In Exercises 53 and 54, explain the mistake that is made.

53. Express $z = -3 - 8i$ in polar form.

Solution:

Find r. $r = \sqrt{x^2 + y^2} = \sqrt{9 + 64} = \sqrt{73}$

Find θ. $\tan\theta = \dfrac{8}{3}$

$$\theta = \tan^{-1}\left(\frac{8}{3}\right) = 69.44°$$

Write the complex number in polar form.

$$z = \sqrt{73}\,(\cos 69.44° + i\sin 69.44°)$$

This is incorrect. What mistake was made?

54. Express $z = -3 + 8i$ in polar form.

Solution:

Find r. $r = \sqrt{x^2 + y^2} = \sqrt{9 + 64} = \sqrt{73}$

Find θ. $\tan\theta = -\dfrac{8}{3}$

$$\theta = \tan^{-1}\left(-\frac{8}{3}\right) = -69.44°$$

Write the complex number in polar form.

$$z = \sqrt{73}\,(\cos[-69.44°] + i\sin[-69.44°])$$

This is incorrect. What mistake was made?

■ CHALLENGE

55. T or F: In the complex plane, any point that lies along the horizontal axis is a real number.

56. T or F: In the complex plane, any point that lies along the vertical axis is an imaginary number.

57. T or F: The modulus of z and the modulus of $\bar{z}$ are equal.

58. T or F: The argument of z and the argument of $\bar{z}$ are equal.

59. Find the argument of $z = a$, where a is a positive real number.

60. Find the argument of $z = bi$, where b is a positive real number.

61. Find the modulus of $z = bi$, where b is a negative real number.

62. Find the modulus of $z = a$, where a is a negative real number.

■ TECHNOLOGY

Graphing calculators are able to convert complex numbers from rectangular to polar form. Use the Abs command to find the modulus and the Angle command to find the angle.

63. Find abs$(1 + i)$. Find angle$(1 + i)$. Write $1 + i$ in polar form.

64. Find abs$(1 - i)$. Find angle$(1 - i)$. Write $1 - i$ in polar form.

A second way of using a graphing calculator to convert between rectangular and polar coordinates is with the Pol and Rec commands.

65. Find Pol $(2, 1)$. Write $2 + i$ in polar form.

66. Find Rec$(3, 45°)$. Write $3(\cos 45° + i\sin 45°)$ in rectangular form.

Skills Objectives

- Find the product of two complex numbers.
- Find the quotient of two complex numbers.
- Raise a complex number to an integer power.
- Find the nth root of a complex number.
- Solve an equation by finding complex roots.

Conceptual Objectives

- Derive the identities for products and quotients of complex numbers.
- Relate DeMoivre's theorem (the power rule) for complex numbers to the product rule for complex numbers.

In this section we will multiply complex numbers, divide complex numbers, raise complex numbers to powers, and find roots of complex numbers.

Products of Complex Numbers

We will first derive a formula for the product of two complex numbers that are given in polar form.

WORDS	MATH
Start with two complex numbers: z_1 and z_2.	$z_1 = r_1(\cos\theta_1 + i\sin\theta_1)$ $z_2 = r_2(\cos\theta_2 + i\sin\theta_2)$
Multiply z_1 and z_2.	$z_1 z_2 = r_1 r_2(\cos\theta_1 + i\sin\theta_1)(\cos\theta_2 + i\sin\theta_2)$
Use the FOIL method to multiply the expressions in parentheses.	$z_1 z_2 = r_1 r_2(\cos\theta_1\cos\theta_2 + i\cos\theta_1\sin\theta_2 + i\sin\theta_1\cos\theta_2 + \underset{-1}{i^2}\sin\theta_1\sin\theta_2)$
Group the real parts and the imaginary parts.	$z_1 z_2 = r_1 r_2[(\cos\theta_1\cos\theta_2 - \sin\theta_1\sin\theta_2) + i(\cos\theta_1\sin\theta_2 + \sin\theta_1\cos\theta_2)]$
Use the cosine and sine sum identities.	$z_1 z_2 = r_1 r_2[\underset{\cos(\theta_1+\theta_2)}{\underbrace{(\cos\theta_1\cos\theta_2 - \sin\theta_1\sin\theta_2)}} + i\underset{\sin(\theta_1+\theta_2)}{\underbrace{(\cos\theta_1\sin\theta_2 + \sin\theta_1\cos\theta_2)}}]$
Simplify.	$z_1 z_2 = r_1 r_2[\cos(\theta_1 + \theta_2) + i\sin(\theta_1 + \theta_2)]$

PRODUCT OF TWO COMPLEX NUMBERS

Let $z_1 = r_1(\cos\theta_1 + i\sin\theta_1)$ and $z_2 = r_2(\cos\theta_2 + i\sin\theta_2)$ be two complex numbers. The **complex product**, $z_1 z_2$, is given by

$$z_1 z_2 = r_1 r_2[\cos(\theta_1 + \theta_2) + i\sin(\theta_1 + \theta_2)]$$

In other words, *when multiplying two complex numbers, the magnitudes are multiplied and the arguments are added.*

STUDY TIP

When two complex numbers are multiplied, the magnitudes are multiplied and the arguments are added.

EXAMPLE 1 Multiplying Complex Numbers

Find the product of $z_1 = 3[\cos 35° + i\sin 35°]$ and $z_2 = 2[\cos 10° + i\sin 10°]$.

Solution:

Set up the product.	$z_1 z_2 = 3[\cos 35° + i\sin 35°] \cdot 2[\cos 10° + i\sin 10°]$
Multiply the magnitudes and add the arguments.	$z_1 z_2 = 3 \cdot 2[\cos(35° + 10°) + i\sin(35° + 10°)]$

Find the product of
$z_1 = 3[\cos 35° + i \sin 35°]$ and
$z_2 = 3[\cos 10° + i \sin 10°]$.

```
3(cos(35)+isin(3
5))*2(cos(10)+is
in(10))
    4.2426+4.2426i
3√(2)+3i√(2)
    4.2426+4.2426i
```

Simplify.

The product is in polar form. To express the product in rectangular form, evaluate the trigonometric functions.

This is the product in polar form:

This is the product in rectangular form:

$z_1 z_2 = 6[\cos 45° + i \sin 45°]$

$z_1 z_2 = 6\left[\dfrac{\sqrt{2}}{2} + i\dfrac{\sqrt{2}}{2}\right] = 3\sqrt{2} + 3i\sqrt{2}$

$z_1 z_2 = 6[\cos 45° + i \sin 45°]$

$z_1 z_2 = 3\sqrt{2} + 3i\sqrt{2}$

YOUR TURN Find the product of $z_1 = 2[\cos 55° + i \sin 55°]$ and $z_2 = 5[\cos 65° + i \sin 65°]$. Express the answer in both polar and rectangular form.

Quotients of Complex Numbers

We now derive a formula for the quotient of two complex numbers.

WORDS	MATH
Start with two complex numbers: z_1 and z_2.	$z_1 = r_1(\cos \theta_1 + i \sin \theta_1)$ $z_2 = r_2(\cos \theta_2 + i \sin \theta_2)$

Divide z_1 by z_2.

$$\frac{z_1}{z_2} = \frac{r_1(\cos \theta_1 + i \sin \theta_1)}{r_2(\cos \theta_2 + i \sin \theta_2)} = \left(\frac{r_1}{r_2}\right)\left(\frac{\cos \theta_1 + i \sin \theta_1}{\cos \theta_2 + i \sin \theta_2}\right)$$

Multiply the second expression in parentheses by the conjugate of the denominator, $\cos \theta_2 - i \sin \theta_2$.

$$\frac{z_1}{z_2} = \left(\frac{r_1}{r_2}\right)\left(\frac{\cos \theta_1 + i \sin \theta_1}{\cos \theta_2 + i \sin \theta_2}\right)\left(\frac{\cos \theta_2 - i \sin \theta_2}{\cos \theta_2 - i \sin \theta_2}\right)$$

Use the FOIL method to multiply the expressions in parentheses in the last two expressions.

$$\frac{z_1}{z_2} = \left(\frac{r_1}{r_2}\right)\left(\frac{\cos \theta_1 \cos \theta_2 - i^2 \sin \theta_1 \sin \theta_2 + i \sin \theta_1 \cos \theta_2 - i \sin \theta_2 \cos \theta_1}{\cos^2 \theta_2 - i^2 \sin^2 \theta_2}\right)$$

Substitute $i^2 = -1$ and group the real parts and the imaginary parts.

$$\frac{z_1}{z_2} = \left(\frac{r_1}{r_2}\right)\left(\frac{[\cos \theta_1 \cos \theta_2 + \sin \theta_1 \sin \theta_2] + i[\sin \theta_1 \cos \theta_2 - \sin \theta_2 \cos \theta_1]}{\underbrace{\cos^2 \theta_2 + \sin^2 \theta_2}_{1}}\right)$$

Simplify.

$$\frac{z_1}{z_2} = \left(\frac{r_1}{r_2}\right)[(\cos \theta_1 \cos \theta_2 + \sin \theta_1 \sin \theta_2) + i(\sin \theta_1 \cos \theta_2 - \sin \theta_2 \cos \theta_1)]$$

Answer: $z_1 z_2 = 10[\cos 120° + i \sin 120°]$ or $z_1 z_2 = -5 + 5i\sqrt{3}$

Use the cosine and sine difference identities.	$$\frac{z_1}{z_2} = \left(\frac{r_1}{r_2}\right)\left[\underbrace{(\cos\theta_1\cos\theta_2 + \sin\theta_1\sin\theta_2)}_{\cos(\theta_1-\theta_2)} + i\underbrace{(\sin\theta_1\cos\theta_2 - \sin\theta_2\cos\theta_1)}_{\sin(\theta_1-\theta_2)}\right]$$
Simplify.	$$z_1 z_2 = \frac{r_1}{r_2}[\cos(\theta_1-\theta_2) + i\sin(\theta_1-\theta_2)]$$

It is important to notice that the argument difference is the argument of the numerator minus the argument of the denominator.

QUOTIENT OF TWO COMPLEX NUMBERS

Let $z_1 = r_1(\cos\theta_1 + i\sin\theta_1)$ and $z_2 = r_2(\cos\theta_2 + i\sin\theta_2)$ be two complex numbers. The **complex quotient,** $\dfrac{z_1}{z_2}$, is given by:

$$\frac{z_1}{z_2} = \frac{r_1}{r_2}[\cos(\theta_1-\theta_2) + i\sin(\theta_1-\theta_2)]$$

In other words, *when dividing two complex numbers, the magnitudes are divided and the arguments are subtracted. It is important to note that the argument difference is the argument of the complex number in the numerator minus the argument of the complex number in the denominator.*

EXAMPLE 2 **Dividing Complex Numbers**

Let $z_1 = 6[\cos 125° + i\sin 125°]$ and $z_2 = 3[\cos 65° + i\sin 65°]$, and find $\dfrac{z_1}{z_2}$.

Solution:

Set up the quotient.	$$\frac{z_1}{z_2} = \frac{6[\cos 125° + i\sin 125°]}{3[\cos 65° + i\sin 65°]}$$	
Divide the magnitudes and subtract the arguments.	$$\frac{z_1}{z_2} = \frac{6}{3}[\cos(125° - 65°) + i\sin(125° - 65°)]$$	**TECHNOLOGY TIP**
Simplify.	$$\frac{z_1}{z_2} = 2[\cos 60° + i\sin 60°]$$	Let $z_1 = 6[\cos 125° + i\sin 125°]$ and $z_2 = 3[\cos 65° + i\sin 65°]$, find $\dfrac{z_1}{z_2}$.
The product is in polar form. To express the product in rectangular form, evaluate the trigonometric functions.	$$\frac{z_1}{z_2} = 2\left[\frac{1}{2} + i\frac{\sqrt{3}}{2}\right] = 1 + i\sqrt{3}$$	Be sure to include parentheses for z_1 and z_2.
This is the product in polar form:	$$\frac{z_1}{z_2} = 2[\cos 60° + i\sin 60°]$$	(6(cos(125)+isin (125)))/(3(cos(6 5)+isin(65))) 1.0000+1.7321i √(3) 1.7321
This is the product in rectangular form:	$$\frac{z_1}{z_2} = 1 + i\sqrt{3}$$	

> ■ **YOUR TURN** Let $z_1 = 10[\cos 275° + i\sin 275°]$ and
>
> $z_2 = 5[\cos 65° + i\sin 65°]$, and find $\dfrac{z_1}{z_2}$. Express the answer in both polar and rectangular form.

When multiplying or dividing complex numbers, we have only considered values of θ: $0° \le \theta \le 360°$. When the value of θ is negative or greater than $360°$, find the coterminal angle in the interval $[0°, 360°]$.

Powers of Complex Numbers

Raising a number to a positive integer power is the same as multiplying that number by itself repeated times.

$$x^3 = x \cdot x \cdot x \qquad (a + b)^2 = (a + b)(a + b)$$

Therefore, raising a complex number to a power that is a positive integer is the same as multiplying the complex number by itself multiple times. Let us illustrate this with the complex number, $z = r(\cos\theta + i\sin\theta)$, and raise it to positive integer powers (n).

WORDS	MATH
Take the case, $n = 2$.	$z^2 = [r(\cos\theta + i\sin\theta)][r(\cos\theta + i\sin\theta)]$
Apply the product rule:	
Multiply the magnitudes and add the arguments.	$z^2 = r^2[\cos 2\theta + i\sin 2\theta]$
Take the case, $n = 3$,	$z^3 = z^2 z = [r^2(\cos 2\theta + i\sin 2\theta)][r(\cos\theta + i\sin\theta)]$
Apply the product rule:	
Multiply the magnitudes and add the arguments.	$z^3 = r^3[\cos 3\theta + i\sin 3\theta]$
Take the case, $n = 4$.	$z^4 = z^3 z = [r^3(\cos 3\theta + i\sin 3\theta)][r(\cos\theta + i\sin\theta)]$
Apply the product rule:	
Multiply the magnitudes and add the arguments.	$z^4 = r^4[\cos 4\theta + i\sin 4\theta]$
The pattern observed is	$z^n = r^n[\cos n\theta + i\sin n\theta]$

Although we will not prove this generalized representation of a complex number raised to a power, it was proved by Abraham De Moivre and hence its name.

DE MOIVRE'S THEOREM

If $z = r(\cos\theta + i\sin\theta)$ is a complex number, then

$$z^n = r^n[\cos n\theta + i\sin n\theta]$$

when n is a positive integer ($n \ge 1$).
In other words, *when raising a complex number to a power (n), raise the magnitude to the same power (n) and multiply the argument by n.*

■ **Answer:** $\dfrac{z_1}{z_2} = 2[\cos 210° + i\sin 210°]$ or $\dfrac{z_1}{z_2} = -\sqrt{3} - i$

Although De Moivre's theorem was proved for all real numbers, n, we will only use it for positive integer values of n and their reciprocals (nth roots). This is a very powerful theorem. For example, if asked to find $(\sqrt{3} - i)^{10}$, you have two choices: (1) multiply out the expression algebraically, which we will call the "long way" or (2) convert to polar coordinates and use De Moivre's theorem, which we will call the "short way." We will use DeMoivre's theorem.

 EXAMPLE 3 Finding a Power of a Complex Number

Find $(\sqrt{3} + i)^{10}$ and express the answer in rectangular form.

Solution:

STEP 1 Convert to polar form.

$$(\sqrt{3} + i)^{10} = [2(\cos 30° + i\sin 30°)]^{10}$$

STEP 2 Use De Moivre's theorem with $n = 10$.

$$(\sqrt{3} + i)^{10} = [2(\cos 30° + i\sin 30°)]^{10}$$
$$= 2^{10}[\cos 10 \cdot 30° + i\sin 10 \cdot 30°]$$

STEP 3 Simplify.

$$(\sqrt{3} + i)^{10} = 2^{10}[\cos 300° + i\sin 300°]$$

Evaluate 2^{10}, sine, and cosine.

$$= 1024\left[\frac{1}{2} - i\frac{\sqrt{3}}{2}\right]$$

Simplify.

$$= 512 - 512i\sqrt{3}$$

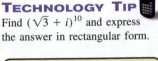
■ YOUR TURN Find $(1 + i\sqrt{3})^{10}$ and express the answer in rectangular form.

Roots of Complex Numbers

De Moivre's theorem is the basis for the *n*th root theorem. Before we proceed, let us motivate it with a problem: solve $x^3 - 1 = 0$.

WORDS	MATH
Add 1 to both sides of the equation.	$x^3 = 1$
Raise both sides to the 1/3 power.	$(x^3)^{1/3} = 1^{1/3}$
Simplify.	$x = 1^{1/3}$

You may think the answer is $x = 1$. There are, however, two more complex solutions. Recall that a polynomial of degree n has n solutions (roots in the complex number system). So the polynomial $P(x) = x^3 - 1$ is degree three and has three solutions (roots). So, how do we find the other two solutions? We use the *n*th root theorem to find the additional cube roots of 1.

WORDS	MATH
Let z be a complex number.	$z = r(\cos\theta + i\sin\theta)$
For k = integer, θ and $\theta + 2k\pi$ are coterminal.	$z = r[\cos(\theta + 2k\pi) + i\sin(\theta + 2k\pi)]$
Apply De Moivre's theorem for $1/n$.	$z^{1/n} = r^{1/n}\left[\cos\left(\dfrac{\theta + 2k\pi}{n}\right) + i\sin\left(\dfrac{\theta + 2k\pi}{n}\right)\right]$
Simplify.	$z^{1/n} = r^{1/n}\left[\cos\left(\dfrac{\theta}{n} + \dfrac{2k\pi}{n}\right) + i\sin\left(\dfrac{\theta}{n} + \dfrac{2k\pi}{n}\right)\right]$

Notice that when $k = n$, the arguments $\dfrac{\theta}{n} + 2\pi$ and $\dfrac{\theta}{n}$ are coterminal. Therefore, to get distinct roots $k = 0, 1, \ldots, n - 1$.

If we let z be a given complex number and w be any complex number that satisfies the relationship $z^{1/n} = w$ or $z = w^n$ where $n \geq 2$, then we say that w is a **complex nth root** of z.

NTH ROOT THEOREM

The **nth roots** of the complex number $z = r(\cos\theta + i\sin\theta)$ are given by

$$w_k = r^{1/n}\left[\cos\left(\dfrac{\theta}{n} + \dfrac{2k\pi}{n}\right) + i\sin\left(\dfrac{\theta}{n} + \dfrac{2k\pi}{n}\right)\right] \quad \theta \text{ in radians}$$

or

$$w_k = r^{1/n}\left[\cos\left(\dfrac{\theta}{n} + \dfrac{k \cdot 360°}{n}\right) + i\sin\left(\dfrac{\theta}{n} + \dfrac{k \cdot 360°}{n}\right)\right] \quad \theta \text{ in degrees}$$

where $k = 0, 1, 2, \ldots, n - 1$.

TECHNOLOGY TIP

Find the three distinct roots of $-4 - 4i\sqrt{3}$.

Caution: If you use a TI calculator to find $(-4 - 4i\sqrt{3})^{1/3}$, the calculator will return only one root.

```
((-4-4i√(3))^(1/3
)
         1.53-1.29i
2(cos(320)+isin(
320))
         1.53-1.29i
```

To find all three distinct roots, you need to change to polar form and apply the nth root theorem.

EXAMPLE 4 Finding Roots of Complex Numbers

Find the three distinct cube roots of $-4 - 4i\sqrt{3}$ and plot the roots in the complex plane.

Solution:

STEP 1 Write $-4 - 4i\sqrt{3}$ in polar form. $\quad 8(\cos 240° + i\sin 240°)$

STEP 2 Find the three cube roots.

$$w_k = r^{1/n}\left[\cos\left(\dfrac{\theta}{n} + \dfrac{k \cdot 360°}{n}\right) + i\sin\left(\dfrac{\theta}{n} + \dfrac{k \cdot 360°}{n}\right)\right]$$

$\theta = 240°, r = 8, n = 3, k = 0, 1, 2$

For $k = 0$:

$$w_0 = 8^{1/3}\left[\cos\left(\dfrac{240°}{3} + \dfrac{0 \cdot 360°}{3}\right) + i\sin\left(\dfrac{240°}{3} + \dfrac{0 \cdot 360°}{3}\right)\right]$$

Simplify.

$$w_0 = 2[\cos(80°) + i\sin(80°)]$$

For $k = 1$:

$$w_1 = 8^{1/3}\left[\cos\left(\frac{240°}{3} + \frac{1 \cdot 360°}{3}\right) + i\sin\left(\frac{240°}{3} + \frac{1 \cdot 360°}{3}\right)\right]$$

Simplify.

$$w_1 = 2\left[\cos(200°) + i\sin(200°)\right]$$

For $k = 2$:

$$w_2 = 8^{1/3}\left[\cos\left(\frac{240°}{3} + \frac{2 \cdot 360°}{3}\right) + i\sin\left(\frac{240°}{3} + \frac{2 \cdot 360°}{3}\right)\right]$$

Simplify.

$$w_2 = 2\left[\cos(320°) + i\sin(320°)\right]$$

STEP 3 Plot the three cube roots in the complex plane.

Notice the following:

- The roots all have a magnitude of 2.

- The roots lie on a circle of radius 2.

- The roots are equally spaced around the circle (120° apart).

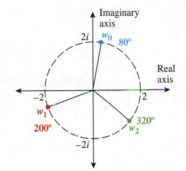

YOUR TURN Find the three distinct cube roots of $4 - 4i\sqrt{3}$ and plot the roots in the complex plane.

Solving Equations Using Roots of Complex Numbers

Let us return to solving the equation: $x^3 - 1 = 0$. As stated, $x = 1$ is the real solution to this cubic equation—however, there are two additional (complex) solutions. Since we are finding the zeros of a third-degree polynomial, we expect three solutions. Furthermore, when complex solutions arise in finding the roots of polynomials, they come in conjugate pairs.

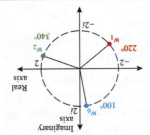

Answer: $w_0 = 2(\cos 100° + i\sin 100°)$
$w_1 = 2(\cos 220° + i\sin 220°)$
$w_2 = 2(\cos 340° + i\sin 340°)$

TECHNOLOGY TIP

The solution to the equation is $x = (1 + 0i)^{1/3}$.

```
abs(1)
                1.00
angle(1)
                0.00
```

```
(cos(0)+isin(0))
                1.00
(cos(0+360/3)+is
in(0+360/3))
            -.50+.87i
■
```

```
cos(0)+isin(0)
                1.00
cos(120)+isin(12
0)
            -.50+.87i
```

```
(cos(0+2*360/3)+
isin(0+2*360/3))

            -.50-.87i
■
```

```
cos(240)+isin(24
0)
            -.50-.87i
```

EXAMPLE 5 Solving Equations Using Complex Roots

Find all complex solutions to $x^3 - 1 = 0$.

Solution: $x^3 = 1$

STEP 1 Write 1 in polar form. $\qquad\qquad$ $1 = 1 + 0i = \cos 0° + i \sin 0°$

STEP 2 Find the three cube roots of 1.

$$w_k = r^{1/n}\left[\cos\left(\frac{\theta}{n} + \frac{k \cdot 360°}{n}\right) + i \sin\left(\frac{\theta}{n} + \frac{k \cdot 360°}{n}\right)\right]$$

$$r = 1, \theta = 0°, n = 3, k = 0, 1, 2$$

For $k = 0$: $\quad w_0 = 1^{1/3}\left[\cos\left(\frac{0°}{3} + \frac{0 \cdot 360°}{3}\right) + i \sin\left(\frac{0°}{3} + \frac{0 \cdot 360°}{3}\right)\right]$

Simplify. $\quad w_0 = \cos 0° + i \sin 0°$

For $k = 1$: $\quad w_1 = 1^{1/3}\left[\cos\left(\frac{0°}{3} + \frac{1 \cdot 360°}{3}\right) + i \sin\left(\frac{0°}{3} + \frac{1 \cdot 360°}{3}\right)\right]$

Simplify. $\quad w_1 = \cos 120° + i \sin 120°$

For $k = 2$: $\quad w_2 = 1^{1/3}\left[\cos\left(\frac{0°}{3} + \frac{2 \cdot 360°}{3}\right) + i \sin\left(\frac{0°}{3} + \frac{2 \cdot 360°}{3}\right)\right]$

Simplify. $\quad w_2 = \cos 240° + i \sin 240°$

STEP 3 Write the roots in rectangular form.

For w_0: $\qquad w_0 = \underbrace{\cos 0°}_{1} + i \underbrace{\sin 0°}_{0} = 1$

For w_1: $\qquad w_1 = \underbrace{\cos 120°}_{-\frac{1}{2}} + i \underbrace{\sin 120°}_{\frac{\sqrt{3}}{2}} = -\frac{1}{2} + i\frac{\sqrt{3}}{2}$

For w_2: $\qquad w_2 = \underbrace{\cos 240°}_{-\frac{1}{2}} + i \underbrace{\sin 240°}_{-\frac{\sqrt{3}}{2}} = -\frac{1}{2} - i\frac{\sqrt{3}}{2}$

STEP 4 Write the solutions to the equation, $x^3 - 1 = 0$.

$$\boxed{x = 1} \qquad \boxed{x = -\frac{1}{2} + i\frac{\sqrt{3}}{2}} \qquad \boxed{x = -\frac{1}{2} - i\frac{\sqrt{3}}{2}}$$

Notice that there is one real solution and two (non-real) complex solutions and that the two complex solutions are complex conjugates.

It is always a good idea to check that the solutions indeed satisfy the equation. The equation $x^3 - 1 = 0$ can also be written as $x^3 = 1$, so the check in this case is to cube the three solutions and confirm that the result is 1.

$x = 1: 1^3 = 1\checkmark$

$$x = -\frac{1}{2} + i\frac{\sqrt{3}}{2}: \left(-\frac{1}{2} + i\frac{\sqrt{3}}{2}\right)^3 = \underbrace{\left(-\frac{1}{2} + i\frac{\sqrt{3}}{2}\right)^2}_{\frac{1}{4} - i\frac{\sqrt{3}}{2} - \frac{3}{4}}\left(-\frac{1}{2} + i\frac{\sqrt{3}}{2}\right)$$

$$= \left(-\frac{1}{2} - i\frac{\sqrt{3}}{2}\right)\left(-\frac{1}{2} + i\frac{\sqrt{3}}{2}\right)$$

$$= \frac{1}{4} + \frac{3}{4}$$

$$= 1\checkmark$$

$$x = -\frac{1}{2} - i\frac{\sqrt{3}}{2}: \left(-\frac{1}{2} - i\frac{\sqrt{3}}{2}\right)^3 = \underbrace{\left(-\frac{1}{2} - i\frac{\sqrt{3}}{2}\right)^2}_{\frac{1}{4} + i\frac{\sqrt{3}}{2} - \frac{3}{4}}\left(-\frac{1}{2} - i\frac{\sqrt{3}}{2}\right)$$

$$= \left(-\frac{1}{2} + i\frac{\sqrt{3}}{2}\right)\left(-\frac{1}{2} - i\frac{\sqrt{3}}{2}\right)$$

$$= \frac{1}{4} + \frac{3}{4}$$

$$= 1\checkmark$$

SECTION 8.3 SUMMARY

In this section we multiplied and divided complex numbers and, using De Moivre's theorem, raised complex numbers to integer powers and found the nth roots of complex numbers, as follows.

Let $z_1 = r_1(\cos\theta_1 + i\sin\theta_1)$ and $z_2 = r_2(\cos\theta_2 + i\sin\theta_2)$ be two complex numbers.

The **product**, $z_1 z_2$, is given by: $z_1 z_2 = r_1 r_2[\cos(\theta_1 + \theta_2) + i\sin(\theta_1 + \theta_2)]$

The **quotient**, $\dfrac{z_1}{z_2}$, is given by: $\dfrac{z_1}{z_2} = \dfrac{r_1}{r_2}[\cos(\theta_1 - \theta_2) + i\sin(\theta_1 - \theta_2)]$

Let $z = r(\cos\theta + i\sin\theta)$ be a complex number. Then for a positive integer, n:

z raised to a **power**, n, is given by: $z^n = r^n[\cos n\theta + i\sin n\theta]$

The **nth roots** of z are given by: $w_k = r^{1/n}\left[\cos\left(\dfrac{\theta}{n} + \dfrac{k \cdot 360°}{n}\right) + i\sin\left(\dfrac{\theta}{n} + \dfrac{k \cdot 360°}{n}\right)\right]$

where $k = 0, 1, 2, \ldots, n - 1$.

SECTION 8.3 EXERCISES

 **■ SKILLS**

In Exercises 1–10, find the product, $z_1 z_2$, and express it in rectangular form.

1. $z_1 = 4[\cos 40° + i \sin 40°]$ and $z_2 = 3[\cos 80° + i \sin 80°]$

2. $z_1 = 2[\cos 100° + i \sin 100°]$ and $z_2 = 5[\cos 50° + i \sin 50°]$

3. $z_1 = 4[\cos 80° + i \sin 80°]$ and $z_2 = 2[\cos 145° + i \sin 145°]$

4. $z_1 = 3[\cos 130° + i \sin 130°]$ and $z_2 = 4[\cos 170° + i \sin 170°]$

5. $z_1 = 2[\cos 10° + i \sin 10°]$ and $z_2 = 4[\cos 80° + i \sin 80°]$

6. $z_1 = 3[\cos 190° + i \sin 190°]$ and $z_2 = 5[\cos 80° + i \sin 80°]$

7. $z_1 = \sqrt{3}\left[\cos \dfrac{\pi}{12} + i \sin \dfrac{\pi}{12}\right]$ and $z_2 = \sqrt{27}\left[\cos \dfrac{\pi}{6} + i \sin \dfrac{\pi}{6}\right]$

8. $z_1 = \sqrt{5}\left[\cos \dfrac{\pi}{15} + i \sin \dfrac{\pi}{15}\right]$ and $z_2 = \sqrt{5}\left[\cos \dfrac{4\pi}{15} + i \sin \dfrac{4\pi}{15}\right]$

9. $z_1 = 4\left[\cos \dfrac{3\pi}{8} + i \sin \dfrac{3\pi}{8}\right]$ and $z_2 = 3\left[\cos \dfrac{\pi}{8} + i \sin \dfrac{\pi}{8}\right]$

10. $z_1 = 6\left[\cos \dfrac{2\pi}{9} + i \sin \dfrac{2\pi}{9}\right]$ and $z_2 = 5\left[\cos \dfrac{\pi}{9} + i \sin \dfrac{\pi}{9}\right]$

In Exercises 11–20, find the quotient, $\dfrac{z_1}{z_2}$, and express it in rectangular form.

11. $z_1 = 6[\cos 100° + i \sin 100°]$ and $z_2 = 2[\cos 40° + i \sin 40°]$

12. $z_1 = 8[\cos 80° + i \sin 80°]$ and $z_2 = 2[\cos 35° + i \sin 35°]$

13. $z_1 = 10[\cos 200° + i \sin 200°]$ and $z_2 = 5[\cos 65° + i \sin 65°]$

14. $z_1 = 4[\cos 280° + i \sin 280°]$ and $z_2 = 4[\cos 55° + i \sin 55°]$

15. $z_1 = \sqrt{12}[\cos 350° + i \sin 350°]$ and $z_2 = \sqrt{3}[\cos 80° + i \sin 80°]$

16. $z_1 = \sqrt{40}[\cos 110° + i \sin 110°]$ and $z_2 = \sqrt{10}[\cos 20° + i \sin 20°]$

17. $z_1 = 9\left[\cos \dfrac{5\pi}{12} + i \sin \dfrac{5\pi}{12}\right]$ and $z_2 = 3\left[\cos \dfrac{\pi}{12} + i \sin \dfrac{\pi}{12}\right]$

18. $z_1 = 8\left[\cos \dfrac{5\pi}{8} + i \sin \dfrac{5\pi}{8}\right]$ and $z_2 = 4\left[\cos \dfrac{3\pi}{8} + i \sin \dfrac{3\pi}{8}\right]$

19. $z_1 = 45\left[\cos \dfrac{22\pi}{15} + i \sin \dfrac{22\pi}{15}\right]$ and $z_2 = 9\left[\cos \dfrac{2\pi}{15} + i \sin \dfrac{2\pi}{15}\right]$

20. $z_1 = 22\left[\cos \dfrac{11\pi}{18} + i \sin \dfrac{11\pi}{18}\right]$ and $z_2 = 11\left[\cos \dfrac{5\pi}{18} + i \sin \dfrac{5\pi}{18}\right]$

In Exercises 21–30, find the result of each expression using De Moivre's theorem. Write the answer in rectangular form.

21. $(-1 + i)^5$ **22.** $(1 - i)^4$ **23.** $(-\sqrt{3} + i)^6$ **24.** $(\sqrt{3} - i)^8$ **25.** $(1 - \sqrt{3}i)^4$

26. $(-1 + \sqrt{3}i)^5$ **27.** $(4 - 4i)^8$ **28.** $(-3 + 3i)^{10}$ **29.** $(4\sqrt{3} + 4i)^7$ **30.** $(-5 + 5\sqrt{3}i)^7$

In Exercises 31–40, find all nth roots of z. Write the answers in polar form, and plot the roots in the complex plane.

31. $2 - 2i\sqrt{3}$, $n = 2$ **32.** $2 + 2\sqrt{3}i$, $n = 2$ **33.** $\sqrt{18} - \sqrt{18}i$, $n = 2$ **34.** $-\sqrt{2} + \sqrt{2}i$, $n = 2$

35. $4 + 4\sqrt{3}i$, $n = 3$ **36.** $-\dfrac{27}{2} + \dfrac{27\sqrt{3}}{2}i$, $n = 3$ **37.** $\sqrt{3} - i$, $n = 3$ **38.** $4\sqrt{2} + 4\sqrt{2}i$, $n = 3$

39. $8\sqrt{2} - 8\sqrt{2}i$, $n = 4$ **40.** $-\sqrt{128} + \sqrt{128}i$, $n = 4$

In Exercises 41–50, find all complex solutions to the given equations.

41. $x^4 - 16 = 0$ **42.** $x^3 - 8 = 0$ **43.** $x^3 + 8 = 0$ **44.** $x^3 + 1 = 0$

45. $x^4 + 16 = 0$ **46.** $x^6 + 1 = 0$ **47.** $x^6 - 1 = 0$ **48.** $4x^2 + 1 = 0$

49. $x^2 + i = 0$ **50.** $x^2 - i = 0$

■ APPLICATIONS

51. Complex Pentagon. When you graph the five fifth roots of $-\dfrac{\sqrt{2}}{2} - \dfrac{\sqrt{2}}{2}i$ and connect the points, you form a pentagon. Find the roots and draw the pentagon.

52. Complex Square. When you graph the four fourth roots of $16i$ and connect the points, you form a square. Find the roots and draw the square.

■ CATCH THE MISTAKE

In Exercises 53–56, explain the mistake that is made.

53. Let $z_1 = 6[\cos 65° + i \sin 65°]$ and
$z_2 = 3[\cos 125° + i \sin 125°]$, and find $\dfrac{z_1}{z_2}$.

Solution:

Use the quotient formula.

$$\frac{z_1}{z_2} = \frac{r_1}{r_2}[\cos(\theta_1 - \theta_2) + i\sin(\theta_1 - \theta_2)]$$

Substitute values.

$$\frac{z_1}{z_2} = \frac{6}{3}[\cos(125° - 65°) + i\sin(125° - 65°)]$$

Simplify. $\dfrac{z_1}{z_2} = 2[\cos(60°) + i\sin(60°)]$

Evaluate the trigonometric functions.

$$\frac{z_1}{z_2} = 2\left[\frac{1}{2} + i\frac{\sqrt{3}}{2}\right] = 1 + i\sqrt{3}$$

This is incorrect. What mistake was made?

54. Let $z_1 = 6[\cos 65° + i \sin 65°]$ and
$z_2 = 3[\cos 125° + i \sin 125°]$, find $z_1 z_2$.

Solution:

Write the product.

$$z_1 z_2 = 6[\cos 65° + i \sin 65°]3[\cos 125° + i \sin 125°]$$

Multiply the magnitudes.

$$z_1 z_2 = 18[\cos 65° + i \sin 65°][\cos 125° + i \sin 125°]$$

Multiply cosine terms and sine terms (add arguments).

$$z_1 z_2 = 18[\cos(65° + 125°) + i^2 \sin(65° + 125°)]$$

Simplify $(i^2 = -1)$.

$$z_1 z_2 = 18[\cos(190°) - \sin(190°)]$$

This is incorrect. What mistake was made?

55. Find $(\sqrt{2} + i\sqrt{2})^6$.

Solution:

Raise each term to the 6th power.　　　$(\sqrt{2})^6 + i^6(\sqrt{2})^6$

Simplify.　　　$8 + 8i^6$

Let $i^6 = i^4 \cdot i^2 = -1$.　　　$8 - 8 = 0$

This is incorrect. What mistake was made?

56. Find all complex solutions to $x^5 - 1 = 0$.

Solution:

Add 1 to both sides.　　　$x^5 = 1$

Raise both sides to the 1/5 power.　　　$x = 1^{1/5}$

Simplify.　　　$x = 1$

This is incorrect. What mistake was made?

■ CHALLENGE

57. T or F: The product of two complex numbers is a complex number.

58. T or F: The quotient of two complex numbers is a complex number.

Use the following identity in Exercises 59–62.

There is an identity you will see in calculus called Euler's formula or identity, $e^{i\theta} = \cos\theta + i\sin\theta$. Notice that when $\theta = \pi$, the identity reduces to $e^{i\pi} + 1 = 0$, which is a beautiful identity in that it relates the fundamental numbers (e, π, 1, and 0) and fundamental operations (multiplication, addition, exponents, and equality) in mathematics.

59. Let $z_1 = r_1(\cos\theta_1 + i\sin\theta_1) = r_1 e^{i\theta_1}$ and $z_2 = r_2(\cos\theta_2 + i\sin\theta_2) = r_2 e^{i\theta_2}$ be two complex numbers, and use properties of exponentials to show that $z_1 z_2 = r_1 r_2 [\cos(\theta_1 + \theta_2) + i\sin(\theta_1 + \theta_2)]$.

60. Let $z_1 = r_1(\cos\theta_1 + i\sin\theta_1) = r_1 e^{i\theta_1}$ and $z_2 = r_2(\cos\theta_2 + i\sin\theta_2) = r_2 e^{i\theta_2}$ be two complex numbers, and use properties of exponentials to show that $\dfrac{z_1}{z_2} = \dfrac{r_1}{r_2}[\cos(\theta_1 - \theta_2) + i\sin(\theta_1 - \theta_2)]$.

61. Let $z = r(\cos\theta + i\sin\theta) = re^{i\theta}$, and use properties of exponents to show that $z^n = r^n[\cos n\theta + i\sin n\theta]$.

62. Let $z = r(\cos\theta + i\sin\theta) = re^{i\theta}$, and use properties of exponents to show that $w_k = r^{1/n}\left[\cos\left(\dfrac{\theta}{n} + \dfrac{2k\pi}{n}\right) + i\sin\left(\dfrac{\theta}{n} + \dfrac{2k\pi}{n}\right)\right]$.

■ TECHNOLOGY

63. Find the fifth roots of $\dfrac{\sqrt{3}}{2} - \dfrac{1}{2}i$ and use the TI to plot the roots.

64. Find the fourth roots of $-\dfrac{\sqrt{2}}{2} + \dfrac{\sqrt{2}}{2}i$ and use the TI to plot the roots.

65. Complex hexagon. Find the sixth roots of $-\dfrac{1}{2} - \dfrac{\sqrt{3}}{2}i$ and use the graphing calculator to draw the hexagon.

66. Complex pentagon. Find the fifth roots of $-4 + 4i$ and use the graphing calculator to draw the pentagon.

SECTION 8.4 Polar Equations and Graphs

Skills Objectives

■ Plot points in the polar coordinate system.
■ Convert between rectangular and polar coordinates.
■ Convert equations between polar form and rectangular form.
■ Graph polar equations.

Conceptual Objectives

■ Relate the rectangular coordinate system to the polar coordinate system.
■ Classify common shapes that arise from plotting some types of polar equations.

We have discussed the rectangular and the trigonometric (polar) form of complex numbers in the complex plane. We now turn our attention back to the familiar Cartesian plane, where the horizontal axis represents the x-variable and the vertical axis represents the *y*-variable and points in this plane represent pairs of real numbers. It is often convenient to instead represent real-number plots in the *polar coordinate system.*

Polar Coordinates

The **polar coordinate system** is anchored by a point, called the **pole** (taken to be the **origin**), and a ray with a vertex at the pole, called the **polar axis**. The polar axis is normally shown where we expect to find the positive *x*-axis in Cartesian coordinates.

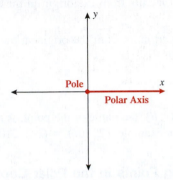

If you align the pole with the origin on the rectangular graph and the polar axis with the positive *x*-axis, you can label a point either with rectangular coordinates, (x, y), or with an ordered pair, (r, θ) in **polar coordinates**.

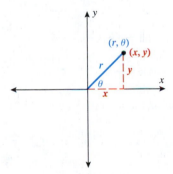

Typically, polar graph paper is used that gives the angles and radii. The graph shown gives the angles in radians (the angle also can be given in degrees) and the radii go from zero through 5.

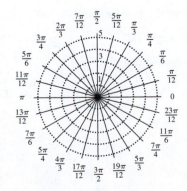

When plotting points in the polar coordinate system, $|r|$ represents the distance from the origin to the point. The following procedure guides us in plotting points in the polar coordinate system.

POINT-PLOTTING POLAR COORDINATES

To plot a point, (r, θ):

1. Start on the polar axis and rotate to the angle θ.

2. If $r > 0$, the point is r units from the origin in the *same direction* of the terminal side of θ.

3. If $r < 0$, the point is $|r|$ units from the origin in the *opposite direction* of the terminal side of θ.

It is important to note that (r, θ), the name of the point, is not unique, whereas in rectangular form (x, y) it is. For example: $(2, 30°) = (-2, 210°)$.

EXAMPLE 1 Plotting Points in the Polar Coordinate System

Plot the points (a) $\left(3, \dfrac{3\pi}{4} \right)$ and (b) $(-2, 60°)$ in the polar coordinate system.

Solution (a):

Start with a pencil along the polar axis.

Rotate the pencil at an angle $\dfrac{3\pi}{4}$.

Go out (in the direction of the pencil) 3 units.

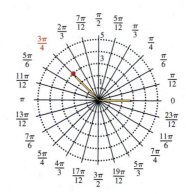

Solution (b):

Start with a pencil along the polar axis.

Rotate the pencil at an angle 60°.

Go out (opposite the direction of the pencil) 2 units.

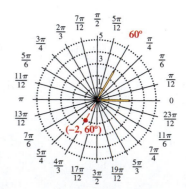

Plot the following points in the polar coordinate system.

a. $\left(-4, \dfrac{3\pi}{2}\right)$ **b.** $(3, 330°)$

The relationships between polar and rectangular coordinates are the familiar relationships:

$$\sin\theta = \frac{y}{r} \qquad \cos\theta = \frac{x}{r} \qquad \tan\theta = \frac{y}{x} \quad (x \neq 0)$$

$$r^2 = x^2 + y^2$$

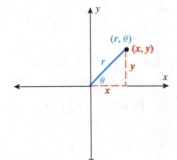

CONVERTING BETWEEN POLAR AND RECTANGULAR COORDINATES

From	To	Identities	
Polar (r, θ)	Rectangular (x, y)	$x = r\cos\theta$	$y = r\sin\theta$
Rectangular (x, y)	Polar (r, θ)	$r = \sqrt{x^2 + y^2}$	$\tan\theta = \dfrac{y}{x},\ x \neq 0$

EXAMPLE 2 Converting Between Polar and Rectangular Coordinates

a. Convert $(-1, \sqrt{3})$ to polar coordinates.
b. Convert $(6\sqrt{2}, 135°)$ to rectangular coordinates.

Solution (a): $(-1, \sqrt{3})$ lies in QII.

Identify x and y. $x = -1 \qquad y = \sqrt{3}$

Find r. $r = \sqrt{x^2 + y^2} = \sqrt{(-1)^2 + (\sqrt{3})^2} = \sqrt{4} = 2$

TECHNOLOGY TIP

```
abs(-1+√(3)i)
                2.00
angle(-1+√(3)i)
              120.00
Ans/180▶Frac
                 2/3
■
```

The solution is $\left(2, \dfrac{2}{3}\pi\right)$.

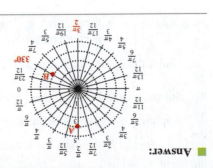

Answer:

Find θ.	$\tan\theta = \dfrac{\sqrt{3}}{-1}$	θ lies in QII
Identify θ from the unit circle.	$\theta = \dfrac{2\pi}{3}$	
Write the point in polar coordinates.	$\left(2, \dfrac{2\pi}{3}\right)$	

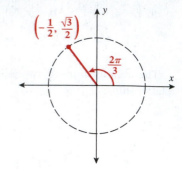

Solution (b): $(6\sqrt{2}, 135°)$ lies in QII.

Identify r and θ.	$r = 6\sqrt{2} \qquad \theta = 135°$
Find x.	$x = r\cos\theta = 6\sqrt{2}\cos 135° = 6\sqrt{2}\left(-\dfrac{\sqrt{2}}{2}\right) = -6$
Find y.	$y = r\sin\theta = 6\sqrt{2}\sin 135° = 6\sqrt{2}\left(\dfrac{\sqrt{2}}{2}\right) = 6$
Write the point in rectangular coordinates.	$(-6, 6)$

Graphs of Polar Equations

We are familiar with equations in rectangular form such as

$$y = 3x + 5 \qquad y = x^2 + 2 \qquad x^2 + y^2 = 9$$
$$\text{(line)} \qquad\quad \text{(parabola)} \qquad \text{(circle)}$$

We now discuss equations in polar form (known as **polar equations**) such as

$$r = 5\theta \qquad r = 2\cos\theta \qquad r = \sin 5\theta$$

which you will learn to recognize in this section as typical equations whose plots are some general shapes.

Our first example deals with two of the simplest forms of polar equations: when r or θ is constant. The results are a circle centered at the origin and a line that passes through the origin, respectively.

EXAMPLE 3 Graphing a Polar Equation (r = constant or θ = constant)

Graph the polar equations **(a)** $r = 3$ and **(b)** $\theta = \dfrac{\pi}{4}$.

Solution (a): Circle centered at the pole (origin)

Approach 1 (polar coordinates): $\qquad\qquad\qquad r = 3$ (θ can take on any value).

Plot points for arbitrary θ and $r = 3$.

Connect the points; they make a circle with radius 3.

Approach 2 (rectangular coordinates):

$$r = 3$$

Square both sides.

$$r^2 = 9$$

Let $r^2 = x^2 + y^2$.

$$x^2 + y^2 = 3^2$$

This is a circle, centered at the origin, with radius 3.

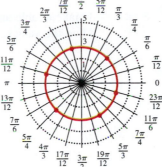

Solution (b): Line passing through the pole (origin)

Approach 1: $\theta = \dfrac{\pi}{4}$ (r can take on any value, positive or negative).

Plot points for arbitrary r and $\theta = \dfrac{\pi}{4}$.

Connect the points; the line pass through the origin with slope $= 1$.

Approach 2:

$$\theta = \frac{\pi}{4}$$

Take the tangent of both sides.

$$\tan \theta = \underbrace{\tan \frac{\pi}{4}}_{1}$$

Use the identity $\tan \theta = \dfrac{y}{x}$.

$$\frac{y}{x} = 1$$

Multiply by x.

$$y = x$$

The result is a line passing through the origin with slope $= 1$.

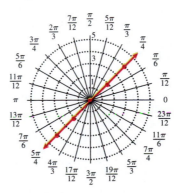

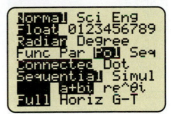
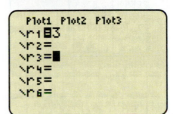
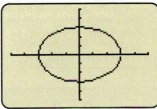
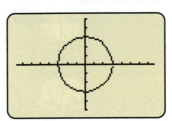
Rectangular equations that depend on variables (not constant) can be graphed by point-plotting (making a table and plotting the points). We will use this same procedure for graphing polar equations that are not constant.

TECHNOLOGY TIP

Graph $r = 4\cos\theta$.

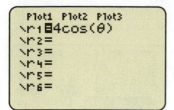

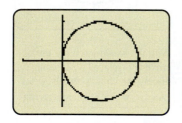

EXAMPLE 4 Graphing a Polar Equation (Circle Not Centered at the Origin)

Graph $r = 4\cos\theta$.

Solution:

STEP 1 Make a table and find the values.

θ	$r = 4\cos\theta$	(r, θ)
0	$4(1) = 4$	$(4, 0)$
$\dfrac{\pi}{4}$	$4\left(\dfrac{\sqrt{2}}{2}\right) \approx 2.8$	$\left(2.8, \dfrac{\pi}{4}\right)$
$\dfrac{\pi}{2}$	$4(0) = 0$	$\left(0, \dfrac{\pi}{2}\right)$
$\dfrac{3\pi}{4}$	$4\left(-\dfrac{\sqrt{2}}{2}\right) \approx -2.8$	$\left(-2.8, \dfrac{3\pi}{4}\right)$
π	$4(-1) = -4$	$(-4, \pi)$

STEP 2 Label the polar coordinates.

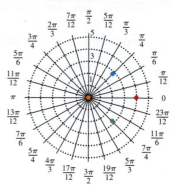

STEP 3 Connect the points with a smooth curve.

Notice that $(4, 0)$ and $(-4, \pi)$ correspond to the same point. There is no need to continue with angles beyond π, as the result would be to go around the same circle again.

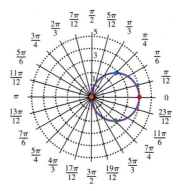

■ **YOUR TURN** Graph $r = 4\sin\theta$.

■ **Answer:**

Compare the result of Example 4, the graph of $r = 4\cos\theta$, with the result of Your Turn, the graph of $r = 4\sin\theta$. Notice that they are 90 degrees out of phase (rotate one graph 90° about the pole to get the other graph).

Graphs of polar equations of the form $r = a\sin\theta$ and $r = a\cos\theta$ are circles.

WORDS	MATH	
Polar Equation	$r = a\sin\theta$	$r = a\cos\theta$
Use trigonometric ratios: $\sin\theta = \dfrac{y}{r}$ and $\cos\theta = \dfrac{x}{r}$.	$r = a\dfrac{y}{r}$	$r = a\dfrac{x}{r}$
Multiply equations by r.	$r^2 = ay$	$r^2 = ax$
Let $r^2 = x^2 + y^2$.	$x^2 + y^2 = ay$	$x^2 + y^2 = ax$
Group x terms together and y terms together.	$x^2 + (y^2 - ay) = 0$	$(x^2 - ax) + y^2 = 0$
Complete the square on the expressions in parentheses.	$x^2 + \left(y^2 - ay + \left(\dfrac{a}{2}\right)^2\right) = \left(\dfrac{a}{2}\right)^2$	$\left(x^2 - ax + \left(\dfrac{a}{2}\right)^2\right) + y^2 = \left(\dfrac{a}{2}\right)^2$
	$x^2 + \left(y - \dfrac{a}{2}\right)^2 = \left(\dfrac{a}{2}\right)^2$	$\left(x - \dfrac{a}{2}\right)^2 + y^2 = \left(\dfrac{a}{2}\right)^2$
The result is a graph of a circle.	Center: $\left(0, \dfrac{a}{2}\right)$ Radius: $\dfrac{a}{2}$	Center: $\left(\dfrac{a}{2}, 0\right)$ Radius: $\dfrac{a}{2}$

EXAMPLE 5 Graphing a Polar Equation (Rose)

Graph $r = 5\sin 2\theta$.

Solution:

STEP 1 Make a table and find the values. Since the argument of sine is doubled, the period is halved. Therefore, instead of steps of $\pi/4$, take steps of $\pi/8$.

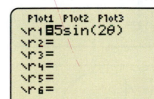

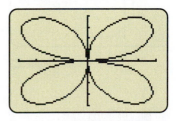

θ	$r = 5\sin 2\theta$	(r, θ)
0	$5(0) = 0$	$(0, 0)$
$\dfrac{\pi}{8}$	$5\left(\dfrac{\sqrt{2}}{2}\right) = 3.5$	$\left(3.5, \dfrac{\pi}{8}\right)$
$\dfrac{\pi}{4}$	$5(1) = 5$	$\left(5, \dfrac{\pi}{4}\right)$
$\dfrac{3\pi}{8}$	$5\left(\dfrac{\sqrt{2}}{2}\right) = 3.5$	$\left(3.5, \dfrac{3\pi}{8}\right)$
$\dfrac{\pi}{2}$	$5(0) = 0$	$(0, \pi)$

STEP 2 Label the polar coordinates.

These values in the table represent what happens in QI. The same pattern repeats in the other three quadrants. The result is a **four-leaved rose**.

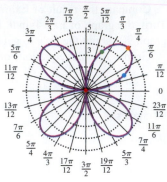

STEP 3 Connect the points with smooth curves.

YOUR TURN Graph $r = 5\cos 2\theta$.

Compare the result of Example 5, the graph of $r = 5\sin 2\theta$, with the result of Your Turn, the graph of $r = 5\cos 2\theta$. Notice that they are 90 degrees out of phase (rotate one graph 90° about the pole to get the other graph).

In general, for $r = a\sin n\theta$ or $r = a\cos n\theta$, the graph has n leaves if n is odd and $2n$ leaves if n is even.

The next class of graphs are called **limaçons**, which have equations of the form $r = a \pm b\cos\theta$ or $r = a \pm b\sin\theta$. When $a = b$, the result is a **cardioid** (heart shape).

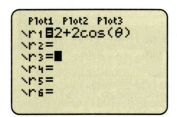
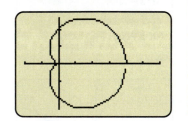
EXAMPLE 6 Graphing a Polar Equation (Cardioid)

Graph $r = 2 + 2\cos\theta$.

Solution:

STEP 1 Make a table and find the values.

This behavior repeats in QIII and QIV because cosine has corresponding values in QI and QIV and in QII and QIII.

θ	$r = 2 + 2\cos\theta$	(r, θ)
0	$2 + 2(1) = 4$	$(4, 0)$
$\dfrac{\pi}{4}$	$2 + 2\left(\dfrac{\sqrt{2}}{2}\right) = 3.4$	$\left(3.4, \dfrac{\pi}{4}\right)$
$\dfrac{\pi}{2}$	$2 + 2(0) = 2$	$\left(2, \dfrac{\pi}{2}\right)$
$\dfrac{3\pi}{4}$	$2 + 2\left(-\dfrac{\sqrt{2}}{2}\right) \approx 0.6$	$\left(0.6, \dfrac{3\pi}{4}\right)$
π	$2 + 2(-1) = 0$	$(0, \pi)$

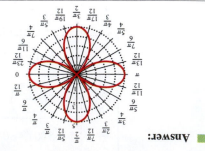

■ **Answer:**

STEP 2 Label the polar coordinates.

STEP 3 Connect the points with a smooth curve.

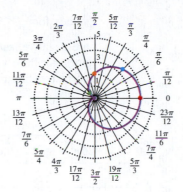

EXAMPLE 7 Graphing a Polar Equation (Spiral)

Graph $r = 0.5\theta$.

Solution:

STEP 1 Make a table and find the values.

θ	$r = 0.5\theta$	(r, θ)
0	$0.5(0) = 0$	$(0, 0)$
$\dfrac{\pi}{2}$	$0.5\left(\dfrac{\pi}{2}\right) = 0.8$	$\left(0.8, \dfrac{\pi}{2}\right)$
π	$0.5(\pi) = 1.6$	$(1.6, \pi)$
$\dfrac{3\pi}{2}$	$0.5\left(\dfrac{3\pi}{2}\right) = 2.4$	$\left(2.4, \dfrac{3\pi}{2}\right)$
2π	$0.5(2\pi) = 3.1$	$(3.1, 2\pi)$

TECHNOLOGY TIP

Graph $r = 0.5\theta$.

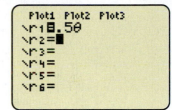

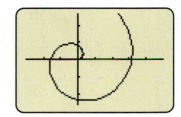

STEP 2 Label the polar coordinates.

STEP 3 Connect the points with a smooth curve.

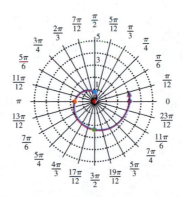

EXAMPLE 8 Graphing a Polar Equation (Lemniscate)

Graph $r^2 = 4\cos 2\theta$.

Solution:

STEP 1 Make a table and find the values.

Solving for r yields $r = \pm 2\sqrt{\cos 2\theta}$. All coordinates $(-r, \theta)$ can be expressed as $(r, \theta + \pi)$. The following table does not have values for $\dfrac{\pi}{4} < \theta < \dfrac{3\pi}{4}$ because the corresponding values of $\cos 2\theta$ are negative, and hence r is an imaginary number. The table also does not have values for $\theta > \pi$ because $2\theta > 2\pi$, and the corresponding points are repeated.

θ	$\cos 2\theta$	$r = \pm 2\sqrt{\cos 2\theta}$	(r, θ)
0	1	$r = \pm 2$	$(2, 0)$ and $(-2, 0) = (2, \pi)$
$\dfrac{\pi}{6}$	0.5	$r = \pm 1.4$	$\left(1.4, \dfrac{\pi}{6}\right)$ and $\left(-1.4, \dfrac{\pi}{6}\right) = \left(1.4, \dfrac{7\pi}{6}\right)$
$\dfrac{\pi}{4}$	0	$r = 0$	$\left(0, \dfrac{\pi}{4}\right)$
$\dfrac{3\pi}{4}$	0	$r = 0$	$\left(0, \dfrac{3\pi}{4}\right)$
$\dfrac{5\pi}{6}$	0.5	$r = \pm 1.4$	$\left(1.4, \dfrac{5\pi}{6}\right)$ and $\left(-1.4, \dfrac{5\pi}{6}\right) = \left(1.4, \dfrac{11\pi}{6}\right)$
π	1	$r = \pm 2$	$(2, \pi)$ and $(-2, \pi) = (2, 2\pi)$

Step 2 Label the polar coordinates.

Step 3 Connect with smooth curve.

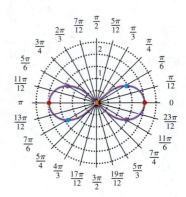

Converting Equations Between Polar and Rectangular Form

It is not always advantageous to plot an equation in the form in which it is given. It is sometimes easier to first convert to rectangular form and then plot. For example, to plot $r = \dfrac{2}{\cos \theta + \sin \theta}$, we could make a table with values. However, as you will see in Example 9, it is much easier to convert to rectangular coordinates.

EXAMPLE 9 Converting an Equation from Polar Form to Rectangular Form

Graph $r = \dfrac{2}{\cos\theta + \sin\theta}$.

Solution:

Multiply the equation by $\cos\theta + \sin\theta$.

$$r(\cos\theta + \sin\theta) = 2$$

Eliminate parentheses.

$$r\cos\theta + r\sin\theta = 2$$

Convert to rectangular form.

$$\underbrace{r\cos\theta}_{x} + \underbrace{r\sin\theta}_{y} = 2$$

Simplify.

$$y = -x + 2$$

Graph the line.

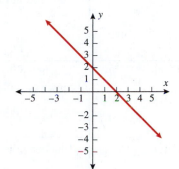

TECHNOLOGY TIP

Graph $r = \dfrac{2}{\cos\theta + \sin\theta}$.

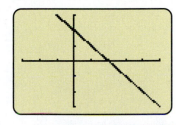

```
Plot1 Plot2 Plot3
\r1■2/(cos(θ)+si
n(θ))
\r2=
\r3=
\r4=
\r5=
\r6=
```

YOUR TURN Graph $r = \dfrac{2}{\cos\theta - \sin\theta}$.

SECTION 8.4 **SUMMARY**

Polar coordinates, (r, θ), are graphed in the polar coordinate system by first rotating the angle and then if r is positive, going out r units from the origin in the direction of the angle and if r is negative, going out $|r|$ units in the opposite direction of the angle. Conversions between polar and rectangular forms are given by:

From	To	Identities	
Polar (r, θ)	Rectangular (x, y)	$x = r\cos\theta$	$y = r\sin\theta$
Rectangular (x, y)	Polar (r, θ)	$r = \sqrt{x^2 + y^2}$	$\tan\theta = \dfrac{y}{x}, x \neq 0$

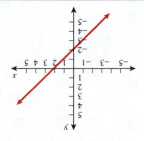

Answer: $y = x - 2$

Polar equations can be graphed by point-plotting. Common shapes that arise are given in the following table. Sine and cosine have the same shapes (just rotated). If more than one equation is given, then the top equation corresponds to the actual graph. In this table a and b are assumed to be positive.

Classification	Special Name	Polar Equations	Graph
Line	Radial line	$\theta = a$	
Circle	Centered at origin	$r = a$	
Circle	Touches pole/ center on polar axis	$r = a\cos\theta$	
Circle	Touches pole/ center on line $\theta = \dfrac{\pi}{2}$	$r = a\sin\theta$	
Limaçon	Cardioid	$r = a + a\cos\theta$ $r = a + a\sin\theta$	
Limaçon	Without inner loop $a > b$	$r = a + b\cos\theta$ $r = a + b\sin\theta$	
Limaçon	With inner loop $a < b$	$r = a + b\sin\theta$ $r = a + b\cos\theta$	
Lemniscate		$r^2 = a^2\cos 2\theta$ $r^2 = a^2\sin 2\theta$	
Rose	Three* petals	$r = a\sin 3\theta$ $r = a\cos 3\theta$	

Classification	Special Name	Polar Equations	Graph
Rose	Four* petals	$r = a\sin 2\theta$ $r = a\cos 2\theta$	
Spiral		$r = a\theta$	

*In the argument, $n\theta$, if n is odd, then there are n petals (leaves), and if n is even then there are $2n$ petals (leaves).

SECTION 8.4 EXERCISES

■ SKILLS

In Exercises 1–10, plot each indicated point in a polar coordinate system.

1. $\left(3, \dfrac{5\pi}{6}\right)$ **2.** $\left(2, \dfrac{5\pi}{4}\right)$ **3.** $\left(4, \dfrac{11\pi}{6}\right)$ **4.** $\left(1, \dfrac{2\pi}{3}\right)$ **5.** $\left(-2, \dfrac{\pi}{6}\right)$

6. $\left(-4, \dfrac{7\pi}{4}\right)$ **7.** $(-4, 270°)$ **8.** $(3, 135°)$ **9.** $(4, 225°)$ **10.** $(-2, 60°)$

In Exercises 11–20, convert each point to exact polar coordinates. Assume $0 \le \theta < 2\pi$.

11. $(2, 2\sqrt{3})$ **12.** $(3, -3)$ **13.** $(-1, -\sqrt{3})$ **14.** $(6, 6\sqrt{3})$ **15.** $(-4, 4)$

16. $(0, \sqrt{2})$ **17.** $(3, 0)$ **18.** $(-7, -7)$ **19.** $(-\sqrt{3}, -1)$ **20.** $(2\sqrt{3}, -2)$

In Exercises 21–30, convert each point to exact rectangular coordinates.

21. $\left(4, \dfrac{5\pi}{3}\right)$ **22.** $\left(2, \dfrac{3\pi}{4}\right)$ **23.** $\left(-1, \dfrac{5\pi}{6}\right)$ **24.** $\left(-2, \dfrac{7\pi}{4}\right)$ **25.** $\left(0, \dfrac{11\pi}{6}\right)$

26. $(6, 0)$ **27.** $(2, 240°)$ **28.** $(-3, 150°)$ **29.** $(-1, 135°)$ **30.** $(5, 315°)$

In Exercises 31–34, match the polar graphs with their corresponding equations.

31. $r = 4\cos\theta$ **32.** $r = 2\theta$ **33.** $r = 3 + 3\sin\theta$ **34.** $r = 3\sin 2\theta$

a.

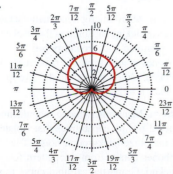

b.

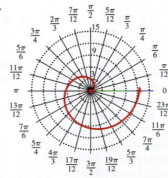

c.

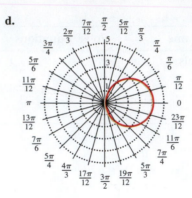

d.

In Exercises 35–50, graph each equation. In Exercises 51–54, convert the equation from polar to rectangular form. Identify the resulting equation as a line, parabola, or circle.

35. $r = 5$

36. $\theta = -\dfrac{\pi}{3}$

37. $r = 2\cos\theta$

38. $r = 3\sin\theta$

39. $r = 4\sin 2\theta$

40. $r = 5\cos 2\theta$

41. $r = 3\sin 3\theta$

42. $r = 4\cos 3\theta$

43. $r^2 = 9\cos 2\theta$

44. $r^2 = 16\sin 2\theta$

45. $r = -2\cos\theta$

46. $r = -3\sin 3\theta$

47. $r = 4\theta$

48. $r = -2\theta$

49. $r = -3 + 2\cos\theta$

50. $r = 2 + 3\sin\theta$

51. $r(\sin\theta + 2\cos\theta) = 1$

52. $r(\sin\theta - 3\cos\theta) = 2$

53. $r^2\cos^2\theta - 2r\cos\theta + r^2\sin^2\theta = 8$

54. $r^2\cos^2\theta - r\sin\theta = -2$

 ■ **APPLICATIONS**

55. Halley's Comet. Halley's comet travels an elliptical path that can be modeled with the polar equation
$r = \dfrac{0.587(1 + 0.967)}{1 - 0.967\cos\theta}$. Sketch the graph of the path of Halley's comet.

56. Dwarf Planet Pluto. The dwarf planet Pluto travels in an elliptical orbit that can be modeled with the polar equation
$r = \dfrac{29.62(1 + 0.249)}{1 - 0.249\cos\theta}$. Sketch the graph of Pluto's orbit.

57. Archimedes Spiral. Spirals are seen in nature—for example, in the swirl of a pine cone. They are also used in machinery to convert motions. An Archimedes spiral has the general equation $r = a\theta$. A more general form for the equation of a spiral is $r = a\theta^{1/n}$, where n is a constant that determines how tightly the spiral is wrapped. Compare the Archimedes spiral $r = \theta$ with the spiral $r = \theta^{1/2}$ by graphing both on the same polar graph.

58. Archimedes Spiral. Spirals are seen in nature—for example, in the swirl of a pine cone. They are also used in machinery to convert motions. An Archimedes spiral has the general equation $r = a\theta$. A more general form for the equation of a spiral is $r = a\theta^{1/n}$, where n is a constant that determines how tightly the spiral is wrapped. Compare the Archimedes spiral $r = \theta$ with the spiral $r = \theta^{4/3}$ by graphing both on the same polar graph.

59. Flapping Wings of Birds. The *lemniscate motion* occurs naturally in the flapping of birds' wings. The bird's vertical lift and wing sweep create the distinctive figure-eight pattern. The patterns vary with different wing profiles. Compare the two possible lemniscate patterns by graphing them on the same polar graph:
$$r^2 = 4\cos 2\theta \text{ and } r^2 = \frac{1}{4}\cos 2\theta.$$

60. Flapping Wings of Birds. The *lemniscate motion* occurs naturally in the flapping of birds' wings. The bird's vertical lift and wing sweep create the distinctive figure-eight pattern. The patterns vary with different wing profiles. Compare the two possible lemniscate patterns by graphing them on the same polar graph:
$r^2 = 4\cos 2\theta$ and $r^2 = 4\cos(2\theta + 2)$.

61. Cardioid Pickup Pattern. Many microphone manufacturers advertise their exceptional pickup capabilities that isolate the sound source and minimize background noise. The name of these microphones comes from the pattern formed by the range of the pickup. Graph the cardioid curve to see what the range looks like: $r = 2 + 2\sin\theta$.

62. Cardioid Pickup Pattern. Many microphone manufacturers advertise their exceptional pickup capabilities that isolate the sound source and minimize background noise. The name of these microphones comes from the pattern formed by the range of the pickup. Graph the cardioid curve to see what the range looks like: $r = -4 - 4\sin\theta$.

For Exercises 63 and 64, refer to the following:

The sword artistry of the samurai is legendary in Japanese folklore and myth. The elegance with which a samurai could wield a sword rivals the grace exhibited by modern figure skaters. In more modern times, such legends have been rendered digitally in many different video games (e.g., *Ominusha*). In order to make the characters realistically move across the screen—and in particular, wield various sword motions true to the legends— trigonometric functions are extensively used in constructing the underlying graphics module. One famous movement is a figure eight, swept out with two hands on the sword. The actual path of the tip of the blade as the movement progresses in this figure-eight motion depends essentially on the length L of the sword and the speed with which the figure is swept out. Such a path is modeled using a polar equation of the form

$$r^2(\theta) = L\cos(A\theta) \ \text{or} \ r^2(\theta) = L\sin(A\theta) \ \theta_1 \le \theta \le \theta_2$$

whose graphs are called *lemniscates*.

63. Video Games.* Graph the following equations:

   a. $r^2(\theta) = 5\cos(\theta), 0 \le \theta \le 2\pi$

   b. $r^2(\theta) = 5\cos(2\theta), 0 \le \theta \le \pi$

   c. $r^2(\theta) = 5\cos(4\theta), 0 \le \theta \le \dfrac{\pi}{2}$

What do you notice about all of these graphs? Suppose that the movement of the tip of the sword in a game is governed by these graphs. Describe what happens if you change the domain in (b) and (c) to $0 \le \theta \le 2\pi$.

64. Video Games.* Write a polar equation that would describe the motion of a sword 12 units long that makes 8 complete motions in $[0, 2\pi]$.

■ CATCH THE MISTAKE

In Exercises 65 and 66, explain the mistake that is made.

65. Convert $(-2, -2)$ to polar coordinates.

   Solution:

   Label x and y. $x = -2, y = -2$

   Find r.
$$r = \sqrt{x^2 + y^2} = \sqrt{4 + 4} = \sqrt{8} = 2\sqrt{2}$$

   Find θ. $\tan\theta = \dfrac{-2}{-2} = 1$

                     $\theta = \tan^{-1}(1) = \dfrac{\pi}{4}$

   Write the point in polar coordinates. $\left(2\sqrt{2}, \dfrac{\pi}{4}\right)$

   This is incorrect. What mistake was made?

66. Convert $(-\sqrt{3}, 1)$ to polar coordinates.

   Solution:

   Label x and y. $x = -\sqrt{3}, y = 1$

   Find r. $r = \sqrt{x^2 + y^2} = \sqrt{3 + 1} = \sqrt{4} = 2$

   Find θ. $\tan\theta = \dfrac{1}{-\sqrt{3}} = -\dfrac{1}{\sqrt{3}}$

            $\theta = \tan^{-1}\left(-\dfrac{1}{\sqrt{3}}\right) = -\dfrac{\pi}{4}$

   Write the point in polar coordinates. $\left(2, -\dfrac{\pi}{4}\right)$

   This is incorrect. What mistake was made?

■ CHALLENGE

67. T or F: All cardioids are limaçons, but not all limaçons are cardioids.

68. T or F: All limaçons are cardioids, but not all cardioids are limaçons.

69. Find the polar equation that is equivalent to a vertical line, $x = a$.

70. Find the polar equation that is equivalent to a horizontal line, $y = b$.

71. Give another pair of polar coordinates for the point (a, θ).

72. Convert $(-a, b)$ to polar coordinates. Assume $a > 0, b > 0$.

*Exercises 63–66 are courtesy of Dr. Mark McKibben, Goucher College.

73. Given: $r = \cos\dfrac{\theta}{2}$. Find the θ-intervals for the inner loop above the x-axis.

74. Given: $r = 2\cos\dfrac{3\theta}{2}$. Find the θ-intervals for the petal in the first quadrant.

75. Given: $r = 1 + 3\cos\theta$. Find the θ-intervals for the inner loop.

76. Given: $r = 1 + \sin 2\theta$ and $r = 1 - \cos 2\theta$. Find all points of intersection.

SECTION 8.5 Parametric Equations and Graphs

Skills Objectives

■ Graph parametric equations.
■ Find an equation (in rectangular form) that corresponds to a graph defined parametrically.
■ Find parametric equations for a graph that is defined by an equation in rectangular form.

Conceptual Objectives

■ Understand that increasing values of the parameter determine the orientation of a curve, or the direction along it.
■ Use time as a parameter in parametric equations.

Parametric Equations of a Curve

Thus far we have always talked about graphs in planes. For example, the equation $x^2 + y^2 = 1$ when graphed in a plane is a unit circle. Similarly, the function $f(x) = \sin x$ when graphed in a plane is a sinusoidal curve. Now, we consider the **path (orientation) along the curve**. For example, if a car is being driven on a circular race-track, we want to see the movement along the circle. We can determine where (position) along the circle the car is at some time, t, using *parametric equations*. Before we define *parametric equations* in general, let us start with a simple example.

Let $x = \sin t$ and $y = \cos t$ and $t \geq 0$. We then can make a table of the corresponding values.

t SECONDS	$x = \cos t$	$y = \sin t$	(x, y)
0	$x = \cos 0 = 1$	$y = \sin 0 = 0$	$(1, 0)$
$\dfrac{\pi}{2}$	$x = \cos\dfrac{\pi}{2} = 0$	$y = \sin\dfrac{\pi}{2} = 1$	$(0, 1)$
π	$x = \cos\pi = -1$	$y = \sin\pi = 0$	$(-1, 0)$
$\dfrac{3\pi}{2}$	$x = \cos\dfrac{3\pi}{2} = 0$	$y = \sin\dfrac{3\pi}{2} = -1$	$(0, -1)$
2π	$x = \cos 2\pi = 1$	$y = \sin 2\pi = 0$	$(1, 0)$

If we plot these points and note the correspondence to time (convert to decimals), we see that the path travels counterclockwise along the curve, which is a unit circle.

Time (seconds)	$t = 0$	$t = 1.57$	$t = 3.14$	$t = 4.71$
Position	$(1, 0)$	$(0, 1)$	$(-1, 0)$	$(0, -1)$

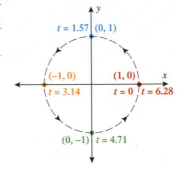

Notice that at time $t = 6.28$ seconds we are back to the point $(1, 0)$.

The path along the curve represents a unit circle, since $x^2 + y^2 = \cos^2 t + \sin^2 t = 1$.

DEFINITION   **PARAMETRIC EQUATIONS**

Let $x = f(t)$ and $y = g(t)$ be functions defined for t on some interval. The collection of points $(x, y) = (f(t), g(t))$ represent a **plane curve**. The equations

$$x = f(t) \qquad \text{and} \qquad y = g(t)$$

are called **parametric equations** of the curve. The variable t is called the **parameter**.

Parametric equations are useful for showing movement along a curve. Arrows show **direction**, or **orientation**, along the curve as t increases.

EXAMPLE 1 Graphing a Curve Defined by Parametric Equations

Graph the curve defined by the parametric equations:

$$x = t^2 \qquad y = (t - 1) \qquad t \text{ in } [-2, 2]$$

Indicate the orientation with arrows.

Solution:

STEP 1 Make a table and find values for t, x, and y.

t	$x = t^2$	$y = (t - 1)$	(x, y)
$t = -2$	$x = (-2)^2 = 4$	$y = (-2 - 1) = -3$	$(4, -3)$
$t = -1$	$x = (-1)^2 = 1$	$y = (-1 - 1) = -2$	$(1, -2)$
$t = 0$	$x = 0^2 = 0$	$y = (0 - 1) = -1$	$(0, -1)$
$t = 1$	$x = 1^2 = 1$	$y = (1 - 1) = 0$	$(1, 0)$
$t = 2$	$x = 2^2 = 4$	$y = (2 - 1) = 1$	$(4, 1)$

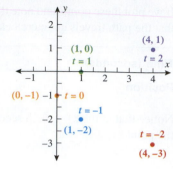

TECHNOLOGY TIP

Graph the curve defined by the parametric equations:
$x = t^2$, $y = t - 1$, t in $[-2, 2]$.
To use a TI calculator, set it in $\boxed{\text{Par}}$ mode.

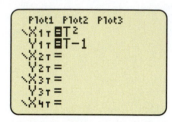

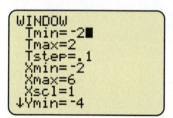

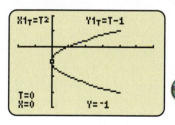

STEP 2 Plot the points in a plane.

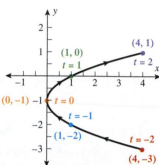

STEP 3 Connect the points with a
smooth curve and use arrows
to indicate direction.

The shape of the graph appears to be a parabola. The parametric equations are $x = t^2$ and $y = (t - 1)$. If we solve the second equation for t, $t = y + 1$, and substitute this expression into $x = t^2$, the result is $x = (y + 1)^2$. The graph of $x = (y + 1)^2$ is a parabola with vertex at the point $(0, -1)$ and opening to the right.

■ **YOUR TURN** Graph the curve defined by the parametric equations:

$$x = t + 1 \qquad y = t^2 \qquad t \text{ in } [-2, 2]$$

Indicate the orientation with arrows.

Sometimes it is easier to show the rectangular equivalent of the curve and eliminate the parameter:

EXAMPLE 2 Graphing a Curve Defined by Parametric Equations by First Finding an Equivalent Rectangular Equation

Graph the curve defined by the parametric equations:

$$x = 4\cos t \qquad y = 3\sin t \qquad t \text{ is any real number}$$

Indicate the orientation with arrows.

■ **Answer:**

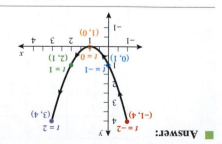

Solution: One approach is to point-plot as in Example 1. A second approach is to find the equivalent rectangular equation that represents the curve.

Use the Pythagorean identity:

$$\sin^2 t + \cos^2 t = 1$$

Find $\sin^2 t$.

$$y = 3\sin t$$

Square both sides.

$$y^2 = 9\sin^2 t$$

Divide by 9.

$$\sin^2 t = \frac{y^2}{9}$$

Find $\cos^2 t$.

Square both sides.

$$x^2 = 16\cos^2 t$$

Divide by 16.

$$\cos^2 t = \frac{x^2}{16}$$

Substitute $\sin^2 t = \dfrac{y^2}{9}$ and $\cos^2 t = \dfrac{x^2}{16}$

into $\sin^2 t + \cos^2 t = 1$.

$$\frac{y^2}{9} + \frac{x^2}{16} = 1$$

The curve is an ellipse.

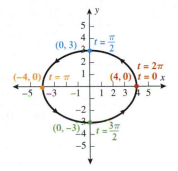

The orientation is counterclockwise. For example, when $t = 0$, the position is $(4, 0)$, and then when $t = \dfrac{\pi}{2}$, the position is $(0, 3)$.

TECHNOLOGY TIP
Graph the curve defined by the parametric equations:
$x = 4\cos t$, $y = 3\sin t$, where t is any real number.

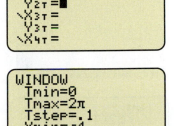

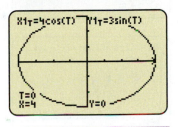

It is important to print out the difference between the graph of the ellipse $\dfrac{y^2}{9} + \dfrac{x^2}{16} = 1$, and the curve defined by $x = 4\cos t$ and $y = 3\sin t$, where t is any real number. The graph is the ellipse. The curve is infinitely many rotations around the ellipse, since t is any real number.

Applications of Parametric Equations

Parametric equations can be used to describe motion in many applications. Two that we will discuss are the *cycloid* and a *projectile*. Suppose you paint a red X on a bicycle tire.

As the bicycle moves in a straight line, if you watch the motion of the red X, it follows the path of a **cycloid**.

The parametric equations that define a cycloid are:

$$x = a\,(t - \sin t) \qquad \text{and} \qquad y = a\,(1 - \cos t)$$

where t is any real number.

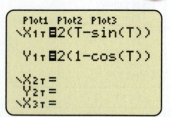

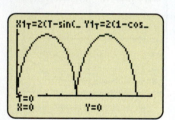

EXAMPLE 3 Graphing a Cycloid

Graph the cycloid given by $x = 2\,(t - \sin t)$ and $y = 2\,(1 - \cos t)$ for t in $[0, 4\pi]$.

Solution:

STEP 1 Make a table and find values for t, x, and y.

t	$x = 2(t - \sin t)$	$y = 2(1 - \cos t)$	(x, y)
$t = 0$	$x = 2(0 - 0) = 0$	$y = 2(1 - 1) = 0$	$(0, 0)$
$t = \pi$	$x = 2(\pi - 0) = 2\pi$	$y = 2(1 - (-1)) = 4$	$(2\pi, 4)$
$t = 2\pi$	$x = 2(2\pi - 0) = 4\pi$	$y = 2(1 - 1) = 0$	$(4\pi, 0)$
$t = 3\pi$	$x = 2(3\pi - 0) = 6\pi$	$y = 2(1 - (-1)) = 4$	$(6\pi, 4)$
$t = 4\pi$	$x = 2(4\pi - 0) = 8\pi$	$y = 2(1 - 1) = 0$	$(8\pi, 0)$

STEP 2 Plot points in a plane and connect them with a smooth curve.

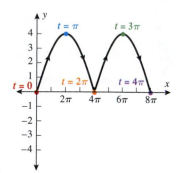

Another example of parametric equations describing real-world phenomena is projectile motion. The accompanying photo of Michelle Wie hitting a golf ball illustrates an example of a projectile.

Let v_0 be the initial velocity of an object, θ be the angle of inclination with the horizontal, and h be the initial height above ground. Then the parametric equations describing the **projectile motion** are:

$$x = (v_0 \cos \theta)t \qquad \text{and} \qquad y = -\frac{1}{2}gt^2 + (v_0 \sin \theta)t + h$$

where t is time and g is the constant acceleration due to gravity (9.8 m/sec^2 or 32 ft/sec^2).

EXAMPLE 4 Graphing Projectile Motion

Suppose Michelle Wie hits her golf ball with an initial velocity of 160 feet per second at an angle of 30° with the ground. How far is her drive, assuming the length of the drive is from the tee to where the ball first hits the ground? Graph the curve representing the path of the golf ball. Assume that she hits the ball straight off the tee and down the fairway.

Solution:

STEP 1 Find the parametric equations that describe the golf ball that Michelle Wie drove. Write the parametric equations for projectile motion:

$$x = (v_0 \cos \theta)t \quad \text{and} \quad y = -\frac{1}{2}gt^2 + (v_0 \sin \theta)t + h$$

Let $g = 32$ ft/sec^2, $v_0 = 160$ ft/sec, $h = 0$, and $\theta = 30°$.

$$x = (160 \cdot \cos 30°)t \quad \text{and} \quad y = -16t^2 + (160 \cdot \sin 30°)t$$

Evaluate the sine and cosine functions and simplify.

$$x = 80\sqrt{3}\,t \quad \text{and} \quad y = -16t^2 + 80t$$

STEP 2 Graph the projectile motion.

t	$x = 80\sqrt{3}\,t$	$y = -16t^2 + 80t$	(x, y)
$t = 0$	$x = 80\sqrt{3}(0) = 0$	$y = -16(0)^2 + 80(0) = 0$	$(0, 0)$
$t = 1$	$x = 80\sqrt{3}(1) \approx 139$	$y = -16(1)^2 + 80(1) = 64$	$(139, 64)$
$t = 2$	$x = 80\sqrt{3}(2) = 277$	$y = -16(2)^2 + 80(2) = 96$	$(277, 96)$
$t = 3$	$x = 80\sqrt{3}(3) \approx 416$	$y = -16(3)^2 + 80(3) = 96$	$(416, 96)$
$t = 4$	$x = 80\sqrt{3}(4) \approx 554$	$y = -16(4)^2 + 80(4) = 64$	$(554, 64)$
$t = 5$	$x = 80\sqrt{3}(5) \approx 693$	$y = -16(5)^2 + 80(5) = 0$	$(693, 0)$

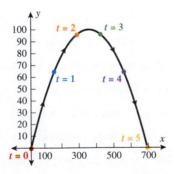

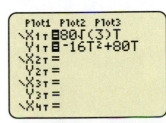

TECHNOLOGY TIP

Graph $x = 80\sqrt{3}\,t$ and $y = -16t^2 + 80t$.

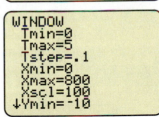

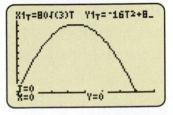

We can see that we selected our time increments well (the last point, $(693, 0)$, corresponds to the ball hitting the ground 693 feet from the tee).

STEP 3   Identify the horizontal distance from the tee to where the ball first hits the ground.

Algebraically, we can determine the distance of the tee shot by setting the height (y) equal to zero. $y = -16t^2 + 80t = 0$

Factor (divide) the common, $-16t$. $-16t(t - 5) = 0$

Solve for t. $t = 0$ or $t = 5$

The ball hits the ground after 5 seconds.

Let $t = 5$ in the horizontal distance,

$x = 80\sqrt{3}\,t$. $x = 80\sqrt{3}(5) \approx$ ┃693┃

The ball hits the ground ┃693 feet┃ from the tee.

SECTION 8.5   SUMMARY

Parametric equations are a way of describing the path an object takes along a curve. Often, t is a parameter used, where $x = f(t)$ and $y = g(t)$ describe the coordinates (x, y) that lie along the curve. Parametric equations have equivalent rectangular equations. However, parametric equations represent a curve which is more than just a graph in the rectangular plane. Two important applications are cycloids and projectiles, whose paths can be traced using parametric equations.

SECTION 8.5 EXERCISES

■ SKILLS

In Exercises 1–20, graph the curve defined by the parametric equations.

1. $x = t + 1$, $y = \sqrt{t}$, $t \geq 0$

2. $x = 3t$, $y = t^2 - 1$, t in $[0, 4]$

3. $x = -3t$, $y = t^2 + 1$, t in $[0, 4]$

4. $x = t^2 - 1$, $y = t^2 + 1$, t in $[-3, 3]$

5. $x = t^2$, $y = t^3$, t in $[-2, 2]$

6. $x = t^3 + 1$, $y = t^3 - 1$, t in $[-2, 2]$

7. $x = \sqrt{t}$, $y = t$, t in $[0, 10]$

8. $x = t$, $y = \sqrt{t^2 + 1}$, t in $[0, 10]$

9. $x = (t + 1)^2$, $y = (t + 2)^3$, t in $[0, 1]$

10. $x = (t - 1)^3$, $y = (t - 2)^2$, t in $[0, 4]$

11. $x = 3\sin t$, $y = 2\cos t$, t in $[0, 2\pi]$

12. $x = \cos 2t$, $y = \sin t$, t in $[0, 2\pi]$

13. $x = \sin t + 1$, $y = \cos t - 2$, t in $[0, 2\pi]$

14. $x = \tan t$, $y = 1$, t in $\left[-\dfrac{\pi}{4}, \dfrac{\pi}{4}\right]$

15. $x = 1$, $y = \sin t$, t in $[-2\pi, 2\pi]$

16. $x = \sin t$, $y = 2$, t in $[0, 2\pi]$

17. $x = \sin^2 t$, $y = \cos^2 t$, t in $[0, 2\pi]$

18. $x = 2\sin^2 t$, $y = 2\cos^2 t$, t in $[0, 2\pi]$

19. $x = 2\sin 3t$, $y = 3\cos 2t$, t in $[0, 2\pi]$

20. $x = 4\cos 2t$, $y = t$, t in $[0, 2\pi]$

In Exercises 21–30, the given parametric equations define a plane curve. Find an equation in rectangular form that also corresponds to the plane curve.

21. $x = \dfrac{1}{t}, y = t^2$

22. $x = t^2 - 1, y = t^2 + 1$

23. $x = t^3 + 1, y = t^3 - 1$

24. $x = 3t, y = t^2 - 1$

25. $x = t, y = \sqrt{t^2 + 1}$

26. $x = \sin^2 t, y = \cos^2 t$

27. $x = 2\sin^2 t, y = 2\cos^2 t$

28. $x = \sec^2 t, y = \tan^2 t$

29. $x = 4(t^2 + 1), y = 1 - t^2$

30. $x = \sqrt{t - 1}, y = \sqrt{t}$

 ■ **APPLICATIONS**

Recall that the flight of a projectile can be modeled with the parametric equations:

$$x = (v_0 \cos \theta)t \qquad y = -16t^2 + (v_0 \sin \theta)t + h$$

where t is in seconds, v_0 is the initial velocity, θ is the angle with the horizontal, and x and y are in feet.

31. Flight of a Projectile. A projectile is launched from the ground at a speed of 400 ft/sec at an angle of 45° with the horizontal. After how many seconds does the projectile hit the ground?

32. Flight of a Projectile. A projectile is launched from the ground at a speed of 400 ft/sec at an angle of 45° with the horizontal. How far does the projectile travel (what is the horizontal distance), and what is its maximum altitude?

33. Flight of a Baseball. A baseball is hit at an initial speed of 105 mph and an angle of 20° at a height of 3 feet above the ground. If home plate is 420 feet from the back fence, which is 15 feet tall, will the baseball clear the back fence for a home run?

34. Flight of a Baseball. A baseball is hit at an initial speed of 105 mph and an angle of 20° at a height of 3 feet above the ground. If there is no back fence or other obstruction, how far does the baseball travel (horizontal distance), and what is its maximum height?

35. Bullet Fired. A gun is fired from the ground at an angle of 60°, and the bullet has an initial speed of 700 ft/sec. How high does the bullet go? What is the horizontal (ground) distance between where the gun was fired and where the bullet hit the ground?

36. Bullet Fired. A gun is fired from the ground at an angle of 60°, and the bullet has an initial speed of 2000 ft/sec. How high does the bullet go? What is the horizontal

(ground) distance between where the gun was fired and where the bullet hit the ground?

37. Missile Fired. A missile is fired from a ship at an angle of 30°, an initial height of 20 feet above the water's surface, and at a speed of 4000 ft/sec. How long will it be before the missile hits the water?

38. Missile Fired. A missile is fired from a ship at an angle of 40°, an initial height of 20 feet above the water's surface, and at a speed of 5000 ft/sec. Will the missile be able to hit a target that is two miles away?

39. Path of a Projectile. A projectile is launched at a speed of 100 ft/sec at an angle of 35° with the horizontal. Plot the path of the projectile on a graph. Assume $h = 0$.

40. Path of a Projectile. A projectile is launched at a speed of 150 ft/sec at an angle of 55° with the horizontal. Plot the path of the projectile on a graph. Assume $h = 0$.

For Exercises 41 and 42, refer to the following:

Modern amusement park rides are often designed to push the envelope in terms of speed, angle, and ultimately gs, and usually take the form of gargantuan roller coasters or skyscraping towers. However, even just a couple of decades ago, such creations were only depicted in fantasy-type drawings, with their creators never truly believing their construction would become a reality. Nevertheless, thrill rides still capable of nauseating any would-be rider were still able to be constructed; one example is the *Calypso*. This ride is a not-too-distant cousin of the better-known *Scrambler*. It consists of four rotating arms (instead of three like the Scrambler), and on each of these arms, four cars (equally spaced around the circumference of a circular frame) are attached. Once in motion, the main piston to which the four arms are connected rotates clockwise, while each of the four arms themselves rotates counterclockwise. The combined motion appears as a blur to any onlooker from the crowd, but the motion of a single rider is much less chaotic. In fact, a single rider's path can be modeled by the following graph:

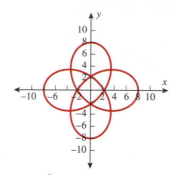

The equation of this graph is defined parametrically by:

$$x(t) = A\cos t + B\cos(-3t)$$
$$y(t) = A\sin t + B\sin(-3t), 0 \le t \le 2\pi$$

41. Amusement Rides.* What is the location of the rider at $t = 0, t = \dfrac{\pi}{2}, t = \pi, t = \dfrac{3\pi}{2}$, and $t = 2\pi$?

42. Amusement Rides.* Suppose the ride conductor was rather sinister and speeded up the ride to twice the speed. How would you modify the parametric equations to model such a change? Now vary the values of A and B. What do you conjecture these parameters are modeling in this problem?

■ **CATCH THE MISTAKE**

In Exercises 43 and 44, explain the mistake that is made.

43. Find the rectangular equation that corresponds to the plane curve defined by the parametric equations $x = t + 1$ and $y = \sqrt{t}$. Describe the plane curve.

Solution:

Square $y = \sqrt{t}$. $\qquad\qquad y^2 = t$

Substitute $t = y^2$ into $x = t + 1$. $\quad x = y^2 + 1$

The graph of $x = y^2 + 1$ is a parabola opening to the right with vertex at $(1, 0)$.

This is incorrect. What mistake was made?

44. Find the rectangular equations that correspond to the plane curve defined by the parametric equations $x = \sqrt{t}$ and $y = t - 1$. Describe the plane curve.

Solution:

Square $x = \sqrt{t}$. $\qquad\qquad x^2 = t$

Substitute $t = x^2$ into $y = t - 1$. $\quad y = x^2 - 1$

The graph of $y = x^2 - 1$ is a parabola opening up with vertex at $(0, -1)$.

This is incorrect. What mistake was made?

■ **CHALLENGE**

45. T or F: Curves given by equations in rectangular form have orientation.

46. T or F: Curves given by parametric equations have orientation.

47. Determine what type of curve the parametric equations $x = \sqrt{t}$ and $y = \sqrt{1 - t}$ define.

48. Determine what type of curve the parametric equations $x = \ln t$ and $y = t$ define.

■ **TECHNOLOGY**

49. Consider the parametric equations: $x = a\sin t - \sin at$ and $y = a\cos t + \cos at$. Use a graphing utility to explore the graphs for $a = 2, 3,$ and 4.

50. Consider the parametric equations: $x = a\cos t - b\cos at$ and $y = a\sin t + \sin at$. Use a graphing utility to explore the graphs for $a = 3$ and $b = 1$, $a = 4$ and $b = 2$, and $a = 6$ and $b = 2$. Find the t-interval that gives one cycle of the curve.

51. Consider the parametric equations: $x = \cos at$ and $y = \sin bt$. Use a graphing utility to explore the graphs for $a = 2$ and $b = 4$, $a = 4$ and $b = 2$, $a = 1$ and $b = 3$, and $a = 3$ and $b = 1$. Find the t-interval that gives one cycle of the curve.

52. Consider the parametric equations: $x = a\sin at - \sin t$ and $y = a\cos at - \cos t$. Use a graphing utility to explore the graphs for $a = 2$ and 3. If $y = a\cos at + \cos t$, explore the graphs for $a = 2$ and 3. Describe the t-interval for each case.

*Exercises 41–42 are courtesy of Dr. Mark McKibben, Goucher College.

Q: Define a coordinate system where x and y are measured in meters, and a cannon is located on top of a fort at the point (500, 30). The origin is located at what is now the entrance to the fort on the opposite side of what used to be the moat.

Guillermo uses this coordinate system to model parametric equations for the trajectory of a cannonball shot from this cannon. The cannon is aimed near the origin. Here are his equations, where t is time measured in seconds:

$$x(t) = 500 - 45t$$

$$y(t) = 30 + 95t - 9.8t^2$$

Guillermo wants to find how far from the fort the cannonball in his model would be when it hit the ground. Thus, he sets $y = 0$ and solves as follows:

$$30 + 95t - 9.8t^2 = 0$$

$$t = \frac{-95 \pm \sqrt{(95)^2 - 4(-9.8)(30)}}{2(-9.8)} = \frac{-95 \pm \sqrt{10201}}{-19.6}$$

(using the quadratic formula)

$$t = 10 \text{ sec} \qquad \text{or} \qquad t \approx -0.3 \text{ sec}$$

He then takes each value and substitutes it into the equation for x:

$$x(10) = 500 - 45(10) = 50 \text{ m}$$

$$x(-0.3) = 500 - 45(-0.3) = 513.5 \text{ m}$$

Guillermo concludes that the cannonball would hit either 50 m or 513.5 m from the fort. However, there are two errors in his calculations. What are they?

A: The variable t is time and cannot be negative. Thus, the value $t \approx -0.3$ sec obtained by solving $30 + 95t - 9.8t^2 = 0$ should be discarded, as should the corresponding value of $x = 513.5$ m. In addition, Guillermo wants to determine how far (horizontally) the cannonball would travel from the fort. Since the fort is 500 meters horizontally from the origin and the ball would hit 50 meters horizontally from the origin (this is what $x(10) = 50$ m tells us), the cannonball would hit $500 - 50 = 450$ m away from the fort.

Now find how far horizontally the cannonball would be from the fort when the cannonball is 10 meters above the ground.

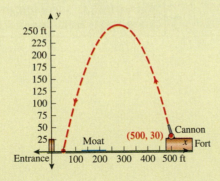

467

TYING IT ALL TOGETHER

A military plane has been flying for 2 hours and is low on fuel. Rather than landing to refuel, it will be met by a tanker plane and refuel in flight. In order to do this, the tanker plane will repeatedly fly around a 10 mile diameter circle in the counterclockwise direction until the military plane meets up with it. The maximum speed of the tanker is 600 miles per hour.

Define a coordinate system with the origin being the location of the military plane at the instant the pilot is told the direction in which to fly in order to meet the tanker (the military plane flies toward the orbit of the tanker). Define this moment as $t = 0$ minutes. Let x and y be measured in miles.

The location of the tanker can be expressed as

$$x(t) = 21 + 5\cos\left(\frac{\pi}{3}t\right), \qquad y(t) = 22 + 5\sin\left(\frac{\pi}{3}t\right)$$

The planes meet 24 miles east and 18 miles north of the origin. At what time t does this occur?

The parametric equations for the military plane's position have the form

$$x(t) = (s\cos\theta)t, \qquad y(t) = (s\sin\theta)t$$

where s is the speed of the plane and θ is the angle that the plane's flight makes with the horizontal. State these equations. (*Note:* You will need to solve for s, and find $\cos\theta$ and $\sin\theta$.)

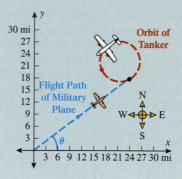

SECTION	TOPIC	PAGES	REVIEW EXERCISES	KEY CONCEPTS
8.1	Complex numbers	416–421	1–16	$i = \sqrt{-1}$, where $i^2 = -1$.
				Complex number: $a + bi$
				Complex conjugate: $a - bi$
				$(a + bi)(a - bi) = a^2 + b^2$
8.2	Polar (trigonometric) form of complex numbers	423–428	17–32	
				Modulus, or magnitude, of a complex number $z = x + iy$ is the distance from the origin to the point (x, y) in the complex plane given by $$\lvert z \rvert = \sqrt{x^2 + y^2}$$
	Polar coordinates	425		**Polar form** of a complex number: $$z = r(\cos \theta + i \sin \theta)$$ where r represents the **modulus** (magnitude) of the complex number and θ represents the **argument** of z.
	Converting from rectangular to polar form	425–428	19–26	**Step 1**: Plot the point, $z = x + yi$, in the complex plane (note the quadrant). **Step 2**: Find r. Use $r = \sqrt{x^2 + y^2}$. **Step 3**: Find θ. Use $\tan \theta = \dfrac{y}{x}$, $x \neq 0$, where θ is in the quadrant found in Step 1.
	Converting from polar to rectangular form	425–428	27–32	
8.3	Products, quotients, powers, and roots of complex numbers	431–439	33–52	
	Product of two complex numbers	431–432	33–36	Let $z_1 = r_1(\cos \theta_1 + i \sin \theta_1)$ and $z_2 = r_2(\cos \theta_2 + i \sin \theta_2)$ be two complex numbers. The product, $z_1 z_2$, is given by $z_1 z_2 = r_1 r_2[\cos(\theta_1 + \theta_2) + i \sin(\theta_1 + \theta_2)]$ *Multiply the magnitudes and add the arguments.*
	Quotient of two complex numbers	432–434	37–40	Let $z_1 = r_1(\cos \theta_1 + i \sin \theta_1)$ and $z_2 = r_2(\cos \theta_2 + i \sin \theta_2)$ be two complex numbers. The quotient, $\dfrac{z_1}{z_2}$, is given by $$\frac{z_1}{z_2} = \frac{r_1}{r_2}[\cos(\theta_1 - \theta_2) + i \sin(\theta_1 - \theta_2)]$$ *Divide the magnitudes and subtract the arguments.*

SECTION	TOPIC	PAGES	REVIEW EXERCISES	KEY CONCEPTS
	Raising a complex number to an integer power	434–435	41–44	**De Moivre's Theorem** If $z = r(\cos\theta + i\sin\theta)$ is a complex number then $z^n = r^n[\cos n\theta + i\sin n\theta]$ $n \geq 1$, where n is an integer.
	Finding the nth root of a complex number	435–437	45–52	The **nth roots** of the complex number $z = r(\cos\theta + i\sin\theta)$ are given by $$w_k = r^{1/n}\left[\cos\left(\frac{\theta}{n} + \frac{k \cdot 360°}{n}\right)\right.$$ $$\left. + i\sin\left(\frac{\theta}{n} + \frac{k \cdot 360°}{n}\right)\right]$$ with θ in degrees, and where $k = 0, 1, 2, \ldots, n-1$
8.4	Polar equations and graphs	442–453	53–68	
	Point-plotting in polar coordinates	443–445	53–64	To plot a point, (r, θ): ■ Start on the polar axis and rotate to the angle θ. ■ If $r > 0$, the point is r units from the origin in the *same direction* of the terminal side of θ. ■ If $r < 0$, the point is $|r|$ units from the origin in the *opposite direction* of the terminal side of θ.
	Converting between rectangular and polar coordinates	445–446	53–58	From polar (r, θ) to rectangular (x, y): $\quad x = r\cos\theta \qquad y = r\sin\theta$ From rectangular (x, y) to polar (r, θ): $\quad r = \sqrt{x^2 + y^2} \qquad \tan\theta = \dfrac{y}{x} \quad x \neq 0$
	Common polar graphs	446–452	65–68	Radial line, circle, spiral, rose petals, lemniscate, and limaçon.
8.5	Parametric equations and graphs	458–464	69–76	Parametric equations: $\quad x = f(t)$ and $y = g(t)$ Plane curve: $\quad (x, y) = (f(t), g(t))$

CHAPTER 8 REVIEW EXERCISES

8.1 Complex Numbers

Simplify.

1. i^{39} **2.** i^{100}

3. $\sqrt{-49}$ **4.** $\sqrt{-50}$

Perform the indicated operation, simplify, and express in standard form.

5. $(2 - 3i) + (4 + 6i)$ **6.** $(-3 - i) - (5 - 2i)$

7. $(1 + 4i)(-6 - 2i)$ **8.** $\dfrac{4 - i}{2 + 3i}$

9. $i(2 + 3i) + 4(-3 + 2i)$ **10.** $-3(2 - 3i) + i(-6 + i)$

Solve the quadratic equations. Write the answers in standard form.

11. $x^2 - 4x + 9 = 0$ **12.** $2x^2 + 2x + 11 = 0$

13. $5x^2 + 6x + 5 = 0$ **14.** $x^2 + 4x + 5 = 0$

Write the negative radicals in terms of imaginary numbers, and then perform the indicated operation and simplify.

15. $\sqrt{-25}\sqrt{-9}$ **16.** $(2 + \sqrt{-100})(-3 + \sqrt{-4})$

8.2 Polar (Trigonometric) Form of Complex Numbers

Graph each complex number.

17. $-6 + 2i$ **18.** $5i$

Express each complex number in polar form.

19. $\sqrt{2} - \sqrt{2}i$ **20.** $\sqrt{3} + i$

21. $-8i$ **22.** $-8 - 8i$

Use a calculator to express each complex number in polar form.

23. $-60 + 11i$ **24.** $9 - 40i$

25. $15 + 8i$ **26.** $-10 - 24i$

Express each complex number in rectangular form.

27. $6(\cos 300° + i\sin 300°)$ **28.** $4(\cos 210° + i\sin 210°)$

29. $\sqrt{2}(\cos 135° + i\sin 135°)$ **30.** $4(\cos 150° + i\sin 150°)$

Use a calculator to express each complex number in rectangular form.

31. $4(\cos 200° + i\sin 200°)$ **32.** $3(\cos 350° + i\sin 350°)$

8.3 Products, Quotients, Powers, and Roots of Complex Numbers

Find the product, z_1z_2.

33. $3(\cos 200° + i\sin 200°)$ and $4(\cos 70° + i\sin 70°)$

34. $3(\cos 20° + i\sin 20°)$ and $4(\cos 220° + i\sin 220°)$

35. $7(\cos 100° + i\sin 100°)$ and $3(\cos 140° + i\sin 140°)$

36. $(\cos 290° + i\sin 290°)$ and $4(\cos 40° + i\sin 40°)$

Find the quotient, $\dfrac{z_1}{z_2}$.

37. $\sqrt{6}(\cos 200° + i\sin 200°)$ and $\sqrt{6}(\cos 50° + i\sin 50°)$

38. $18(\cos 190° + i\sin 190°)$ and $2(\cos 100° + i\sin 100°)$

39. $24(\cos 290° + i\sin 290°)$ and $4(\cos 110° + i\sin 110°)$

40. $\sqrt{200}(\cos 93° + i\sin 93°)$ and $\sqrt{2}(\cos 48° + i\sin 48°)$

Find the result of each expression using De Moivre's theorem. Write the answer in rectangular form.

41. $(3 + 3i)^4$ **42.** $(3 + \sqrt{3}i)^4$

43. $(1 + \sqrt{3}i)^5$ **44.** $(-2 - 2I)^7$

Find all nth roots of z. Write the answers in polar form, and plot the roots in the complex plane.

45. $2 + 2\sqrt{3}i, n = 2$ **46.** $-8 + 8\sqrt{3}i, n = 4$

47. $-256, n = 4$ **48.** $-18i, n = 2$

Find all complex solutions to the given equations.

49. $x^3 + 216 = 0$ **50.** $x^4 - 1 = 0$

51. $x^4 + 1 = 0$ **52.** $x^3 - 125 = 0$

8.4 Polar Equations and Graphs

Convert each point to exact polar coordinates (assuming that $0 \le \theta < 2\pi$), and then graph the point in the polar coordinate system.

53. $(-2, 2)$ **54.** $(4, -4\sqrt{3})$ **55.** $(-5\sqrt{3}, -5)$

56. $(\sqrt{3}, \sqrt{3})$ **57.** $(0, -2)$ **58.** $(11, 0)$

Convert each point to exact rectangular coordinates.

59. $\left(-3, \dfrac{5\pi}{3}\right)$ **60.** $\left(4, \dfrac{5\pi}{4}\right)$ **61.** $\left(2, \dfrac{\pi}{3}\right)$

62. $\left(6, \dfrac{7\pi}{6}\right)$ **63.** $\left(1, \dfrac{4\pi}{3}\right)$ **64.** $\left(-3, \dfrac{7\pi}{4}\right)$

Graph each equation.

65. $r = 4\cos 2\theta$

66. $r = \sin 3\theta$

67. $r = -\theta$

68. $r = 4 - 3\sin\theta$

8.5 Parametric Equations and Graphs

Graph the curve defined by the parametric equations.

69. $x = \sin t, y = 4\cos t$ for t in $[-\pi, \pi]$

70. $x = 5\sin^2 t, y = 2\cos^2 t$ for t in $[-\pi, \pi]$

71. $x = 4 - t^2, y = t^2$ for t in $[-3, 3]$

72. $x = t + 3, y = 4$ for t in $[-4, 4]$

The given parametric equations define a plane curve. Find an equation in rectangular form that also corresponds to the plane curve.

73. $x = 4 - t^2, y = t$

74. $x = 5\sin^2 t, y = 2\cos^2 t$

75. $x = 2\tan^2 t, y = 4\sec^2 t$

76. $x = 3t^2 + 4, y = 3t^2 - 5$

1. Simplify: $i^{20} - i^{13} + i^{17} - i^{1000}$.

2. Plot the point $-3 - 5i$ in the complex plane.

3. Convert the point $3\sqrt{2}(1 - i)$ to polar form.

4. Find the modulus and argument of $-5 - 5i$.

For the complex numbers $z_1 = 4[\cos 75° + i \sin 75°]$ and $z_2 = 2[\cos 15° + i \sin 15°]$,

5. Find $z_1 \cdot z_2$.

6. Find $\dfrac{z_1}{z_2}$.

For the complex number $z = 16[\cos 120° + i \sin 120°]$,

7. Find z^4.

8. Find the four distinct 4th roots of z.

9. Give another pair of polar coordinates for the point (a, θ).

10. Convert the point $(3, 210°)$ to rectangular coordinates.

11. Use a calculator to convert the point $(-3, -4)$ to polar coordinates.

12. Graph $r = 6 \sin 2\theta$.

13. Graph $r^2 = 9 \cos 2\theta$.

14. Describe (classify) the plane curve defined by the parametric equations $x = \sqrt{1 - t}$ and $y = \sqrt{t}$ for t in $[0, 1]$.

15. A golf ball is hit with an initial speed of 120 feet per second at an angle of 45° with the ground. How long will the ball stay in the air? How far will the ball travel (horizontal distance) before it hits the ground?

Algebraic Prerequisites and Review

Skills Objectives

- Plot points in the Cartesian plane.
- Calculate the distance between two points.
- Calculate the midpoint between two points.

Conceptual Objective

- Expand the concept of a one-dimensional number line to a two-dimensional plane.

Cartesian Plane

HIV infection rates, circular orbits, stock prices, and temperature conversions are all examples of relationships between two quantities that can be expressed in a two-dimensional graph, which is called a **plane**.

The **axes** are two perpendicular real number lines in the plane that intersect at their origins. Typically, the horizontal axis is called the **x-axis** and the vertical axis is denoted as the **y-axis**. The point where the axes intersect is called the **origin**. The axes divide the plane into four **quadrants**, numbered by roman numerals and ordered counterclockwise.

Each point in the plane is called an **ordered pair**, denoted (x, y). The first number of the ordered pair indicates the position in the horizontal direction and is often called the *x*-coordinate or **abscissa**. The second number indicates the position in the vertical direction and is often called the *y*-coordinate or **ordinate**. The origin is denoted $(0, 0)$. Examples of other coordinates are given on the graph below:

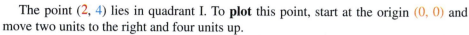

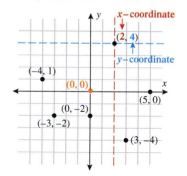

The point $(2, 4)$ lies in quadrant I. To **plot** this point, start at the origin $(0, 0)$ and move two units to the right and four units up.

All points in quadrant I have positive coordinates and all points in quadrant III have negative coordinates. Quadrant II has negative *x*-coordinates and positive *y*-coordinates; quadrant IV has positive *x*-coordinates and negative *y*-coordinates.

This representation is called the **rectangular coordinate system** or **Cartesian coordinate system**, named after the French mathematician René Descartes.

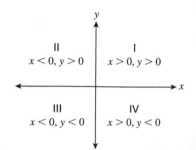

EXAMPLE 1 Plotting Points in a Cartesian Plane

a. Plot and label the points $(-1, -4)$, $(2, 2)$, $(-2, 3)$, $(2, -3)$, $(0, 5)$, and $(-3, 0)$ in the Cartesian plane.

b. List the points and corresponding quadrant or axis in a table.

Solution:

a.

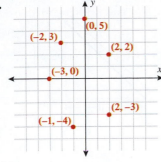

b.

Point	Quadrant
$(2, 2)$	I
$(-2, 3)$	II
$(-1, -4)$	III
$(2, -3)$	IV
$(0, 5)$	y-axis
$(-3, 0)$	x-axis

Distance Between Two Points

Suppose you want to find the distance between any two points in the plane. In the previous graph, to find the distance between the points $(2, -3)$ and $(2, 2)$, count the units between the two points. The distance is 5. What if the two points do not lie along a horizontal or vertical line? Example 2 uses the Pythagorean theorem to help find the distance between any two points.

EXAMPLE 2 Finding the Distance Between Two Points

Find the distance between the points $(-2, -1)$ and $(1, 3)$.

Solution:

STEP 1 Plot and label the two points in the Cartesian plane and draw a segment indicating the distance, d, between the two points.

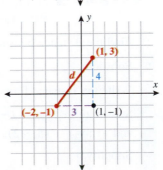

STEP 2 Form a right triangle by connecting the points to a third point, $(1, -1)$.

STEP 3 Calculate the length of the horizontal segment. $3 = |1 - (-2)|$
Calculate the length of the vertical segment. $4 = |3 - (-1)|$

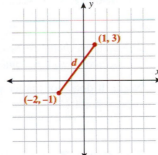

STEP 4 Use the Pythagorean theorem to calculate the distance, d.

$$d^2 = 3^2 + 4^2$$
$$d^2 = 9 + 16 = 25$$
$$d = 5$$

WORDS	MATH
For any two points, (x_1, y_1) and (x_2, y_2).	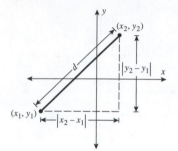

The distance along the horizontal segment is the absolute value.	$\lvert x_2 - x_1 \rvert$
The distance along the vertical segment is the absolute value.	$\lvert y_2 - y_1 \rvert$
Use the Pythagorean theorem to calculate the distance, d.	$d^2 = \lvert x_2 - x_1 \rvert^2 + \lvert y_2 - y_1 \rvert^2$
$\lvert a \rvert^2 = a^2$ for all real numbers a.	$d^2 = (x_2 - x_1)^2 + (y_2 - y_1)^2$
Use the square root property.	$d = \pm\sqrt{(x_2 - x_1)^2 + (y_2 - y_1)^2}$
Distance can only be positive.	$d = \sqrt{(x_2 - x_1)^2 + (y_2 - y_1)^2}$

DISTANCE FORMULA

The **distance**, d, between two points $P_1 = (x_1, y_1)$ and $P_2 = (x_2, y_2)$ is given by

$$d = \sqrt{(x_2 - x_1)^2 + (y_2 - y_1)^2}$$

The distance between two points is the square root of the distance between the x-coordinates squared plus the distance between the y-coordinates squared.

Note: It does not matter which point is taken to be the first point or the second point.

EXAMPLE 3 Using the Distance Formula to Find the Distance Between Two Points

Find the distance between $(-3, 7)$ and $(5, -2)$.

Solution:

Write the distance formula.	$d = \sqrt{[x_2 - x_1]^2 + [y_2 - y_1]^2}$
Substitute $(x_1, y_1) = (-3, 7)$ and $(x_2, y_2) = (5, -2)$.	$d = \sqrt{[5 - (-3)]^2 + [-2 - 7]^2}$
Simplify.	$d = \sqrt{[5 + 3]^2 + [-2 - 7]^2}$
	$d = \sqrt{8^2 + (-9)^2} = \sqrt{64 + 81} = \sqrt{145}$
Solve for d.	$d = \sqrt{145}$

■ **YOUR TURN**   Find the distance between $(4, -5)$ and $(-3, -2)$.

Midpoint of a Segment Joining Two Points

The **midpoint**, (x, y), of a segment connecting two points (x_1, y_1) and (x_2, y_2) is defined as the point that lies on the segment which has the same distance, d, from both points.

   In other words, the midpoint of a segment lies halfway between the given endpoints. The coordinates of the midpoint are found by *averaging* the x-coordinates and averaging the y-coordinates.

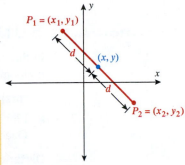

MIDPOINT FORMULA

The **midpoint**, (x, y), between two points (x_1, y_1) and (x_2, y_2) is given by

$$(x, y) = \left(\frac{x_1 + x_2}{2}, \frac{y_1 + y_2}{2} \right)$$

The midpoint can be found by averaging the x-coordinates and averaging the y-coordinates.

EXAMPLE 4 Finding the Midpoint of a Segment

Find the midpoint of the segment joining points $(2, 6)$ and $(-4, -2)$.

Solution:

Write the midpoint formula.

$$(x, y) = \left(\frac{x_1 + x_2}{2}, \frac{y_1 + y_2}{2} \right)$$

Substitute $(x_1, y_1) = (2, 6)$ and $(x_2, y_2) = (-4, -2)$.

$$(x, y) = \left(\frac{2 + (-4)}{2}, \frac{6 + (-2)}{2} \right)$$

Simplify.

$$(x, y) = (-1, 2)$$

One way to check your answer is to plot the points and midpoint to make sure your answer looks reasonable.

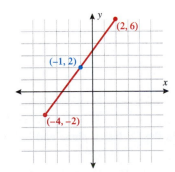

CONCEPT CHECK What quadrant would you expect the midpoint of the segment joining the points $(3, -4)$ and $(5, 8)$ to lie in?

YOUR TURN Compute the midpoint of the segment joining points $(3, -4)$ and $(5, 8)$.

SECTION A.1   SUMMARY

In this section we discussed point plotting in a plane. Formulas for the distance and midpoint between two points were developed.

Cartesian Plane

- Plotting coordinates: (x, y)
- Quadrants: I, II, III, and IV

Distance Between Two Points

$$d = \sqrt{(x_2 - x_1)^2 + (y_2 - y_1)^2}$$

Midpoint of Segment Joining Two Points

$$\text{Midpoint} = (x, y) = \left(\frac{x_1 + x_2}{2}, \frac{y_1 + y_2}{2} \right)$$

SECTION A.1   EXERCISES

■ SKILLS

In Exercises 1–6, give the coordinates for each point labeled.

1. Point A
2. Point B
3. Point C
4. Point D
5. Point E
6. Point F

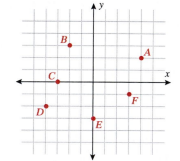

In Exercises 7 and 8, plot each point in the Cartesian plane and indicate in which quadrant or on which axis the point lies.

7. A: $(-2, 3)$ B: $(1, 4)$ C: $(-3, -3)$ D: $(5, -1)$ E: $(0, -2)$ F: $(4, 0)$

8. A: $(-1, 2)$ B: $(1, 3)$ C: $(-4, -1)$ D: $(3, -2)$ E: $(0, 5)$ F: $(-3, 0)$

■ **Answer:** Midpoint $= (4, 2)$

9. Plot the points $(-3, 1)$, $(-3, 4)$, $(-3, -2)$, $(-3, 0)$, $(-3, -4)$. Describe the line containing points of the form $(-3, y)$.

10. Plot the points $(-1, 2)$, $(-3, 2)$, $(0, 2)$, $(3, 2)$, $(5, 2)$. Describe the line containing points of the form $(x, 2)$.

In Exercises 11–24, calculate the distance and midpoint between the segment joining the given points.

11. $(1, 3)$ and $(5, 3)$

12. $(-2, 4)$ and $(-2, -4)$

13. $(-1, 4)$ and $(3, 0)$

14. $(-3, -1)$ and $(1, 3)$

15. $(-10, 8)$ and $(-7, -1)$

16. $(-2, 12)$ and $(7, 15)$

17. $\left(-\frac{1}{2}, \frac{1}{3}\right)$ and $\left(\frac{7}{2}, \frac{10}{3}\right)$

18. $\left(\frac{1}{5}, \frac{7}{3}\right)$ and $\left(\frac{9}{5}, -\frac{2}{3}\right)$

19. $\left(-\frac{2}{3}, -\frac{1}{5}\right)$ and $\left(\frac{1}{4}, \frac{1}{3}\right)$

20. $\left(\frac{7}{5}, \frac{1}{9}\right)$ and $\left(\frac{1}{2}, -\frac{7}{3}\right)$

21. $(-1.5, 3.2)$ and $(2.1, 4.7)$

22. $(-1.2, -2.5)$ and $(3.7, 4.6)$

23. $(-14.2, 15.1)$ and $(16.3, -17.5)$

24. $(1.1, 2.2)$ and $(3.3, 4.4)$

In Exercises 25 and 26, calculate (to two decimal places) the perimeter of the triangle with the following vertices:

25. Points A, B, and C

26. Points C, D, and E

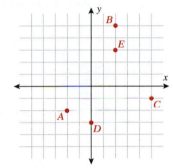

In Exercises 27–30, determine if the triangle with the given vertices is a right triangle, isosceles triangle, neither, or both. (Recall that a right triangle satisfies the Pythagorean theorem and an isosceles triangle has at least two sides of equal length.)

27. $(0, -3)$, $(3, -3)$, and $(3, 5)$

28. $(0, 2)$, $(-2, -2)$, and $(2, -2)$

29. $(1, 1)$, $(3, -1)$, and $(-2, -4)$

30. $(-3, 3)$, $(3, 3)$, and $(-3, -3)$

■ **APPLICATIONS**

31. Cell Phones. A cellular phone company currently has three towers: one in Tampa, one in Orlando, and one in Gainesville to serve the central Florida region. If Orlando is 80 miles east of Tampa and Gainesville is 100 miles north of Tampa, what is the distance from Orlando to Gainesville?

32. Cell Phones. The same cellular phone company in Exercise 31 has decided to add additional towers "halfway" between each city. How many miles from Tampa is each "halfway" tower?

33. Travel. A retired couple who lives in Columbia, South Carolina, decides to take their motor home and visit two children who live in Atlanta and in Savannah, Georgia. Savannah is 160 miles south of Columbia and Atlanta is 215 miles west of Columbia. How far apart do the children live from each other?

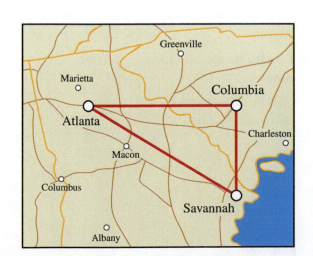

34. Sports. In the 1984 Orange Bowl, Doug Flutie, the 5 foot 9 inch quarterback for Boston College, shocked the world as he threw a "hail Mary" pass that was caught in the end zone with no time left on the clock, defeating the Miami Hurricanes 47–45. Although the record books have it listed as a 48 yard pass, what was the actual distance the ball was thrown?

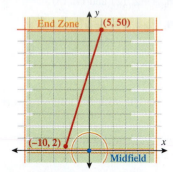

35. NASCAR Revenue. Action Performance Inc., the leading seller of NASCAR merchandise, recorded $260 million in revenue in 2002 and $400 million in revenue in 2004. Calculate the midpoint to estimate the revenue Action Performance Inc. recorded in 2003.

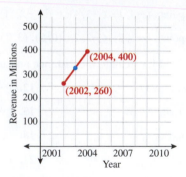

36. Ticket Price. In 1993 a Miami Dolphins average ticket price was $28 and in 2001 the average price was $56. Find the midpoint of the segment joining these two points to estimate the ticket price in 1997.

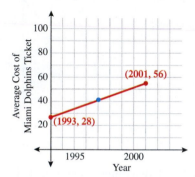

■ CATCH THE MISTAKE

In Exercises 37–40, explain the mistake that is made.

37. Calculate the distance between $(2, 7)$ and $(9, 10)$.

Solution:

Write the distance formula. $d = \sqrt{(x_2 - x_1)^2 + (y_2 - y_1)^2}$

Substitute $(2, 7)$ and $(9, 10)$. $d = \sqrt{(7 - 2)^2 + (10 - 9)^2}$

Simplify. $d = \sqrt{(5)^2 + (1)^2} = \sqrt{26}$

38. Calculate the distance between $(-2, 1)$ and $(3, -7)$.

Solution:

Write the distance formula. $d = \sqrt{(x_2 - x_1)^2 + (y_2 - y_1)^2}$

Substitute $(-2, 1)$ and $(3, -7)$. $d = \sqrt{(3 - 2)^2 + (-7 - 1)^2}$

Simplify. $d = \sqrt{(1)^2 + (-8)^2} = \sqrt{65}$

39. Compute the midpoint between the points $(-3, 4)$ and $(7, 9)$.

Solution:

Write the midpoint formula. $(x, y) = \left(\dfrac{x_1 + x_2}{2}, \dfrac{y_1 + y_2}{2} \right)$

Substitute $(-3, 4)$ and $(7, 9)$. $(x, y) = \left(\dfrac{-3 + 4}{2}, \dfrac{7 + 9}{2} \right)$

Simplify. $(x, y) = \left(\dfrac{1}{2}, \dfrac{16}{2} \right) = \left(\dfrac{1}{2}, 4 \right)$

40. Compute the midpoint between the points $(-1, -2)$ and $(-3, -4)$.

Solution:

Write the midpoint formula. $(x, y) = \left(\dfrac{x_1 - x_2}{2}, \dfrac{y_1 - y_2}{2} \right)$

Substitute $(-1, -2)$ and $(-3, -4)$. $(x, y) = \left(\dfrac{-1 - (-3)}{2}, \dfrac{-2 - (-4)}{2} \right)$

Simplify. $(x, y) = (1, 1)$

■ CHALLENGE

41. T or F: The distance from the origin to the point (a, b) is $d = \sqrt{a^2 + b^2}$.

42. T or F: The midpoint between the origin and the point (a, a) is $\left(\frac{a}{2}, \frac{a}{2}\right)$.

43. T or F: The midpoint of any segment joining two points in quadrant I also lies in quadrant I.

44. T or F: The midpoint of any segment joining a point in quadrant I to a point in quadrant 3 also lies in either quadrant I or quadrant III.

45. Calculate the length and the midpoint for the segment joining the points (a, b) and (b, a).

46. Calculate the length and the midpoint for the segment joining the points (a, b) and $(-a, -b)$.

47. Assume two points, (x_1, y_1) and (x_2, y_2), are connected by a segment. Prove that the distance from the midpoint of the segment to either of the two points is the same.

48. Prove that the diagonals of a parallelogram in the figure intersect at their midpoints.

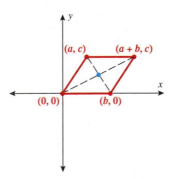

■ TECHNOLOGY

In Exercises 49–52, calculate the distance between the two points. Use a graphing utility to graph the segment joining the two points and find the length of the segment.

49. $(-2.3, 4.1)$ and $(3.7, 6.2)$

50. $(-4.9, -3.2)$ and $(5.2, 3.4)$

51. $(1.1, 2.2)$ and $(3.3, 4.4)$

52. $(-1.3, 7.2)$ and $(2.3, -4.5)$

SECTION A.2 Graphing Equations: Point Plotting and Symmetry

Skills Objectives

■ Plot equations by plotting points.
■ Conduct a test for symmetry about x-axis, y-axis, and origin.

Conceptual Objective

■ Relate symmetry graphically and algebraically.

We will learn how to graph equations in this section using point plotting. Later we will learn other techniques. All equations can be graphed by plotting points. However, as we discuss graphing aids in Section A.4, you will see that other techniques can be more efficient.

Point Plotting

An equation in two variables, such as $y = x^2$, has an infinite number of ordered pairs which are its solutions. For example, $(0, 0)$ is a solution to $y = x^2$ because when $x = 0$ and $y = 0$ the equation is true. Two other solutions are $(-1, 1)$ and $(1, 1)$.

The **graph of an equation** in two variables, x and y, consists of all the points in the xy-plane whose coordinates (x, y) satisfy the equation. A procedure for plotting the graphs of equations is outlined below and is illustrated with the above example, $y = x^2$.

WORDS

MATH

Step 1: In a table, list several pairs of coordinates that make the equation true.

x	$y = x^2$	(x, y)
0	0	$(0, 0)$
-1	1	$(-1, 1)$
1	1	$(1, 1)$
-2	4	$(-2, 4)$
2	4	$(2, 4)$

Step 2: Plot these points on a graph and connect the points with a smooth curve. Use arrows to indicate that the graph continues.

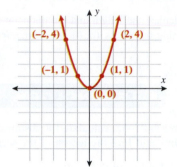

EXAMPLE 1 Graphing an Equation of a Line by Plotting Points

Graph the equation $y = 2x - 1$.

Solution:

STEP 1 In a table, list several pairs of coordinates that make the equation true.

x	$y = 2x - 1$	(x, y)
0	-1	$(0, -1)$
-1	-3	$(-1, -3)$
1	1	$(1, 1)$
-2	-5	$(-2, -5)$
2	3	$(2, 3)$

STEP 2 Plot these points on a graph and connect the points, resulting in a line. Arrows indicate that the graph continues.

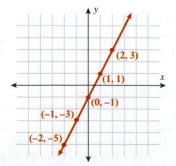

CONCEPT CHECK How many points are necessary to draw a straight line?

■ **YOUR TURN** The graph of the equation $y = -x + 1$ is a line. Graph the line.

EXAMPLE 2 Graphing an Equation of a Parabola by Plotting Points

Graph the equation $y = x^2 - 5$.

Solution:

STEP 1 In a table, list several pairs of coordinates that make the equation true.

x	$y = x^2 - 5$	(x, y)
0	-5	$(0, -5)$
-1	-4	$(-1, -4)$
1	-4	$(1, -4)$
-2	-1	$(-2, -1)$
2	-1	$(2, -1)$
-3	4	$(-3, 4)$
3	4	$(3, 4)$

STEP 2 Plot these points on a graph and connect the points with a smooth curve, indicating with arrows that the curve continues.

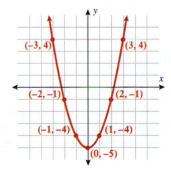

Recall from algebra that this graph is called a *parabola*.

■ **YOUR TURN** Graph $x = y^2 - 1$.

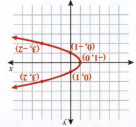

Answer: ■

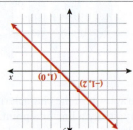

Answer: ■

EXAMPLE 3 Graphing an Equation by Plotting Points

Graph the equation $y = x^3$.

Solution:

STEP 1 In a table, list several pairs of coordinates in a table that satisfy the equation.

x	$y = x^3$	(x, y)
0	0	$(0, 0)$
-1	-1	$(-1, -1)$
1	1	$(1, 1)$
-2	-8	$(-2, -8)$
2	8	$(2, 8)$

STEP 2 Plot these points on a graph and connect the points with a smooth curve, indicating with arrows that the curve continues in both the positive and negative directions.

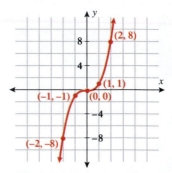

Symmetry

The word **symmetry** conveys balance. Suppose you have two pictures to hang on a wall. If you space them equally apart on the wall, then you prefer a symmetric décor. This is an example of symmetry about a line. The word (water) written below is identical if you rotate the word 180 degrees. This is an example of symmetry about a point.

Symmetric graphs have the characteristic that their mirror image can be obtained about a reference, typically a line or a point.

In Example 2, the points $(-2, -1)$ and $(2, -1)$ both lie on the graph as do the points $(-1, -4)$ and $(1, -4)$. Notice that the graph on the right side of the y-axis is a mirror image of the part of the graph to the left of the *y*-axis. This graph illustrates *symmetry with respect to the y-axis* (the line $x = 0$).

In the *Your Turn* following Example 2, the points $(0, 1)$ and $(0, -1)$ both lie on the graph as do the points $(3, 2)$ and $(3, -2)$. Notice that the part of the graph above the *x*-axis is a mirror image of the part of the graph below the *x*-axis. This graph illustrates *symmetry* with respect to the *x-axis* (the line $y = 0$).

In Example 3, the points $(-1, -1)$ and $(1, 1)$ both lie on the graph. Notice that rotating this graph 180 degrees results in an identical graph. This is an example of *symmetry* with respect to the *origin* $(0, 0)$.

Symmetry aids in graphing by giving information "for free." For example, if a graph is symmetric about the *y*-axis, then once the graph to the right of the *y*-axis is found, the left side of the graph is the mirror image of that. If a graph is symmetric about the origin, then once the graph is known in quadrant I the graph in quadrant III is found by rotating the known graph 180 degrees.

It would be beneficial to know if a graph of an equation is symmetric about a line or point before the equation is plotted. Although a graph can be symmetric about any line or point, we will only discuss symmetry about the *x*-axis, *y*-axis, or origin. These types of symmetry and the algebraic procedure for testing for symmetry are outlined in the box below.

Types and Tests for Symmetry

TYPE OF SYMMETRY	GRAPH	IF THE POINT (a, b) IS ON THE GRAPH, THEN THE POINT . . .	ALGEBRAIC TEST FOR SYMMETRY
Symmetric with respect to the **x-axis**		$(a, -b)$ is on the graph.	Replacing y with $-y$ leaves the equation unchanged.
Symmetric with respect to the **y-axis**		$(-a, b)$ is on the graph.	Replacing x with $-x$ leaves the equation unchanged.
Symmetric with respect to the **origin**		$(-a, -b)$ is on the graph.	Replacing x with $-x$ and y with $-y$ leaves the equation unchanged.

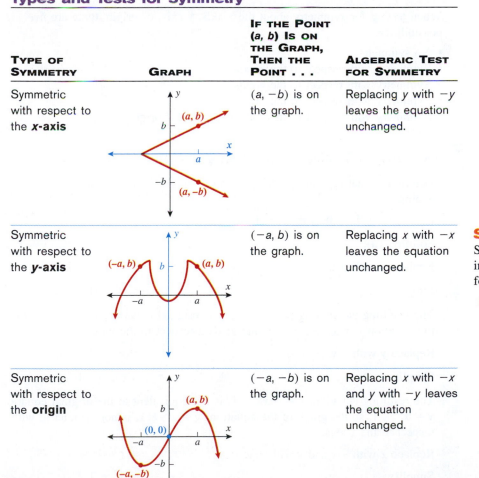

STUDY TIP

Symmetry gives us information about the graph for "free."

TECHNOLOGY TIP
To enter the graph of $y^2 = x^3$, solve for y first. The graphs of $y_1 = \sqrt[3]{x^3}$ and $y_2 = -\sqrt[3]{x^3}$ are shown.

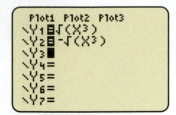

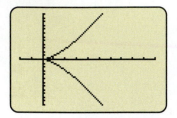

EXAMPLE 4 **Testing for Symmetry with Respect to the Axes**

Test the equation $y^2 = x^3$ for symmetry with respect to the axes.

Solution:

Test for symmetry with respect to the x-axis.

Replace y with $-y$. $\hspace{3cm} (-y)^2 = x^3$

Simplify. $\hspace{5cm} y^2 = x^3$

The resulting equation is the same as the original equation, $y^2 = x^3$. Therefore,

$y^2 = x^3$ is **symmetric with respect to the x-axis** .

Test for symmetry with respect to the y-axis.

Replace x with $-x$. $\hspace{3cm} y^2 = (-x)^3$

Simplify. $\hspace{5cm} y^2 = -x^3$

The resulting equation, $y^2 = -x^3$, is not the same as the original equation, $y^2 = x^3$. Therefore, $y^2 = x^3$ is **not** symmetric with respect to the y-axis .

When testing for symmetry about the x-axis, y-axis, or origin, there are *five* possibilities:

• No symmetry
• Symmetry with respect to the x-axis
• Symmetry with respect to the y-axis
• Symmetry with respect to the origin
• Symmetry with respect to the x-axis, y-axis, and origin

EXAMPLE 5 **Testing for Symmetry**

Determine what type of symmetry (if any) the graphs of the equations exhibit:

a. $y = x^2 + 1$ $\hspace{1cm}$ **b.** $y = x^3 + 1$

Solution (a):

Replace x with $-x$. $\hspace{4cm} y = (-x)^2 + 1$

Simplify. $\hspace{6cm} y = x^2 + 1$

The resulting equation is equivalent to the original equation, so the graph of the equation $y = x^2 + 1$ is symmetric with respect to the y-axis.

Replace y with $-y$. $\hspace{4cm} (-y) = x^2 + 1$

Simplify. $\hspace{6cm} y = -x^2 - 1$

The resulting equation, $y = -x^2 - 1$, is not equivalent to the original equation, $y = x^2 + 1$, so the graph of the equation, $y = x^2 + 1$, is not symmetric with respect to the x-axis.

Replace x with $-x$ and y with $-y$. $\hspace{2cm} (-y) = (-x)^2 + 1$

Simplify. $\hspace{6cm} -y = x^2 + 1$

$\hspace{7.5cm} y = -x^2 - 1$

The resulting equation, $y = -x^2 - 1$, is not equivalent to the original equation, $y = x^2 + 1$, so the graph of the equation, $y = x^2 + 1$, is not symmetric with respect to the origin.

The equation $y = x^2 + 1$ is **symmetric with respect to the y-axis**.

Solution (b):

Replace x with $-x$. $\qquad\qquad\qquad\qquad\qquad y = (-x)^3 + 1$

Simplify. $\qquad\qquad\qquad\qquad\qquad\qquad\quad y = -x^3 + 1$

The resulting equation, $y = -x^3 + 1$, is not equivalent to the original equation, $y = x^3 + 1$. Therefore, the graph of the equation, $y = x^3 + 1$, is not symmetric with respect to the y-axis.

Replace y with $-y$. $\qquad\qquad\qquad\qquad\qquad (-y) = x^3 + 1$

Simplify. $\qquad\qquad\qquad\qquad\qquad\qquad\quad y = -x^3 - 1$

The resulting equation, $y = -x^3 - 1$, is not equivalent to the original equation, $y = x^3 + 1$. Therefore, the graph of the equation, $y = x^3 + 1$, is not symmetric with respect to the x-axis.

Replace x with $-x$ and y with $-y$. $\qquad\quad (-y) = (-x)^3 + 1$

Simplify. $\qquad\qquad\qquad\qquad\qquad\qquad -y = -x^3 + 1$

$$y = x^3 - 1$$

The resulting equation, $y = x^3 - 1$, is not equivalent to the original equation, $y = x^3 + 1$. Therefore, the graph of the equation, $y = x^3 + 1$, is not symmetric with respect to the origin.

The equation $y = x^3 + 1$ exhibits **no symmetry**.

■ **YOUR TURN**   Determine the symmetry (if any) for $x = y^2 - 1$.

Using Symmetry as a Graphing Aid

How can we use symmetry to assist us in graphing? Look back at Example 2, $y = x^2 - 5$. We selected seven x-coordinates and solved the equation to find the corresponding y-coordinates. If we had known that this graph was symmetric with respect to the y-axis, then we would have only had to find the solutions to the positive x-coordinates, since we get the negative x-coordinates for free. For example, we found the point $(1, -4)$ to be a solution to the equation. The rules of symmetry tell us that $(-1, -4)$ is also on the graph.

EXAMPLE 6 Using Symmetry as a Graphing Aid

For the equation, $x^2 + y^2 = 25$, use symmetry to help you graph the equation using the point plotting technique.

Solution:

Test for symmetry with respect to the y-axis.

Replace x with $-x$. $\qquad\qquad\qquad\qquad\qquad (-x)^2 + y^2 = 25$

Simplify. $\qquad\qquad\qquad\qquad\qquad\qquad\qquad\qquad x^2 + y^2 = 25$

The resulting equation is equivalent to the original, so the graph of $x^2 + y^2 = 25$ is symmetric with respect to the y-axis.

Test for symmetry with respect to the x-axis.

Replace y with $-y$. $\qquad\qquad\qquad\qquad\qquad x^2 + (-y)^2 = 25$

Simplify. $\qquad\qquad\qquad\qquad\qquad\qquad\qquad\qquad x^2 + y^2 = 25$

The resulting equation is equivalent to the original, so the graph of $x^2 + y^2 = 25$ is symmetric with respect to the x-axis.

Test for symmetry with respect to the origin.

Replace x with $-x$ and y with $-y$. $\qquad\quad (-x)^2 + (-y)^2 = 25$

Simplify. $\qquad\qquad\qquad\qquad\qquad\qquad\qquad\qquad x^2 + y^2 = 25$

The resulting equation is equivalent to the original, so the graph of $x^2 + y^2 = 25$ is symmetric with respect to the origin.

Since the graph is symmetric with respect to the y-axis, x-axis, and origin, we need to determine solutions to the equation on only the positive x- and y-axes and in quadrant I because of the following symmetries:

- Symmetry with respect to the y-axis gives the solutions in quadrant II.
- Symmetry with respect to the origin gives the solutions in quadrant III.
- Symmetry with respect to the x-axis yields solutions in quadrant IV.

Solutions to $x^2 + y^2 = 25$.

Axes: $(0, 5)$ and $(5, 0)$

Quadrant I: $(3, 4)$, $(4, 3)$

Additional points due to symmetry:

Axes: $(0, -5)$ and $(-5, 0)$

Quadrant II: $(-3, 4)$, $(-4, 3)$

Quadrant III: $(-3, -4)$, $(-4, -3)$

Quadrant IV: $(3, -4)$, $(4, -3)$

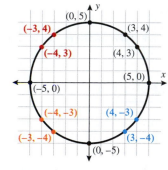

Connecting the points with a smooth curve yields a **circle**. We discuss circles in more detail in Chapter 3.

TECHNOLOGY TIP

To enter the graph of $x^2 + y^2 = 25$, solve for y first. The graphs of $y_1 = \sqrt{25 - x^2}$ and $y_2 = -\sqrt{25 - x^2}$ are shown.

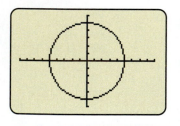

SECTION A.2 **SUMMARY**

In this section you graphed equations using a point plotting technique, tested equations for symmetry, and used symmetry about the x-axis, y-axis, and origin as a graphing aid.

SECTION A.2 EXERCISES

■ SKILLS

In Exercises 1–4, determine whether each point lies on the graph of the equation.

1. $y = x^2 - 2x + 1$ **a.** $(-1, 4)$ **b.** $(0, -1)$ **2.** $y = x^3 - 1$ **a.** $(-1, 0)$ **b.** $(-2, -9)$

3. $y = \sqrt{x + 2}$ **a.** $(7, 3)$ **b.** $(-6, 4)$ **4.** $y = 2 + |3 - x|$ **a.** $(9, -4)$ **b.** $(-2, 7)$

In Exercises 5–8, complete the table and use the table to sketch a graph of the equation.

5.

x	$y = 2 + x$	(x, y)
-2		
0		
1		

6.

x	$y = 3x - 1$	(x, y)
-1		
0		
2		

7.

x	$y = x^2 - x$	(x, y)
-1		
0		
$\frac{1}{2}$		
1		
2		

8.

x	$y = -\sqrt{x + 2}$	(x, y)
-2		
-1		
2		
7		

In Exercises 9–14, graph the equation by plotting points.

9. $y = -3x + 2$ **10.** $y = x^2 - x - 2$ **11.** $x = y^2 - 1$

12. $x = |y + 1| + 2$ **13.** $y = \dfrac{1}{2}x - \dfrac{3}{2}$ **14.** $y = 0.5|x - 1|$

In Exercises 15–20, match the graph with the corresponding symmetry.

a. No symmetry   **b.** Symmetry with respect to the x-axis   **c.** Symmetry with respect to the y-axis
d. Symmetry with respect to the origin   **e.** Symmetry with respect to the x-axis, y-axis, and origin

15.

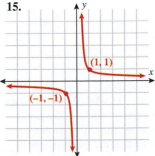

16.

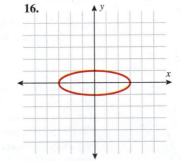

17.

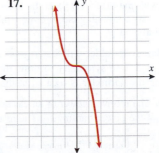

18. 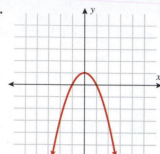 **19.** 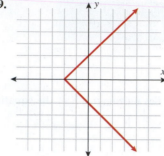 **20.**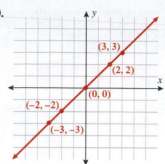

In Exercises 21–26, a point that lies on a graph is given along with that graph's symmetry. State the other points that must also lie on the graph.

POINT ON A GRAPH	THE GRAPH IS SYMMETRIC ABOUT THE	POINT ON A GRAPH	THE GRAPH IS SYMMETRIC ABOUT THE
21. $(-1, 3)$	x-axis	**22.** $(-2, 4)$	y-axis
23. $(7, -10)$	origin	**24.** $(-1, -1)$	origin
25. $(3, -2)$	x-axis, y-axis, and origin	**26.** $(-1, 7)$	x-axis, y-axis, and origin

In Exercises 27–40, use algebraic tests to determine whether the equation's graph is symmetric with respect to the x-axis, y-axis, or origin.

27. $x = y^2 + 4$
28. $x = 2y^2 + 3$
29. $y = x^3 + x$
30. $y = x^5 + 1$
31. $x = |y|$

32. $x = |y| - 2$
33. $x^2 - y^2 = 100$
34. $x^2 + 2y^2 = 30$
35. $y = x^{2/3}$
36. $x = y^{2/3}$

37. $x^2 + y^3 = 1$
38. $y = \sqrt{1 + x^2}$
39. $y = \dfrac{2}{x}$
40. $xy = 1$

In Exercises 41–52, use symmetry to help you graph the given equations.

41. $y = x$
42. $y = x^2 - 1$
43. $y = \dfrac{x^3}{2}$
44. $x = y^2 + 1$

45. $y = \dfrac{1}{x}$
46. $xy = -1$
47. $y = |x|$
48. $|x| = |y|$

49. $x^2 + y^2 = 16$
50. $\dfrac{x^2}{4} + \dfrac{y^2}{9} = 1$
51. $x^2 - y^2 = 16$
52. $x^2 - \dfrac{y^2}{25} = 1$

 ■ **APPLICATIONS**

53. Bomb. A particular bomb has a destruction pattern governed by the equation $x^2 + y^2 = 9$ where the xy-plane represents a town with the origin (0, 0) as the town center. Use symmetry as a graphing aid to draw a picture of the destruction area.

54. Sprinkler. A sprinkler will water grass in the shape of $x^2 + \frac{y^2}{9} = 1$. Use symmetry to draw the watered area assuming the sprinkler is located at the origin.

55. Signals. The received power of an electromagnetic signal is a fraction of the power transmitted. The relationship is given by

$$P_{\text{received}} = P_{\text{transmitted}} \cdot \dfrac{1}{R^2}$$

where R is the distance that the signal has traveled in meters. Plot the points showing the percentage of

transmitted power that is received for $R = 100$ m, 1 km, and 10,000 km.

56. Signals. The wavelength, λ, and the frequency, f, of a signal are related by the equation

$$f = \dfrac{c}{\lambda}$$

where c is the speed of light in a vacuum, $c = 3.0 \times 10^8$ meters per second. For the values, $\lambda = 0.001, \lambda = 1$, and $\lambda = 100$ mm, plot the points corresponding to frequency, f. What do you notice about the relationship between frequency and wavelength? Note that the frequency will have units Hz = 1/seconds.

■ CATCH THE MISTAKE

In Exercises 57–60, explain the mistake that is made.

57. Graph the equation $y = x^2 + 1$.

Solution:

x	$y = x^2 + 1$	(x, y)
0	1	(0, 1)
1	2	(1, 2)

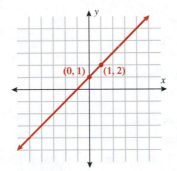

58. Test $y = -x^2$ for symmetry with respect to the y-axis.

Solution:

Replace x with $-x$. $y = -(-x)^2$

Simplify. $y = x^2$

The resulting equation, $y = x^2$, is not equivalent to the original equation. $y = -x^2$ is not symmetric with respect to the y-axis.

59. Test $x = |y|$ for symmetry with respect to the y-axis.

Solution:

Replace y with $-y$. $x = |-y|$

Simplify. $x = |y|$

The resulting equation is equivalent to the original equation. $x = |y|$ is symmetric with respect to the y-axis.

60. Use symmetry to help you graph $x^2 = y - 1$.

Solution:

Replace x with $-x$. $(-x)^2 = y - 1$

Simplify. $x^2 = y - 1$

$x^2 = y - 1$ is symmetric with respect to the x-axis.

Determine points that lie on the graph in quadrant I.

y	$x^2 = y - 1$	(x, y)
1	0	(0, 1)
2	1	(1, 2)
5	2	(2, 5)

Symmetry with respect to the x-axis implies that $(0, -1)$, $(1, -2)$, and $(2, -5)$ are also points that lie on the graph.

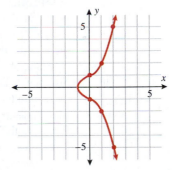

■ CHALLENGE

61. T or F: If the point (a, b) lies on a graph that is symmetric about the x-axis, then the point $(-a, b)$ also must lie on the graph.

62. T or F: If the point (a, b) lies on a graph that is symmetric about the y-axis, then the point $(-a, b)$ also must lie on the graph.

63. T or F: If the point $(a, -b)$ lies on a graph that is symmetric about the x-axis, y-axis, and origin, then the points (a, b), $(-a, -b)$ and $(-a, b)$ must also lie on the graph.

64. Determine if the graph of $y = \dfrac{ax^2 + b}{cx^3}$ has any symmetry, where a, b, and c are real numbers.

■ TECHNOLOGY

In Exercises 65–68, graph the equation using a graphing utility and state if there is any symmetry.

65. $y = 16.7x^4 - 3.3x^2 + 7.1$ **66.** $y = 0.4x^5 + 8.2x^3 - 1.3x$ **67.** $2.3x^2 = 5.5\,|y|$ **68.** $3.2x^2 - 5.1y^2 = 1.3$

Introduction to Functions

What do the following relations have in common?

- Every person has a blood type.
- Every real number has an absolute value.
- Every person has a DNA sequence.
- Every working household phone in the United States has a 10-digit phone number.

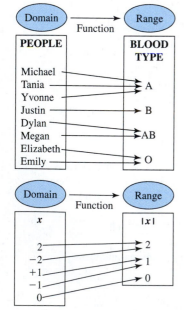

They all describe a particular correspondence between two groups. A **relation** is a correspondence between two sets. A relation is a **function** when each element in one group or set corresponds to *only* one element in another group or set. The first set (or input set) is called the **domain**, and the corresponding second set (or output set) is called the **range**. If each member of the domain has *exactly one* corresponding member in the range, then that relation is called a function. All of the above examples are functions.

In the blood-type example on the left, each person can have only *one* blood type. For instance, Michael cannot be both type A and type O. Because each person in the domain corresponds to a single blood type in the range, this relation is a function. Notice that two elements in the domain may correspond to the same element in the range. In this example, both Elizabeth and Emily have type O blood.

One way of determining if this example constitutes a function is to ask the question: If I know the person's name, do I know his or her blood type? And the answer is yes. Before DNA testing revolutionized the forensic sciences, blood types were used to narrow down the list of suspects in investigations. Because no two people have the same DNA, forensic scientists now can match DNA to a specific suspect. The DNA example also constitutes a function because every person corresponds to a unique DNA structure.

In the absolute value example on the left, each number in the domain corresponds to a single value. For instance, the absolute value of -1 is 1.

Notice that two numbers in the domain may correspond to the same number in the range. For example, both -1 and 1 in the domain correspond to 1 in the range. The key to a relation being classified as a function is that each element in the domain has *one and only one* element in the range that it maps to. If an element in the domain corresponds to two or more elements in the range, then the relationship is *not* a function.

In the following examples, elements in the domain map to more than one element in the range. Hence the two examples are not functions.

At a university, four primary sports typically overlap in the late fall: football, volleyball, soccer, and basketball. The start times are 1 P.M., 3 P.M., and 7 P.M. The 1 P.M. start time corresponds to one and only one event: football. However, the 3 P.M. start time corresponds to *both* volleyball and soccer. Because 3 P.M. corresponds to *both* volleyball and soccer, this is not a unique relation and, hence, is *not* a function.

If the temperature is measured throughout the day, there can be more than one time during the day when the same temperature is recorded. If we know the temperature, do we know the time of day? No, because each temperature does not correspond to one and only one time.

In the example on the right, although 85 °F corresponds to only 11 A.M. and 91 °F corresponds to only 4 P.M., 87 °F corresponds to two different times (2 P.M. and 6 P.M.). Since one element in the domain corresponds to two elements in the range, this is *not* a function.

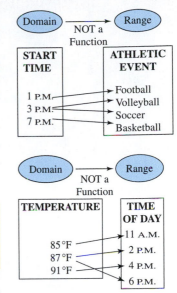

FUNCTION

A **function** is a relation that maps each element in the domain to *one and only one* element in the range.

If we let x represent an element in the domain and y represent an element in the range, then we can think of a function as a set of ordered pairs (x, y). Because a function maps each element in the domain to one and only one element in the range, then functions are ordered pairs that have no common x values. For example, here are three sets of ordered pairs. Which sets represent a function and why?

■ $\{(1, 2), (3, 4), (5, 6), (7, 8)\}$ Function
■ $\{(1, 2), (3, 2), (5, 6), (7, 6)\}$ Function
■ $\{(1, 2), (1, 3), (5, 6), (5, 7)\}$ Not a function

In the first set, all of the pairs are distinct with no first elements repeated. Therefore, it is a function. In the second set, each first element corresponds to one and only one second element. Although two first elements map to the same second element, this still represents a function. The last set does not represent a function because there are repeated first elements, indicating multiple mappings for each element in the domain.

EXAMPLE 1 Classifying Relationships as Functions

Classify the following relationships as functions or not functions and justify your answer.

a. $\{(-3, 4), (2, 4), (3, 5), (6, 4)\}$ **b.** $\{(-3, 4), (2, 4), (3, 5), (2, 2)\}$

c. Domain = set of all items for sale in a grocery store; Range = price

Solution:

a. Function: No x value is repeated. Therefore, each x maps to one and only one y.

b. Not a function: $x = 2$ maps to *both* $y = 2$ and $y = 4$.

c. Function: Each item has one and only one price. It is still a function even though some items have the same price (or elements in the domain map to the same element in the range).

All of the examples we have discussed thus far are finite, discrete sets. They represent a finite number of distinct pairs of (x, y). For instance, for a set consisting of certain people, we have their corresponding blood type: (Michael, A), (Tania, A) (Yvonne, A), (Justin, B), (Dylan, AB), (Megan, AB), (Elizabeth, O), and (Emily, O).

Using ordered pairs of numbers, we described the set $\{(1, 2), (3, 4), (5, 6), (7, 8)\}$ as a function. If we plot this set of ordered pairs, we have 4 distinct points on the graph.

Domain	Range
x	y
1	2
3	4
5	6
7	8

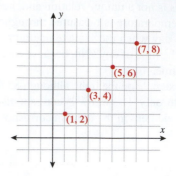

There are only four elements in the domain and four elements in the range.

Notice that these points all lie on the line, $y = x + 1$. What if we wanted other points that lie on this line? We could use this equation to map x values to corresponding y values. Some additional points are $(0, 1)$, $(-2, -1)$, and $(8, 9)$.

Function: $y = x + 1$.

Domain	Range
x	y
1	2
3	4
5	6
7	8
-2	-1
0	1
8	9

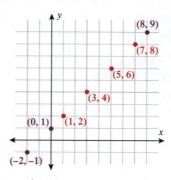

We now have seven distinct pairs of numbers. What if we want *all* of the points that lie on this line? The equation, $y = x + 1$, describes the correspondence between the x and y values. To describe the mapping from all real values of x to the corresponding real values of y, we no longer use a discrete set of points but rather indicate an infinite set of points, or continuum, described by the equation: $y = x + 1$.

Function: $y = x + 1$.

Domain	Range
x	y
$\mathbb{R}$	$\mathbb{R}$

The symbol $\mathbb{R}$ denotes all real numbers.

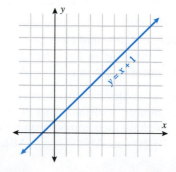

This function maps an infinite set of real numbers to a corresponding infinite set of real numbers and forms a continuum of ordered pairs as opposed to discrete points.

Functions Defined by Equations

Displaying a function as a table or a set of ordered pairs only works if the domain and range are finite sets. If the domain and/or range are infinite sets of real numbers, then functions must be defined using equations.

Let's start with the equation $y = x^2 - 3x$, where x can be any real number. This equation assigns each x *one and only one* corresponding value for y.

x	$y = x^2 - 3x$	y
1	$y = (1)^2 - 3(1)$	-2
5	$y = (5)^2 - 3(5)$	10
$-\dfrac{2}{3}$	$y = \left(-\dfrac{2}{3}\right)^2 - 3\left(-\dfrac{2}{3}\right)$	$\dfrac{22}{9}$
1.2	$y = (1.2)^2 - 3(1.2)$	-2.16

Since the variable y *depends* on what value of x is selected, we denote y as the **dependent variable**. The variable x can be any number in the domain; therefore we denote x as the **independent variable**. Although it is customary to use the variables x and y, any variables can be used in a function relation. The variable in the domain is the independent variable, and the corresponding variable in the range is the dependent variable.

Words that are synonymous with *domain* are *input* or *independent variable*. Words that are synonymous with *range* are *output* or *dependent variable*. Typically, the variable x is used to denote the domain or input, and the variable y is often used to denote the range or output. The coordinates (x, y) represent a corresponding pair of elements: one in the domain and one in the range.

Function

Domain	Range
Input	Output
x	y
Independent variable	Dependent variable

Although functions are defined by equations, it is important to recognize that *not all equations are functions*. The requirement for an equation to define a function is that each element in the domain corresponds to only one element in the range.

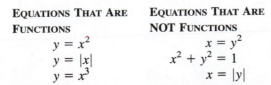

EQUATIONS THAT ARE FUNCTIONS	EQUATIONS THAT ARE NOT FUNCTIONS
$y = x^2$	$x = y^2$
$y = |x|$	$x^2 + y^2 = 1$
$y = x^3$	$x = |y|$

In the "equations that are functions," every x corresponds to only one y. Some points that correspond to these functions are

$$y = x^2: \quad (-1, 1)\ (0, 0)\ (1, 1)$$

$$y = |x|: \quad (-1, 1)\ (0, 0)\ (1, 1)$$

$$y = x^3: \quad (-1, -1)\ (0, 0)\ (1, 1)$$

The fact that $x = -1$ and $x = 1$ both correspond to $y = 1$ in the first two examples does not violate the definition of a function.

In the "equations that are NOT functions," some x values correspond to *more than one* y value. Some points that correspond to these equations are

$$x = y^2: \quad (\mathbf{1}, \mathbf{-1})\ (0, 0)\ (\mathbf{1}, \mathbf{1}) \qquad x = \mathbf{1} \text{ maps to } \mathbf{both}\ y = \mathbf{-1} \text{ and } y = \mathbf{1}$$

$$x^2 + y^2 = 1: \quad (\mathbf{0}, \mathbf{-1})\ (\mathbf{0}, \mathbf{1})\ (-1, 0)\ (1, 0) \qquad x = \mathbf{0} \text{ maps to } \mathbf{both}\ y = \mathbf{-1} \text{ and } y = \mathbf{1}$$

$$x = |y|: \quad (\mathbf{1}, \mathbf{-1})\ (0, 0)\ (\mathbf{1}, \mathbf{1}) \qquad x = \mathbf{1} \text{ maps to } \mathbf{both}\ y = \mathbf{-1} \text{ and } y = \mathbf{1}$$

Let's look at the graphs of the three **functions:**

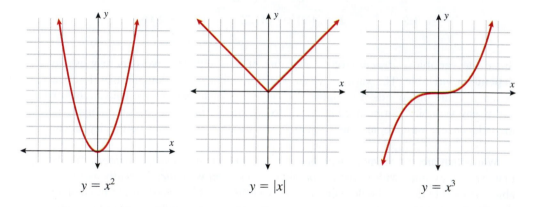

$$y = x^2 \qquad\qquad y = |x| \qquad\qquad y = x^3$$

Let's take any value for x, say $x = a$. This corresponds to a vertical line. A function maps one x to only one y; therefore there should be only one point of intersection with any vertical line. We see in the three graphs of functions above that if a vertical line is drawn at any value of x on any of the three graphs, the vertical line only intersects the graph in one place. Look at the graphs of the three equations that are **not functions**.

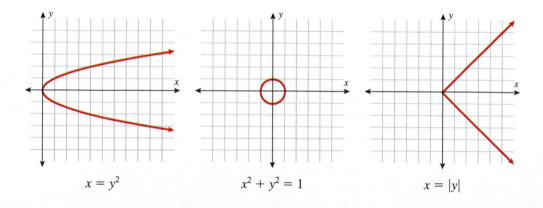

$$x = y^2 \qquad\qquad x^2 + y^2 = 1 \qquad\qquad x = |y|$$

If a vertical line is drawn at $x = \frac{1}{2}$ in any of the three graphs, that vertical line will intersect each of these graphs at two points. Thus, there are two y values that correspond to each x in the domain, which is why these equations do not define functions.

EXAMPLE 2 Using the Vertical Line Test

Use the vertical line test to determine if the graphs of equations define functions.

a.

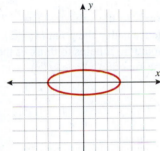

b.

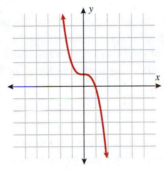

Solution:

Apply the vertical line test.

a.

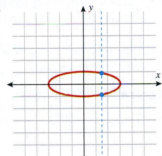

b.

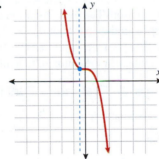

a. Because the vertical line intersects the graph of the equation at two points, this equation is *not a function.*

b. Because the vertical line intersects the graph of the equation at *one and only one* point for any x-value, this equation *is a function.*

YOUR TURN   Determine if the equation $(x - 3)^2 + (y + 2)^2 = 16$ is a function.

■ Answer: The graph of the equation is a circle which does not pass the vertical line test. Therefore, the equation does not define a function.

Function Notation

It was mentioned earlier that functions can be defined by equations such as $y = x^2 - 3x$, where y is the dependent variable. We often use an equivalent **function notation** instead:

$$f(x) = x^2 - 3x$$

The symbol $f(x)$ is read "f evaluated at x" or "f of x" and represents the value in the range that the function corresponds to, given a specified value of x in the domain. For instance, $f(1)$ represents the value of the function when $x = 1$.

$$f(x) = x^2 - 3x$$
$$f(1) = (1)^2 - 3(1)$$
$$f(1) = 1 - 3 = -2$$

STUDY TIP

When using function notation, it helps to think of the independent variable as a placeholder.

Therefore, the value $x = 1$ in the domain corresponds to the value $f(1) = -2$ in the range. A graphical interpretation of this correspondence is that the point $(1, -2)$ lies on the graph of the function, f. When using function notation, it helps to think of the independent variable as a placeholder. For instance, the function above, $f(x) = x^2 - 3x$, can be thought of as:

$$f(\ ) = (\ )^2 - 3(\ )$$

It is important to note that $f(x)$ **does not** represent f "times" x.

In the above example, we have used x to represent the independent variable. Because $y = f(x)$, we say that y and $f(x)$ both represent the dependent variable. We say that "f maps x into $f(x)$." In other words, the function, f, maps the independent variable, x in the domain, to a dependent variable, $f(x)$ in the range.

Although f is the most logical letter to represent a function, other letters can be used to denote a function. The most common are F, G, or g, but any letter can be used to represent either the independent or dependent variable.

We state that "G maps t into $G(t)$." In other words, the function, G, maps the independent variable, t in the domain, to a dependent variable, $G(t)$ in the range.

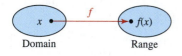

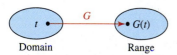

EXAMPLE 3 Evaluating Functions: Interpreting Mapping

Given the function $f(x) = 2x^3 - 3x^2 + 6$ find $f(-1)$.

Solution:

The independent variable, x, is a placeholder. $\qquad f(\ ) = 2(\ )^3 - 3(\ )^2 + 6$

To find $f(-1)$, substitute $x = -1$
into the function. $\qquad\qquad\qquad\qquad f(-1) = 2(-1)^3 - 3(-1)^2 + 6$

Evaluate the right side. $\qquad\qquad\qquad\qquad f(-1) = -2 - 3 + 6$

Simplify. $\qquad\qquad\qquad\qquad\qquad\qquad f(-1) = 1$

We can interpret $f(-1) = 1$ as "the function $f(x) = 2x^3 - 3x^2 + 6$ maps -1 in the domain to 1 in the range.

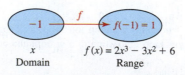

The independent variable is also referred to as the **argument** of a function. In the case of the function $f(x) = x^2 - 3x$, we say that x is the argument of f. The function $f(x) = x^2 - 3x$ can be thought of as $f(\ ) = (\ )^2 - 3(\ )$, because the argument is a placeholder. A way to describe this function is "the argument squared minus 3 times the argument." *Any* expression can be substituted in for the argument.

$$f(1) = (1)^2 - 3(1)$$
$$f(x + 1) = (x + 1)^2 - 3(x + 1)$$
$$f(-x) = (-x)^2 - 3(-x)$$

 EXAMPLE 4 Evaluating Functions with Variable Arguments

Evaluate $f(x + 1)$, given that $f(x) = x^2 - 3x$.

COMMON MISTAKE

A common misunderstanding is to interpret the notation $f(x + 1)$ as a sum: $f(x + 1) \neq f(x) + f(1)$.

 CORRECT

Write original function.
$$f(x) = x^2 - 3x$$

Replace the argument, x, with a placeholder.
$$f(\ ) = (\ )^2 - 3(\ )$$

Substitute $x + 1$ for the argument.
$$f(x + 1) = (x + 1)^2 - 3(x + 1)$$

Eliminate parentheses.
$$f(x + 1) = x^2 + 2x + 1 - 3x - 3$$

Combine like terms.
$$f(x + 1) = x^2 - x - 2$$

 INCORRECT

The **ERROR** is interpreting as a sum.
$$f(x + 1) = f(x) + f(1)$$

Substituting $f(x) = x^2 - 3x$ and $f(1) = -2$ into the right side leads to the **INCORRECT** answer.
$$f(x + 1) = x^2 - 3x - 2$$

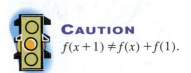 **CAUTION**
$f(x+1) \neq f(x) + f(1)$.

■ **YOUR TURN**   Evaluate $g(x - 1)$ given that $g(x) = x^2 - 2x + 3$.

EXAMPLE 5 Evaluating Functions: Sums

For the given function $H(x) = x^2 + 2x$ evaluate:

a. $H(x + 1)$ **b.** $H(x) + H(1)$

Solution (a):

Write the function H in placeholder notation.
$$H(\ ) = (\ )^2 + 2(\ )$$

Substitute $x + 1$ in for the argument of H.
$$H(x + 1) = (x + 1)^2 + 2(x + 1)$$

■ **Answer:** $g(x - 1) = x^2 - 4x + 6$

TECHNOLOGY TIP

Use a graphing utility to display graphs of $y_1 = H(x + 1) = (x + 1)^2 + 2(x + 1)$ and $y_2 = H(x) + H(1) = x^2 + 2x + 3$ in a $[-6, 3]$ by $[-5, 10]$ viewing rectangle.

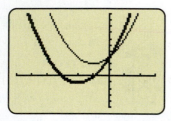

```
Plot1 Plot2 Plot3
\Y1◘(X+1)²+2(X+1
)
\Y2◘X²+2X+3
\Y3=█
```

The graphs are not the same. $H(x + 1) \neq H(x) + H(1)$

Eliminate the parentheses on the right side.

$$H(x + 1) = x^2 + 2x + 1 + 2x + 2$$

Combine like terms on the right side.

$$H(x + 1) = x^2 + 4x + 3$$

Solution (b):

Write $H(x)$.

$$H(x) = x^2 + 2x$$

Evaluate H at $x = 1$.

$$H(1) = (1)^2 + 2(1) = 3$$

Evaluate the sum $H(x) + H(1)$.

$$H(x) + H(1) = x^2 + 2x + 3$$

$$H(x) + H(1) = x^2 + 2x + 3$$

Note: Comparing the results of part a and part b, we see that $H(x + 1) \neq H(x) + H(1)$.

EXAMPLE 6 Evaluating Functions: Negatives

For the given function $G(t) = t^2 - t$ evaluate:

a. $G(-t)$ **b.** $-G(t)$

Solution (a):

Write the function G in placeholder notation.

$$G(\ ) = (\ )^2 - (\ )$$

Substitute $-t$ in for the argument of G.

$$G(-t) = (-t)^2 - (-t)$$

Eliminate the parentheses on the right side.

$$G(-t) = t^2 + t$$

Solution (b):

Write $G(t)$.

$$G(t) = t^2 - t$$

Multiply both sides by -1.

$$-G(t) = -(t^2 - t)$$

Eliminate parentheses on right side.

$$-G(t) = -t^2 + t$$

Note: Comparing the results of part a and part b, we see that $G(-t) \neq -G(t)$.

EXAMPLE 7 Evaluating Functions: Quotients

For the given function $F(x) = 3x + 5$ evaluate:

a. $F\left(\dfrac{1}{2}\right)$ **b.** $\dfrac{F(1)}{F(2)}$

Solution (a):

Write F in placeholder notation.

$$F(\ ) = 3(\ ) + 5$$

Replace the argument with $\frac{1}{2}$.

$$F\left(\frac{1}{2}\right) = 3\left(\frac{1}{2}\right) + 5$$

Simplify the right side.

$$F\left(\frac{1}{2}\right) = \frac{13}{2}$$

Solution (b):

Evaluate $F(1)$.

$$F(1) = 3(1) + 5 = 8$$

Evaluate $F(2)$.

$$F(2) = 3(2) + 5 = 11$$

Divide $F(1)$ by $F(2)$.

$$\frac{F(1)}{F(2)} = \frac{8}{11}$$

CAUTION

$$f\left(\frac{a}{b}\right) \neq \frac{f(a)}{f(b)}$$

Note: Comparing the results of part a and part b, we see that $F\left(\dfrac{1}{2}\right) \neq \dfrac{F(1)}{F(2)}$.

■ **YOUR TURN** Given the function, $G(t) = 3t - 4$, evaluate:

a. $G(t - 2)$ **b.** $\dfrac{G(1)}{G(3)}$

Domain of a Function

The word **domain** has been used to represent the input or independent variables that are mapped to a range via the function correspondence. What elements are allowed in the domain? Let's investigate three different functions and ask the question: Are there any restrictions on what elements are allowed in the domain?

$$f(x) = x^3 - 4x^2 + 17 \qquad g(x) = \sqrt{x} \qquad h(x) = \frac{1}{x}$$

The first function, $f(x) = x^3 - 4x^2 + 17$, maps each value in the domain to one and only one corresponding value in the range. Any real number can be cubed, squared, and combined with other constants, so there are no restrictions on what numbers can be "input" into this function. Therefore, we say the domain of the function f is all real numbers, $\mathbb{R}$.

The second function, $g(x) = \sqrt{x}$, maps numbers from the domain to the range by taking the square root. Are there any real numbers on which the square root is taken that yield something other than a real number? In other words, what can x be? The square root of a positive real number yields some other positive real number. The square root of zero is zero. The square root of a negative real number, however, is an imaginary number. Therefore, in order to have real numbers in the range, we must restrict the domain to nonnegative real numbers, $[0, \infty)$.

The third function, $h(x) = \frac{1}{x}$, maps quantities in the domain to unique quantities in the range by taking the reciprocal of the input to yield the output. We again ask the question: What can x be? The variable x can be any real number except zero. Therefore, we must restrict the domain to all real numbers except zero, which can be written as *all real numbers except 0, $x \neq 0$,* or in interval notation as $(-\infty, 0) \cup (0, \infty)$.

EXAMPLE 8 Determining the Domain of a Function

State the domain of the given functions.

a. $F(x) = \dfrac{3}{x^2 - 25}$ **b.** $H(x) = \sqrt{9 - 2x}$ **c.** $G(t) = |4 - t|$

Solution (a):

Write the original equation.

$$F(x) = \frac{3}{x^2 - 25}$$

Determine any restrictions on the values of x.

$$x^2 - 25 = 0$$

■ **Answer: a.** $G(t-2) = 3t - 10$ **b.** $\dfrac{G(1)}{G(3)} = -\dfrac{1}{5}$

Solve the restriction equation.

$x^2 = 25$ or $x = \pm\sqrt{25} = \pm 5$

State the domain restrictions.

$x \neq \pm 5$

Write domain in interval notation.

$(-\infty, -5) \cup (-5, 5) \cup (5, \infty)$

The domain is restricted to all real numbers except -5 and 5.

Solution (b):

Write the original equation.

$H(x) = \sqrt{9 - 2x}$

Determine any restrictions on the values of x.

$9 - 2x \geq 0$

Solve the restriction equation.

$9 \geq 2x$

State the domain restrictions.

$\dfrac{9}{2} \geq x$ or $x \leq \dfrac{9}{2}$

Write domain in interval notation.

$\left(-\infty, \dfrac{9}{2}\right]$

The domain is restricted to all real numbers less than or equal to $\frac{9}{2}$.

Solution (c):

Write the original equation.

$G(t) = |4 - t|$

Determine any restrictions on the values of t.

no restrictions

State the domain.

$\mathbb{R}$

Write domain in interval notation.

$(-\infty, \infty)$

The domain is the set of all real numbers.

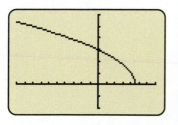

TECHNOLOGY TIP
Graph of $H(x) = \sqrt{9 - 2x}$.

Notice that $H(x)$ is defined for all x less than or equal to $\frac{9}{2}$.

CONCEPT CHECK Is the function $f(x) = \sqrt{x - 3}$ defined for $x = 0$?

Is the function $g(x) = \dfrac{1}{x^2 - 4}$ defined for $x = 2$?

■ **YOUR TURN** State the domain of the given functions.

a. $f(x) = \sqrt{x - 3}$ **b.** $g(x) = \dfrac{1}{x^2 - 4}$

Applications

Functions that are used in applications often have restrictions on the domains due to physical constraints. For example, the volume of a square cube is given by the function $V(x) = x^3$ where x is the length of a side. The function $f(x) = x^3$ has no restrictions on x, and therefore the domain is the set of all real numbers. However, the volume of a box has the restriction that the length of a side can never be negative or zero.

■ **Answer: a.** The domain is the set of all real numbers greater than or equal to 3, which can be written as $x \geq 3$ or $[3, \infty)$. **b.** The domain is the set of all real numbers except -2 and 2, which can be written as $x \neq \pm 2$ or $(-\infty, -2) \cup (-2, 2) \cup (2, \infty)$.

EXAMPLE 9 **Price of Gasoline**

Following the capture of Saddam Hussain in Iraq in 2003, gas prices in the United States escalated and then finally returned to their precapture prices. Over a 6-month period the average price of a gallon of 87 octane gasoline was given by the function $C(x) = -0.05x^2 + 0.3x + 1.7$ where C is the cost function and x represents the number of months after the capture.

a. Determine the domain of the cost function.

b. What was the average price of gas per gallon 3 months after the capture?

Solution (a):

Since the cost function $C(x) = -0.05x^2 + 0.3x + 1.7$ modeled the price of gas only for 6 months after the capture, the domain is $0 \le x \le 6$ or $[0, 6]$.

Solution (b):

Write the cost function.

$$C(x) = -0.05x^2 + 0.3x + 1.7 \quad 0 \le x \le 6$$

Find the value of the function when $x = 3$.

$$C(3) = -0.05(3)^2 + 0.3(3) + 1.7$$

Simplify.

$$C(3) = 2.15$$

The average price per gallon 3 months after the capture was $\$2.15$.

EXAMPLE 10 **The Dimensions of a Pool**

Express the volume of a 30 foot $\times$ 10 foot rectangular swimming pool as a function of its depth.

Solution:

The volume of any rectangular box is $V = lwh$ where V is the volume, l is the length, w is the width, and h is the height. In this example the length is 30 feet, the width is 10 feet, and the height is the depth, d.

Write the volume as a function of depth, d. $V(d) = (30)(10)d$

Simplify. $V(d) = 300d$

Determine any restrictions on the domain. $d > 0$

SECTION A.3 SUMMARY

Determining if a relation is a function can be done both algebraically and graphically.

- Algebraically: Each element in the domain maps to one and only one element in the range. It is still a function if two elements in the domain map to the same element in the range.

- Graphically: Vertical line test—If a vertical line is drawn anywhere on a graph and intersects the graph of an equation in more than one point, then the graph does not represent a function. If they intersect in at most one point, then it is a function.

When evaluating functions the independent variable acts like a placeholder.

To determine the domain of a function, ask the question: Are there any restrictions on what the input can be?

SECTION A.3 EXERCISES

■ SKILLS

In Exercises 1–24, determine whether each relation is a function. Assume that the coordinate pair (x, y) represents the independent variable x and the dependent variable y.

1.

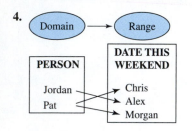

2.

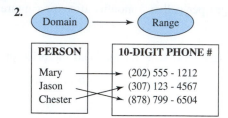

3.

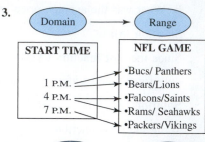

4.

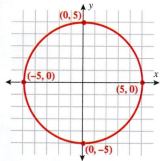

5.

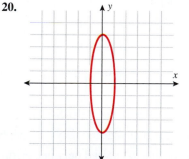

6.

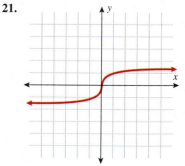

7. $\{(0, -3), (0, 3), (-3, 0), (3, 0)\}$

8. $\{(2, -2), (2, 2), (5, -5), (5, 5)\}$

9. $\{(0, 0), (9, -3), (4, -2), (4, 2), (9, 3)\}$

10. $\{(0, 0), (-1, -1), (-2, -8), (1, 1), (2, 8)\}$

11. $\{(0, 1), (1, 0), (2, 1), (-2, 1), (5, 4), (-3, 4)\}$

12. $\{(0, 1), (1, 1), (2, 1), (3, 1)\}$

13. $x^2 + y^2 = 9$

14. $x = |y|$

15. $x = y^2$

16. $y = x^3$

17. $y = |x - 1|$

18. $y = 3$

19.

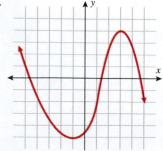

20.

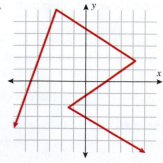

21.

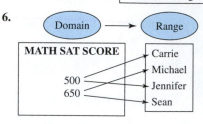

22.

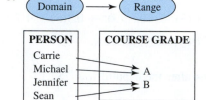

23.

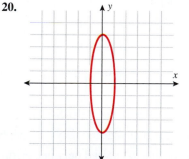

24.

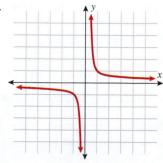

In Exercises 25–48, evaluate the given quantities using the following four functions.

$$f(x) = 2x - 3 \qquad F(t) = 4 - t^2 \qquad g(t) = 5 + t \qquad G(x) = x^2 + 2x - 7$$

25. $f(-2)$

26. $G(-3)$

27. $g(1)$

28. $F(-1)$

29. $f(-2) + g(1)$

30. $G(-3) - F(-1)$

31. $3f(-2) - 2g(1)$

32. $2F(-1) - 2G(-3)$

33. $\dfrac{f(-2)}{g(1)}$

34. $\dfrac{G(-3)}{F(-1)}$

35. $\dfrac{f(0) - f(-2)}{g(1)}$

36. $\dfrac{G(0) - G(-3)}{F(-1)}$

37. $f(x + 1) - f(x - 1)$

38. $F(t + 1) - F(t - 1)$

39. $g(x + a) - f(x + a)$

40. $G(x + b) + F(b)$

41. $\dfrac{f(x + h) - f(x)}{h}$

42. $\dfrac{F(t + h) - F(t)}{h}$

43. $\dfrac{g(t + h) - g(t)}{h}$

44. $\dfrac{G(x + h) - G(x)}{h}$

45. $\dfrac{f(-2 + h) - f(-2)}{h}$

46. $\dfrac{F(-1 + h) - F(-1)}{h}$

47. $\dfrac{g(1 + h) - g(1)}{h}$

48. $\dfrac{G(-3 + h) - G(-3)}{h}$

In Exercises 49–66, find the domain of the given function. Express the domain in interval notation.

49. $f(x) = 2x - 5$

50. $f(x) = -2x - 5$

51. $g(t) = t^2 + 3t$

52. $h(x) = 3x^4 - 1$

53. $P(x) = \dfrac{x + 5}{x - 5}$

54. $Q(t) = \dfrac{2 - t^2}{t + 3}$

55. $T(x) = \dfrac{2}{x^2 - 4}$

56. $R(x) = \dfrac{1}{x^2 - 1}$

57. $F(x) = \dfrac{1}{x^2 + 1}$

58. $G(t) = \dfrac{2}{t^2 + 4}$

59. $q(x) = \sqrt{7 - x}$

60. $k(t) = \sqrt{t - 7}$

61. $f(x) = \sqrt{2x + 5}$

62. $g(x) = \sqrt{5 - 2x}$

63. $G(t) = \sqrt{t^2 - 4}$

64. $F(x) = \sqrt{x^2 - 25}$

65. $F(x) = \dfrac{1}{\sqrt{x - 3}}$

66. $G(x) = \dfrac{2}{\sqrt{5 - x}}$

 ■ **APPLICATIONS**

67. Budget. The cost associated with a catered wedding reception is $45 per person for a reception for more than 75 people. Write the cost of the reception in terms of the number of guests and state any domain restrictions.

68. Budget. The cost of a local home phone plan is $35 for basic service and $.10 per minute for any domestic long-distance calls. Write the cost of monthly phone service in terms of the number of monthly long-distance minutes and state any domain restrictions.

69. Temperature. The average temperature in Tampa, Florida, in the springtime is given by the function $T(x) = -0.7x^2 + 16.8x - 10.8$ where T is the temperature in degrees Fahrenheit and x is the time of day in military time and is restricted to $6 \le x \le 18$ (sunrise to sunset). What is the temperature at 6 A.M.? What is the temperature at noon?

70. Firecrackers. A firecracker is launched straight up, and its height is a function of time, $h(t) = -16t^2 + 128t$ where h is the height in feet and t is the time in seconds with $t = 0$ corresponding to the instant it launches. What is the height 4 seconds after launch? What is the domain of this function?

71. Collector Card. The price of a signed Alex Rodriguez card is a function of how many are for sale. When he was traded from the Texas Rangers to the New York Yankees, the going rate for a signed card on eBay in 2004 was $P(x) = 10 + \sqrt{400,000 - 100x}$ where x represents the number of signed cards for sale. What was the value of the card when there were 10 signed cards for sale? What was the value of the card when there were 100 signed cards for sale?

72. Collector Card. In Exercise 71, what was the lowest price on eBay, and how many cards were available then? What was the highest price on eBay, and how many cards were available then?

73. Volume. An open box is constructed from a square 10 inch piece of cardboard by cutting squares of length x out of each corner and folding the sides up. Express the volume of the box as a function of x and state the domain.

74. Volume. A cylindrical water basin will be built to harvest rainwater. The basin is limited in that the largest radius it can have is 10 feet. Write a function representing the volume of water, V, as a function of height, h. How many additional gallons of water will be collected if you increase the height by 2 feet? (Hint: 1 cubic foot = 7.48 gallons.)

■ CATCH THE MISTAKE

In Exercises 75–80, explain the mistake that is made.

75. Determine if the relationship is a function.

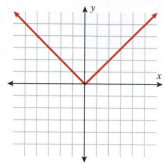

Solution: Apply the horizontal line test.

Because the horizontal line intersects the graph in two places, this is not a function.

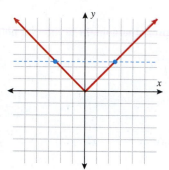

76. Given the function, $H(x) = 3x - 2$, evaluate the quantity $H(3) - H(-1)$.

Solution: $H(3) - H(-1) = H(3) + H(1)$

$H(3) = 7$ and $H(1) = 1$

$H(3) + H(1) = 7 + 1 = 8$

77. Given the function, $f(x) = x^2 - x$, evaluate the quantity $f(x + 1)$.

Solution: $f(x + 1) = f(x) + f(1)$

$f(x) = x^2 - x$ and $f(1) = 0$

$f(x + 1) = x^2 - x$

78. Determine the domain of the function $g(t) = \sqrt{3 - t}$, and express in interval notation.

Solution:

What can t be? Any nonnegative real number.

$3 - t > 0$

$3 > t$ or $t < 3$

Domain: $(-\infty, 3)$

79. Given the function, $G(x) = x^2$, evaluate

$$\frac{G(-1 + h) - G(-1)}{h}.$$

Solution:

$$\frac{G(-1 + h) - G(-1)}{h} = \frac{G(-1) + G(h) - G(-1)}{h} = \frac{G(h)}{h}$$

$$G(x) = x^2 \Rightarrow G(\ ) = (\ )^2 \Rightarrow G(h) = (h)^2$$

$$\frac{G(h)}{h} = \frac{h^2}{h} = h$$

80. Given the function, $f(x) = |x - A| - 1$ and $f(1) = -1$, find A.

Solution:

Since $f(1) = -1$, the point $(-1, 1)$ must satisfy the function. $-1 = |-1 - A| - 1$

Add 1 to both sides of the equation. $|-1 - A| = 0$

The absolute value of zero is zero so there is no need for the absolute value signs.

$$-1 - A = 0 \Rightarrow A + 1 = 0 \Rightarrow A = -1$$

■ CHALLENGE

81. T or F: If a vertical line does not intersect the graph of an equation, then that equation does not represent a function.

82. T or F: If a horizontal line intersects a graph of an equation more than once, the equation does not represent a function.

83. T or F: If $f(-a) = f(a)$, then f does not represent a function.

84. T or F: If $f(-a) = f(a)$, then f may or may not represent a function.

85. If $f(x) = Ax^2 - 3x$ and $f(1) = -1$, find A.

86. If $g(x) = \dfrac{1}{b - x}$, and $g(3)$ is undefined, find b.

87. If $F(x) = \dfrac{C - x}{D - x}$, $F(-2)$ is undefined and $F(-1) = 4$, find C and D.

88. Construct a function that is undefined at $x = 5$ and the point $(1, -1)$ lies on the graph of the function.

■ **TECHNOLOGY**

89. Using a graphing utility, graph the temperature function in Exercise 69. What time of day is it the warmest? What is the temperature? Looking at this function, explain why this model for Tampa, Florida, is only valid from sunrise to sunset (6 A.M. to 6 P.M. [1800]).

90. Using a graphing utility, graph the height of the firecracker in Exercise 70. How long after liftoff is the firecracker airborne? What is the maximum height that the firecracker attains? Explain why this height model is only valid for the first 8 seconds.

91. Using a graphing utility, graph the price function in Exercise 71. What are the lowest and highest prices of the cards? Does this agree with what you found in Exercise 72?

92. The makers of malted milk balls are considering increasing the size of the spherical treats. The thin chocolate coating on a malted milk ball can be approximated by the surface area, $S(r) = 4\pi r^2$. If the radius is increased 3 mm, what is the resulting increase in required chocolate for the thin outer coating?

SECTION A.4 Graphs of Functions: Common Functions, Transformations, and Symmetry

Skills Objectives

■ Recognize and graph common functions.
■ Classify functions as even, odd, or neither.
  ■ Algebraically
  ■ By graphing
■ Determine if intervals are increasing, decreasing, or constant.
■ Graph horizontal and vertical shifting of common functions.
■ Graph reflections of common functions about the *x*-axis or *y*-axis.
■ Graph expansion and contraction of common functions.

Conceptual Objectives

■ Identify common functions.
■ Use multiple transformations of common functions to obtain graphs of functions.

Common Functions

In Section A.2, point-plotting techniques were discussed, and it was mentioned that we would discuss more efficient ways of graphing functions later in this chapter. There are nine main functions that we will discuss that will constitute a "library" of functions. We will draw on this library of functions when graphing aids are discussed. Several of these functions you already know, and we will classify them specifically by name and discuss properties that each function exhibits.

In Section A.2 we discussed equations and graphs of lines. All lines (with the exception of vertical lines) pass the vertical line test, and hence are classified as functions. Instead of the traditional notation of a line, $y = mx + b$, we use function notation and classify a function whose graph is a *line* as a *linear* function.

LINEAR FUNCTION

$$f(x) = mx + b \qquad m \text{ and } b \text{ are real numbers.}$$

The domain of a linear function, $f(x) = mx + b$, is the set of all real numbers, $\mathbb{R}$. The graph of this function has slope m and y-intercept $(0, b)$. Two specific linear functions are the constant function (when $m = 0$) and the identity function (when $b = 0$ and $m = 1$).

When the slope is zero, $m = 0$, the function reduces to the special case of a constant function.

CONSTANT FUNCTION

$$f(x) = b \qquad b \text{ is any real number.}$$

The graph of a constant function, $f(x) = b$, is a horizontal line. The y-intercept is the point $(0, b)$. The domain of a constant function is the set of all real numbers, $\mathbb{R}$. The range, however, is a single value, b. In other words, all x values correspond to a single y value.

Points that lie on the graph of a constant function, $f(x) = b$, are

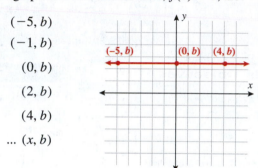

$$(-5, b)$$
$$(-1, b)$$
$$(0, b)$$
$$(2, b)$$
$$(4, b)$$
$$\dots (x, b)$$

Another specific example of a linear function is when the slope is one, $m = 1$, and the y-intercept is zero, $b = 0$. This special case is called the identity function.

IDENTITY FUNCTION

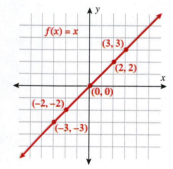

$$f(x) = x$$

The graph of the identity function has the following properties: It passes through the origin, and every point that lies on the line has equal x- and y-coordinates. Both the domain and the range of the identity function are the set of all real numbers, $\mathbb{R}$.

A function that squares the input is called the square function.

SQUARE FUNCTION

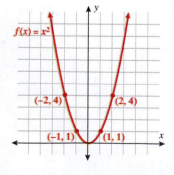

$$f(x) = x^2$$

The graph of the square function is called a parabola. The domain of the square function is the set of all real numbers, $\mathbb{R}$. Because squaring a real number always yields a positive number or zero, the range of the square function is the set of all nonnegative numbers. Notice that the intercept is the origin and the square function is symmetric about the y-axis. This graph is contained in quadrants I and II.

A function that cubes the input is called the cube function.

CUBE FUNCTION

$$f(x) = x^3$$

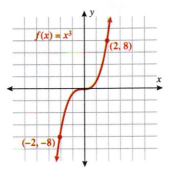

The domain of the cube function is the set of all real numbers, $\mathbb{R}$. Because cubing a negative number yields a negative number, the range of the cube function is also the set of all real numbers, $\mathbb{R}$. Notice that the intercept is the origin and the cube function is symmetric about the origin. This graph extends only into quadrants I and III.

The next two functions have similar names to the previous two functions: square root and cube root. When a function takes the square root of the input or the cube root of the input, the function is called the square root function or the cube root function, respectively. In the following discussion we restrict ourselves to real numbers.

SQUARE ROOT FUNCTION

$$f(x) = \sqrt{x} \qquad \text{or} \qquad f(x) = x^{1/2}$$

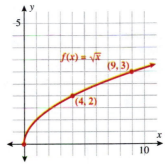

What values for x will yield real values for the function? All negative numbers must be eliminated from the domain. Therefore, we state the domain to be all nonnegative real numbers. The output of the function will be all real numbers greater than or equal to zero. The graph of this function will be contained in quadrant I.

CUBE ROOT FUNCTION

$$f(x) = \sqrt[3]{x} \qquad \text{or} \qquad f(x) = x^{1/3}$$

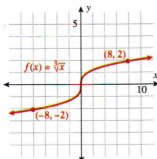

The cube root function does not have any restrictions on the domain and, hence, the range. In fact, the domain and range both are the set of all real numbers, $\mathbb{R}$. This graph is contained in quadrants I and III and passes through the origin. This function is symmetric about the origin.

ABSOLUTE VALUE FUNCTION

$$f(x) = |x|$$

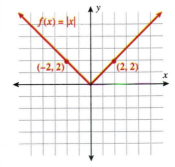

Some points that are on the graph of the absolute value function are $(-1, 1)$, $(0, 0)$, and $(1, 1)$. The domain of the absolute value function is the set of all real numbers, $\mathbb{R}$, yet the range is the set of nonnegative real numbers. The graph of this function is symmetric with respect to the y-axis and is contained in quadrants I and II.

A function that takes the reciprocal of the input is called the reciprocal function.

RECIPROCAL FUNCTION

$$f(x) = \frac{1}{x} \qquad x \neq 0$$

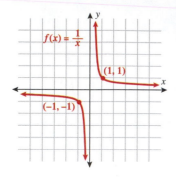

The only restriction on the domain of the reciprocal function is that $x \neq 0$. Hence, we say the domain is the set of all real numbers excluding zero. The range is the set of all real numbers except zero. This function is symmetric with respect to the origin and is contained in quadrants I and III.

Even and Odd Functions

Of the nine functions discussed above, several have similar properties of symmetry. The constant function, square function, and absolute value function are all symmetric with respect to the y-axis. The identity function, cube function, cube root function, and reciprocal function are all symmetric with respect to the origin. The term **even** is used to describe functions that are symmetric with respect to the y-axis, or vertical axis, and the term **odd** is used to describe functions that are symmetric with respect to the origin. Recall from Section 0.2 that symmetry can be determined both graphically and algebraically. The box below summarizes the graphic and algebraic characteristics of even and odd functions.

TECHNOLOGY TIP

Graph $f(x) = |x|$.

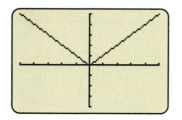

EVEN AND ODD FUNCTIONS

FUNCTION	SYMMETRIC WITH RESPECT TO	ON REPLACING x WITH $-x$
Even	y-axis or vertical axis	$f(-x) = f(x)$
Odd	origin	$f(-x) = -f(x)$

The algebraic method for determining symmetry with respect to the y-axis, or vertical axis, is to substitute in $-x$ for x. If it results in an equivalent equation, the function is symmetric with respect to the y-axis. Some examples of even functions are $f(x) = b$, $f(x) = x^2$, $f(x) = x^4$, and $f(x) = |x|$. In any of these equations, if $-x$ is substituted for x, the result is the same, $f(-x) = f(x)$. Also note that, with the exception of the absolute value function, these examples are all even-degree polynomial equations. One common mistake is assuming the constant function is odd if the constant is odd or even if the constant is even. All constant functions are even functions.

The algebraic method for determining symmetry with respect to the origin is to substitute $-x$ for x. If the result is the negative of the original function, $f(-x) = -f(x)$, then the function is symmetric with respect to the origin and, hence, classified as an odd function. Examples of odd functions are $f(x) = x$, $f(x) = x^3$, $f(x) = x^5$, and $f(x) = x^{1/3}$. In any of these functions, if $-x$ is substituted in for x, the result is the negative of the original function. Notice that with the exception of the cube root function, these equations are odd-degree polynomials.

Be careful, though, because functions that are combinations of even- and odd-degree polynomials can turn out to be neither even nor odd, as we will see in the next example.

EXAMPLE 1 Determining If a Function Is Even, Odd, or Neither

Determine if the functions are even, odd, or neither.

a. $f(x) = x^2 - 3$ **b.** $g(x) = x^5 + x^3$ **c.** $h(x) = x^2 - x$

Solution (a):

Original function.

$$f(x) = x^2 - 3$$

Replace x with $-x$.

$$f(-x) = (-x)^2 - 3$$

Simplify.

$$f(-x) = x^2 - 3 = f(x)$$

Because $f(-x) = f(x)$, we say that $f(x)$ is an *even* function .

Solution (b):

Original function:

$$g(x) = x^5 + x^3$$

Replace x with $-x$.

$$g(-x) = (-x)^5 + (-x)^3$$

Simplify.

$$g(-x) = -x^5 - x^3 = -(x^5 + x^3) = -g(x)$$

Because $g(-x) = -g(x)$, we say that $g(x)$ is an *odd* function .

Solution (c):

Original function:

$$h(x) = x^2 - x$$

Replace x with $-x$.

$$h(-x) = (-x)^2 - (-x)$$

Simplify.

$$h(-x) = x^2 + x$$

$h(-x)$ is neither $-h(x)$ nor $h(x)$; therefore the function $h(x)$ is neither even nor odd .

In parts (a), (b), and (c), we classified these functions as either even, odd, or neither, using the algebraic test. Look back at them now and reflect on whether these classifications agree with your intuition. In part (a), we combined two functions: the square function and constant function. Both of these functions are even, and adding even functions yields another even function. In part (b), we combined two odd functions: the fifth-power function and the cube function. Both of these functions are odd, and adding two odd functions yields another odd function. In part (c), we combined two functions: the square function and the identity function. The square function is even, and the identity function is odd. In this part, combining an even function with an odd function yields a function that is neither even nor odd and, hence, has no symmetry with respect to the vertical axis or the origin.

■ **YOUR TURN** Classify the functions as even, odd, or neither.

a. $f(x) = |x| + 4$ **b.** $f(x) = x^3 - 1$

Increasing and Decreasing Functions

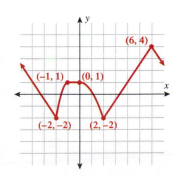

Functions can be described as increasing, decreasing, or constant. There is no fourth option. Look at the figure on the right. If we start at the left side of the graph and trace the red curve with our pen, we see that the function values (values in the vertical direction) are decreasing until arriving at the point $(-2, 2)$. Then, the function values increase until arriving at the point $(-1, 1)$. The value then remains constant ($y = 1$) between the points $(-1, 1)$ and $(0, 1)$. Proceeding beyond the point $(0, 1)$, the function values decrease again until the point $(2, -2)$. Beyond the point $(2, -2)$ the function values increase again until the point (6, 4). Finally, the function values decrease and continue to do so.

When specifying intervals of increasing, decreasing, and constant function values, the intervals are classified according to the x-coordinate. For instance, in this graph, we say the function is increasing when x is between $x = -2$ and $x = -1$ and again when x is between $x = 2$ and $x = 6$. The graph is classified as decreasing when x is less than $x = -2$ and again when x is between $x = 0$ and $x = 2$ and again when x is greater than $x = 6$. The graph is classified as constant when x is between $x = -1$ and $x = 0$. In interval notation, this is summarized as

DECREASING	INCREASING	CONSTANT
$(-\infty, -2) \cup (0, 2) \cup (6, \infty)$	$(-2, -1) \cup (2, 6)$	$(-1, 0)$

An algebraic test for increasing, decreasing, or constant is to compare the value, $f(x)$, of the function for particular points in intervals.

INCREASING, DECREASING, AND CONSTANT INTERVALS

If $x_1 < x_2$ then

Increasing interval: $f(x_1) < f(x_2)$

Decreasing interval: $f(x_1) > f(x_2)$

Constant interval: $f(x_1) = f(x_2)$

Let's now classify our library of functions according to their intervals of increasing, decreasing, or constant. The constant function stays constant for all x. Therefore, the interval that is constant is $(-\infty, \infty)$, which accounts for all real numbers—there are no intervals of increasing or decreasing. The identity function, cube function, and cube root function always increase from left to right on the graph. Therefore, we classify those intervals of increasing as $(-\infty, \infty)$, and, hence, those functions have no intervals of decreasing or constant values. The square root function always increases. Since it is only defined for nonnegative numbers, we say the square root function has an increasing interval of $(0, \infty)$. The square and absolute value functions, however, both decrease until reaching the origin, then increase. Therefore, we classify those functions as having no intervals of constant value, and the interval, $(-\infty, 0)$, is decreasing and the interval, $(0, \infty)$, is increasing.

Horizontal and Vertical Shifting

The focus of this section to now has been to learn the graphs that correspond to particular functions such as identity, square, cube, square root, cube root, absolute value, and reciprocal. Therefore, at this point, you should be able to recognize and generate the graphs of $y = x$, $y = x^2$, $y = x^3$, $y = \sqrt{x}$, $y = \sqrt[3]{x}$, $y = |x|$, and $y = \frac{1}{x}$. We will now discuss how to graph functions that look almost like these functions. For instance, a common function may be shifted (horizontally or vertically), reflected, or stretched (or compressed). Collectively, these techniques are called **transformations**.

Let's take the absolute value function as an example. The graph of $f(x) = |x|$ was given in the last section. Now look at two examples that are much like this function:

$g(x) = |x| + 2$ and $h(x) = |x - 1|$. Graphing these functions by point plotting yields

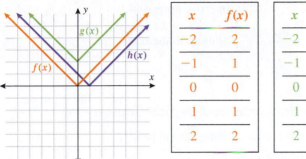

x	$f(x)$
-2	2
-1	1
0	0
1	1
2	2

x	$g(x)$
-2	4
-1	3
0	2
1	3
2	4

x	$h(x)$
-2	3
-1	2
0	1
1	0
2	1

Instead of point plotting the function, $g(x) = |x| + 2$, we could have started with the function, $f(x) = |x|$, and shifted the entire graph *up* 2 units. Similarly, we could have generated the graph of the function, $h(x) = |x - 1|$, by shifting the function $f(x) = |x|$ to the *right* 1 unit. In both cases, the base or starting function is $f(x) = |x|$. Why did we go up for $g(x)$ and to the right for $h(x)$?

Notice that we could rewrite the functions $g(x)$ and $h(x)$ in terms of $f(x)$:

$$g(x) = |x| + 2 \ = f(x) + 2$$

$$h(x) = |x - 1| \ = f(x - 1)$$

In the case of $g(x)$, the shift ($+2$) occurs *outside* the function. Therefore, the output for $g(x)$ is 2 more than the typical output for $f(x)$. Because the output corresponds to the vertical axis, this results in a shift *upward* of two units. In general, shifts that occur outside the function correspond to a vertical shift that corresponds to the sign of the shift. For instance, had the function been $G(x) = |x| - 2$, this graph would have started with the function $f(x)$ and shifted down 2 units.

In the case of $h(x)$, the shift occurs *inside* the function. Notice that the point $(0, 0)$ that lies on $f(x)$ was shifted to the point $(1, 0)$ on the function $h(x)$. The y value remained the same, but the x value shifted to the right 1 unit. Similarly, the points $(-1, 1)$ and $(1, 1)$ were shifted to the points $(0, 1)$ and $(2, 1)$, respectively. In general, shifts that occur inside the function correspond to a horizontal shift that corresponds to opposite the sign. In this case, $h(x) = |x - 1|$, shifted the function $f(x)$ to the right 1 unit. If instead, we had the function, $H(x) = |x + 1|$, this graph would have started with the function $f(x)$ and shifted to the left 1 unit.

VERTICAL SHIFTS

Assuming c is a positive constant, shifts **outside** the function correspond to a **vertical** shift that goes **with the sign**.

TO GRAPH	SHIFT THE FUNCTION $f(x)$
$y = f(x) + c$	c units upward
$y = f(x) - c$	c units downward

HORIZONTAL SHIFTS

Assuming c is a positive constant, shifts **inside** the function correspond to a **horizontal** shift that goes **opposite the sign**.

TO GRAPH	SHIFT THE FUNCTION $f(x)$
$y = f(x + c)$	c units to the left
$y = f(x - c)$	c units to the right

EXAMPLE 2 Plotting Horizontal and Vertical Translations

Plot the given functions using graph-shifting techniques:

a. $g(x) = x^2 - 1$ **b.** $H(x) = (x + 1)^2$

Solution: In both cases, the function to start with is $f(x) = x^2$.

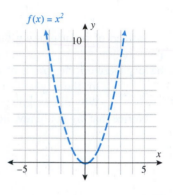

a. $g(x) = x^2 - 1$ can be rewritten as $g(x) = f(x) - 1$.

1. The shift (1 unit) occurs *outside* of the function. Therefore, we expect a vertical shift that goes with the sign.

2. Since the sign is *negative,* this corresponds to a *downward* shift.

3. Shifting the function $f(x) = x^2$ down 1 unit yields the graph of $g(x) = x^2 - 1$.

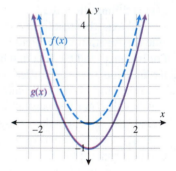

b. $H(x) = (x + 1)^2$ can be rewritten as $H(x) = f(x + 1)$.

1. The shift (1 unit) occurs *inside* of the function. Therefore, we expect a horizontal shift that goes *opposite* the sign.

2. Since the sign is *positive,* this corresponds to a *left* shift.

3. Shifting the function $f(x) = x^2$ to the left 1 unit yields the graph of $H(x) = (x + 1)^2$.

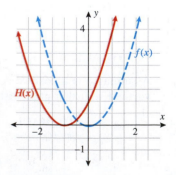

⭕ **CONCEPT CHECK**   What does the graph of $y = x^2$ look like?

EXAMPLE 3 Horizontal and Vertical Shifts, and Changes in Domain and Range

Graph the functions using graph-shifting techniques, and state the domain and range of each function.

a. $g(x) = \sqrt{x} + 1$ **b.** $G(x) = \sqrt{x} - 2$

Solution:

In both cases the function to start with is $f(x) = \sqrt{x}$.

Domain: $[0, \infty)$

Range: $[0, \infty)$

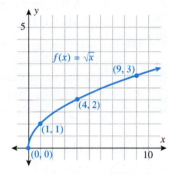

a.

1. $g(x) = \sqrt{x} + 1$ can be rewritten as $g(x) = f(x + 1)$.

2. The shift (1 unit) is *inside* the function, which corresponds to a *horizontal* shift *opposite the sign.*

3. Shifting $f(x) = \sqrt{x}$ to the *left* 1 unit yields the graph of $g(x) = \sqrt{x} + 1$. Notice that the point $(0, 0)$, which lies on $f(x)$, gets shifted to the point $(-1, 0)$ on the graph of $g(x)$.

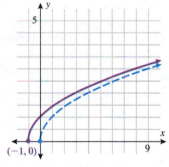

Although the original function, $f(x) = \sqrt{x}$, had an implicit restriction on the domain: $[0, \infty)$, the function, $g(x) = \sqrt{x} + 1$, has the implicit restriction that

---

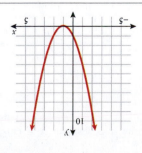

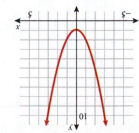

$x \geq -1$. We see that the output or range of $g(x)$ is the same as the output of the original function, $f(x)$.

Domain: $[-1, \infty)$ **Range:** $[0, \infty)$

b.

1. $G(x) = \sqrt{x} - 2$ can be rewritten as $G(x) = f(x) - 2$.

2. The shift (2 units) is *outside* the function, which corresponds to a *vertical* shift *with the sign*.

3. $G(x) = \sqrt{x} - 2$ is found by shifting $f(x) = \sqrt{x}$ to the *down* 2 units. Notice that the point $(0, 0)$, which lies on $f(x)$, gets shifted to the point $(0, -2)$ on the graph of $G(x)$.

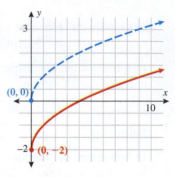

The original function, $f(x) = \sqrt{x}$, has an implicit restriction on the domain: $[0, \infty)$. The function, $G(x) = \sqrt{x} - 2$, also has the implicit restriction that $x \geq 0$. The output or range of $G(x)$ is always 2 units less than the output of the original function, $f(x)$.

Domain: $[0, \infty)$ **Range:** $[-2, \infty)$

■ **YOUR TURN** Plot the graph of the functions using graph-shifting techniques.

a. $G(x) = x^3 + 1$ **b.** $h(x) = (x + 2)^3$

The previous examples have involved graphing functions by shifting a known function either in the horizontal or vertical direction. Let us now look at combinations of horizontal and vertical translations.

EXAMPLE 4 Combination of Horizontal and Vertical Translations

Graph the function: $F(x) = (x + 1)^2 - 2$.

Solution:

The base function is $y = x^2$.

1. The shift 1 unit is *inside* the function, so it represents a *horizontal* shift *opposite the sign.*

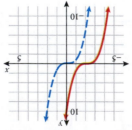

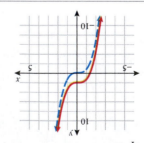

■ **Answer: a.** $G(x) = x^3 + 1$ is $y = x^3$ shifted up 1.   **b.** $h(x) = (x + 2)^3$ is $y = x^3$ shifted to the left 2.

2. The -2 shift is *outside* the function, which represents a *vertical* shift *with the sign.*

3. Therefore, we shift the graph of $y = x^2$ to the left 1 unit and down 2 units. For instance, the point $(0, 0)$ on $y = x^2$ shifts to the point $(-1, -2)$ on $F(x) = (x + 1)^2 - 2$.

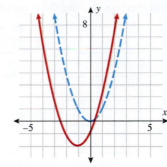

YOUR TURN Graph the function $f(x) = |x - 2| + 1$.

All of the previous transformation examples involve starting with a common function and shifting the function in either the horizontal or vertical direction (or a combination of both). Now, let's investigate *reflections* of functions about the *x*-axis or *y*-axis.

Reflection About the Axes

Plot $f(x) = x^2$ and $g(x) = -x^2$ on the same graph. Start by first listing points that are on each of the graphs and then connecting the points with smooth curves.

x	$f(x)$
-2	4
-1	1
0	0
1	1
2	4

x	$g(x)$
-2	-4
-1	-1
0	0
1	-1
2	-4

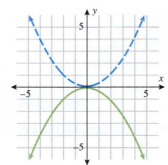

Notice that if the graph of $f(x) = x^2$ is reflected around the *x*-axis, the result is the graph of $g(x) = -x^2$. Also note the function $g(x)$ can be written as the negative of the function $f(x)$, $g(x) = -f(x)$. In general, **reflection about the x-axis** is produced by multiplying a function by -1.

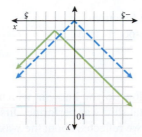

Answer: $f(x) = |x - 2| + 1$
$f(x) = |x|$

Let's now investigate reflection about the y-axis. Plot the functions $f(x) = \sqrt{x}$ and $g(x) = \sqrt{-x}$ on the same graph. Start by listing points that are on each of the graphs and then connecting the points with smooth curves.

x	$f(x)$
0	0
1	1
4	2
9	3

x	$g(x)$
-9	3
-4	2
-1	1
0	0

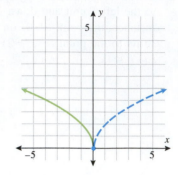

Notice that if the graph of $f(x) = \sqrt{x}$ is reflected around the y-axis, the result is the graph of $g(x) = \sqrt{-x}$. Also note that the function $g(x)$ can be written as $g(x) = f(-x)$. In general, **reflection about the y-axis** is produced by replacing x with $-x$ in the function.

> **REFLECTION ABOUT THE AXES**
>
> The graph of $-f(x)$ is obtained by reflecting the function $f(x)$ about the x-axis.
>
> The graph of $f(-x)$ is obtained by rotating the function $f(x)$ about the y-axis.

Combinations of Shifting and Reflection Techniques

When multiple steps are involved, some students find it easier to use function notation, $f(x)$, and others find it easier to use y in place of $f(x)$. In Example 5, function notation is used, and more detailed explanations are given. In Example 6 the y notation is used, and a step-by-step approach is described. Either way of describing the transformations is acceptable.

EXAMPLE 5 **Graph Using Combinations of Translations and Reflections Relying on Function Notation**

Graph the function $G(x) = -\sqrt{x + 1}$.

Solution:

1. Start with the square root function $f(x) = \sqrt{x}$.
2. The shift 1 unit is *inside* the function, which corresponds to a *horizontal* shift *opposite the sign*.
3. Shift the graph of $f(x) = \sqrt{x}$ to the left 1 unit to yield $f(x + 1) = \sqrt{x + 1}$.
4. The negative outside the function $G(x) = -\sqrt{x + 1}$ corresponds to a reflection of $f(x + 1) = \sqrt{x + 1}$ about the x-axis.

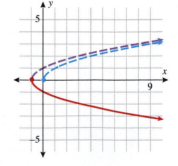

The question always arises: How do you know in which order to do the transformations? In Example 5, the horizontal shift was performed first. The reflection about the x-axis was performed second. Had we reflected about the x-axis and then shifted to the left instead, the same graph would have resulted.

EXAMPLE 6 **Graph Using Combinations of Translations and Reflections Relying on *y* Notation**

Graph the function $y = \sqrt{2 - x} + 1$.

Solution:

1. Start with the square root function. $y = \sqrt{x}$

2. Shift $y = \sqrt{x}$ 2 units to the left. $y = \sqrt{x + 2}$

3. Replacement of x with $-x$ results in reflection about the y-axis. $y = \sqrt{-x + 2}$

4. Shift 1 unit upward. $y = \sqrt{2 - x} + 1$

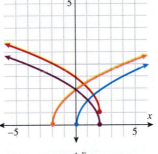

The graph of the function $y = \sqrt{2 - x} + 1$ is:

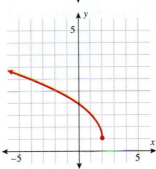

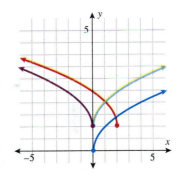

What graph would have resulted in Example 6 had we performed the shifts and reflection in another order? The same graph. For instance, if we start with the function $y = \sqrt{x}$, then vertically shift up 1 unit, the result is the function $y = \sqrt{x} + 1$. Then that function is reflected about the y-axis, resulting in the function $y = \sqrt{-x} + 1$. Shifting that function to the left 2 leads to the graph of $y = \sqrt{-x + 2} + 1$ or $y = \sqrt{2 - x} + 1$. This is exactly what we found before.

■ **YOUR TURN** Use shifts and reflection to graph the function $f(x) = -\sqrt{x - 1} + 2$. State the domain and range of $f(x)$.

Expansion and Contraction

In addition to shifts and reflections, another graphing aid is expansion or contraction. An expansion or a contraction of a graph occurs when the function is multiplied by a positive

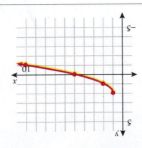

■ **Answer:** Domain: $[1, \infty)$
Range: $(-\infty, 2]$

constant. For example, plot the function $f(x) = x^2$ on the same graph with the functions $g(x) = 2f(x) = 2x^2$ and $h(x) = \frac{1}{2}f(x) = \frac{1}{2}x^2$. Depending on whether or not the constant is larger than 1 or smaller than 1 will determine if it corresponds to a stretch or compression in the vertical direction.

x	$f(x)$
-2	4
-1	1
0	0
1	1
2	4

x	$g(x)$
-2	8
-1	2
0	0
1	2
2	8

x	$h(x)$
-2	2
-1	$\frac{1}{2}$
0	0
1	$\frac{1}{2}$
2	2

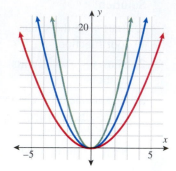

Notice that when the function is multiplied by 2, the result is a graph expanded (stretched) in the vertical direction. When the function is multiplied by $\frac{1}{2}$, the result is a graph that is contracted (compressed) in the vertical direction. Let the function $f(x) = x^2$ represent this year's profit for your company. If profit doubles next year, $g(x) = 2f(x) = 2x^2$, you have *expanded* your profit. If the profit is cut in half next year, $h(x) = \frac{1}{2}f(x) = \frac{1}{2}x^2$, you have *contracted* your profit.

EXPANSION AND CONTRACTION OF GRAPHS

The graph of $cf(x)$ is found by

    Expanding the graph of $f(x)$ if $c > 1$
    Contracting the graph of $f(x)$ if $0 < c < 1$

EXAMPLE 7 Expanding and Contracting Graphs

Graph the function $h(x) = \frac{1}{4}x^3$.

Solution:

1. Start with the cube function. $f(x) = x^3$

2. *Contraction* is expected because $\frac{1}{4}$ is less than 1. $h(x) = \frac{1}{4}x^3$

3. Determine a few points that lie on the graph of h. $(0,0) \; (2,2) \; (-2,-2)$

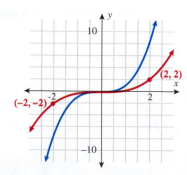

■ **YOUR TURN**   Graph the function $g(x) = 4x^3$.

Combining Graphing Aids

Combinations of shifting, reflecting, and expansion or contraction can be used to generate graphs from our library of common functions.

EXAMPLE 8 Combination of Graphing Aids

Use the appropriate graphing aids to plot the function $f(x) = -2(x - 3)^2$.

1. Start with $y = x^2$.

2. Shift to the right 3 units $y = (x - 3)^2$.

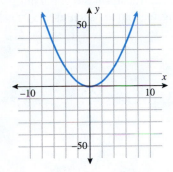

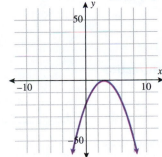

3. Reflect about the x-axis.

$$y = -(x - 3)^2$$

4. Expand in vertical direction.

$$y = -2(x - 3)^2$$

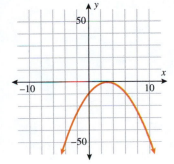

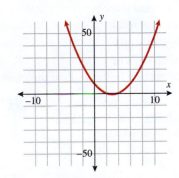

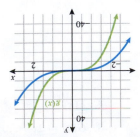

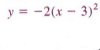

■ **Answer:** *Expansion of the graph* $f(x) = x^3$

SECTION A.4 **SUMMARY**

Common Functions

- Linear function
- Constant function
- Identity function
- Square function
- Cube function
- Square root function
- Cube root function
- Absolute value function
- Reciprocal function

Even and Odd Functions

- Even function: Symmetry with respect to the y-axis and $f(-x) = f(x)$
- Odd function: Symmetry with respect to the origin and $f(-x) = -f(x)$

Intervals When a Function Is Increasing, Decreasing, or Constant

In this section, we discussed three types of graphing aids that transform the graph of a known function into a graph of a similar function using techniques such as

- Horizontal and vertical shifts
- Reflection about the x-axis or y-axis
- Expansion and contraction in the vertical direction

Vertical Shifts

To graph	Shift the function $f(x)$
$y = f(x) + c$	c units upward
$y = f(x) - c$	c units downward

Horizontal Shifts

To graph	Shift the function $f(x)$
$y = f(x + c)$	c units to the left
$y = f(x - c)$	c units to the right

Reflection About the Axes

To graph	Reflect the function $f(x)$ about the
$-f(x)$	x-axis
$f(-x)$	y-axis

Expansion and Contraction

To graph	The graph of $f(x)$ is
$cf(x)$ if $c > 1$	expanded in the vertical direction
$cf(x)$ if $0 < c < 1$	contracted in the vertical direction

A series or combination of these techniques can be used. *They can be used in any order.*

SECTION A.4 **EXERCISES**

 **SKILLS**

In Exercises 1–24, determine if the function is even, odd, or neither.

1. $G(x) = x + 4$

2. $h(x) = 3 - x$

3. $f(x) = 3x^2 + 1$

4. $F(x) = x^4 + 2x^2$

5. $g(t) = 5t^3 - 3t$

6. $f(x) = 3x^5 + 4x^3$

7. $h(x) = x^2 + 2x$

8. $G(x) = 2x^4 - 3x^3$

9. $h(x) = x^{1/3} - x$

10. $g(x) = x^{-1} + x$

11. $f(x) = |x| + 5$

12. $f(x) = |x| + x^2$

13. $f(x) = |x|$

14. $f(x) = |x^3|$

15. $G(t) = |t - 3|$

16. $g(t) = |t + 2|$

17. $G(t) = \sqrt{t - 3}$

18. $f(x) = \sqrt{2 - x}$

19. $g(x) = \sqrt{x^2 + x}$

20. $f(x) = \sqrt{x^2 + 2}$

21. $h(x) = \dfrac{1}{x} + 3$

22. $h(x) = \dfrac{1}{x} - 2x$

23.

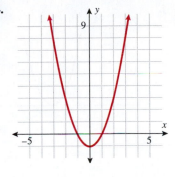

24.

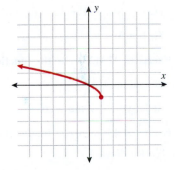

In Exercises 25–36, match the function to the graph.

a.

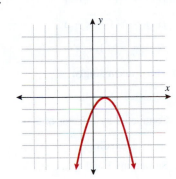

b.

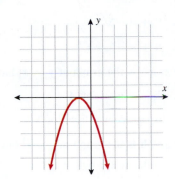

c.

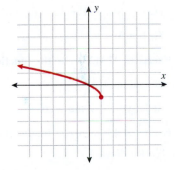

d.

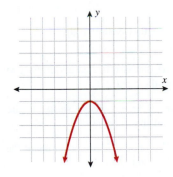

e.

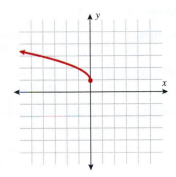

f.

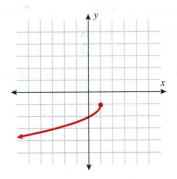

g.

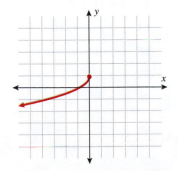

h.

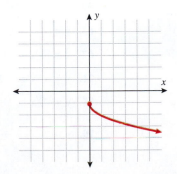

i.

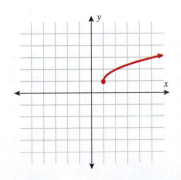

j.

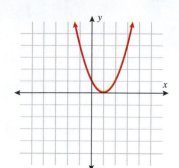

k.

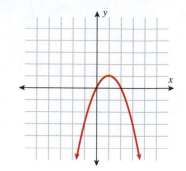

l.

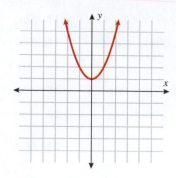

25. $f(x) = x^2 + 1$

26. $f(x) = (x - 1)^2$

27. $f(x) = -(1 - x)^2$

28. $f(x) = -x^2 - 1$

29. $f(x) = -(x + 1)^2$

30. $f(x) = -(1 - x)^2 + 1$

31. $f(x) = \sqrt{x - 1} + 1$

32. $f(x) = -\sqrt{x} - 1$

33. $f(x) = \sqrt{1 - x} - 1$

34. $f(x) = \sqrt{-x} + 1$

35. $f(x) = -\sqrt{-x} + 1$

36. $f(x) = -\sqrt{1 - x} - 1$

In Exercises 37–42, write the function whose graph is the graph of $y = |x|$ but is transformed accordingly.

37. Shifted up 3

38. Shifted to the left 4

39. Reflected about the y-axis

40. Reflected about the x-axis

41. Expanded by a factor of 3

42. Contracted by a factor of 3

In Exercises 43–48, write the function whose graph is the graph of $y = x^3$ but is transformed accordingly.

43. Shifted down 4

44. Shifted to the right 3

45. Shifted up 3 and to the left 1

46. Reflected about the x-axis

47. Reflected about the y-axis

48. Reflected about both the x-axis and the y-axis

In Exercises 49–56, write the function that results from the transformation.

49. The function $f(x) = x^2$ that is shifted to the right 1 and up 2 and reflected about the x-axis.

50. The function $f(x) = \sqrt[3]{x}$ that is shifted up 10 and reflected about the x-axis.

51. The function $f(x) = \sqrt{x}$ that is reflected about the y-axis and shifted to the right 1 and down 2.

52. The function $f(x) = |x|$ that is reflected about the y-axis and shifted up 3 and to the right 2.

53. The function $f(x) = \frac{1}{x}$ that is shifted to the right 2 and up 5 and reflected about the x-axis.

54. The function $f(x) = x^2 + 3x - 1$ that is shifted to the left 2 and reflected about the y-axis.

55. The function $f(x) = x^5 + x$ that is reflected about the y-axis and shifted up 2.

56. The function $f(x) = x^4 + x^3$ that is reflected about the x-axis and shifted to the right 2 and down 3.

In Exercises 57–82, graph the function using graphing aids.

57. $y = x^2 - 2$

58. $y = x^2 + 3$

59. $y = (x + 1)^2$

60. $y = (x - 2)^2$

61. $y = (x - 3)^2 + 2$

62. $y = (x + 2)^2 + 1$

63. $y = -(1 - x)^2$

64. $y = -(x + 2)^2$

65. $y = |-x|$

66. $y = -|x|$

67. $y = -|x + 2| - 1$

68. $y = |1 - x| + 2$

69. $y = 2x^2 + 1$

70. $y = 2|x| + 1$

71. $y = -\sqrt{x - 2}$

72. $y = \sqrt{2 - x}$

73. $y = -\sqrt{2 + x} - 1$

74. $y = \sqrt{2 - x} + 3$

75. $y = \sqrt[3]{x - 1} + 2$

76. $y = \sqrt[3]{x + 2} - 1$

77. $y = \dfrac{1}{x + 3} + 2$

78. $y = \dfrac{1}{3 - x}$

79. $y = 2 - \dfrac{1}{x + 2}$

80. $y = 2 - \dfrac{1}{1 - x}$

81. $y = 5\sqrt{-x}$

82. $y = -\dfrac{1}{5}\sqrt{x}$

■ APPLICATIONS

83. Salary. A manager hires an employee at a rate of $10 per hour. Write the function that describes the current salary of the employee as a function of number of hours worked per week, x. After a year, the manager decides to award the employee a raise equivalent to paying him for an additional 5 hours per week. Write a function that describes the salary of the employee after the raise.

84. Profit. The profit associated with St. Augustine sod in Florida is typically $P(x) = -x^2 + 14{,}000x - 48{,}700{,}000$ where x is the number of pallets sold per year in a normal year. In rainy years Sod King gives away 10 free pallets per year. Write the function that describes the profit of x pallets of sod in rainy years.

85. Taxes. Every year in the United States each working American typically pays in taxes a percentage of his or her earnings (minus the standard deduction). Karen's 2005 taxes were calculated based on the formula $T(x) = 0.22(x - 6500)$. That year the standard deduction was $6,500 and her tax bracket paid 22%. Write the function that will determine her 2006 taxes assuming she receives a raise that places her in a 33% bracket.

86. Medication. The amount of medication that an infant requires is typically a function of the baby's weight. The number of milliliters of an antiseizure medication, A, is given by $A(x) = \sqrt{x} + 2$ where x is the weight of the infant in ounces. In emergencies there is often not enough time to weigh the infant so nurses have to estimate the baby's weight. What is the function that is the actual amount the infant is given if his weight is overestimated by 3 ounces?

■ CATCH THE MISTAKE

In Exercises 87–90, explain the mistake that is made.

87. Describe a procedure for graphing the function $f(x) = \sqrt{x - 3} + 2$.

Solution:

a. Start with the function $f(x) = \sqrt{x}$.

b. Shift the function to the left 3.

c. Shift the function up 2.

88. Describe a procedure for graphing the function $f(x) = -\sqrt{x + 2} - 3$.

Solution:

a. Start with the function $f(x) = \sqrt{x}$.

b. Shift the function to the left 2.

c. Reflect the function about the y-axis.

d. Shift the function down 3.

89. Describe a procedure for graphing the function $f(x) = |3 - x| + 1$.

Solution:

a. Start with the function $f(x) = |x|$.

b. Reflect graph about the y-axis.

c. Shift to the left 3.

d. Shift up 1.

90. Describe a procedure for graphing the function $f(x) = -2x^2 + 1$.

Solution:

a. Start with the function $f(x) = x^2$.

b. Reflect graph about the y-axis.

c. Shift up 1 unit.

d. Expand in the vertical direction by a factor of 2.

■ CHALLENGE

91. T or F: The graph of $y = |-x|$ is the same as the graph of $y = |x|$.

92. T or F: The graph of $y = \sqrt{-x}$ is the same as the graph of $y = \sqrt{x}$.

93. T or F: If the graph of an odd function is reflected around the x-axis and then the y-axis, the result is the graph of the original odd function.

94. T or F: If the graph of $y = \frac{1}{x}$ is reflected around the x-axis, it produces the same graph as if it had been reflected about the y-axis.

95. The point (a, b) lies on the graph of the function $y = f(x)$. What point is guaranteed to lie on the graph of $f(x - 3) + 2$?

96. The point (a, b) lies on the graph of the function $y = f(x)$. What point is guaranteed to lie on the graph of $-f(-x) + 1$?

■ **TECHNOLOGY**

97. Use a graphing utility to graph

  a. $y = x^2 - 2$ and $y = |x^2 - 2|$

  b. $y = x^3 + 1$ and $y = |x^3 + 1|$

  What is the relationship between $f(x)$ and $|f(x)|$?

98. Use a graphing utility to graph

  a. $y = x^2 - 2$ and $y = |x|^2 - 2$

  b. $y = x^3 + 1$ and $y = |x|^3 + 1$

  What is the relationship between $f(x)$ and $f(|x|)$?

99. Use a graphing utility to graph

  a. $y = \sqrt{x}$ and $y = \sqrt{0.1x}$ **b.** $y = \sqrt{x}$ and $y = \sqrt{10x}$

  What is the relationship between $f(x)$ and $f(ax)$ assuming a is positive?

100. Use a graphing utility to graph

  a. $y = \sqrt{x}$ and $y = 0.1\sqrt{x}$ **b.** $y = \sqrt{x}$ and $y = 10\sqrt{x}$

  What is the relationship between $f(x)$ and $af(x)$ assuming a is positive?

SECTION A.5 Operations on Functions and Composition of Functions

Skills Objectives

■ Add, subtract, multiply, and divide functions.
■ Evaluate composite functions.
■ Determine domain of functions resulting from operations and composite functions.

Conceptual Objectives

■ Understand domain restrictions when dividing functions.
■ Determine domain restrictions during composition of functions.

Two different functions can be combined using mathematical operations such as addition, subtraction, multiplication, and division. Also, there is an operation on functions called composition, which can be thought of as a function of a function. When we combine functions, we do so algebraically. Special attention must be paid to the domain and range of the combined functions.

Operations on Functions

Consider the two functions $f(x) = x^2 + 2x - 3$ and $g(x) = x + 1$. The domain of both of these functions is the set of all real numbers. Therefore, we can add, subtract, or multiply these functions for any real number x.

$$\text{Addition: } f(x) + g(x) = x^2 + 2x - 3 + x + 1 = x^2 + 3x - 2$$

The result is in fact a new function, which we denote:

$$(f + g)(x) = x^2 + 3x - 2 \quad \text{This is the \textbf{sum function}.}$$

$$\text{Subtraction: } f(x) - g(x) = x^2 + 2x - 3 - (x + 1) = x^2 + x - 4$$

The result is in fact a new function, which we denote:

$$(f - g)(x) = x^2 + x - 4 \quad \text{This is the \textbf{difference function}.}$$

$$\text{Multiplication: } f(x) \cdot g(x) = (x^2 + 2x - 3)(x + 1) = x^3 + 3x^2 - x - 3$$

The result is in fact a new function, which we denote:

$$(f \cdot g)(x) = x^3 + 3x^2 - x - 3 \quad \text{This is the \textbf{product function}.}$$

Although both f and g are defined for all real numbers x, we must restrict x such that $x \neq -1$ to form the quotient f/g.

$$\text{Division:} \quad \frac{f(x)}{g(x)} = \frac{x^2 + 2x - 3}{x + 1}, \quad x \neq -1$$

The result is in fact a new function, which we denote:

$$\left(\frac{f}{g}\right)(x) = \frac{x^2 + 2x - 3}{x + 1}, \quad x \neq -1 \quad \text{This is called the \textbf{quotient function}.}$$

The new function that arises from adding, subtracting, multiplying, or dividing two functions is defined for all real values that both f and g are defined for. The exception is the quotient function, which eliminates values based on what numbers make the value of the denominator equal to zero.

The previous examples involved polynomials, whose domain is all real numbers. Adding, subtracting, and multiplying polynomials result in other polynomials, which have domains of all real numbers. Let's now investigate operations applied to functions that have a restricted domain.

The domain of the sum function, difference function, or product function is the *intersection* of the individual domains of the two functions. The quotient function has a similar domain in that it is the intersection of the two domains. But any values that make the denominator zero must also be eliminated.

FUNCTION	NOTATION	DOMAIN
Sum	$(f + g)(x) = f(x) + g(x)$	{domain of f} $\cap$ {domain of g}
Difference	$(f - g)(x) = f(x) - g(x)$	{domain of f} $\cap$ {domain of g}
Product	$(f \cdot g)(x) = f(x) \cdot g(x)$	{domain of f} $\cap$ {domain of g}
Quotient	$\left(\dfrac{f}{g}\right)(x) = \dfrac{f(x)}{g(x)}$	{domain of f} $\cap$ {domain of g} $\cap$ {$g(x) \neq 0$}

We can think of this in the following way. Any number that is in the domain of *both* of the functions is in the domain of the combined function. The exception to this is the quotient function, which also eliminates values that make the denominator equal to zero.

EXAMPLE 1 Operations on Functions: Determining Domains of New Functions

For the functions $f(x) = \sqrt{x - 1}$ and $g(x) = \sqrt{4 - x}$, determine the sum function, difference function, product function, and quotient function. State the domain of these four new functions.

Solution:

Sum function: $\qquad f(x) + g(x) = \sqrt{x - 1} + \sqrt{4 - x}$

Difference function: $\quad f(x) - g(x) = \sqrt{x - 1} - \sqrt{4 - x}$

Product function:
$$f(x) \cdot g(x) = \sqrt{x-1} \cdot \sqrt{4-x}$$
$$= \sqrt{(x-1)(4-x)} = \sqrt{-x^2 + 5x - 4}$$

Quotient function:
$$\frac{f(x)}{g(x)} = \frac{\sqrt{x-1}}{\sqrt{4-x}} = \sqrt{\frac{x-1}{4-x}}$$

The domain of the square root function is determined by setting the argument under the radical greater than or equal to zero.

$$\text{Domain of } f(x): \quad [1, \infty)$$

$$\text{Domain of } g(x): \quad (-\infty, 4]$$

Domain of the sum, difference, and product functions is

$$[1, \infty) \cap (-\infty, 4] = [1, 4]$$

The quotient function has the additional constraint that the denominator cannot be zero. This implies that $x \neq 4$, so the domain of the quotient function is $[1, 4)$.

CONCEPT CHECK Are there any restrictions on the domain of the functions $f(x) = \sqrt{x+3}$ and $g = \sqrt{1-x}$?

■ **YOUR TURN** Given the function $f(x) = \sqrt{x+3}$ and $g = \sqrt{1-x}$ find $(f+g)(x)$ and state its domain.

EXAMPLE 2 Quotient Function and Domain Restrictions

Given the functions $F(x) = \sqrt{x}$ and $G(x) = |x-3|$, find the quotient function, $\left(\frac{F}{G}\right)(x)$, and state its domain.

Solution:

The quotient function is written as

$$\left(\frac{F}{G}\right)(x) = \frac{F(x)}{G(x)} = \frac{\sqrt{x}}{|x-3|}$$

Domain of $F(x)$: $[0, \infty)$ Domain of $G(x)$: $(-\infty, \infty)$

The real numbers that are in both the domain for $F(x)$ and the domain for $G(x)$ are represented by the intersection: $[0, \infty) \cap (-\infty, \infty) = [0, \infty)$. Also, the denominator of the quotient function is equal to zero when $x = 3$, so we must eliminate this value from the domain.

$$\text{Domain of } \left(\frac{F}{G}\right)(x): \quad [0, 3) \cup (3, \infty)$$

■ **YOUR TURN** For the functions given in Example 2, determine the quotient function $\left(\frac{G}{F}\right)(x)$, and state its domain.

■ **Answer:** $(f+g)(x) = \sqrt{x+3} + \sqrt{1-x}$ domain: $[-3, 1]$

■ **Answer:** $\left(\frac{G}{F}\right)(x) = \frac{G(x)}{F(x)} = \frac{|x-3|}{\sqrt{x}}$ domain: $(0, \infty)$

Composition of Functions

Recall that a function maps every element in the domain to one and only one corresponding element in the range as shown in the figure on the right.

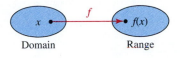

Suppose there is a sales rack of clothes in the juniors department at Macy's. Let x correspond to the original price of each item on the rack. These clothes have recently been marked down 20%. Therefore, the function, $f(x) = 0.80x$, represents the current sale price of each item. You have been invited to a special sale that lets you take 10% off the current sale price and an additional \$5 off every item at checkout. The function, $g(f(x)) = 0.90f(x) - 5$, determines the checkout price. Note that the input of the function g is the output of the function f as shown in the figure below.

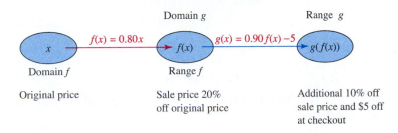

Original price   Sale price 20%   Additional 10% off
                  off original price   sale price and \$5 off
                                         at checkout

This is an example of a **composition of functions**, when the output of one function is the input of another function. It is commonly referred to as a function of a function.

An algebraic example of this is the function $y = \sqrt{x^2 - 2}$. Suppose we let $g(x) = x^2 - 2$ and $f(x) = \sqrt{x}$. Recall that the independent variable in function notation is a placeholder. Since $f(\ ) = \sqrt{(\ )}$, then $f(g(x)) = \sqrt{(g(x))}$. Substituting the expression for $g(x)$, we find $f(g(x)) = \sqrt{x^2 - 2}$. The function $y = \sqrt{x^2 - 2}$ is said to be a composite function, $y = f(g(x))$.

Note that the domain of $g(x)$ is the set of all real numbers, and the domain of $f(x)$ is the set of all nonnegative numbers. The domain of a composite function is the set of all x such that $g(x)$ is in the domain of f. For instance, in the composite function $y = f(g(x))$, we know that the allowable inputs into f are all numbers greater than or equal to zero. Therefore, we restrict the outputs of $g(x) \geq 0$ and find the corresponding x values. Those x values are the only allowable inputs and constitute the domain of the composite function $y = f(g(x))$.

The symbol that represents composition of functions is a small open circle, $(f \circ g)(x) = f(g(x))$ and is read aloud as "f of g." It is important not to confuse this with the multiplication sign, $(f \cdot g)(x) = f(x)g(x)$.

CAUTION
$f \circ g \neq f \cdot g$

COMPOSITE FUNCTIONS

Given two functions f and g, the composite function is denoted by $f \circ g$ and defined by

$$(f \circ g)(x) = f(g(x))$$

The domain of $f \circ g$ is the set of all numbers x in the domain of g where $g(x)$ is in the domain of f.

It is important to realize that there are two "filters" that allow certain values of x into the domain. The first filter is $g(x)$. If x is not in the domain of $g(x)$, it cannot be in the domain of $(f \circ g)(x) = f(g(x))$. Of those values for x that are in the domain of $g(x)$, only some pass through, because we restrict the output of $g(x)$ to values that are allowable as input into f. This adds an additional filter.

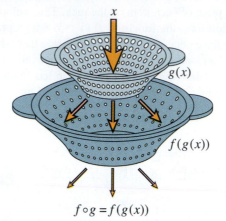

$$f \circ g = f(g(x))$$

STUDY TIP

The domain of $f \circ g$ is always a subset of the domain of g, and the range of $f \circ g$ is always a subset of the range of f.

The domain of $f \circ g$ is always a subset of the domain of g, and the range of $f \circ g$ is always a subset of the range of f.

EXAMPLE 3 Finding a Composite Function

Given the functions $f(x) = x^2 + 1$ and $g(x) = x - 3$, find the composite function, $f \circ g$.

Solution:

Write $f(x)$ using placeholder notation. $f(\ ) = (\ )^2 + 1$

Substitute $g(x) = x - 3$ into f. $f(g(x)) = (x - 3)^2 + 1$

Eliminate parentheses on right side. $f(g(x)) = x^2 - 6x + 10$

$$(f \circ g)(x) = f(g(x)) = x^2 - 6x + 10$$

■ **YOUR TURN** Given the functions in Example 3, find $g \circ f$.

EXAMPLE 4 Determining the Domain of a Composite Function

Given the functions $f(x) = \dfrac{1}{x - 1}$ and $g(x) = \dfrac{1}{x}$, determine the composite function, $f \circ g$, and state its domain.

Solution:

Write $f(x)$ using placeholder notation. $f(\ ) = \dfrac{1}{(\ ) - 1}$

Substitute $g(x) = \dfrac{1}{x}$ into f. $f(g(x)) = \dfrac{1}{\dfrac{1}{x} - 1}$

Answer: $g \circ f = g(f(x)) = x^2 - 2$

Multiply right side by $\dfrac{x}{x}$.

$$f(g(x)) = \cfrac{1}{\dfrac{1}{x} - 1} \cdot \frac{x}{x} = \frac{x}{1-x}$$

$$(f \circ g) = f(g(x)) = \frac{x}{1-x}$$

What is the domain of $(f \circ g)(x) = f(g(x))$? By inspecting the final result of $f(g(x))$, we see that the denominator is zero when $x = 1$. Therefore, $x \neq 1$. Are there any other values for x that are not allowed? The function $g(x)$ has the domain $x \neq 0$; therefore we must also exclude zero.

The domain of $(f \circ g)(x) = f(g(x))$ is $x \neq 0$ and $x \neq 1$ or in interval notation

$$(-\infty, 0) \cup (0, 1) \cup (1, \infty)$$

The domain of the composite function cannot always be determined by examining the final form of $f \circ g$.

CAUTION

The domain of the composite function cannot always be determined by examining the final form of $f \circ g$.

YOUR TURN   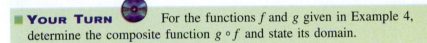   For the functions f and g given in Example 4, determine the composite function $g \circ f$ and state its domain.

EXAMPLE 5   Evaluating a Composite Function

Given the functions $f(x) = x^2 - 7$ and $g(x) = 5 - x^2$, evaluate

a. $f(g(1))$ **b.** $f(g(-2))$ **c.** $g(f(3))$ **d.** $g(f(-4))$

Solution:

One way of evaluating these composite functions is to calculate the two individual composites in terms of x: $f(g(x))$ and $g(f(x))$. Once those functions are known, the values can be substituted for x and evaluated.

Another way of proceeding is as follows:

a. Write desired quantity. $f(g(1))$

Find the value of the inner function, g. $g(1) = 5 - 1^2 = 4$

Substitute $g(1) = 4$ into f. $f(g(1)) = f(4)$

Evaluate $f(4)$. $f(4) = 4^2 - 7 = 9$

$$f(g(1)) = 9$$

b. Write desired quantity. $f(g(-2))$

Find the value of the inner function, g. $g(-2) = 5 - (-2)^2 = 1$

Substitute $g(-2) = 1$ into f. $f(g(-2)) = f(1)$

Evaluate $f(1)$. $f(1) = 1^2 - 7 = -6$

$$f(g(-2)) = -6$$

c. Write desired quantity. $\quad\quad\quad\quad\quad\quad g(f(3))$

Find the value of the inner function, f. $\quad f(3) = 3^2 - 7 = 2$

Substitute $f(3) = 2$ into g. $\quad\quad\quad g(f(3)) = g(2)$

Evaluate $g(2)$. $\quad\quad\quad\quad\quad\quad\quad g(2) = 5 - 2^2 = 1$

$$g(f(3)) = 1$$

d. Write desired quantity. $\quad\quad\quad\quad\quad\quad g(f(-4))$

Find the value of the inner function, f. $\quad f(-4) = (-4)^2 - 7 = 9$

Substitute $f(-4) = 9$ into g. $\quad\quad\quad g(f(-4)) = g(9)$

Evaluate $g(9)$. $\quad\quad\quad\quad\quad\quad\quad g(9) = 5 - 9^2 = -76$

$$g(f(-4)) = -76$$

YOUR TURN Given the functions $f(x) = x^3 - 3$ and $g(x) = 1 + x^3$, evaluate $f(g(1))$ and $g(f(1))$.

Application Problems

Recall the first example in this section regarding the clothes that are on sale. Often, real-world applications are modeled with composite functions. In the clothes example, x is the original price of each item. The first function maps the input (original price) to an output (sale price). The second function maps the input (sale price) to an output (check-out price). The next example is another real-world application of composite functions.

Three temperature scales are commonly used

- The degree Celsius (°C) scale

  - This scale was devised by dividing the range between the freezing and boiling of pure water at sea level into 100 equal parts. This scale is used in science and is one of the standards of the "metric" (SI) system of measurements.

  - The **degrees Celsius** range from 0 °C (freezing) to 100 °C (boiling).

- The Kelvin (K) temperature scale

  - This scale extends the Celsius scale down to absolute zero, a hypothetical temperature at which there is a complete absence of heat energy.

  - Temperatures on this scale are called **kelvins**, *not* degrees kelvin, and kelvin is not capitalized. The symbol for kelvin is K.

- The degree Fahrenheit (°F) scale

  - This scale evolved over time and is still widely used mainly in the United States, although Celsius is the preferred "metric" scale.

  - With respect to pure water at sea level, the **degrees Fahrenheit** correspond to 32 °F (freezing) and 212 °F (boiling).

The equations that relate these temperature scales are

$$F = \frac{9}{5}C + 32 \quad\quad C = K - 273.15$$

■ Answer: $f(g(1)) = 5$ and $g(f(1)) = -7$

EXAMPLE 6 Applications Involving Composite Functions

Determine degrees Fahrenheit as a function of kelvins.

Solution:

Degrees Fahrenheit is a function of degrees Celsius.

$$F = \frac{9}{5}C + 32$$

Now substitute $C = K - 273.15$ into the equation for F.

$$F = \frac{9}{5}(K - 273.15) + 32$$

Simplify.

$$F = \frac{9}{5}K - 491.67 + 32$$

$$F = \frac{9}{5}K - 459.67$$

SECTION A.5 SUMMARY

Operations on Functions

FUNCTION	NOTATION
Sum	$(f + g)(x) = f(x) + g(x)$
Difference	$(f - g)(x) = f(x) - g(x)$
Product	$(f \cdot g)(x) = f(x) \cdot g(x)$
Quotient	$\left(\dfrac{f}{g}\right)(x) = \dfrac{f(x)}{g(x)} \qquad g(x) \neq 0$

The domain of the sum, difference, and product functions is the intersection, or common, domain shared by both $f(x)$ and $g(x)$. The domain of the quotient function is also the intersection of the domain shared by both $f(x)$ and $g(x)$ with an additional restriction that $g(x) \neq 0$.

Composition of Functions

$$(f \circ g)(x) = f(g(x))$$

The domain restrictions cannot be determined only by inspecting the final form of $f(g(x))$. The domain of the composite function is a subset of the domain of $g(x)$. Values for x must be eliminated if their corresponding values, $g(x)$, are not in the domain of f.

SECTION A.5 EXERCISES

■ SKILLS

In Exercises 1–10, given the functions f and g, find $f + g$, $f - g$, $f \cdot g$, and $\dfrac{f}{g}$, and state the domain of each.

1. $f(x) = 2x + 1$
$g(x) = 1 - x$

2. $f(x) = 3x + 2$
$g(x) = 2x - 4$

3. $f(x) = 2x^2 - x$
$g(x) = x^2 - 4$

4. $f(x) = 3x + 2$
$g(x) = x^2 - 25$

5. $f(x) = \dfrac{1}{x}$
$g(x) = x$

6. $f(x) = \dfrac{2x + 3}{x - 4}$

$g(x) = \dfrac{x - 4}{3x + 2}$

7. $f(x) = \sqrt{x}$

$g(x) = 2\sqrt{x}$

8. $f(x) = \sqrt{x - 1}$

$g(x) = 2x^2$

9. $f(x) = \sqrt{4 - x}$

$g(x) = \sqrt{x + 3}$

10. $f(x) = \sqrt{1 - 2x}$

$g(x) = \dfrac{1}{x}$

In Exercises 11–20, for the given functions f and g, find the composite functions $f \circ g$ and $g \circ f$, and state their domains.

11. $f(x) = 2x + 1$

$g(x) = x^2 - 3$

12. $f(x) = x^2 - 1$

$g(x) = 2 - x$

13. $f(x) = \dfrac{1}{x - 1}$

$g(x) = x + 2$

14. $f(x) = \dfrac{2}{x - 3}$

$g(x) = 2 + x$

15. $f(x) = |x|$

$g(x) = \dfrac{1}{x - 1}$

16. $f(x) = |x - 1|$

$g(x) = \dfrac{1}{x}$

17. $f(x) = \sqrt{x - 1}$

$g(x) = x + 5$

18. $f(x) = \sqrt{2 - x}$

$g(x) = x^2 + 2$

19. $f(x) = x^3 + 4$

$g(x) = (x - 4)^{1/3}$

20. $f(x) = \sqrt[3]{x^2 - 1}$

$g(x) = x^{2/3} + 1$

In Exercises 21–28, evaluate the functions for the specified values, if possible.

$$f(x) = x^2 + 10 \qquad g(x) = \sqrt{x - 1}$$

21. $(f + g)(2)$

22. $(f - g)(5)$

23. $(f \cdot g)(4)$

24. $\left(\dfrac{f}{g}\right)(2)$

25. $f(g(2))$

26. $g(f(4))$

27. $f(g(0))$

28. $g(f(0))$

In Exercises 29–34, evaluate $f(g(1))$ and $g(f(2))$, if possible.

29. $f(x) = \dfrac{1}{x}$ $g(x) = 2x + 1$

30. $f(x) = x^2 + 1$ $g(x) = \dfrac{1}{2 - x}$

31. $f(x) = \sqrt{1 - x}$ $g(x) = x^2 + 2$

32. $f(x) = \sqrt{3 - x}$ $g(x) = x^2 + 1$

33. $f(x) = \dfrac{1}{|x - 1|}$ $g(x) = x + 3$

34. $f(x) = \dfrac{1}{x}$ $g(x) = |2x - 3|$

In Exercises 35–40, show that $f(g(x)) = x$ and $g(f(x)) = x$.

35. $f(x) = 2x + 1$ $g(x) = \dfrac{x - 1}{2}$

36. $f(x) = \dfrac{x - 2}{3}$ $g(x) = 3x + 2$

37. $f(x) = \sqrt{x - 1}$ $g(x) = x^2 + 1$ for $x \ge 1$

38. $f(x) = 2 - x^2$ $g(x) = \sqrt{2 - x}$ for $x \le 2$

39. $f(x) = \dfrac{1}{x}$ $g(x) = \dfrac{1}{x}$ for $x \ne 0$

40. $f(x) = (5 - x)^{1/3}$ $g(x) = 5 - x^3$

In Exercises 41–46, write the function as a composite of two functions f and g. (More than one answer is correct.)

41. $f(g(x)) = 2(3x - 1)^2 + 5(3x - 1)$

42. $f(g(x)) = \dfrac{1}{1 + x^2}$

43. $f(g(x)) = \dfrac{2}{|x - 3|}$

44. $f(g(x)) = \sqrt{1 - x^2}$

45. $f(g(x)) = \dfrac{3}{\sqrt{x + 1} - 2}$

46. $f(g(x)) = \dfrac{\sqrt{x}}{3\sqrt{x} + 2}$

■ APPLICATIONS

Exercises 47 and 48 depend on the relationship between degrees Fahrenheit, degrees Celsius, and kelvins:

$$F = \tfrac{9}{5}C + 32 \quad C = K - 273.15$$

47. **Temperature.** Write a composite function that converts degrees kelvins into Fahrenheit.

48. **Temperature.** Convert the following degrees Fahrenheit to kelvins: 32 °F and 212 °F.

49. Dog Run. Suppose you want to build a *square* fenced-in area for your dog. Fence is purchased in linear feet.

   a. Write a composite function that determines the area of your dog pen as a function of how many linear feet are purchased.

   b. If you purchase 100 linear feet, what is the area of your dog pen?

   c. If you purchase 200 linear feet, what is the area of your dog pen?

50. Dog Run. Suppose you want to build a *circular* fenced-in area for your dog. Fence is purchased in linear feet.

   a. Write a composite function that determines the area of your dog pen as a function of how many linear feet are purchased.

   b. If you purchase 100 linear feet, what is the area of your dog pen?

   c. If you purchase 200 linear feet, what is the area of your dog pen?

51. Market Price. Typical supply and demand relationships state that as the number of units for sale increases, the market price decreases. Assume the market price, p, and the number of units for sale, x, are related by the demand equation:

$$p = 3000 - \frac{1}{2}x$$

Assume the cost, $C(x)$, of producing x items is governed by the equation

$$C(x) = 2000 + 10x$$

And the revenue, $R(x)$, generated by selling x units is governed by

$$R(x) = 100x$$

   a. Write the cost as a function of price, p.

   b. Write the revenue as a function of price, p.

   c. Write the profit as a function of price, p.

52. Market Price. Typical supply and demand relationships state that as the number of units for sale increases, the market price decreases. Assume the market price, p, and the number of units for sale, x, are related by the demand equation:

$$p = 10,000 - \frac{1}{4}x$$

Assume the cost, $C(x)$, of producing x items is governed by the equation

$$C(x) = 30000 + 5x$$

And the revenue, $R(x)$, generated by selling x units is governed by

$$R(x) = 1000x$$

   a. Write the cost as a function of price, p.

   b. Write the revenue as a function of price, p.

   c. Write the profit as a function of price, p.

53. Oil Spill. An oil spill makes a circular pattern around a ship. If the radius, in feet, grows as a function of time, in hours: $r(t) = 150\sqrt{t}$. Find the area of the spill as a function of time.

54. Pool Volume. A 20 foot $\times$ 10 foot rectangular pool has been built. If 50 cubic feet of water is pumped into the pool per hour, write the water level height (feet) as a function of time (hours).

55. Fireworks. A family is watching a fireworks display. If they are 2 miles from where the fireworks are being launched and the fireworks travel vertically, what is the distance between the family and the fireworks as a function of height above ground?

56. Real Estate. A couple are about to put their house up for sale. They bought the house for $172,000 a few years ago, and when they list it with a realtor they will pay 6% commission. Write a function that represents the amount of money they will make on their home as a function of asking price, p.

■ CATCH THE MISTAKE

In Exercises 57–60, for the functions $f(x) = x + 2$ and $g(x) = x^2 - 4$, find the composite function and state its domain. Explain the mistake made in each problem.

57. $\dfrac{g}{f}$

   Solution:
   $$\frac{g(x)}{f(x)} = \frac{x^2 - 4}{x + 2}$$
   $$= \frac{(x - 2)(x + 2)}{(x + 2)}$$
   $$= x - 2$$

   Domain: All real numbers

58. $\dfrac{f}{g}$

   Solution:
   $$\frac{f(x)}{g(x)} = \frac{x + 2}{x^2 - 4}$$
   $$= \frac{x + 2}{(x - 2)(x + 2)} = \frac{1}{x - 2}$$
   $$= \frac{1}{x - 2}$$

   Domain: All real numbers except $x = 2$

59. $f \circ g$

Solution: $f \circ g = f(x)g(x)$

$$= (x + 2)(x^2 - 4)$$
$$= x^3 + 2x^2 - 4x - 8$$

Domain: All real numbers

60. Given the function $f(x) = x^2 + 7$ and $g(x) = \sqrt{x - 3}$, find $f \circ g$, and state the domain.

Solution: $f \circ g = f(g(x)) = (\sqrt{x - 3})^2 + 7$

$$= f(g(x)) = x - 3 + 7$$
$$= x - 4$$

Domain: All real numbers

■ CHALLENGE

61. T or F: When adding, subtracting, multiplying, or dividing two functions, the domain of the resulting function is the union of the domains of the individual functions.

62. T or F: For any functions f and g, $f(g(x)) = g(f(x))$ for all values of x that are in the domain of both f and g.

63. T or F: For any functions f and g, $(f \circ g)(x)$ exists for all values of x that are in the domain of $g(x)$ provided the range of g is a subset of the domain of f.

64. T or F: The domain of a composite function can be found by inspection, without knowledge of the domain of the individual functions.

65. For the functions $f(x) = x + a$ and $g(x) = \dfrac{1}{x - a}$, find $g \circ f$ and state its domain.

66. For the functions $f(x) = ax^2 + bx + c$ and $g(x) = \dfrac{1}{x - c}$, find $g \circ f$ and state its domain.

67. For the functions $f(x) = \sqrt{x + a}$ and $g(x) = x^2 - a$, find $g \circ f$ and state its domain.

68. For the functions $f(x) = \dfrac{1}{x^a}$ and $g(x) = \dfrac{1}{x^b}$, find $g \circ f$ and state its domain.

■ TECHNOLOGY

69. Using a graphing utility plot $y_1 = \sqrt{x + 7}$ and $y_2 = \sqrt{9 - x}$. Plot $y_3 = y_1 + y_2$. Explain what you see for y_3.

70. Using a graphing utility plot $y_1 = \sqrt{1 - x}$, $y_2 = x^2 + 2$, and $y_3 = y_1^2 + 2$. If y_1 represents a function f and y_2 represents a function g, then y_3 represents the composite function $g \circ f$. The graph of y_3 is only defined for the domain of $g \circ f$.

SECTION A.6 One-to-One Functions and Inverse Functions

Skills Objectives

■ Determine if a function is a one-to-one function
  ■ Algebraically
  ■ Graphically
■ Find the inverse of a function.
■ Graph the inverse function given the graph of the function.

Conceptual Objectives

■ Visualize the relationships between domain and range of a function and the domain and range of its inverse.
■ Understand why functions and their inverses are symmetric about $y = x$.

Every human being has a blood type, and every human being has a DNA sequence. These are examples of functions when a person is the input and the output is blood type or DNA sequence. These relationships are classified as functions because each person can have one and only one blood type or DNA strand. The difference between these functions is that many people have the same blood type, but DNA is unique to each individual. Could we map backwards? For instance, if you know the blood type, do you know specifically which person it came from? No. But, if you know the DNA sequence, you know exactly which person it corresponds to. When a function has a one-to-one correspondence, like the DNA example, then mapping backwards is possible. The map back is called the *inverse function*.

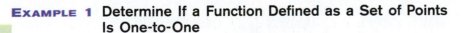

Square function: $f(x) = x^2$

One-to-One Functions

In Section A.3, we defined a function as a relationship that maps an input (domain) to one and only one output (range). Algebraically, each value for x can only correspond to a single value for y. Recall the square, identity, absolute value, and reciprocal functions from our library of functions in Section A.4 (See graphs on the right.)

All of the graphs of these functions satisfy the vertical line test. Although the square function and the absolute value function map each value of x to one and only one value for y, these two functions map two values of x to the same value for y. For example $(-1, 1)$ and $(1, 1)$ lie on both graphs. The identity and reciprocal functions, on the other hand, map each x to a single value for y, and no two x values map to the same y value. These two functions are examples of one-to-one functions.

Identity function: $f(x) = x$

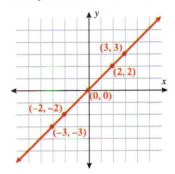

ONE-TO-ONE FUNCTION

A function $f(x)$ is **one-to-one** if every two distinct values for x in the domain, $x_1 \neq x_2$, correspond to two distinct values of the function, $f(x_1) \neq f(x_2)$.

Absolute value function: $f(x) = |x|$

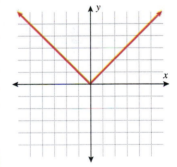

In other words, no two inputs map to the same output. Just as there is a graphical test for functions, the vertical line test, there is a graphical test for one-to-one functions, the *horizontal line test*. Note that a horizontal line can be drawn on the square and absolute value functions so that it intersects the graph of each function at two points. The identity and reciprocal functions, however, will only intersect a horizontal line in at most one point. This leads us to the horizontal line test for one-to-one functions.

HORIZONTAL LINE TEST

If every horizontal line intersects the graph of a function in at most one point, then the function is classified as a one-to-one function.

Reciprocal function: $f(x) = 1/x$

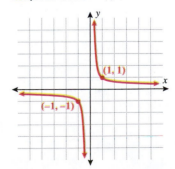

EXAMPLE 1 **Determine If a Function Defined as a Set of Points Is One-to-One**

For each of the three relationships, determine if the relationship is a function. If it is a function, determine if it is a one-to-one function.

$$f = \{(0, 0), (1, 1), (1, -1)\}$$

$$g = \{(-1, 1), (0, 0), (1, 1)\}$$

$$h = \{(-1, -1), (0, 0), (1, 1)\}$$

Solution:

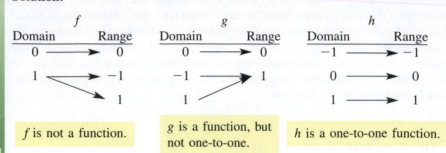

| f is not a function. | g is a function, but not one-to-one. | h is a one-to-one function. |

EXAMPLE 2 Determine If a Function Defined by an Equation Is One-to-One

For each of the three relationships, determine if the relationship is a function. If it is a function, determine if it is a one-to-one function. Assume that x is the independent variable and y is the dependent variable.

$$x = y^2 \qquad y = x^2 \qquad y = x^3$$

Solution:

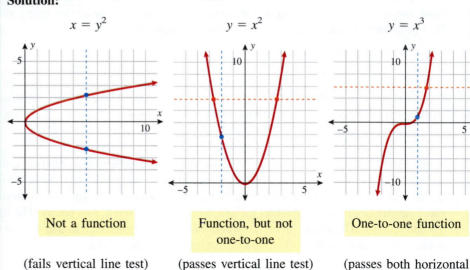

| Not a function | Function, but not one-to-one | One-to-one function |
| (fails vertical line test) | (passes vertical line test) (fails horizontal line test) | (passes both horizontal and vertical line tests) |

✔️**CONCEPT CHECK**   Does a parabola pass the horizontal line test?

■ **YOUR TURN**   Determine if the functions are one to one.

a. $f(x) = x + 2$ **b.** $f(x) = x^2 + 1$

Inverse Functions

If a function is one-to-one, then the function maps each x to one and only one y, and no two x values map to the same y value. This implies that there is a one-to-one correspondence between the input (domain) and output (range) of a one-to-one function $f(x)$. In

■ **Answer: a.** yes   **b.** no

the special case of a one-to-one function, it would be possible to map from the output (range of f) back to the input (domain of f), and this mapping would also be a function. The function that maps the output back to the input of a function, f, is called the **inverse function** and is denoted $f^{-1}(x)$.

A one-to-one function, f, maps every x in the domain to a unique and distinct corresponding y in the range. Therefore, the inverse, f^{-1}, maps every y back to a unique and distinct x.

The function notation $f(x) = y$ and $f^{-1}(y) = x$ indicates that if the point (x, y) satisfies the function, then the point (y, x) satisfies the inverse function.

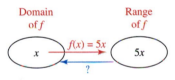

For example, the function $h(x) = \{(-1, 0), (1, 2), (3, 4)\}$

$$h = \{(-1, 0)\ (1, 2)\ (3, 4)\}$$

Domain   Range

$$-1 \rightleftarrows 0$$

$$1 \rightleftarrows 2 \qquad \begin{array}{l} h \text{ is a one-to-one} \\ \text{function} \end{array}$$

$$3 \rightleftarrows 4$$

Range   Domain

$$h^{-1} = \{(0, -1), (2, 1), (4, 3)\}$$

The inverse function undoes whatever the function does. For example, if $f(x) = 5x$, then the function f maps any value x in the domain to a value $5x$ in the range.

If we want to map backwards or undo the $5x$, we develop a function called the inverse function that takes $5x$ as input and maps back to x as output.

The inverse function is $f^{-1}(x) = \frac{1}{5}x$. Notice that if we input $5x$ into the inverse function, the output is x: $f^{-1}(5x) = \frac{1}{5}(5x) = x$.

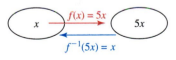

DEFINITION   **INVERSE FUNCTIONS**

If f and g denote two one-to-one functions such that

$$f(g(x)) = x \text{ for every } x \text{ in the domain of } g$$

and

$$g(f(x)) = x \text{ for every } x \text{ in the domain of } f,$$

then the function g is the **inverse** of the function f. The function g is denoted by f^{-1} (read "f-inverse").

Note: f^{-1} is used to denote the inverse of f. The -1 is not used as an exponent and, therefore, does not represent the reciprocal of f, $\frac{1}{f}$.

CAUTION

$$f^{-1} \neq \frac{1}{f}$$

Two properties hold true relating one-to-one functions to their inverses: (1) the range of the function is the domain of the inverse, and the range of the inverse is the domain of the function and (2) the composite function that results with a function and its inverse (and vice versa) yields x.

Domain of f = range of f^{-1} and range of f = domain of f^{-1}

$$f^{-1}(f(x)) = x \quad \text{and} \quad f(f^{-1}(x)) = x$$

EXAMPLE 3 Verifying Inverse Functions

Verify that $f^{-1}(x) = \frac{1}{2}x - 2$ is the inverse of $f(x) = 2x + 4$.

Solution:

Show that $f^{-1}(f(x)) = x$ and $f(f^{-1}(x)) = x$.

Write f^{-1} using placeholder notation.

$$f^{-1}(\ ) = \frac{1}{2}(\ ) - 2$$

Substitute $f(x) = 2x + 4$ into f^{-1}.

$$f^{-1}(f(x)) = \frac{1}{2}(2x + 4) - 2$$

Simplify.

$$f^{-1}(f(x)) = x + 2 - 2 = x$$
$$f^{-1}(f(x)) = x$$

Write f using placeholder notation.

$$f(\ ) = 2(\ ) + 4$$

Substitute $f^{-1}(x) = \frac{1}{2}x - 2$ into f.

$$f(f^{-1}(x)) = 2\left(\frac{1}{2}x - 2\right) + 4$$

Simplify.

$$f(f^{-1}(x)) = x - 4 + 4 = x$$
$$f(f^{-1}(x)) = x$$

Note the relationship between the domain and range of f and f^{-1}.

	Domain	Range
$f(x) = 2x + 4$	$\mathbb{R}$	$\mathbb{R}$
$f^{-1}(x) = \frac{1}{2}x - 2$	$\mathbb{R}$	$\mathbb{R}$

EXAMPLE 4 Verifying Inverse Functions with Domain Restrictions

Verify that $f^{-1}(x) = x^2, x \geq 0$ is the inverse of $f(x) = \sqrt{x}$.

Solution:

Show that $f^{-1}(f(x)) = x$ and $f(f^{-1}(x)) = x$.

Write f^{-1} using placeholder notation.

$$f^{-1}(\ ) = (\ )^2$$

Substitute $f(x) = \sqrt{x}$ into f^{-1}.

$$f^{-1}(f(x)) = (\sqrt{x})^2 = x$$
$$f^{-1}(f(x)) = x \text{ for } x \geq 0$$

Write f using placeholder notation.

$$f(\ ) = \sqrt{(\ )}$$

Substitute $f^{-1}(x) = x^2, x \geq 0$ into f.

$$f(f^{-1}(x)) = \sqrt{x^2} = x, \ x \geq 0$$
$$f(f^{-1}(x)) = x \text{ for } x \geq 0$$

	Domain	Range
$f(x) = \sqrt{x}$	$[0, \infty)$	$[0, \infty)$
$f^{-1}(x) = x^2, x \geq 0$	$[0, \infty)$	$[0, \infty)$

Graphical Interpretation of Inverse Functions

In Example 3, we showed that $f^{-1}(x) = \frac{1}{2}x - 2$ is the inverse of $f(x) = 2x + 4$. Let's now investigate the graphs that correspond to the function f and its inverse f^{-1}.

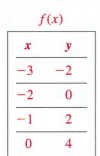

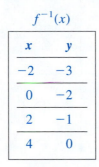

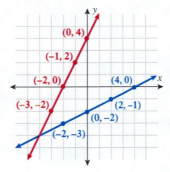

Notice that the point $(-3, -2)$ lies on the function and the point $(-2, -3)$ lies on the inverse. In fact, every point (a, b) that lies on the function corresponds to a point (b, a) that lies on the inverse.

Draw the line $y = x$ on the graph. In general, the point (b, a) on the inverse, $f^{-1}(x)$, is the reflection (about $y = x$) of the point (a, b) on the function, $f(x)$.

In general, if the point (a, b) is on the graph of a function, then the point (b, a) is on the graph of its inverse.

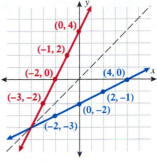

EXAMPLE 5 Graphing the Inverse Function

Given the graph of the function, $f(x)$, plot the graph of its inverse, $f^{-1}(x)$.

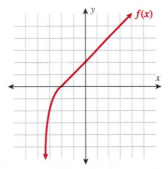

Solution:

Because the points $(-3, -2)$, $(-2, 0)$, $(0, 2)$, and $(2, 4)$ lie on the graph of f, then the points $(-2, -3)$, $(0, -2)$, $(2, 0)$ and $(4, 2)$ lie on the graph of f^{-1}.

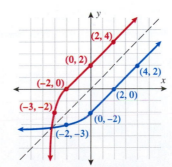

CONCEPT CHECK If the point (2, 0) lies on the graph of a one-to-one function, what point lies on the graph of its inverse?

■ **YOUR TURN** Given the graph of a function, f, plot the inverse function.

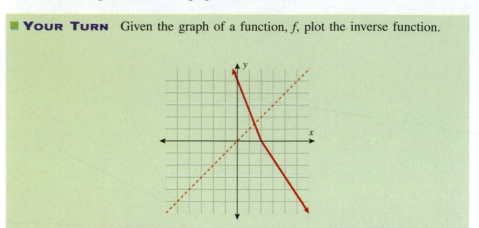

We have developed the definition of an inverse function, and properties of inverses. At this point you should be able to verify if two functions are inverses of one another. Let's turn our attention to another problem: How do you find the inverse of a function?

Finding the Inverse Function

Recall that if the point (a, b) lies on the graph of a function, then the point (b, a) lies on the graph of the inverse function. The symmetry about the line $y = x$ tells us that the roles of x and y interchange. Therefore, if we start with every point (x, y) that lies on the graph of a function, every point (y, x) lies on the graph of its inverse. Algebraically, this corresponds to interchanging x and y. Finding the inverse of a finite set of ordered pairs is easy: Simply interchange the x and y coordinates. Earlier, we found that if $h(x) = \{(-1, 0),(1, 2),(3, 4)\}$ then $h^{-1}(x) = \{(0, -1),(2, 1),(4, 3)\}$. But how do we find the inverse of a function defined by an equation?

Recall the mapping relationship if f is a one-to-one function. This relationship implies that $f(x) = y$ and $f^{-1}(y) = x$. Let's use these two identities in finding the inverse. Consider the function defined by $f(x) = 3x - 1$. To find f^{-1}, we let $f(x) = y$, which yields $y = 3x - 1$. Solve for the variable, x: $x = \frac{1}{3}y + \frac{1}{3}$.

Recall that $f^{-1}(y) = x$, so we have found the inverse to be $f^{-1}(y) = \frac{1}{3}y + \frac{1}{3}$. It is customary to write the independent variable as x, so we write the inverse as

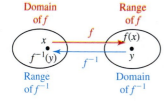

Domain of f · Range of f · x · f · $f(x)$ · $f^{-1}(y)$ · f^{-1} · y · Range of f^{-1} · Domain of f^{-1}

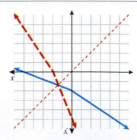

Answer:

$f^{-1}(x) = \frac{1}{3}x + \frac{1}{3}$. Now that we have found the inverse, let's confirm that the property $f^{-1}(f(x)) = x$ and $f(f^{-1}(x)) = x$, holds.

$$f(f^{-1}(x)) = 3\left(\frac{1}{3}x + \frac{1}{3}\right) - 1 = x + 1 - 1 = x$$

$$f^{-1}(f(x)) = \frac{1}{3}(3x - 1) + \frac{1}{3} = x - \frac{1}{3} + \frac{1}{3} = x$$

Finding inverses is often summarized with a seven-step approach.

FINDING THE INVERSE OF A FUNCTION

Step 1: Verify that $f(x)$ is a one-to-one function.
Step 2: Let $y = f(x)$.
Step 3: Interchange x and y.
Step 4: Solve for y.
Step 5: Let $y = f^{-1}(x)$.
Step 6: Note any domain restrictions on $f^{-1}(x)$.
Step 7: Check.

EXAMPLE 6 **The Inverse of a Square Root Function Is a Square Function**

Find the inverse of the function $f(x) = \sqrt{x + 2}$.

Solution:

STEP 1 $f(x)$ is a one-to-one function because it passes the horizontal line test.

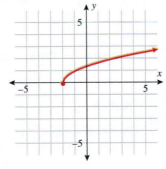

STEP 2 Let $y = f(x)$. $\qquad\qquad y = \sqrt{x + 2}$

STEP 3 Interchange x and y. $\qquad x = \sqrt{y + 2}$

STEP 4 Solve for y.

Square both sides of the equation. $\quad x^2 = y + 2$

Subtract 2 from both sides. $\qquad x^2 - 2 = y$ or $y = x^2 - 2$

STEP 5 Let $y = f^{-1}(x)$. $\qquad\qquad f^{-1}(x) = x^2 - 2$

STEP 6 Note any domain restrictions.

$\qquad f:$ Domain: $[-2, \infty)$ $\qquad$ Range: $[0, \infty)$

$\qquad f^{-1}:$ Domain: $[0, \infty)$ $\qquad$ Range: $[-2, \infty)$

The inverse of $f(x) = \sqrt{x + 2}$ is $\boxed{f^{-1}(x) = x^2 - 2 \text{ for } x \geq 0}$.

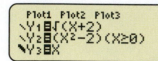
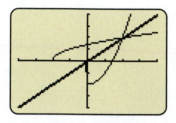
STEP 7 Check.

$f^{-1}(f(x)) = x$ for all x in the domain of f.

$$f^{-1}(f(x)) = (\sqrt{x + 2})^2 - 2$$
$$= x + 2 - 2 \quad \text{for } x \ge -2$$
$$= x$$

$f(f^{-1}(x)) = x$ for all x in the domain of f^{-1}.

$$f(f^{-1}(x)) = \sqrt{(x^2 - 2) + 2}$$
$$= \sqrt{x^2} \quad \text{for } x \ge 0$$
$$= x$$

Note that the function $f(x) = \sqrt{x + 2}$ and its inverse $f^{-1}(x) = x^2 - 2$ for $x \ge 0$ are symmetric about the line $y = x$.

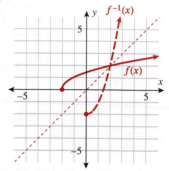

■ **YOUR TURN** Find the inverse of the given function. State the domain and range of the inverse function.

a. $f(x) = 7x - 3$ **b.** $g(x) = \sqrt{x - 1}$

EXAMPLE 7 When the Inverse of a Function Does Not Exist

Find the inverse of the function $f(x) = |x|$.

Solution:

The function $f(x) = |x|$ fails the horizontal line test and therefore is not a one-to-one function. Because f is not a one-to-one function, its inverse does not exist.

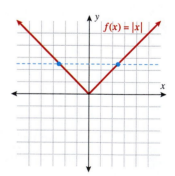

EXAMPLE 8 Finding the Inverse of a Function

The function $f(x) = \dfrac{2}{x+3}$, $x \neq -3$ is a one-to-one function. Find its inverse.

Solution:

STEP 1 The problem states that $f(x)$ is a one-to-one function.

STEP 2 Let $y = f(x)$.

$$y = \frac{2}{x+3}$$

STEP 3 Interchange x and y.

$$x = \frac{2}{y+3}$$

STEP 4 Solve for y.

Multiply equation by $(y+3)$.　　　　$x(y+3) = 2$

Eliminate parentheses.　　　　　　　$xy + 3x = 2$

Subtract $3x$ from both sides.　　　　$xy = -3x + 2$

Divide equation by x.　　　　$y = \dfrac{-3x+2}{x} = -3 + \dfrac{2}{x}$

STEP 5 Let $y = f^{-1}(x)$.　　　　$f^{-1}(x) = -3 + \dfrac{2}{x}$

STEP 6 Note any domain restrictions on $f^{-1}(x)$.　　$x \neq 0$

The inverse of the function $f(x) = \dfrac{2}{x+3}$, $x \neq -3$ is $\boxed{f^{-1}(x) = -3 + \dfrac{2}{x}, x \neq 0}$.

STEP 7 Check.

$$f^{-1}(f(x)) = -3 + \frac{2}{\left(\dfrac{2}{x+3}\right)} = -3 + (x+3) = x \qquad x \neq -3$$

$$f(f^{-1}(x)) = \frac{2}{\left(-3+\dfrac{2}{x}\right)+3} = \frac{2}{\left(\dfrac{2}{x}\right)} = x \qquad x \neq 0$$

■ **YOUR TURN** The function $f(x) = \dfrac{4}{x-1}$, $x \neq 1$ is a one-to-one function. Find its inverse.

SECTION A.6 **SUMMARY**

One-to-One Functions

- Algebraic test: No two x values map to the same y value.
- Graph test: Horizontal line test

Properties of Inverses

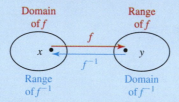

1. If f is one-to-one, then f^{-1} exists.
2. Domain of f^{-1} = range of f.
   Range of f^{-1} = domain of f.
3. $f^{-1}(f(x)) = x$ (for all x in domain of $f(x)$) and $f(f^{-1}(y)) = y$ (for all y in domain of $f^{-1}(y)$).
4. f and f^{-1} are symmetric with respect to the line $y = x$.

There is a seven-step procedure for finding the inverse of a function.

SECTION A.6 EXERCISES

▪ SKILLS

In Exercises 1–20, determine whether the given function is a one-to-one function.

1.

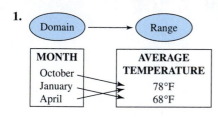

2.

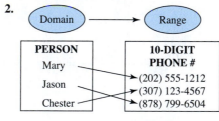

3.

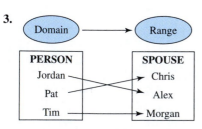

4.

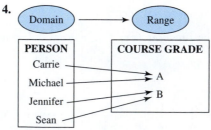

5. $\{(0, 1),(1, 2),(2, 3),(3, 4)\}$

6. $\{(0, -2),(2, 0),(5, 3),(-5, -7)\}$

7. $\{(0, 0),(9, -3),(4, -2),(4, 2),(9, 3)\}$

8. $\{(0, 1),(1, 1),(2, 1),(3, 1)\}$

9. $\{(0, 1),(1, 0),(2, 1),(-2, 1),(5, 4),(-3, 4)\}$

10. $\{(0, 0),(-1, -1),(-2, -8),(1, 1),(2, 8)\}$

11.

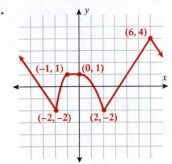

12.

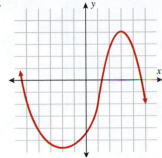

13.

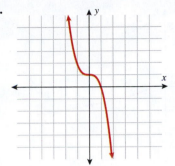

14.

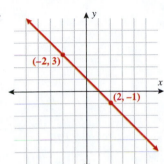

15.

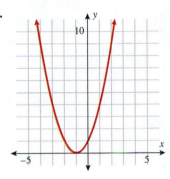

16.

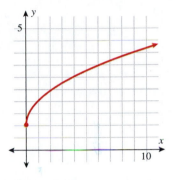

17. $f(x) = |x - 3|$　　**18.** $f(x) = (x - 2)^2 + 1$　　**19.** $f(x) = \dfrac{1}{x - 1}$　　**20.** $f(x) = \sqrt[3]{x}$

In Exercises 21–30, verify that the function, $f^{-1}(x)$, is the inverse of $f(x)$ by showing that $f(f^{-1}(x)) = x$ and $f^{-1}(f(x)) = x$. Graph $f(x)$ and $f^{-1}(x)$ on the same axes to show the symmetry about the line $y = x$.

21. $f(x) = 2x + 1$　$f^{-1}(x) = \dfrac{x - 1}{2}$

22. $f(x) = \dfrac{x - 2}{3}$　$f^{-1}(x) = 3x + 2$

23. $f(x) = \sqrt{x - 1}$　$x \geq 1$　$f^{-1}(x) = x^2 + 1$　$x \geq 0$

24. $f(x) = 2 - x^2$　$x \geq 0$　$f^{-1}(x) = \sqrt{2 - x}$　$x \leq 2$

25. $f(x) = \dfrac{1}{x}$　$f^{-1}(x) = \dfrac{1}{x}$　$x \neq 0$

26. $f(x) = (5 - x)^{1/3}$　$f^{-1}(x) = 5 - x^3$

27. $f(x) = \dfrac{1}{2x + 6}$　$x \neq -3$　$f^{-1}(x) = \dfrac{1}{2x} - 3$　$x \neq 0$

28. $f(x) = \dfrac{3}{4 - x}$　$x \neq 4$　$f^{-1}(x) = 4 - \dfrac{3}{x}$　$x \neq 0$

29. $f(x) = \dfrac{x + 3}{x + 4}$　$x \neq -4$　$f^{-1}(x) = \dfrac{3 - 4x}{x - 1}$　$x \neq 1$

30. $f(x) = \dfrac{x - 5}{3 - x}$　$x \neq 3$　$f^{-1}(x) = \dfrac{3x + 5}{x + 1}$　$x \neq -1$

In Exercises 31–38, a graph of a one-to-one function is given; plot its inverse.

31.

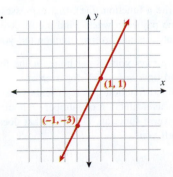

32.

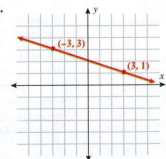

33.

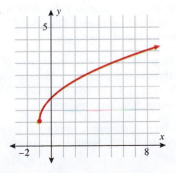

34.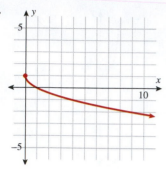

35.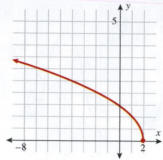

36.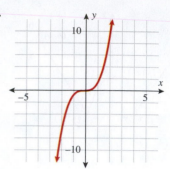

37.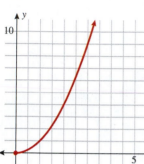

38.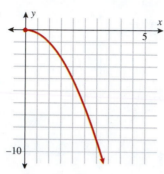

In Exercises 39–56, the function f is one-to-one. Find its inverse, and check your answer. State the domain and range of both f and f^{-1}.

39. $f(x) = x - 1$

40. $f(x) = 7x$

41. $f(x) = -3x + 2$

42. $f(x) = 2x + 3$

43. $f(x) = x^3 + 1$

44. $f(x) = x^3 - 1$

45. $f(x) = \sqrt{x - 3}$

46. $f(x) = \sqrt{3 - x}$

47. $f(x) = x^2 - 1, x \geq 0$

48. $f(x) = -x^2 + 1, x \geq 0$

49. $f(x) = (x + 2)^2 - 3, x \geq -2$

50. $f(x) = (x - 3)^2 - 2, x \geq 3$

51. $f(x) = \dfrac{2}{x}$

52. $f(x) = -\dfrac{3}{x}$

53. $f(x) = \dfrac{2}{3 - x}$

54. $f(x) = \dfrac{7}{x + 2}$

55. $f(x) = \dfrac{7x + 1}{5 - x}$

56. $f(x) = \dfrac{2x + 5}{7 + x}$

■ APPLICATIONS

57. Temperature. The equation used to convert from degrees Celsius, x, to degrees Fahrenheit is $f(x) = \frac{9}{5}x + 32$. Determine the inverse function, $f^{-1}(x)$. What does the inverse function represent?

58. Temperature. The equation used to convert from degrees Fahrenheit, x, to degrees Celsius is $C(x) = \frac{5}{9}(x - 32)$. Determine the inverse function, $C^{-1}(x)$. What does the inverse function represent?

59. Salary. A student is working at Target making $7 per hour and the weekly number of hours worked per week, x,

varies. If Target withholds 25% of his earnings for taxes and social security, write a function, $E(x)$, that expresses the student's take-home pay each week. Find the inverse function, $E^{-1}(x)$. What does the inverse function tell you?

60. Salary. A grocery store pays you $8 per hour for the first 40 hours per week and time and a half for overtime. Write a piecewise-defined function that represents your weekly earnings, $E(x)$, as a function of number of hours worked, x. Find the inverse function, $E^{-1}(x)$. What does the inverse function tell you?

■ CATCH THE MISTAKE

In Exercises 61–64, explain the mistake that is made.

61. Is $x = y^2$ a one-to-one function?

Solution:

Yes, this graph represents a one-to-one function because it passes the horizontal line test.

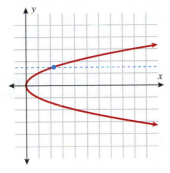

62. A linear one-to-one function is graphed below. Draw its inverse.

Solution:

Note that the points $(3, 3)$ and $(0, -4)$ lie on the graph of the function.

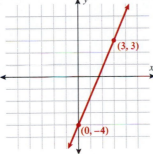

By symmetry, the points $(-3, -3)$ and $(0, 4)$ lie on the graph of the inverse.

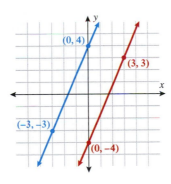

63. Given the function, $f(x) = x^2$, find the inverse function, $f^{-1}(x)$.

Solution:

Step 1: Let $y = f(x)$. $\qquad y = x^2$
Step 2: Solve for x. $\qquad x = \sqrt{y}$
Step 3: Interchange x and y. $\qquad y = \sqrt{x}$
Step 4: Let $y = f^{-1}(x)$. $\qquad f^{-1}(x) = \sqrt{x}$

Check: $f(f^{-1}(x)) = (\sqrt{x})^2 = x$ and $f^{-1}(f(x)) = \sqrt{x^2} = x$.

The inverse of $f(x) = x^2$ is $f^{-1}(x) = \sqrt{x}$.

64. Given the function, $f(x) = \sqrt{x} - 2$, find the inverse function, $f^{-1}(x)$, and state the domain restrictions on $f^{-1}(x)$.

Solution:

Step 1: Let $y = f(x)$. $\qquad y = \sqrt{x} - 2$
Step 2: Interchange x and y. $\qquad x = \sqrt{y} - 2$
Step 3: Solve for y. $\qquad y = x^2 + 2$
Step 4: Let $f^{-1}(x) = y$. $\qquad f^{-1}(x) = x^2 + 2$
Step 5: Domain restrictions $f(x) = \sqrt{x} - 2$ has the domain restriction that $x \geq 2$.

The inverse of $f(x) = \sqrt{x} - 2$ is $f^{-1}(x) = x^2 + 2$.

The domain of $f^{-1}(x)$ is $x \geq 2$.

■ CHALLENGE

65. T or F: Every even function is a one-to-one function.

66. T or F: Every odd function is a one-to-one function.

67. T or F: It is not possible for $f = f^{-1}$.

68. T or F: A function f has an inverse. If the function lies in quadrant II, then its inverse lies in quadrant IV.

69. If $(0, b)$ is the y-intercept of a one-to-one function f, what is the x-intercept of the inverse, f^{-1}?

70. The unit circle is not a function. If we restrict ourselves to the semicircle that lies in quadrants I and II, the graph represents a function, but it is not a one-to-one function. If we further restrict ourselves to the quarter circle lying in quadrant I, the graph does represent a one-to-one function. Determine the equations of both the one-to-one function and its inverse. State the domain restrictions of both.

71. Under what conditions is the linear function $f(x) = mx + b$ a one-to-one function?

72. Assuming the conditions found in Exercise 71 are met, determine the inverse of the linear function.

■ TECHNOLOGY

In Exercises 73–76, graph the following functions and determine if they are one-to-one.

73. $f(x) = |4 - x^2|$

75. $f(x) = x^{1/3} - x^5$

74. $f(x) = \dfrac{3}{x^3 + 2}$

76. $f(x) = \dfrac{1}{x^{1/2}}$

In Exercises 77 and 78, graph the functions f and g and the line $y = x$ in the same screen. Are the two function inverses of each other?

77. $f(x) = \sqrt{3x - 5}$ $g(x) = \dfrac{x^2}{3} + \dfrac{5}{3}$

78. $f(x) = (x - 7)^{1/3} + 2$ $g(x) = x^3 - 6x^2 + 12x - 1$

SECTION	TOPIC	PAGES	REVIEW EXERCISES	KEY CONCEPTS
A.1	Cartesian plane	476–480	1–14	Two points in the xy plane: (x_1, y_1) and (x_2, y_2).
	Distance	477–479	5–8	$d = \sqrt{(x_2 - x_1)^2 + (y_2 - y_1)^2}$
	Midpoint	479–480	9–12	$(x, y) = \left(\dfrac{x_1 + x_2}{2}, \dfrac{y_1 + y_2}{2} \right)$
	Applications	481–482	13–14	
A.2	Graphing equations: Point-plotting and symmetry	483–490	15–26	
	Plotting points	483–486	15–26	List a table with several coordinates that are solutions to the equation, plot, and connect with a smooth curve.
	Symmetry	486–489	15–18	The graph of an equation can be symmetric about the x-axis, y-axis, or origin.
	Using symmetry as a graphing aid	489–490	19–26	If (a, b) is on the graph of the equation, then ■ $(-a, b)$ is on the graph if it is symmetric about the y-axis. ■ $(a, -b)$ is on the graph if it is symmetric about the x-axis. ■ $(-a, -b)$ is on the graph if it is symmetric about the origin.
A.3	Functions	494–505	27–52	
	Determine if a relationship is a function	494–499	27–36	Each x corresponds to one and only one y.
	Vertical line test	499	35–36	A vertical line can only intersect a function in at most one point.
	Evaluate a function	500–503	37–44	Placeholder notation.
	Domain of a function	503–504	45–50	Are there any restrictions on x?
A.4	Graphs of functions: Common functions, transformations, and symmetry	509–524	53–74	
	Common functions	509–512		$f(x) = mx + b$, $f(x) = x$, $f(x) = x^2$, $f(x) = x^3$, $f(x) = \sqrt{x}$, $f(x) = \sqrt[3]{x}$, $f(x) = |x|$, $f(x) = 1/x$
	Even functions	512–513	53–60	$f(-x) = f(x)$ Symmetry about the y-axis
	Odd functions	512–513	53–60	$f(-x) = -f(x)$ Symmetry about the origin
	Horizontal shifts	514–519	61–74	$y = f(x + c)$ c units to the left $y = f(x - c)$ c units to the right
	Vertical shifts	514–519	61–74	$y = f(x) + c$ c units upward $y = f(x) - c$ c units downward

A.1 Cartesian Plane

Plot each point and indicate which quadrant the point lies in.

1. $(-4, 2)$ **2.** $(4, 7)$ **3.** $(-1, -6)$ **4.** $(2, -1)$

Calculate the distance between the two points.

5. $(-2, 0)$ and $(4, 3)$ **6.** $(1, 4)$ and $(4, 4)$

7. $(-4, -6)$ and $(2, 7)$ **8.** $\left(\frac{1}{4}, \frac{1}{12}\right)$ and $\left(\frac{1}{3}, -\frac{7}{3}\right)$

Calculate the midpoint of the segment joining the two points.

9. $(2, 4)$ and $(3, 8)$ **10.** $(-2, 6)$ and $(5, 7)$

11. $(2.3, 3.4)$ and $(5.4, 7.2)$ **12.** $(-a, 2)$ and $(a, 4)$

Applications

13. Sports. A quarterback drops back to pass. At the point $(-5, -20)$ he throws the ball to his wide receiver located at $(10, 30)$. Find the distance the ball has traveled. Assume the width of the football field is $[-15, 15]$ and the length is $[-50, 50]$.

14. Sports. Suppose in the above exercise a defender was midway between the quarterback and the receiver. At what point was the defender located when the ball was thrown over his head?

A.2 Graphing Equations: Point Plotting and Symmetry

Use algebraic tests to determine symmetry with respect to the x-axis, y-axis, or origin.

15. $x^2 + y^3 = 4$ **16.** $y = x^2 - 2$

17. $xy = 4$ **18.** $y^2 = 5 + x$

Use symmetry as a graphing aid and point plot the given equations.

19. $y = x^2 - 3$ **20.** $y = |x| - 4$ **21.** $y = \sqrt[3]{x}$

22. $x = y^2 - 2$ **23.** $y = x\sqrt{9 - x^2}$ **24.** $x^2 + y^2 = 36$

Applications

25. Track. A track around a high school football field is in the shape of the graph of $8x^2 + y^2 = 8$. Graph using symmetry and by plotting points.

26. Highways. A "bypass" around a town follows the graph of $y = x^3 + 2$ where the origin is the center of town. Graph the equation.

A.3 Functions

Determine whether each relationship is a function.

27.

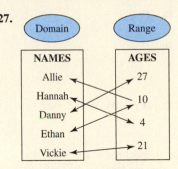

28. $\{(1, 2), (3, 4), (2, 4), (3, 7)\}$

29. $\{(-2, 3), (1, -3), (0, 4), (2, 6)\}$

30. $\{(4, 7), (2, 6), (3, 8), (1, 7)\}$ **31.** $x^2 + y^2 = 36$

32. $x = 4$ **33.** $y = |x + 2|$ **34.** $y = \sqrt{x}$

35.

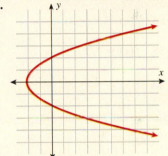

36.

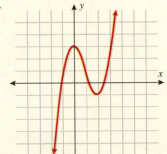

Evaluate the given quantities using the following three functions.

$$f(x) = 4x - 7 \quad F(t) = t^2 + 4t - 3 \quad g(x) = |x^2 + 2x + 4|$$

37. $f(3)$ **38.** $F(4)$ **39.** $f(-7) \cdot g(3)$

40. $\dfrac{F(0)}{g(0)}$ **41.** $\dfrac{f(2) - F(2)}{g(0)}$ **42.** $f(3 + h)$

43. $\dfrac{f(3 + h) - f(3)}{h}$ **44.** $\dfrac{F(t + h) - F(t)}{h}$

Find the domain of the given function. Express the domain in interval notation.

45. $f(x) = -3x - 4$ **46.** $g(x) = x^2 - 2x + 6$

47. $h(x) = \dfrac{1}{x + 4}$ **48.** $F(x) = \dfrac{7}{x^2 + 3}$

49. $G(x) = \sqrt{x - 4}$ **50.** $H(x) = \dfrac{1}{\sqrt{2x - 6}}$

Challenge

51. If $f(x) = \dfrac{D}{x^2 - 16}$, $f(4)$ and $f(-4)$ are undefined and $f(5) = 2$, find D.

52. Construct a function that is undefined at $x = -3$ and $x = 2$, and the point $(0, -4)$ lies on the graph of the function.

A.4 Graphs of Functions: Common Functions, Transformations, and Symmetry

Determine if the function is even, odd, or neither.

53. $f(x) = 2x - 7$ **54.** $g(x) = 7x^5 + 4x^3 - 2x$

55. $h(x) = x^3 - 7x$ **56.** $f(x) = x^4 + 3x^2$

57. $f(x) = x^{1/4} + x$ **58.** $f(x) = \sqrt{x + 4}$

59. $f(x) = \dfrac{1}{x^3} + 3x$ **60.** $f(x) = \dfrac{1}{x^2} + 3x^4 + |x|$

Graph the following functions using graphing aids.

61. $y = -(x - 2)^2 + 4$ **62.** $y = |-x + 5| - 7$

63. $y = \sqrt[3]{x - 3} + 2$ **64.** $y = \dfrac{1}{x - 2} - 4$

65. $y = \dfrac{-1}{2}x^3$ **66.** $y = 2x^2 + 3$

Write the function whose graph is the graph of $y = \sqrt{x}$ but is transformed accordingly, and state the domain of the resulting function.

67. shifted to left 3 units

68. shifted down 4 units

69. shifted to the right 2 units and up 3 units

70. reflected about the y-axis

71. expanded by a factor of 5 and shifted down 6 units

72. compressed by a factor of 2 and shifted up 3 units

Transform the function into the form $f(x) = c(x - h)^2 + k$ by completing the square and graph the resulting function using transformations.

73. $y = x^2 + 4x - 8$ **74.** $y = 2x^2 + 6x - 5$

A.5 Operations on Functions and Composition of Functions

Given the functions g and h, find $g + h$, $g - h$, $g \cdot h$, $\frac{g}{h}$, and state the domain.

75. $g(x) = -3x - 4$ **76.** $g(x) = 2x + 3$
$h(x) = x - 3$ $h(x) = x^2 + 6$

77. $g(x) = \dfrac{1}{x^2}$ **78.** $g(x) = \dfrac{x + 3}{2x - 4}$
$h(x) = \sqrt{x}$ $h(x) = \dfrac{3x - 1}{x - 2}$

79. $g(x) = \sqrt{x - 4}$ **80.** $g(x) = x^2 - 4$
$h(x) = \sqrt{2x + 1}$ $h(x) = x + 2$

For the given functions f and g, find the composite functions $f \circ g$ and $g \circ f$, and state the domains.

81. $f(x) = 3x - 4$ **82.** $f(x) = x^3 + 2x - 1$
$g(x) = 2x + 1$ $g(x) = x + 3$

83. $f(x) = \dfrac{2}{x + 3}$ **84.** $f(x) = \sqrt{2x^2 - 5}$
$g(x) = \dfrac{1}{4 - x}$ $g(x) = \sqrt{x + 6}$

Evaluate $f(g(3))$ and $g(f(-1))$, if possible.

85. $f(x) = 4x^2 - 3x + 2$ **86.** $f(x) = \sqrt{4 - x}$
$g(x) = 6x - 3$ $g(x) = x^2 + 5$

87. $f(x) = \dfrac{x}{|2x - 3|}$ **88.** $f(x) = \dfrac{1}{x - 1}$
$g(x) = |5x + 2|$ $g(x) = x^2 - 1$

Write the function as a composite of two functions f and g.

89. $h(x) = 3(x - 2)^2 + 4(x - 2) + 7$

90. $h(x) = \dfrac{\sqrt[3]{x}}{1 - \sqrt[3]{x}}$

91. $h(x) = \dfrac{1}{\sqrt{x^2 + 7}}$

92. $h(x) = \sqrt{|3x + 4|}$

Applications

93. Rain. A rain drop hitting a lake makes a circular ripple. If the radius, in inches, grows as a function of time, in minutes: $r(t) = 25\sqrt{t + 2}$, find the area of the ripple as a function of time.

94. Geometry. Let the area of a rectangle be given by: $42 = l \cdot w$ and let the perimeter be $36 = 2 \cdot l + 2 \cdot w$. Express the perimeter in terms of w.

A.6 One-to-One Functions and Inverse Functions

Determine whether the given function is a one-to-one function.

95.

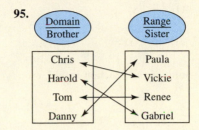

96. $\{(2, 3),(-1, 2),(3, 3),(-3, -4),(-2, 1)\}$

97. $\{(-2, 0),(4, 5),(3, 7)\}$

98. $\{(-8, -6),(-4, 2),(0, 3),(2, -8),(7, 4)\}$

99. $y = \sqrt{x}$

100. $y = x^2$

Verify that the function $f^{-1}(x)$ is the inverse of $f(x)$ by showing that $f(f^{-1}(x)) = x$. Graph $f(x)$ and $f^{-1}(x)$ on the same graph and show the symmetry about the line $y = x$.

101. $f(x) = 3x + 4 \qquad f^{-1}(x) = \dfrac{x - 4}{3}$

102. $f(x) = \dfrac{1}{4x - 7} \qquad f^{-1}(x) = \dfrac{1 + 7x}{4x}$

103. $f(x) = \sqrt{x + 4} \qquad f^{-1}(x) = x^2 - 4 \qquad x \geq 0$

104. $f(x) = \dfrac{x + 2}{x - 7} \qquad f^{-1}(x) = \dfrac{7x + 2}{x - 1}$

The function f is one-to-one. Find its inverse and check your answer. State the domain and range of both f and f^{-1}.

105. $f(x) = 2x + 1$

106. $f(x) = x^5 + 2$

107. $f(x) = \sqrt{x + 4}$

108. $f(x) = (x + 4)^2 + 3 \quad x \geq -4$

109. $f(x) = \dfrac{x + 6}{x + 3}$

110. $f(x) = 2\sqrt[3]{x - 5} - 8$

Applications

111. Salary. A pharmaceutical salesperson makes $22,000 base salary a year plus 8% of the total products sold. Write a function $S(x)$ that represents her yearly salary as a function of the total dollars worth of products sold, x. Find $S^{-1}(x)$. What does this inverse function tell you?

112. Volume. Express the volume of a rectangular box, V, that has a square base of length s and is 3 feet high as a function of the square length. Find V^{-1}. If a certain volume is desired what does the inverse tell you?

1. Find the distance between the points $(-7, -3)$ and $(2, -2)$.

2. Find the midpoint between $(-3, 5)$ and $(5, -1)$.

3. Determine the length and the midpoint of a segment that joins the points $(-2, 4)$ and $(3, 6)$.

4. **Research Triangle.** The Research Triangle in North Carolina was established as a collaborative research center among Duke University (Durham, NC), North Carolina State University (Raleigh, NC), and the University of North Carolina (Chapel Hill, NC).

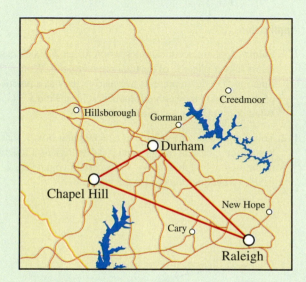

Durham is 10 miles north and 8 miles east of Chapel Hill, and Raleigh is 28 miles east and 15 miles south of Chapel Hill. What is the perimeter of the research triangle? Round your answer to the nearest mile.

5. If the point $(3, -4)$ is on a graph that is symmetric with respect to the y-axis, what point must also be on the graph?

6. If the point $(1, -1)$ is on the graph that is symmetric with respect to the x-axis, y-axis, and origin, what other points also must lie on the graph?

In Exercises 7–9, assuming x represents the independent variable and y represents the dependent variable, classify the relationships as

a. not a function   b. function, but not one-to-one

c. one-to-one function

7. $f(x) = |2x + 3|$ 8. $x = y^2 + 2$ 9. $y = \sqrt[3]{x + 1}$

In Exercises 10–13, use $f(x) = \sqrt{x - 2}$ and $g(x) = x^2 + 11$, and determine the desired quantity or expression. In the case of an expression, state the domain.

10. $f(11) - 2g(-1)$ 11. $\left(\dfrac{f}{g}\right)(x)$

12. $\left(\dfrac{g}{f}\right)(x)$ 13. $g(f(x))$

In Exercises 14–16, determine if the function is odd, even, or neither.

14. $f(x) = |x| - x^2$

15. $f(x) = 9x^3 + 5x - 3$

16. $f(x) = \dfrac{2}{x}$

In Exercises 17 and 18, graph the functions. State the domain and range of each function.

17. $f(x) = -\sqrt{x - 3} + 2$

18. $f(x) = -2(x - 1)^2$

In Exercises 19 and 20, given the function f, find the inverse if it exists. State the domain and range of both f and f^{-1}.

19. $f(x) = \sqrt{x - 5}$ 20. $f(x) = x^2 + 5$

21. What domain restriction can be made so that $f(x) = x^2$ has an inverse?

22. If the point $(-2, 5)$ lies on the graph of a function, what point lies on the graph of its inverse function?

23. **Discount.** Suppose a suit has been marked down 40% off the original price. An advertisement in the newspaper has an "additional 30% off the sale price" coupon. Write a function that determines the "checkout" price of the suit.

24. **Temperature.** Degrees Fahrenheit (F), degrees Celsius (C), and kelvins (K) are related by the two equations: $F = \frac{9}{5}C + 32$ and $K = C + 273.15$. Write a function whose input is kelvins and output is degrees Fahrenheit.

ANSWERS TO ODD NUMBERED EXERCISES*

Chapter 1

Section 1.1
1. $180°$
3. $-120°$
5. $300°$
7. $-288°$
9. **a.** $72°$ **b.** $162°$
11. **a.** $48°$ **b.** $138°$
13. **a.** $1°$ **b.** $91°$
15. $54°/36°$
17. $120°/60°$
19. $\gamma = 30°$
21. $\alpha = 120°$, $\beta = 30°$, $\gamma = 30°$
23. $c = 5$
25. $b = 8$
27. $c = \sqrt{89}$
29. $10\sqrt{2} \approx 14.14$ inches
31. 2 centimeters
33. other leg: $5\sqrt{3} \approx 8.66$ meters   hypotenuse: 10 meters
35. other leg: $4\sqrt{3} \approx 6.93$ yards   hypotenuse: $8\sqrt{3} \approx 13.9$ yards
37. short leg: 5 inches long   leg: $5\sqrt{3} \approx 8.66$ inches
39. $120°$
41. $144°$
43. 60 minutes or 1 hour
45. $\sqrt{7300} \approx 85$ feet
47. 241 feet
49. 9.8 feet
51. 17.3 feet
53. 48 feet by 28 feet
55. The length opposite the $60°$ angle is $\sqrt{3}$ times (not twice) the length opposite the $30°$ angle.
57. false   59. true   61. $110°$ 63. $DC = 3$ 65. 25
67. The triangles are isosceles right triangles, $45°$-$45°$-$90°$.

Section 1.2
1. $B = 80°$ 3. $F = 80°$ 5. $B = 75°$
7. $x = 15$, $A = 120°$, $D = 120°$
9. $x = 8$, $A = 110°$, $G = 70°$
11. b   13. a   15. d   17. $f = 3$
19. $a = 15$ 21. $c = 11.55$ kilometers   23. 38 feet
25. 225 feet   27. 120 meters   29. 6 feet 5 inches
31. The similar triangle ratio should be $\dfrac{A}{D} = \dfrac{B}{E}$, not $\dfrac{A}{E} = \dfrac{B}{D}$.
33. true   35. false   37. $x = 2$
39. The triangles 1 and 2 share vertical angles, and it is assumed that they are right triangles. Therefore they are similar triangles, since the triangles have three angles with equal measure. The same is true for triangles 3 and 4.

Section 1.3
1. $\dfrac{4}{5}$ 3. $\dfrac{5}{4}$ 5. $\dfrac{4}{3}$ 7. $\dfrac{\sqrt{5}}{5}$ 9. $\sqrt{5}$ 11. 2
13. $30°$ 15. $90° - x$ 17. $60°$
19. $\cos(90° - x - y)$ 21. $\sin(70° - A)$
23. $\tan(45° + x)$ 25. 10 miles
27. Opposite of angle y is 3 (not 4).
29. Secant is the reciprocal of cosine (not sine).
31. true
33. $\sin 30° = \dfrac{1}{2}$ and $\cos 30° = \dfrac{\sqrt{3}}{2}$
35. $\tan 30° = \dfrac{\sqrt{3}}{3}$ and $\tan 60° = \sqrt{3}$
37. $\sec 45° = \sqrt{2}$ and $\csc 45° = \sqrt{2}$

Section 1.4
1. a   3. b   5. c
7. $\dfrac{\sqrt{3}}{3}$ 9. $\sqrt{3}$ 11. $\dfrac{2\sqrt{3}}{3}$

13. $\dfrac{2\sqrt{3}}{3}$ 15. $\dfrac{\sqrt{3}}{3}$ 17. $\sqrt{2}$
19. 0.6018   21. 0.1392   23. 1.3764
25. 1.0098   27. 1.0002   29. 0.7002
31. $68°\,46'\,4''$ 33. $79°\,40'\,5''$ 35. $47°\,17'\,5''$
37. $84°42'31''$ 39. $33.33°$ 41. $59.45°$
43. $27.754°$ 45. $42.470°$ 47. $15°\,45'$
49. $22°\,21'$ 51. $30°\,10'30''$ 53. $77°\,32'\,6''$
55. 0.1808   57. 0.4091   59. 2.1007
61. 2.405   63. 1.335
65. $\cos 60° = \dfrac{1}{2}$, not $\dfrac{\sqrt{3}}{2}$.
67. 0
69. 0
71. **a.** 2.92398 **b.** 2.92380   b is more accurate
73. 3.240

Section 1.5
1. $a = 14$ in.   3. $a = 18$ ft   5. $a = 5.50$ miles
7. $c = 12$ km   9. $c = 20.60$ cm   11. $\alpha = 50°$
13. $\alpha = 62°$ 15. $a = 82.12$ yards
17. $c = 10.6$ km   19. $c = 19{,}293$ km
21. $\beta = 58°$, $a = 6.4$ ft, $b = 10$ ft
23. $\alpha = 18°$, $a = 3.0$ mm, $b = 9.2$ mm
25. $\beta = 35.8°$, $b = 80.1$ miles, $c = 137$ miles
27. $\beta = 61.62°$, $a = 936.9$ ft, $c = 1971$ ft
29. $\alpha = 56.0°$, $\beta = 34.0°$, $c = 51.3$ ft
31. $\alpha = 55.480°$, $\beta = 34.520°$, $b = 24{,}235$ km
33. 286 ft
35. $a = 88$ ft
37. 260 ft (262 rounded to two significant digits)
39. $11.0°$ (she is too low)   41. 80 ft
43. 170 m   45. $0.000016°$
47. 26 ft   49. 4414 ft
51. $8.32°$ 53. $138.3°$
55. $\tan^{-1}$ should have been used (not tan)
57. true   59. false   61. 4000 ft   63. $40°$
65. 0.8   67. θ

OOPS!
Divide the triangle into two right triangles by drawing a vertical line down the center of the triangle as in the diagram below.

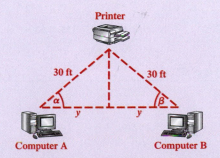

---

*Selected answers that require a proof, graph, or otherwise lengthy solution are not included.

The distance between the computers is x, so the distance from each computer to the dashed line is $\frac{x}{2}$. Let $y = \frac{x}{2}$. Thus,

$$\cos 40° = \frac{y}{30}$$

$$y = 30\cos 40° \approx 23 \text{ feet}$$

Since $y = \frac{x}{2}$, then $x \approx 46$ feet.

Tying It All Together

Angles α and β are alternate interior angles, so their measure is the same. Call the length of the partition x. Thus, $\tan 51° = \frac{x}{22}$ and $x \approx 27$ feet. Angle β can also be found using the fact that angles β and γ are supplementary.

Review Exercises

1. a. 62° **b.** 152°   **3. a.** 55° **b.** 145°
5. $\gamma = 25°$ **7.** $\alpha = 140°, \beta = 20°, \gamma = 20°$
9. $b = \sqrt{128} = 8\sqrt{2}$ **11.** $c = \sqrt{65}$
13. $12\sqrt{2}$ yards   **15.** leg: $3\sqrt{3}$ feet, hypotenuse: 6 feet
17. 150° **19.** 75°   **21.** 75°   **23.** 75°
25. $F = 4$ **27.** $C = 147.6$ km **29.** 32 m   **31.** 48 inches
33. $\frac{2\sqrt{13}}{13}$ **35.** $\frac{\sqrt{13}}{2}$ **37.** $\frac{3}{2}$ **39.** $\cos 60°$
41. $\cot 45°$ **43.** $\cos(x + 60°)$ **45.** $\sec(x + 45°)$ **47.** b
49. b   **51.** c   **53.** $\frac{\sqrt{3}}{3}$ **55.** $\sqrt{3}$
57. $\sqrt{2}$ **59.** $\frac{2\sqrt{3}}{3}$ **61.** 2   **63.** 1
65. 0.6691   **67.** 0.9548   **69.** 1.5399
71. 1.5477   **73.** 39.28°   **75.** 29.507°
77. 42° 15′   **79.** 30° 10′ 30″   **81.** 1.50
83. 14 inches   **85.** 14.5 miles   **87.** 92.91 yards
89. $\beta = 60°, a = 11$ ft, $b = 18$ ft
91. $\beta = 41.5°, b = 190$ miles, $c = 287$ miles
93. $\alpha = 33.7°, \beta = 56.3°, c = 54.9$ feet
95. 75 feet

Practice Test

1. 20°, 60°, and 100°   **3.** 6000 feet
5. a. $\frac{3\sqrt{10}}{10}$ **b.** $\sqrt{10}$ **c.** $\frac{1}{3}$ **7.** 1.3629
9. 33.756°   **11.** $\beta = 66°$ **13.** 1.53

Chapter 2

Section 2.1
1. QI   **3.** QII   **5.** QIV
7. y-axis   **9.** x-axis   **11.** QIII
13. QI   **15.** QIII   **17.** QII
19. QII

21.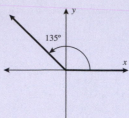
23.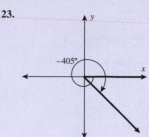

25.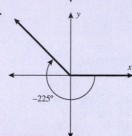
27. 330°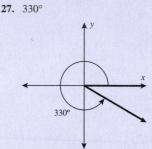

29. 510°
31. 840°

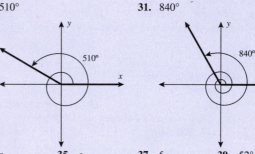

33. c   **35.** e   **37.** f   **39.** 52°
41. 268°   **43.** 330°   **45.** 150°   **47.** −1200°
49. Yes, these are coterminal angles, so the steer were in approximately the same place when roped.
51. Don and Ron end up in the same place, 180° from where they started.
53. 1440°
55. Coterminal angles are not complementary angles.
57. true
59. false
61. a. $30° + n\,360°, n = 0, 1, 2,\ldots$
    b. $-330° - n\,360°, n = 0, 1, 2,\ldots$

Section 2.2

	$\sin \theta$	$\cos \theta$	$\tan \theta$	$\cot \theta$	$\sec \theta$	$\csc \theta$
1.	$\frac{2\sqrt{5}}{5}$	$\frac{\sqrt{5}}{5}$	2	$\frac{1}{2}$	$\sqrt{5}$	$\frac{\sqrt{5}}{2}$
3.	$\frac{2\sqrt{5}}{5}$	$\frac{\sqrt{5}}{5}$	2	$\frac{1}{2}$	$\sqrt{5}$	$\frac{\sqrt{5}}{2}$
5.	$\frac{4\sqrt{41}}{41}$	$\frac{5\sqrt{41}}{41}$	$\frac{4}{5}$	$\frac{5}{4}$	$\frac{\sqrt{41}}{5}$	$\frac{\sqrt{41}}{4}$
7.	$\frac{2\sqrt{5}}{5}$	$-\frac{\sqrt{5}}{5}$	-2	$-\frac{1}{2}$	$-\sqrt{5}$	$\frac{\sqrt{5}}{2}$

	sin θ	cos θ	tan θ	cot θ	sec θ	csc θ
9.	$-\dfrac{7\sqrt{65}}{65}$	$-\dfrac{4\sqrt{65}}{65}$	$\dfrac{7}{4}$	$\dfrac{4}{7}$	$-\dfrac{\sqrt{65}}{4}$	$\dfrac{\sqrt{65}}{7}$
11.	$\dfrac{\sqrt{15}}{5}$	$-\dfrac{\sqrt{10}}{5}$	$-\dfrac{\sqrt{6}}{2}$	$-\dfrac{\sqrt{6}}{3}$	$-\dfrac{\sqrt{10}}{2}$	$\dfrac{\sqrt{15}}{3}$
13.	$-\dfrac{\sqrt{6}}{4}$	$-\dfrac{\sqrt{10}}{4}$	$\dfrac{\sqrt{15}}{5}$	$\dfrac{\sqrt{15}}{3}$	$\dfrac{2\sqrt{10}}{5}$	$\dfrac{2\sqrt{6}}{3}$
15.	$-\dfrac{2\sqrt{29}}{29}$	$-\dfrac{5\sqrt{29}}{29}$	$\dfrac{2}{5}$	$\dfrac{5}{2}$	$-\dfrac{\sqrt{29}}{5}$	$\dfrac{\sqrt{29}}{2}$
17.	$\dfrac{2\sqrt{5}}{5}$	$\dfrac{\sqrt{5}}{5}$	2	$\dfrac{1}{2}$	$\sqrt{5}$	$\dfrac{\sqrt{5}}{2}$
19.	$\dfrac{\sqrt{5}}{5}$	$\dfrac{2\sqrt{5}}{5}$	$\dfrac{1}{2}$	2	$\dfrac{\sqrt{5}}{2}$	$\sqrt{5}$
21.	$-\dfrac{\sqrt{10}}{10}$	$\dfrac{3\sqrt{10}}{10}$	$-\dfrac{1}{3}$	-3	$\dfrac{\sqrt{10}}{3}$	$-\sqrt{10}$
23.	$\dfrac{2\sqrt{13}}{13}$	$-\dfrac{3\sqrt{13}}{13}$	$-\dfrac{2}{3}$	$-\dfrac{3}{2}$	$\dfrac{\sqrt{13}}{3}$	$\dfrac{\sqrt{13}}{2}$
25.	1	0	undefined	0	undefined	1
27.	-1	0	undefined	0	undefined	-1
29.	1	0	undefined	0	undefined	1
31.	-1	0	undefined	0	undefined	-1
33.	-1	0	undefined	0	undefined	-1
35.	1	0	undefined	0	undefined	1

37. $\tan\theta = \dfrac{7}{24}$
39. 120 feet
41. $r = \sqrt{5}$ (not 5)
43. false
45. true
47. $-\dfrac{3}{5}$
49. $m = \tan\theta$
51. $y = (x - a)\tan\theta$
53. -1
55. because tangent is undefined $\left(\dfrac{1}{0}\right)$

Section 2.3

1. QIV
3. QII
5. QI
7. $\sin\theta = -\dfrac{4}{5}$
9. $\tan\theta = -\dfrac{60}{11}$
11. $\sin\theta = -\dfrac{84}{85}$
13. $\tan\theta = \sqrt{3}$
15. $\tan\theta = -\dfrac{\sqrt{11}}{5}$
17. 1
19. -1
21. 0
23. 1
25. 1
27. possible
29. not possible
31. possible
33. possible
35. $-\dfrac{1}{2}$
37. $-\dfrac{\sqrt{3}}{2}$
39. $\dfrac{\sqrt{3}}{3}$
41. 1
43. -2
45. 30° and 330°
47. 210° and 330°
49. 90° and 270°
51. 270°
53. -0.8387
55. -11.4301
57. 2.0627
59. -1.0711
61. -2.6051
63. 110°
65. 143°
67. 322°
69. 140°
71. 340°
73. 1.3
75. 12°

77. The reference angle is made with the terminal side and the x-axis (not the y-axis).
79. true   81. false   83. $-\dfrac{a}{\sqrt{a^2 + b^2}}$ 85. QI   87. QI

Section 2.4

1. $\dfrac{8}{7}$
3. $-\dfrac{5}{3}$
5. $-\dfrac{1}{5}$
7. $\dfrac{2\sqrt{5}}{5}$
9. $-\dfrac{5\sqrt{7}}{7}$
11. $-\dfrac{\sqrt{3}}{3}$
13. $\dfrac{4}{3}$
15. $\dfrac{\sqrt{11}}{5}$
17. $\dfrac{b}{a}$ $a \neq 0$
19. $\dfrac{5}{64}$
21. -8
23. $-\dfrac{\sqrt{3}}{2}$
25. $-\dfrac{\sqrt{21}}{5}$
27. $-\sqrt{17}$
29. $-\sqrt{5}$
31. $-\dfrac{8\sqrt{161}}{161}$
33. $-\dfrac{15\sqrt{11}}{44}$
35. $-\dfrac{2\sqrt{13}}{13}$
37. $\cos\theta = -\dfrac{3}{5},\ \sin\theta = \dfrac{4}{5}$
39. $\cos\theta = -\dfrac{\sqrt{5}}{5},\ \sin\theta = -\dfrac{2\sqrt{5}}{5}$
41. $\cos\theta = -\dfrac{5\sqrt{34}}{34},\ \sin\theta = -\dfrac{3\sqrt{34}}{34}$
43. $\dfrac{1}{\sin\theta}$
45. -1
47. $\dfrac{\cos^2\theta}{\sin\theta}$
49. $\dfrac{1}{\sin\theta}$
51. $1 + 2\sin\theta\cos\theta$
53. $r = 8\sin\theta(1 - \sin^2\theta) = 8\sin\theta - 8\sin^3\theta$
$\theta = 30°, r = 3$
$\theta = 60°, r = \sqrt{3}$
$\theta = 90°, r = 0$
55. Cosine is negative in quadrant III.   57. F   59. 90°
61. $\cot\theta = \pm\dfrac{\sqrt{1 - \sin^2\theta}}{\sin\theta}$
63. $8|\cos\theta|$ 65. Yes

OOPS!

$\theta = \cos^{-1}(0.52) \approx 59°$
$\alpha = \sin^{-1}(0.95) \approx 72°$
$\gamma = \tan^{-1}(-27) \approx -88°$

Tying It All Together

$\alpha = 40°, \beta = 55°$. The monument's coordinates are (172.1, 245.7).

Review Exercises

1. QIV
3. QII
5.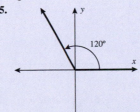
7.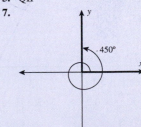
9. 280°
11. 120°

561

	$\sin\theta$	$\cos\theta$	$\tan\theta$	$\cot\theta$	$\sec\theta$	$\csc\theta$
13.	$-\dfrac{4}{5}$	$\dfrac{3}{5}$	$-\dfrac{4}{3}$	$-\dfrac{3}{4}$	$\dfrac{5}{3}$	$-\dfrac{5}{4}$
15.	$\dfrac{\sqrt{10}}{10}$	$-\dfrac{3\sqrt{10}}{10}$	$-\dfrac{1}{3}$	-3	$-\dfrac{\sqrt{10}}{3}$	$\sqrt{10}$
17.	$\dfrac{1}{2}$	$\dfrac{\sqrt{3}}{2}$	$\dfrac{\sqrt{3}}{3}$	$\sqrt{3}$	$\dfrac{2\sqrt{3}}{3}$	2
19.	$-\dfrac{\sqrt{2}}{2}$	$-\dfrac{\sqrt{2}}{2}$	1	1	$-\sqrt{2}$	$-\sqrt{2}$
21.	$-\dfrac{3}{5}$	$\dfrac{4}{5}$	$-\dfrac{3}{4}$	$-\dfrac{4}{3}$	$\dfrac{5}{4}$	$\dfrac{5}{3}$
23.	-1	0	undefined	0	undefined	-1
25.	0	1	0	undefined	1	undefined

27. QIII **29.** QIV **31.** $\dfrac{4}{5}$

33. $-2\sqrt{2}$ **35.** 1 **37.** 0

39. impossible **41.** possible **43.** $-\dfrac{1}{2}$

45. $-\dfrac{\sqrt{3}}{3}$ **47.** $-\dfrac{2\sqrt{3}}{3}$ **49.** -0.2419

51. 1.0355 **53.** -2.7904 **55.** -0.6494

57. $-\dfrac{11}{7}$ **59.** $-\dfrac{8}{15}$ **61.** $\dfrac{4}{27}$

63. $3\sqrt{3}$ **65.** $-\dfrac{7}{24}$ **67.** $-\dfrac{5}{12}$

69. $\sin\theta = -\dfrac{\sqrt{3}}{2}, \cos\theta = -\dfrac{1}{2}$ **71.** $\dfrac{\cos\theta}{\sin^2\theta}$

73. $\sin\theta$ **75.** $\dfrac{1}{\sin\theta} + \cos\theta$

Practice Test
1.

	0°	QI	90°	QII	180°	QIII	270°	QIV	360°
$\sin\theta$	0	+	1	+	0	−	−1	−	0
$\cos\theta$	1	+	0	−	−1	−	0	+	1

3. $\theta = 90°$ and $\theta = 270°$

5. 220° **7.** $\dfrac{\sqrt{2}}{2}$ **9.** 1 **11.** 1.1106

13. $\sin\theta = -\dfrac{\sqrt{35}}{6}$ and $\tan\theta = -\sqrt{35}$

15. a

Chapter 3

Section 3.1
1. $\dfrac{1}{5}$ or 0.2 **3.** $\dfrac{2}{11}$ or 0.18 **5.** $\dfrac{1}{50}$ or 0.02

7. $\dfrac{1}{8}$ or 0.125 **9.** $\dfrac{1}{5}$ or 0.2 **11.** $\dfrac{\pi}{6}$

13. $\dfrac{\pi}{4}$ **15.** $\dfrac{7\pi}{4}$ **17.** $\dfrac{5\pi}{12}$

19. $\dfrac{17\pi}{18}$ **21.** $\dfrac{13\pi}{3}$ **23.** $-\dfrac{7\pi}{6}$

25. 30° **27.** 135° **29.** 67.5°

31. 75° **33.** 1620° **35.** 171°

37. $-84°$ **39.** 229.18° **41.** 48.70°

43. $-160.37°$ **45.** 0.820 **47.** 1.95

49. 0.986 **51.** $\dfrac{\pi}{3}$ or 60° **53.** $\dfrac{\pi}{4}$ or 45°

55. $\dfrac{5\pi}{12}$ or 75° **57.** $\dfrac{\pi}{3}$ or 60° **59.** $\dfrac{\sqrt{2}}{2}$

61. $-\dfrac{\sqrt{2}}{2}$ **63.** $-\dfrac{\sqrt{3}}{3}$ **65.** $\dfrac{\sqrt{3}}{3}$

67. 0 **69.** -1 **71.** $\dfrac{3\pi}{2}$

73. $\dfrac{\pi}{5}$ **75.** 5π **77.** $\dfrac{\pi}{16}$

79. Radius and arc length must be converted to the same units before using the radian formula.

81. The arguments are radians (not degrees).

83. false **85.** $\dfrac{\pi}{2}$

87. $\dfrac{9}{40} = 0.225$ radians **89.** $\dfrac{13\pi}{18}$

91. -0.917 for both. It makes sense because the conversion from radians to degrees involved multiplying by $\dfrac{180°}{\pi}$.

Section 3.2
1. 12 mm **3.** $\dfrac{2\pi}{3}$ ft **5.** $\dfrac{11\pi}{5}\ \mu m$

7. $\dfrac{200\pi}{3}$ km **9.** 25 ft **11.** 8 in.

13. 4 yd **15.** 12 mi **17.** 1.3 mm

19. 1.7 yd **21.** 2.2 μm **23.** 1300 km

25. 12.8 sq. ft **27.** 2.85 sq. km **29.** 8.62 sq. cm

31. 0.0236 sq. ft **33.** 5262 km **35.** 37 ft

37. $200\pi \approx 628$ ft **39.** 50° **41.** 60 in. (5 ft)

43. 911 mi **45.** 157 sq. ft **47.** 78 sq. in.

49. The arc length formula for radian measure was used (should have used the formula for degree measure).

51. true **53.** $\theta_2° = \left(\dfrac{r_1}{r_2}\right)\theta_1°$

55. 11,800 sq. ft **57.** 133 sq. ft

Section 3.3
1. $\dfrac{2}{5}$ m/sec **3.** 272 km/hr **5.** 7 nm/ms

7. $\dfrac{1}{64}$ in./min **9.** 9.8 m **11.** 1.5 mi

13. 15 mi **15.** 4,320,000 km **17.** $\dfrac{5\pi}{2}\dfrac{\text{rad}}{\text{sec}}$

19. $20\pi\dfrac{\text{rad}}{\text{sec}}$ **21.** $\dfrac{2\pi}{9}\dfrac{\text{rad}}{\text{sec}}$ **23.** $\dfrac{13\pi}{9}\dfrac{\text{rad}}{\text{sec}}$

25. 6π inches per second **27.** $\dfrac{\pi}{4}$ mm per second

29. $\dfrac{2\pi}{3}$ inches per second **31.** 26.2 cm

33. 653 in. (or 54.5 ft) **35.** 6.69 mi

37. 69.82 mph

39. 1037.51 mph; 1040.35 mph

41. 15.71 ft/sec

43. $66\frac{2}{3}\pi$ rad/min

45. 12.05 mph

47. 640 rev/min

49. 10.11 rad/sec or 1.6 rotations per second

51. 17.59 m/s

53. Angular speed needs to be in radians (not degrees) per second.

55. false

57. $v_2 = \left(\dfrac{r_2}{r_1}\right)v_1$

59. $2\pi \approx 6.3$ cm/sec

Section 3.4

1. $-\dfrac{\sqrt{3}}{2}$

3. $-\dfrac{\sqrt{3}}{2}$

5. $\dfrac{\sqrt{2}}{2}$

7. -1

9. $-\sqrt{2}$

11. $\sqrt{3}$

13. $-\dfrac{\sqrt{3}}{2}$

15. $-\dfrac{\sqrt{3}}{2}$

17. $-\dfrac{\sqrt{2}}{2}$

19. $-\dfrac{\sqrt{3}}{2}$

21. $\dfrac{\sqrt{2}}{2}$

23. 1

25. $\dfrac{\sqrt{2}}{2}$

27. 0

29. $\dfrac{\pi}{6}, \dfrac{11\pi}{6}$

31. $\dfrac{4\pi}{3}, \dfrac{5\pi}{3}$

33. $0, \pi, 2\pi, 3\pi, 4\pi$

35. $\pi, 3\pi$

37. $\dfrac{3\pi}{4}, \dfrac{7\pi}{4}$

39. $\dfrac{3\pi}{4}, \dfrac{5\pi}{4}$

41. $0, \pi, 2\pi$

43. $\dfrac{\pi}{2}, \dfrac{3\pi}{2}$

45. 22.9°F

47. 99.1°F

49. 2.6 ft

51. 135 lb

53. 10,000 guests

55. Used the x coordinate for sine and the y coordinate for cosine. Should have done the opposite.

57. true

59. false

61. odd

63. $\dfrac{\pi}{4}, \dfrac{5\pi}{4}$

65. $\sin(423°) = 0.891$ and $\sin(-423°) = -0.891$

67. 0.5

OOPS!

Arc length: $s = r\theta = 6\left(\dfrac{4\pi}{3}\right) = 8\pi$

Time to fly 8π miles: $(8\pi \text{ mi})\left(\dfrac{1 \text{ hr}}{92 \text{ mi}}\right)\left(\dfrac{60 \text{ min}}{1 \text{ hr}}\right) \approx 16$ min

Tying It All Together

If $r = 30$ mm, $\omega = \dfrac{v}{r} = \dfrac{1200 \text{ mm/s}}{30 \text{ mm}} = 40$ rad/s

If $r = 42$ mm, $\omega = \dfrac{v}{r} = \dfrac{1200 \text{ mm/s}}{42 \text{ mm}} = \dfrac{200}{7}$ rad/s ≈ 28.6 rad/s

The angular speed at which the disk is being played for the song closer to the center of the CD is faster.

The length of the tracks is $D = rt = \dfrac{1.2 \text{ m}}{s}(4440 \text{ s}) = 5328 \text{ m} \approx 5.3$ km

(There are 4440 seconds in 74 minutes.)

Review Exercises

1. $\dfrac{3\pi}{4}$

3. $\dfrac{11\pi}{6}$

5. $\dfrac{6\pi}{5}$

7. $\dfrac{14\pi}{5}$

9. $-\dfrac{5\pi}{6}$

11. 60°

13. 225°

15. 100°

17. 585°

19. $-50°$

21. $\dfrac{5\pi}{3}$ cm ≈ 5.24 cm

23. $\dfrac{25\pi}{9}$ in. ≈ 8.73 in.

25. 0.5 radians

27. $\dfrac{2\pi}{3}$

29. $\dfrac{1}{6}$ radian

31. 1.6 in.

33. 16.8 m

35. 5 yd

37. 96π sq. mi ≈ 302 sq. mi

39. 600π sq. m ≈ 1885 sq. m

41. $\dfrac{1}{3}$ ft/sec

43. 5 mi/min

45. 360 mi

47. 20 mi

49. $\dfrac{2\pi}{3}$ rad/sec

51. $\dfrac{\pi}{16}$ rad/sec

53. 10π m/sec

55. 75π ft

57. 240π yd

59. 240π in./min ≈ 754 in./min

61. $-\dfrac{\sqrt{3}}{3}$

63. $-\dfrac{1}{2}$

65. 1

67. -1

69. -1

71. $\dfrac{1}{2}$

73. $-\dfrac{1}{2}$

75. $-\dfrac{1}{2}$

77. $\dfrac{7\pi}{6}, \dfrac{11\pi}{6}$

79. $0, \pi, 2\pi, 3\pi, 4\pi$

Practice Test

1. 0.02 radians

3. $\dfrac{13\pi}{9}$

5. $\dfrac{5\pi}{12}$

7. 164 sq. ft

9. 52°

11. -1

13. 0

15. $\dfrac{23\pi}{36}$

Chapter 4

Section 4.1

1. c

3. a

5. h

7. b

9. e

11. $A = \dfrac{3}{2}, p = \dfrac{2\pi}{3}$

13. $A = 1, p = \dfrac{2\pi}{5}$

15. $A = \dfrac{2}{3}, p = \dfrac{4\pi}{3}$

17. $A = 3, p = 2$

19. $A = 5, p = 6$

21.

23.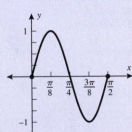

25.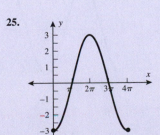

27.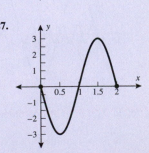

29.

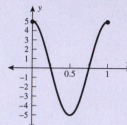

31.

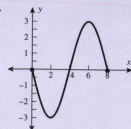

77. As t increases, the amplitude goes to zero for Y_1 and Y_3.

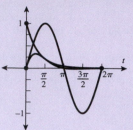

33.

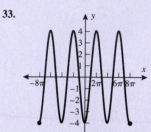

35.

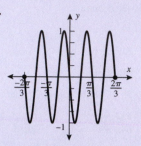

Section 4.2

1. c **3.** a **5.** e **7.** f

9.

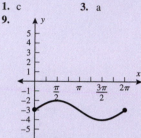

11.

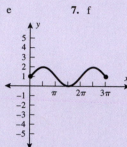

37.

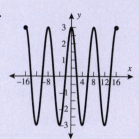

39.

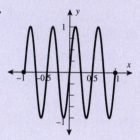

13.

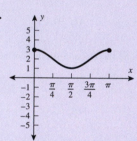

15.

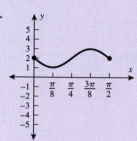

41. $y = -\sin 2x$

43. $y = \cos \pi x$

45. $y = -2\sin\left(\dfrac{\pi}{2}x\right)$

47. $y = \sin 8\pi x$

49. amplitude: 4 cm; mass: 4 g **51.** $\dfrac{1}{4\pi}$ cycles per second

53. amplitude: 0.005 cm; frequency: 256 hertz

55. amplitude: 0.008 cm; frequency: 375 hertz

57. 660 m/sec

59. 660 m/sec

61. forgot to reflect about the x-axis

63. true **65.** false **67.** $(0, A)$

69. $x = \dfrac{n\pi}{B}$, where n is an integer.

17.

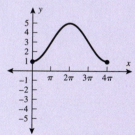

19.

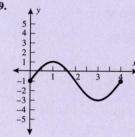

71. no **73.** They coincide.

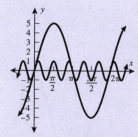

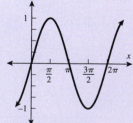

21. amplitude: 2; period: 2; phase shift: $\dfrac{1}{\pi}$

23. amplitude: 5; period: $\dfrac{2\pi}{3}$; phase shift: $-\dfrac{2}{3}$

25. amplitude: 6; period: 2; phase shift: -2

27.

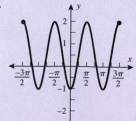

29.

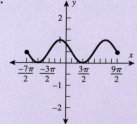

75. a. Y_2 is Y_1 shifted to the left $\dfrac{\pi}{3}$.

 b. Y_2 is Y_1 shifted to the right $\dfrac{\pi}{3}$.

31.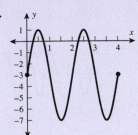

33.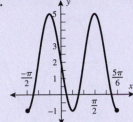

35.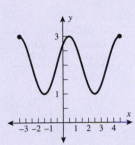

37. $y = 1 + \sin(\pi x)$ or $y = 1 + \cos\left[\pi\left(x - \frac{1}{2}\right)\right]$

39. $y = \cos(\pi(x + 1))$ or $y = \sin\left[\pi\left(x - \frac{1}{2}\right)\right]$

41. maximum current: 220 A; minimum current: -220 A
period: 0.1 seconds; phase shift: 0.01 seconds

43. Charlotte Monthly Temperature is $^\circ$F =
$59.3 + 20\sin\left(\frac{\pi}{6}x + \frac{4\pi}{3}\right)$, where $x = 1$ is January.

45. height: 50 ft; 2400 ft in length

47. 400 to 600 deer; 8 years

49. $y = 3\sin\left(\frac{\pi}{11}t\right) + 9$

51. $y = 25 + 25\sin\left[\frac{2\pi}{4}(t - 1)\right]$

53. The mistake was that the phase shift of $y = k + A\sin(Bx + C)$
is $-\frac{C}{B}$, (not $-C$). Had we either factored out the common 2 or
set the arguments equal to 0 and 2π, we would have found that
the shift was $\frac{\pi}{2}$ units.

55. false   **57.** true   **59.** $\left[\frac{\phi}{\omega}, \frac{\phi}{\omega} + \frac{2\pi}{\omega}\right]$

Section 4.3

1. b   **3.** h   **5.** c   **7.** d

9.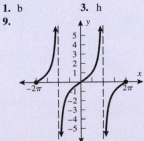

11.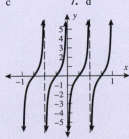

13.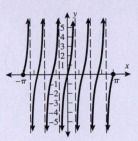

15.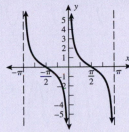

17.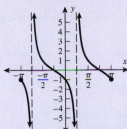

19.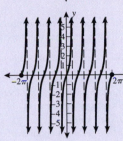

21.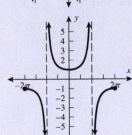

23.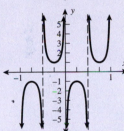

25.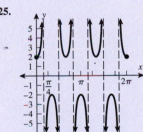

27.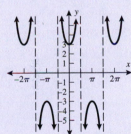

29.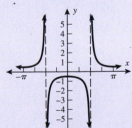

31.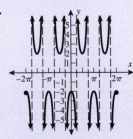

33.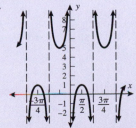

35.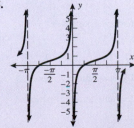

37.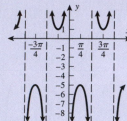

39.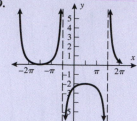

17.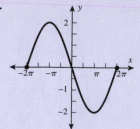

19.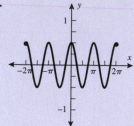

41. domain: all real numbers such that $x \neq n$, where n is an integer
range: all real numbers

43. domain: all real numbers such that $x \neq \dfrac{2n+1}{10}\pi$, where n is an
integer range: $(-\infty, -2] \cup [2, \infty)$

45. domain: all real numbers such that $x \neq 2n\pi$, where n is an
integer range: $(-\infty, 1] \cup [3, \infty)$

47. 3.3 miles **49.** 55.4 sq. in.

51. forgot the amplitude of 3 **53.** true

55. n = integer **57.** $x = -\pi, -\dfrac{\pi}{2}, 0, \dfrac{\pi}{2}, \pi$

59. $A = \sqrt{2}$

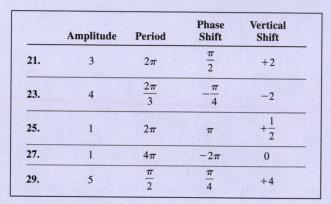

	Amplitude	Period	Phase Shift	Vertical Shift
21.	3	2π	$\dfrac{\pi}{2}$	$+2$
23.	4	$\dfrac{2\pi}{3}$	$-\dfrac{\pi}{4}$	-2
25.	1	2π	π	$+\dfrac{1}{2}$
27.	1	4π	-2π	0
29.	5	$\dfrac{\pi}{2}$	$\dfrac{\pi}{4}$	$+4$

OOPS!

$$A = \frac{16.17 - 8.28}{2} = 3.945$$

$$k = \frac{16.17 + 8.28}{2} = 12.225$$

$$12 = \frac{2\pi}{B}, \text{ so } B = \frac{\pi}{6}.$$

$$y = 12.225 + 3.945 \sin\left(\frac{\pi}{6}x + C\right)$$

If we let $x = 1$, then $y = 8.28$. Solve for C:

$$8.28 = 12.225 + 3.945 \sin\left(\frac{\pi}{6}(1) + C\right)$$

$$-1 = \sin\left(\frac{\pi}{6} + C\right)$$

$$\frac{\pi}{6} + C = \frac{3\pi}{2}$$

$$C = \frac{4\pi}{3}$$

$$y = 12.225 + 3.945 \sin\left(\frac{\pi}{6}x + \frac{4\pi}{3}\right)$$

Tying It All Together

$$x(t) = 5 \sin \pi t$$

$$x\left(\frac{1}{2}\right) = 5 \text{ cm}$$

Review Exercises

1. 2π **3.** $y = 4\cos x$ **5.** 5

7. 2π **9.** $y = \dfrac{1}{4}\sin x$ **11.** 2

13. amplitude: 2; period: 1

15. amplitude: $\dfrac{1}{5}$; period: $\dfrac{2\pi}{3}$

31.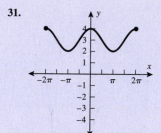

33.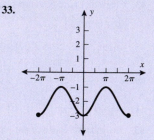

35.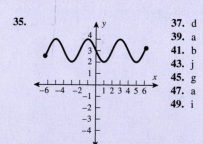

37. d
39. a
41. b
43. j
45. g
47. a
49. i

51. domain: all real numbers such that $x \neq n\pi$, where n is an
integer; range: all real numbers

53. domain: all real numbers such that $x \neq \dfrac{2n+1}{4}\pi$, where n is an
integer; range: $(-\infty, -3] \cup [3, \infty)$

55.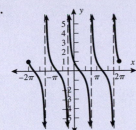

57.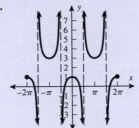

59.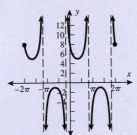

Practice Test

1. Amplitude: 5; period: $\dfrac{2\pi}{3}$

3.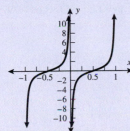

5.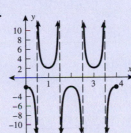

7.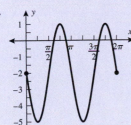

9.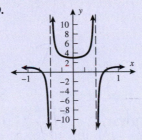

11. period: $\dfrac{2\pi}{\omega}$; phase shift: $\dfrac{\phi}{\omega}$ units to the left

13. $x = \dfrac{n\pi}{2}$ or $\left(\dfrac{n\pi}{2}, 0\right)$ where n is an integer.

15. a, c, e

Chapter 5

Section 5.1

1. 1   **3.** $\csc x$ **5.** 1
7. $\sin^2 x - \cos^2 x$ **9.** $\sec x$ **11.** $\sin^2 x$
13. $-\cos x$ **15.** 1   **39.** conditional

41. identity   **43.** conditional   **45.** conditional
47. conditional   **49.** identity   **51.** $|a|\cos\theta$
53. The $\cos x$ and $\sin x$ terms do not cancel. The numerators become $\cos^2 x$ and $\sin^2 x$, respectively.
55. This is a conditional equation. Just because the equation is true for $\dfrac{\pi}{4}$ does not mean it is true for all x.
57. false   **59.** QI, QIV   **61.** $a^2 + b^2$
63. No. Let $A = 30°$ and $B = 60°$.
65. $\cos(A + B) = \cos A \cos B - \sin A \sin B$
67. $\sin(A + B) = \sin A \cos B + \cos A \sin B$

Section 5.2

1. $\dfrac{\sqrt{6} - \sqrt{2}}{4}$ **3.** $\dfrac{\sqrt{6} - \sqrt{2}}{4}$ **5.** $-2 + \sqrt{3}$

7. $\dfrac{\sqrt{2} + \sqrt{6}}{4}$ **9.** $2 + \sqrt{3}$ **11.** $2 + \sqrt{3}$

13. $\sqrt{2} - \sqrt{6}$ **15.** $\cos x$ **17.** $-\sin x$
19. $-2\cos(A - B)$ **21.** $-2\sin(A + B)$ **23.** $\tan(26°)$
25. $\dfrac{1 + 2\sqrt{30}}{12}$ **27.** $\dfrac{-6\sqrt{6} + 4}{25}$

29. $\dfrac{3 - 4\sqrt{15}}{4 + 3\sqrt{15}} = \dfrac{192 - 25\sqrt{15}}{-119}$

31. identity   **41.** $y = \sin\left(x + \dfrac{\pi}{3}\right)$
33. conditional
35. identity
37. conditional
39. identity

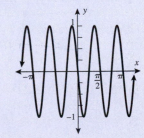

43. $y = \cos\left(x - \dfrac{\pi}{4}\right)$ **45.** $y = -\sin 4x$

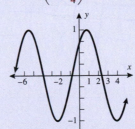 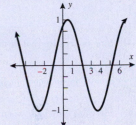

51. $\cos(kz - ct) = \cos kz \cos ct + \sin kz \sin ct$; when $\dfrac{z}{\lambda} = $ integer then $kz = 2\pi n$, and the $\sin kz$ term goes to zero.
53. Tangent of a sum is not the sum of the tangent. Needed to use the tangent of a sum identity.
55. false
59. $A = n\pi$ and $B = m\pi$ where n and m are integers.

61. a.

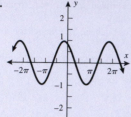

b.

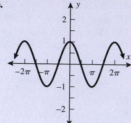

c.

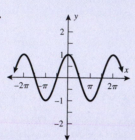

The difference quotients of $y = \sin x$ better approximate $y = \cos x$ as h goes to zero.

Section 5.3

1. $-\dfrac{4}{5}$ **3.** $\dfrac{120}{119}$ **5.** $\dfrac{120}{169}$

7. $-\dfrac{4}{3}$ **9.** $\dfrac{\sqrt{19}}{10}$ **11.** $\dfrac{119}{120}$

13. $\tan 30° = \dfrac{\sqrt{3}}{3}$ **15.** $\dfrac{1}{2}\sin\dfrac{\pi}{4} = \dfrac{\sqrt{2}}{4}$ **17.** $\cos 4x$

35. $y = \cot x$ **37.** $y = \sec 2x$ **39.** 22,565,386 lb.

41. Sine is negative (not positive).

43. false

45. false

47. $\tan 4x = \dfrac{4\tan x(1 - \tan^2 x)}{(1 - \tan^2 x)^2 - 4\tan^2 x}$

49. Identities only hold when functions are defined.

51. good approximation

53. not an identity

Section 5.4

1. $\dfrac{\sqrt{2 - \sqrt{3}}}{2}$ **3.** $-\dfrac{\sqrt{2 + \sqrt{3}}}{2}$ **5.** $\dfrac{\sqrt{2 - \sqrt{3}}}{2}$

7. $\sqrt{3 + 2\sqrt{2}}$ **9.** $-\dfrac{2}{\sqrt{2 + \sqrt{2}}}$ **11.** $1 - \sqrt{2}$ or $\dfrac{1}{\sqrt{3 + 2\sqrt{2}}}$

13. $\dfrac{2\sqrt{13}}{13}$ **15.** $\dfrac{3\sqrt{13}}{13}$

17. $\dfrac{\sqrt{5} - 1}{2}$ or $\sqrt{\dfrac{3 - \sqrt{5}}{2}}$ **19.** $\sqrt{\dfrac{3 + 2\sqrt{2}}{6}}$ or $\dfrac{1 + \sqrt{2}}{\sqrt{6}}$

21. $-\dfrac{\sqrt{15}}{5}$ **23.** $\cos\dfrac{5\pi}{12}$ **25.** $\tan 75°$

37.

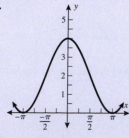

39.

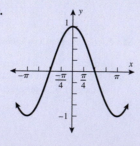

41. $= a^2\sqrt{\dfrac{(1 - \cos\theta)(1 + \cos\theta)}{4}} = a^2\dfrac{\sqrt{1 - \cos^2\theta}}{\sqrt{4}} = a^2\dfrac{\sqrt{\sin^2\theta}}{2}$

$= \dfrac{1}{2}a^2\sin\theta = \dfrac{a^2}{2}\sin\theta$

43. Wrong identity, and $\sin\dfrac{x}{2}$ is positive, not negative, in this case.

45. false   **47.** false   **51.** yes   **53.** identity

Section 5.5

1. $\dfrac{1}{2}[\sin 3x + \sin x]$ **3.** $\dfrac{5}{2}[\cos 2x - \cos 10x]$

5. $2[\cos x + \cos 3x]$ **7.** $\dfrac{1}{2}[\cos x - \cos 4x]$

9. $\dfrac{1}{2}\left[\cos\dfrac{2x}{3} + \cos 2x\right]$ **11.** $2\cos 4x\cos x$

13. $2\sin x\cos 2x$ **15.** $-2\sin x\cos\dfrac{3}{2}x$

17. $2\cos\dfrac{3}{2}x\cos\dfrac{5}{6}x$ **19.** $2\sin 0.5x\cos 0.1x$

21. $-\tan x$ **23.** $\tan 2x$

31. $2\cos(886\pi t)\cos(102\pi t)$

     average frequency: 443 Hz; beat frequency: 102 Hz

33. $2\sin\left[\dfrac{2\pi ct}{2}\left(\dfrac{1}{1.55} + \dfrac{1}{0.63}\right)10^6\right]\cos\left[\dfrac{2\pi ct}{c}\left(\dfrac{1}{1.55} - \dfrac{1}{0.63}\right)10^6\right]$

35. $2\sin\left[\dfrac{2\pi(1979)t}{2}\right]\cos\left[\dfrac{2\pi(439)t}{2}\right] = 2\sin[1979\pi t]\cos[439\pi t]$

37. $\dfrac{25(\sqrt{6} - \sqrt{3})}{3} \approx 5.98$ square feet

39. $\cos A\cos B \neq \cos AB$ and $\sin A\sin B \neq \sin AB$. Should have used product-to-sum identities.

41. false

43. true

45. Answers will vary. Here is one approach.

$\underset{\frac{1}{2}[\cos(A - B) - \cos(A + B)]}{\underline{\sin A\sin B}}\ \sin C = \dfrac{1}{2}\left[\underset{\frac{1}{2}[\sin(A-B+C)+\sin(C-A+B)]}{\underline{\cos(A - B)\sin C}} - \underset{\frac{1}{2}[\sin(A+B+C)+\sin(C-A-B)]}{\underline{\cos(A + B)\sin C}}\right]$

$\sin A\sin B\sin C = \dfrac{1}{4}[\sin(A - B + C) + \sin(C - A + B)$

$- \sin(A + B + C) - \sin(C - A - B)]$

47. $y = \sin 4x$

OOPS!

$E = 10\sin[100 + 300{,}000(5)] = 10\sin(1{,}500{,}100)$

    ≈ 8.8 volts per meter

$E = 10[(\sin 100)(\cos 1{,}500{,}000) + (\cos 100)(\sin 1{,}500{,}000)]$

    ≈ 8.8 volts per meter

Tying It All Together

Let $x = 3\sin\theta$.

$\dfrac{x}{\sqrt{9 - x^2}} = \dfrac{3\sin\theta}{\sqrt{9 - (3\sin\theta)^2}} = \dfrac{3\sin\theta}{\sqrt{9 - 9\sin^2\theta}}$

$= \dfrac{3\sin\theta}{\sqrt{9(1 - \sin^2\theta)}} = \dfrac{3\sin\theta}{\sqrt{9\cos^2\theta}} = \dfrac{3\sin\theta}{3|\cos\theta|}$

Since $-\dfrac{\pi}{2} \leq \theta \leq \dfrac{\pi}{2}$, $\dfrac{3\sin\theta}{3|\cos\theta|} = \dfrac{3\sin\theta}{3\cos\theta} = \tan\theta$.

Review Exercises

1. $\sec^2 x$ **3.** $\sec^2 x$ **5.** $\cos^2 x$

13. identity   **15.** Conditional   **17.** $\dfrac{\sqrt{2}-\sqrt{6}}{4}$

19. $\dfrac{\sqrt{3}-3}{3+\sqrt{3}}=\sqrt{3}-2$ **21.** $\sin x$ **23.** $\tan x$

25. $\dfrac{117}{44}$ **27.** $-\dfrac{897}{1025}$ **29.** identity

31.

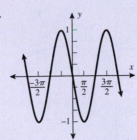

33. $\dfrac{7}{25}$

35. $\dfrac{671}{1800}$

37. $\dfrac{336}{625}$

39. $\dfrac{\sqrt{3}}{2}$

41. $\dfrac{3}{2}$ **49.** $-\dfrac{\sqrt{2-\sqrt{2}}}{2}$

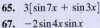

51. $\sqrt{2}-1$ **53.** $\dfrac{7\sqrt{2}}{10}$

55. $-\dfrac{5}{4}$ **57.** $\sin\dfrac{\pi}{12}$

63.

65. $3[\sin 7x + \sin 3x]$

67. $-2\sin 4x \sin x$

69. $2\sin\dfrac{x}{3}\cos x$

71. $\cot 3x$

61. $-\dfrac{\pi}{4}$ **63.** $\dfrac{\sqrt{7}}{4}$ **65.** $\dfrac{12}{13}$

67. $\dfrac{3}{4}$ **69.** $\dfrac{5\sqrt{23}}{23}$ **71.** $\dfrac{4\sqrt{15}}{15}$

73. $\dfrac{11}{60}$ **75.** $t = 0.026476$ sec or 26 ms

77. 173.4; June 22   **79.** $11.3 \approx 11$ years

81. $\tan\theta = \dfrac{\dfrac{7}{x}+\dfrac{1}{x}}{1-\dfrac{7}{x}\cdot\dfrac{1}{x}} = \dfrac{8x}{x^2-7}$ **83.** 0.70 m; 0.24 m

85. $\theta = \pi - \tan^{-1}\left(\dfrac{150}{x}\right) - \tan^{-1}\left(\dfrac{300}{200-x}\right)$

87. The wrong interval for the identity was used. The correct domain is $\left[-\dfrac{\pi}{2}, \dfrac{\pi}{2}\right]$.

89. $\cot^{-1}x \neq \dfrac{1}{\tan^{-1}x}$

91. false

93. $\dfrac{1}{2}$ is not in the domain of the inverse secant function.

95. $\dfrac{\sqrt{6}-\sqrt{2}}{4}$

97. (a) $0 \le x \le \pi$, (b) $f^{-1}(x) = \dfrac{\pi}{2} + \sin^{-1}\left(\dfrac{2-x}{4}\right)$; domain: $[-2, 6]$

99. The identity $\sin[\sin^{-1}x] = x$ only holds on $-1 \le x \le 1$.

101. $\left[-\dfrac{\pi}{2}, 0\right) \cup \left(0, \dfrac{\pi}{2}\right]$

Practice Test

3. conditional   **5.** $-2-\sqrt{3}$ **7.** $\dfrac{23}{25}$

9. $-\tan 2x$ **11.** $\sin x + \sin 3$ **13.** $3|\cos x|$

Chapter 6

Section 6.1

1. $\dfrac{\pi}{4}$ **3.** $-\dfrac{\pi}{3}$ **5.** $\dfrac{3\pi}{4}$

7. $\dfrac{\pi}{6}$ **9.** $\dfrac{\pi}{6}$ **11.** $-\dfrac{\pi}{3}$

13. 60°   **15.** 45°   **17.** 120°

19. 30°   **21.** −30°   **23.** 135°

25. 57.10°   **27.** 62.18°   **29.** 48.10°

31. −15.30°   **33.** 166.70°   **35.** −0.63

37. 1.43   **39.** 0.92   **41.** 2.09

43. 0.31   **45.** $\dfrac{5\pi}{12}$ **47.** not possible

49. $\dfrac{\pi}{6}$ **51.** $\dfrac{2\pi}{3}$ **53.** $\sqrt{3}$

55. $\dfrac{\pi}{3}$ **57.** not possible   **59.** 0

Section 6.2

1. $\dfrac{3\pi}{4}, \dfrac{5\pi}{4}$ **3.** $\dfrac{7\pi}{6}, \dfrac{11\pi}{6}, \dfrac{19\pi}{6}, \dfrac{23\pi}{6}$

5. $n\pi$, where n is any integer   **7.** $\dfrac{7\pi}{12}, \dfrac{11\pi}{12}, \dfrac{19\pi}{12}, \dfrac{23\pi}{12}$

9. $\dfrac{(7+12n)\pi}{3}$ or $\dfrac{(11+12n)\pi}{3}$, where n is any integer

11. $-\dfrac{4\pi}{3}, \dfrac{5\pi}{6}, -\dfrac{11\pi}{6}, -\dfrac{\pi}{3}, \dfrac{\pi}{6}, \dfrac{2\pi}{3}, \dfrac{7\pi}{6}, \dfrac{5\pi}{3}$

13. $-\dfrac{4\pi}{3}, -\dfrac{2\pi}{3}$ **15.** $\dfrac{\pi}{6}, \dfrac{\pi}{3}, \dfrac{7\pi}{6}, \dfrac{4\pi}{3}$

17. $\dfrac{\pi}{12}, \dfrac{7\pi}{12}, \dfrac{13\pi}{12}, \dfrac{19\pi}{12}$ **19.** $\dfrac{\pi}{3}, \dfrac{2\pi}{3}, \dfrac{4\pi}{3}, \dfrac{5\pi}{3}$

21. $\dfrac{\pi}{3}$ **23.** $\dfrac{\pi}{4}, \dfrac{3\pi}{4}, \dfrac{5\pi}{4}, \dfrac{7\pi}{4}$

25. $\dfrac{\pi}{2}, \dfrac{3\pi}{2}, \dfrac{\pi}{3}, \dfrac{5\pi}{3}$ **27.** $\dfrac{3\pi}{2}, \dfrac{7\pi}{6}, \dfrac{11\pi}{6}$

29. $\dfrac{\pi}{2}$ **31.** 115.83°, 154.17°, 295.83°, 334.17°

33. 333.63°   **35.** 29.05°, 209.05°

37. 200.70°, 339.30°   **39.** 41.41°, 318.59°

41. 56.31°, 126.87°, 236.31°, 306.87°

43. 9.74°, 80.26°, 101.79°, 168.21°, 189.74°, 260.26°, 281.79°, 348.21°

45. 80.12°, 279.88°

47. 64.93°, 121.41°, 244.93°, 301.41°

49. March

51. $A = \frac{1}{2}h(b_1 + b_2)$

$= \frac{1}{2}(x\sin\theta)(x + (x\cos\theta + x + x\cos\theta))$

$= x^2(\sin\theta + \sin\theta\cos\theta)$

$= x^2\sin\theta(1 + \cos\theta)$

53. 2001

55. 24°

57. Found answer in QI; what about answer in QIV?

59. Extraneous solution. Forgot to check.

61. false

63. $\frac{\pi}{6}, \frac{5\pi}{6}, \frac{7\pi}{6}, \frac{11\pi}{6}$

65. $x = \frac{\pi}{6} = 0.524$ and $x = \frac{5\pi}{6} = 2.618$

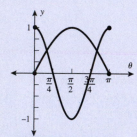

67. no solution   **69.** no solution

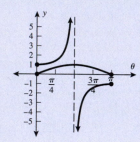

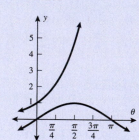

Section 6.3

1. $\frac{\pi}{4}, \frac{5\pi}{4}$

3. π

5. $\frac{\pi}{6}$

7. $\frac{\pi}{3}$

9. $\frac{\pi}{4}, \frac{3\pi}{4}, \frac{5\pi}{4}, \frac{7\pi}{4}$

11. $\frac{\pi}{2}, \frac{3\pi}{2}$

13. $0, \frac{\pi}{4}, \pi, \frac{7\pi}{4}$

15. $\frac{\pi}{6}, \frac{5\pi}{6}, \frac{7\pi}{6}, \frac{11\pi}{6}, \frac{\pi}{2}, \frac{3\pi}{2}$

17. $\frac{\pi}{6}, \frac{\pi}{3}, \frac{7\pi}{6}, \frac{4\pi}{3}$

19. $\frac{\pi}{6}, \frac{5\pi}{6}, \frac{7\pi}{6}, \frac{11\pi}{6}$

21. $\frac{3\pi}{2}$

23. $\frac{2\pi}{3}, \frac{4\pi}{3}$

25. $\frac{\pi}{3}, \frac{5\pi}{3}, \pi$

27. 57.47°, 122.53°, 323.62°, 216.38°

29. 30°, 150°, 199.47°, 340.53°

31. 14.48°, 165.52°, 270°

33. 111.47°, 248.53°

35. $\frac{3}{4}$ second

37. $(0, 1), \left(\frac{\pi}{3}, \frac{3}{2}\right), (\pi, -3), \left(\frac{5\pi}{3}, \frac{3}{2}\right)$

39. Can't divide by $\cos x$. Must factor.

41. true

43. $\frac{\pi}{6}$ or 30°

45.

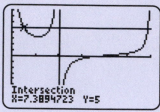

47. $x \approx 1.3$

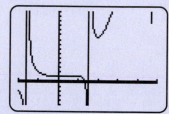

OOPS!

$\tan\theta = \frac{4}{-4} = -1$

Since $\tan^{-1}(-1) = -\frac{\pi}{4}$ is an angle in the fourth quadrant, add π to it to get the correct angle in the second quadrant.

$\theta = \pi + \tan^{-1}(-1) = \frac{3\pi}{4}$

Tying It All Together

$20\sin(10{,}000 - 300{,}000t) = 10$

$\sin(10{,}000 - 300{,}000t) = \frac{1}{2}$

$10{,}000 - 300{,}000t = \frac{\pi}{6}$

$300{,}000t = 10{,}000 - \frac{\pi}{6}$

$t = \dfrac{10{,}000 - \dfrac{\pi}{6}}{300{,}000} \approx 0.03 \text{ seconds}$

Review Exercises

1. $\frac{\pi}{4}$

3. $\frac{\pi}{2}$

5. $-90°$

7. 60°

9. $-37.50°$

11. 22.50°

13. 1.75

15. -0.10

17. $-\frac{\pi}{4}$

19. $\sqrt{3}$

21. $\frac{\pi}{3}$

23. $\frac{60}{61}$

25. $\frac{7}{6}$

27. $\frac{6\sqrt{35}}{35}$

29. July

31. $\frac{2\pi}{3}, \frac{5\pi}{6}, \frac{5\pi}{3}, \frac{11\pi}{6}$

33. $-\frac{3\pi}{2}, -\frac{\pi}{2}$

35. $\frac{\pi}{3}, \frac{2\pi}{3}, \frac{4\pi}{3}, \frac{5\pi}{3}$

37. $\frac{3\pi}{8}, \frac{7\pi}{8}, \frac{11\pi}{8}, \frac{15\pi}{8}$

39. $0, \pi, \frac{3\pi}{4}, \frac{7\pi}{4}$

41. 80.46°, 170.46°, 260.46°, 350.46°

43. 90°, 138.59°, 221.41°, 270° **45.** 17.62°, 162.38°

47. $\dfrac{\pi}{4}, \dfrac{5\pi}{4}$ **49.** $\dfrac{\pi}{3}, \pi$

51. $0, \dfrac{\pi}{6}, \pi, \dfrac{11\pi}{6}$ **53.** $\dfrac{3\pi}{2}$

55. π **57.** 90°, 135°, 270°, 315°

59. 0°   **61.** 60°, 90°, 270°, 300°

Practice Test

1. $-1 \le x \le 1$ **3.** $0 \le x \le \pi$

5. $\dfrac{5\pi}{6}$ **7.** $\dfrac{\sqrt{5}}{5}$

9. $\theta = \begin{cases} \dfrac{4\pi}{3} + 2n\pi \\[2mm] \dfrac{5\pi}{3} + 2n\pi \end{cases}$, where n is an integer.

11. $\dfrac{\pi}{3}, \dfrac{5\pi}{3}, \pi$

13. 14.48°, 90°, 165.52°, 270°

15. no solution

Chapter 7

Section 7.1

1. SSA   **3.** SSS   **5.** ASA

7. $\gamma = 75°, b = 5\sqrt{6}$ m, $c = 13.66$ m

9. $\beta = 62°, a = 163$ cm, $c = 215$ cm

11. $\beta = 116.1°, a = 80.2$ yd, $b = 256.6$ yd

13. $\gamma = 120°, a = 7$ m, $b = 7$ m

15. $\alpha = 97°, a = 118$ yd, $b = 52$ yd

17. $\beta_1 = 20°, \gamma_1 = 144°, c_1 = 9; \beta_2 = 160°, \gamma_2 = 4°, c_2 = 1$

19. $\alpha = 40°, \beta = 100°, b = 18$

21. no triangle

23. $\beta_1 = 77°, \alpha_1 = 63°, a = 457; \beta_2 = 103°, \alpha_2 = 37°, a = 309$

25. $\alpha = 31°, \gamma = 43°, c = 2$

27. 1246 ft   **29.** 1.7 mi   **31.** 1.3 mi

33. 42 ft   **35.** No triangle because $\sin B \ne 1.13 > 1$

37. true   **39.** true

Section 7.2

1. C   **3.** S   **5.** S   **7.** C

9. $b = 5, \alpha = 47°, \gamma = 33°$ **11.** $a = 5, \beta = 158°, \gamma = 6°$

13. $a = 2, \beta = 80°, \gamma = 80°$ **15.** $b = 5, \alpha = 43°, \gamma = 114°$

17. $b = 7, \alpha = 30°, \gamma = 90°$ **19.** $\alpha = 93°, \beta = 39°, \gamma = 48°$

21. $\alpha = 51.32°, \beta = 51.32°, \gamma = 77.36°$

23. $\alpha = 75°, \beta = 57°, \gamma = 48°$ **25.** no triangle

27. $\alpha = 67°, \beta = 23°, \gamma = 90°$ **29.** $\gamma = 105°, b = 5, c = 9$

31. $\beta = 12°, \gamma = 137°, c = 16$ **33.** $\alpha = 66°, \beta = 77°, \gamma = 37°$

35. $\gamma = 2°, \alpha = 168°, a = 13$ **37.** 2710 miles

39. 63.7 ft   **41.** 16 ft

43. 0.8 miles   **45.** should have used the smaller angle, β, in step 2

47. false   **49.** true

Section 7.3

1. 55.4   **3.** 0.5   **5.** 23.6

7. 6.4   **9.** 4408.4   **11.** 9.6

13. 97.4   **15.** 25.0   **17.** 26.7

19. 111.64   **21.** 111,632,076   **23.** no triangle

25. 312,297 sq. nm   **27.** 10,591 sq. ft   **29.** 16°

31. 312.4 sq. mi   **33.** 12 in.   **35.** 47,128 sq. ft

37. 23.38 sq. ft   **39.** Semiperimeter is *half* the perimeter.   **41.** true

Section 7.4

1. $\sqrt{13}$ **3.** $5\sqrt{2}$ **5.** 25

7. $\sqrt{73}; \theta = 69.4°$ **9.** $\sqrt{26}; \theta = 348.7°$ **11.** $\sqrt{17}; \theta = 166.0°$

13. $8; \theta = 180°$ **15.** $2\sqrt{3}; \theta = 60°$ **17.** $\langle -2, -2 \rangle$

19. $\langle -12, 9 \rangle$ **21.** $\langle 0, -14 \rangle$ **23.** $\langle -36, 48 \rangle$

25. $\langle 6.3, 3.0 \rangle$ **27.** $\langle -2.8, 15.8 \rangle$ **29.** $\langle 2.6, -3.1 \rangle$

31. $\langle 8.2, -3.8 \rangle$ **33.** $\langle -1, 1.7 \rangle$ **35.** $\left\langle -\dfrac{5}{13}, -\dfrac{12}{13} \right\rangle$

37. $\left\langle \dfrac{60}{61}, \dfrac{11}{61} \right\rangle$ **39.** $\left\langle \dfrac{24}{25}, -\dfrac{7}{25} \right\rangle$ **41.** $\left\langle -\dfrac{3}{5}, -\dfrac{4}{5} \right\rangle$

43. $\left\langle \dfrac{\sqrt{10}}{10}, \dfrac{3\sqrt{10}}{10} \right\rangle$ **45.** $7i + 3j$ **47.** $5i - 3j$

49. $-i + 0j$ **51.** $2i + 0j$ **53.** $-5i + 5j$

55. $7i + 0j$

57. vertical: 1100 ft/sec; horizontal: 1905 ft/sec

59. 2801 lb

61. 11.7 mph; 31° west of due north

63. 52.41°; 303 mph.

65. 250 lb

67. 51.4 ft/sec; 61.3 ft/sec

69. 29.93 yd   **71.** 10.9°   **73.** 1156 lb

75. Magnitude is never negative. Should not have factored out the negative but instead squared it in finding the magnitude.

77. false   **79.** true   **81.** vector

83. $\sqrt{a^2 + b^2}$

Section 7.5

1. 2   **3.** -3 **5.** 42   **7.** 11   **9.** $-13a$ **11.** -1.4

13. 98°   **15.** 109°   **17.** 3°   **19.** 30°   **21.** 105°   **23.** 180°

25. no   **27.** yes   **29.** no   **31.** yes   **33.** yes   **35.** yes

37. 400 ft-lb   **39.** 80,000 ft-lb   **41.** 1299 ft-lb

43. 148 ft-lb   **45.** 1607 lb   **47.** 3,939,231 ft-lb

49. The dot product of two vectors is a scalar (not a vector). Should have summed the products of components.

51. false   **53.** true   **55.** 17

61. -1083 **63.** 31.4°

OOPS!

$\mathbf{d} = \langle 12, 0 \rangle$

$\mathbf{F} = \langle 35\cos 60°, 35\sin 60° \rangle = \left\langle 35\left(\dfrac{1}{2}\right), 35\left(\dfrac{\sqrt{3}}{2}\right) \right\rangle = \left\langle \dfrac{35}{2}, \dfrac{35\sqrt{3}}{2} \right\rangle$

$W = \mathbf{F} \cdot \mathbf{d} = \left\langle \dfrac{35}{2}, \dfrac{35\sqrt{3}}{2} \right\rangle \cdot \langle 12, 0 \rangle$

$= 12\left(\dfrac{35}{2}\right) + 0\left(\dfrac{35\sqrt{3}}{2}\right)$

$= 210$ ft-lb

Tying It All Together

$|\mathbf{v}| = \sqrt{3+1} = 2$, so $\mathbf{u} = \dfrac{\mathbf{v}}{|\mathbf{v}|} = \dfrac{1}{2}\langle \sqrt{3}, 1 \rangle = \left\langle \dfrac{\sqrt{3}}{2}, \dfrac{1}{2} \right\rangle$ is a unit vector in the same direction as $\mathbf{v}$. The velocity of escalator A is

$\mathbf{e}_A = 90\left\langle \dfrac{\sqrt{3}}{2}, \dfrac{1}{2} \right\rangle = \langle 45\sqrt{3}, 45 \rangle$,

and the velocity of escalator B is

$\mathbf{e}_B = 96\left\langle \dfrac{\sqrt{3}}{2}, \dfrac{1}{2} \right\rangle = \langle 48\sqrt{3}, 48 \rangle$.

Jessica's resultant velocity is

$$\mathbf{J} = (440 - 90)\left\langle \frac{\sqrt{3}}{2}, \frac{1}{2} \right\rangle = \langle 175\sqrt{3}, 175 \rangle,$$

and Mike's resultant velocity is

$$\mathbf{M} = (432 - 96)\left\langle \frac{\sqrt{3}}{2}, \frac{1}{2} \right\rangle = \langle 168\sqrt{3}, 168 \rangle.$$

Mike's actual speed is 336 feet per minute and Jessica's is 350 feet per minute. Since they are covering the same distance, Jessica will reach the bottom of the escalators first.

Review Exercises

1. $\gamma = 150°, c = 12, b = 8$ 3. $\gamma = 130°, a = 1, b = 9$
5. $\beta = 158°, a = 11, b = 22$ 7. $\beta = 90°, a = \sqrt{2}, c = \sqrt{2}$
9. $\beta = 146°, b = 266, c = 178$
11. $\beta = 26°, \gamma = 134°, c = 15$ or $\beta = 154°, \gamma = 6°, c = 2$
13. $\beta = 127°, \gamma = 29°, b = 20$ or $\beta = 5°, \gamma = 151°, b = 2$
15. no triangle
17. $\beta = 15°, \gamma = 155°, c = 10$ or $\beta = 165°, \gamma = 5°, c = 2$
19. 6.2 mi   21. $\alpha = 42°, \beta = 88°, c = 46$
23. $\alpha = 51°, \beta = 54°, \gamma = 75°$ 25. $\alpha = 42°, \beta = 48°, \gamma = 90°$
27. $\beta = 28°, \gamma = 138°, a = 4$ 29. $\beta = 68°, \gamma = 22°, a = 11$
31. $\alpha = 51°, \beta = 59°, \gamma = 70°$ 33. $\beta = 37°, \gamma = 43°, a = 26$
35. $\beta = 4°, \gamma = 166°, a = 28$ 37. no triangle
39. $\beta = 10°, \gamma = 155°, c = 10.3$ 41. 141.8
43. 51.5   45. 89.8
47. 41.7   49. 5.2 in.
51. 13   53. 13
55. 26; 112.6°   57. 20; $-323.1°$
59. $\langle 2, 11 \rangle$ 61. $\langle 38, -7 \rangle$
63. $\langle 2.6, 9.7 \rangle$ 65. $\langle -3.1, 11.6 \rangle$
67. $\left\langle \frac{\sqrt{2}}{2}, -\frac{\sqrt{2}}{2} \right\rangle$ 69. $5\mathbf{i} + \mathbf{j}$
71. -6 73. -9 75. 16   77. 59°
79. 49°   81. 166°   83. no   85. yes
87. no   89. no

Practice Test

1. $a = 7.8, c = 14.6,$ and $\gamma = 110°$
3. $\alpha = 35.4°, \beta = 48.2°,$ and $\gamma = 96.4°$
5. 57
7. magnitude $= 13, \theta = 112.6°$
9. a. $\langle -14, 5 \rangle$ b. -16
11. 59.5°
13. Graphically, because the magnitude is a length, and length is always nonnegative. Algebraically, the magnitude is the square root of a quantity (principal root is nonnegative).
15. scalar; the same (equal)

Chapter 8

Section 8.1

1. $-i$ 3. 1   5. $4i$
7. $2i\sqrt{5}$ 9. $2 - 9i$ 11. $2 - 2i$
13. $5 - i$ 15. 65   17. $\frac{3}{10} + \frac{1}{10}i$
19. $\frac{18}{53} - \frac{43}{53}i$ 21. $\frac{3}{4} \pm \frac{\sqrt{7}}{4}i$ 23. $-\frac{5}{2} \pm \frac{5\sqrt{3}}{2}i$
25. $\frac{1}{4} \pm \frac{\sqrt{19}}{4}i$ 27. $5i$ 29. $13 + i$
31. $R = R_1 + R_2$

33. $R = \dfrac{(X_1R_2 + X_2R_1)(X_1 + X_2) + (R_1R_2 - X_1X_2)(R_1 + R_2)}{(R_1 + R_2)^2 + (X_1 + X_2)^2}$

35. $\sqrt{-4} = 2i$. Convert square roots to imaginary numbers before multiplication.
37. Multiplied by the denominator, $4 - i$, instead of the conjugate, $4 + i$. Also multiplied $(4 - i)(4 - i)$ incorrectly.
39. true   41. true   43. $-2 + 2i$

Section 8.2

1. 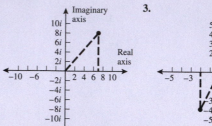   3.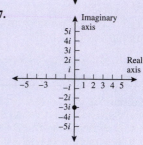

5. 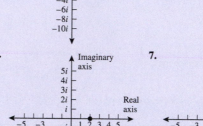   7.

9. $\sqrt{2}\left(\cos\frac{7\pi}{4} + i\sin\frac{7\pi}{4}\right) = \sqrt{2}(\cos 315° + i\sin 315°)$
11. $2\left(\cos\frac{\pi}{3} + i\sin\frac{\pi}{3}\right) = 2(\cos 60° + i\sin 60°)$
13. $4\sqrt{2}\left(\cos\frac{3\pi}{4} + i\sin\frac{3\pi}{4}\right) = 4\sqrt{2}(\cos 135° + i\sin 135°)$
15. $2\sqrt{3}\left(\cos\frac{5\pi}{3} + i\sin\frac{5\pi}{3}\right) = 2\sqrt{3}(\cos 300° + i\sin 300°)$
17. $3(\cos 0 + i\sin 0) = 3(\cos 0° + i\sin 0°)$
19. $\sqrt{58}(\cos 293.2° + i\sin 293.2°)$
21. $\sqrt{61}(\cos 140.2° + i\sin 140.2°)$
23. $13(\cos 112.6° + i\sin 112.6°)$
25. $10(\cos 323.1° + i\sin 323.1°)$
27. $a\sqrt{5}(\cos 296.6° + i\sin 296.6°)$
29. -5 31. $\sqrt{2} - \sqrt{2}i$
33. $-2 - 2\sqrt{3}i$ 35. $-\frac{3}{2} + \frac{\sqrt{3}}{2}i$
37. $1 + i$ 39. $2.1131 - 4.5315i$
41. $-0.5209 + 2.9544i$ 43. $5.3623 - 4.4995i$
45. $-2.8978 + 0.7765i$ 47. $0.6180 - 1.9021i$
49. $100(\cos 0° + i\sin 0°) + 120(\cos 30° + i\sin 30°)$; 212.5 lb
51. $80(\cos 0° + i\sin 0°) + 150(\cos 30° + i\sin 30°)$; 19.7°
53. The point is in QIII (not QI).
55. true   57. true   59. 0°   61. $|b|$

Section 8.3

1. $-6 + 6\sqrt{3}i$ 3. $-4\sqrt{2} - 4\sqrt{2}i$
5. $0 + 8i$ 7. $\frac{9\sqrt{2}}{2} + \frac{9\sqrt{2}}{2}i$

9. $0 + 12i$

11. $\dfrac{3}{2} + \dfrac{3\sqrt{3}}{2}i$

13. $-\sqrt{2} + \sqrt{2}i$

15. $0 - 2i$

17. $\dfrac{3}{2} + \dfrac{3\sqrt{3}}{2}i$

19. $-\dfrac{5}{2} - \dfrac{5\sqrt{3}}{2}i$

21. $4 - 4i$

23. $-64 + 0i$

25. $-8 + 8\sqrt{3}i$

27. $1{,}048{,}576 + 0i$

29. $-1{,}048{,}576\sqrt{3} - 1{,}048{,}576i$

31. $2[\cos 150° + i \sin 150°]$ and $2[\cos 330° + i \sin 330°]$

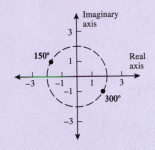

33. $\sqrt{6}\,[\cos 157.5° + i \sin 157.5°], \sqrt{6}\,[\cos 337.5° + i \sin 337.5°]$

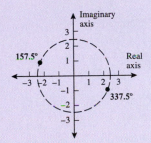

35. $2[\cos 20° + i \sin 20°], 2[\cos 140° + i \sin 140°],$
$2[\cos 260° + i \sin 260°]$

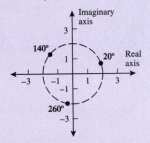

37. $\sqrt[3]{2}\,[\cos 110° + i \sin 110°], \sqrt[3]{2}\,[\cos 230° + i \sin 230°],$
$\sqrt[3]{2}\,[\cos 350° + i \sin 350°]$

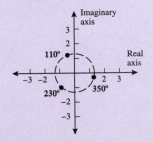

39. $2[\cos 78.75° + i \sin 78.75°], 2[\cos 168.75° + i \sin 168.75°],$
$2[\cos 258.75° + i \sin 258.75°], 2[\cos 348.75° + i \sin 348.75°]$

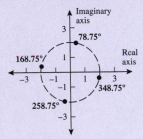

41. $x = \pm 2, x = \pm 2i$

43. $-2, 1 - \sqrt{3}i, 1 + \sqrt{3}i$

45. $\sqrt{2} - \sqrt{2}i, \sqrt{2} + \sqrt{2}i, -\sqrt{2} - \sqrt{2}i, -\sqrt{2} + \sqrt{2}i$

47. $1, -1, \dfrac{1}{2} + \dfrac{\sqrt{3}}{2}i, \dfrac{1}{2} - \dfrac{\sqrt{3}}{2}i, -\dfrac{1}{2} + \dfrac{\sqrt{3}}{2}i, -\dfrac{1}{2} - \dfrac{\sqrt{3}}{2}i$

49. $-\dfrac{\sqrt{2}}{2} + \dfrac{\sqrt{2}}{2}i, \dfrac{\sqrt{2}}{2} - \dfrac{\sqrt{2}}{2}i$

51. $[\cos 45° + i \sin 45°], [\cos 117° + i \sin 117°],$
$[\cos 189° + i \sin 189°], [\cos 261° + i \sin 261°],$
$[\cos 333° + i \sin 333°]$

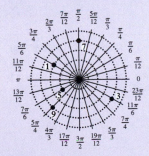

53. Reversed order of angles being subtracted.

55. Use De Moivre's theorem. In general, $(a + b)^6 \neq a^6 + b^6$.

57. true

Section 8.4

Exercises 1, 3, 5, 7, and 9 are all plotted on same graph below:

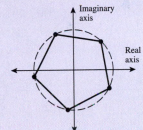

11. $\left(4, \dfrac{\pi}{3}\right)$

13. $\left(2, \dfrac{4\pi}{3}\right)$

15. $\left(4\sqrt{2}, \dfrac{3\pi}{4}\right)$

17. $(3, 0)$

19. $\left(2, \dfrac{7\pi}{6}\right)$

21. $(2, -2\sqrt{3})$

23. $\left(\dfrac{\sqrt{3}}{2}, -\dfrac{1}{2}\right)$

25. $(0, 0)$

27. $(-1, -\sqrt{3})$

29. $\left(\dfrac{\sqrt{2}}{2}, -\dfrac{\sqrt{2}}{2}\right)$

31. d

33. a

35.

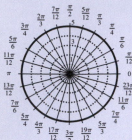

37.

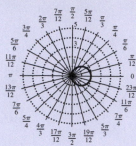

57.

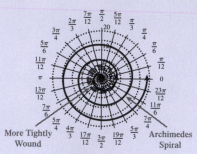

More Tightly Wound   Archimedes Spiral

39.

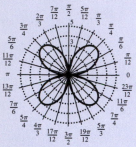

41.

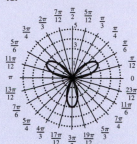

59.

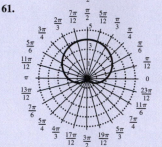

Multiplier 4

Multiplier $\frac{1}{4}$

43.

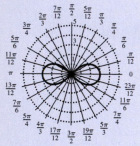

45.

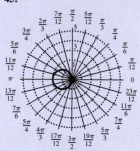

61.

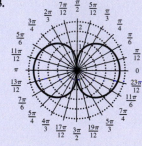

47.

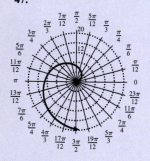

49.

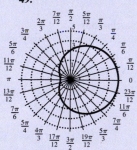

63.

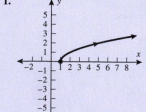

All three graphs are figure eights. Extending the domain in (b) results in movement that is twice as fast. Extending the domain in (c) results in movement that is four times as fast.

51. line: $y = -2x + 1$
53. circle: $(x - 1)^2 + y^2 = 9$

55.

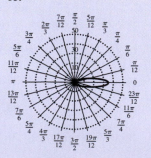

65. Point is in QIII; the angle found was the reference angle (needed to add π).

67. true   **69.** $r = \dfrac{a}{\cos\theta}$ **71.** $(-a, \theta \pm 180°)$

Section 8.5

1.

3.

5.

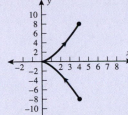

7.

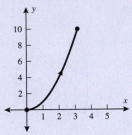

9.

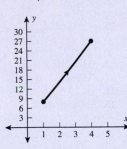

11.

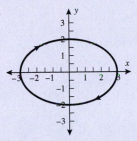

13.

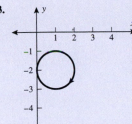

15.

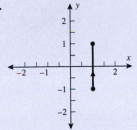

17. Arrow in different directions, depending on t.

19.

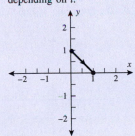

21. $y = \dfrac{1}{x^2}$ **23.** $y = x - 2$ **25.** $y = \sqrt{x^2 + 1}$

27. $x + y = 2$ **29.** $x + 4y = 8$ **31.** 17.7 seconds

33. Yes. The height will be over 20 feet at that time.

35. Height: 5742 ft; horizontal distance: 13,261 ft

37. 125 seconds

39.

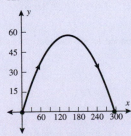

41.

t	x	y
0	$A + B$	0
$\dfrac{\pi}{2}$	0	$A + B$
π	$-A - B$	0
$\dfrac{3\pi}{2}$	0	$-A - B$
2π	$A + B$	0

43. The original domain must be $t \geq 0$; therefore only the part of the parabola where $y \geq 0$ is part of the plane curve.

45. false

47. quarter circle in QI

OOPS!

$30 + 95t - 9.8t^2 = 10$

$20 + 95t - 9.8t^2 = 0$

$t = \dfrac{-95 \pm \sqrt{9025 - 4(-9.8)(20)}}{2(-9.8)} = \dfrac{-95 \pm \sqrt{9809}}{-19.6}$

$t \approx -0.2$ sec (discard this negative value of time), $t \approx 9.9$ sec

$x(9.9) = 500 - 45(9.9) = 54.5$ m

The cannonball would be $500 - 54.5 = 445.5$ m horizontally from the fort.

Tying It All Together

$21 + 5\cos\left(\dfrac{\pi}{3}t\right) = 24$

$\cos\left(\dfrac{\pi}{3}t\right) = \dfrac{3}{5}$

$\dfrac{\pi}{3}t = 2\pi - \cos^{-1}\left(\dfrac{3}{5}\right)$

$t = \dfrac{3}{\pi}\left(2\pi - \cos^{-1}\left(\dfrac{3}{5}\right)\right)$

$t \approx 5.1$ min

$22 + 5\sin\left(\dfrac{\pi}{3}t\right) = 18$

$\sin\left(\dfrac{\pi}{3}t\right) = -\dfrac{4}{5}$

$\dfrac{\pi}{3}t = 2\pi + \sin^{-1}\left(-\dfrac{4}{5}\right)$

$t = \dfrac{3}{\pi}\left(2\pi + \sin^{-1}\left(-\dfrac{4}{5}\right)\right)$

$t \approx 5.1$ min

$\cos\theta = \dfrac{24}{30} = \dfrac{4}{5}$

$x(t) = (s\cos\theta)t$

$24 = s\left(\dfrac{4}{5}\right)(5.1)$

$s \approx 5.9$ mi/min

$x(t) = 5.9\left(\dfrac{4}{5}\right)t$

$y(t) = 5.9\left(\dfrac{3}{5}\right)t$

$\sin\theta = \dfrac{18}{30} = \dfrac{3}{5}$

Review Exercises

1. $-i$ **3.** $7i$ **5.** $6 + 3i$

7. $2 - 26i$ **9.** $-15 + 10i$ **11.** $2 + \sqrt{5}\,i, 2 - \sqrt{5}\,i$

13. $-\dfrac{3}{5} + \dfrac{4}{5}i, -\dfrac{3}{5} - \dfrac{4}{5}i$ **15.** -15

17.

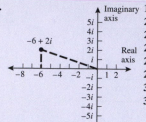

19. $2(\cos 315° + i \sin 315°)$
21. $8(\cos 270° + i \sin 270°)$
23. $61(\cos 169.6° + i \sin 169.6°)$
25. $17(\cos 28.1° + i \sin 28.1°)$
27. $3 - 3\sqrt{3}i$
29. $-1 + i$
31. $-3.7588 - 1.3681i$
33. $-12i$

35. $-\dfrac{21}{2} - \dfrac{21\sqrt{3}}{2}i$

37. $-\dfrac{\sqrt{3}}{2} + \dfrac{1}{2}i$

39. -6 **41.** -324

43. $16 - 16\sqrt{3}i$

45. $2(\cos 30° + i \sin 30°)$,
$2(\cos 210° + i \sin 210°)$

47. $4(\cos 45° + i \sin 45°)$,
$4(\cos 135° + i \sin 135°)$,
$4(\cos 225° + i \sin 225°)$,
$4(\cos 315° + i \sin 315°)$

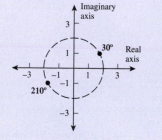

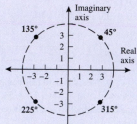

49. $-6, 3 + 3\sqrt{3}i, 3 - 3\sqrt{3}i$

51. $\dfrac{\sqrt{2}}{2} + \dfrac{\sqrt{2}}{2}i, \dfrac{\sqrt{2}}{2} - \dfrac{\sqrt{2}}{2}i, -\dfrac{\sqrt{2}}{2} + \dfrac{\sqrt{2}}{2}i, -\dfrac{\sqrt{2}}{2} - \dfrac{\sqrt{2}}{2}i$

All the points from Exercises 53, 55, and 57 are graphed on the single graph below:

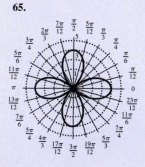

53. $\left(2\sqrt{2}, \dfrac{3\pi}{4}\right)$ **55.** $\left(10, \dfrac{7\pi}{6}\right)$ **57.** $\left(2, \dfrac{3\pi}{2}\right)$

59. $\left(-\dfrac{3}{2}, \dfrac{3\sqrt{3}}{2}\right)$ **61.** $(1, \sqrt{3})$ **63.** $\left(-\dfrac{1}{2}, -\dfrac{\sqrt{3}}{2}\right)$

65.

67.

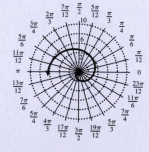

69.

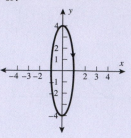

71.

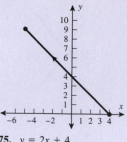

73. $x = 4 - y^2$

75. $y = 2x + 4$

Practice Test

1. 0
5. $8i$
9. $(-a, \theta \pm 180°)$
13.

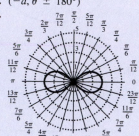

3. $6[\cos 315° + i \sin 315°]$
7. $32{,}768[-1 + i\sqrt{3}]$
11. $(5, 233°)$
15. 5.3 sec; 450 ft

Appendix

Section A.1

1. $(4, 2)$ **3.** $(-3, 0)$ **5.** $(0, -3)$

7.

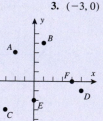

9. The line being described is $x = -3$.

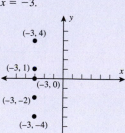

11. $d = 4, (3, 3)$ **13.** $d = 4\sqrt{2}, (1, 2)$

15. $d = 3\sqrt{10}, \left(-\dfrac{17}{2}, \dfrac{7}{2}\right)$ **17.** $d = 5, \left(\dfrac{3}{2}, \dfrac{11}{6}\right)$

19. $d = \dfrac{\sqrt{4049}}{60}, \left(-\dfrac{5}{24}, \dfrac{1}{15}\right)$ **21.** $d = 3.9, (0.3, 3.95)$

23. $d = \sqrt{1993.01} \approx 44.643, (1.05, -1.2)$

25. The perimeter of the triangle rounded to two decimal places is: 21.84.

27. right triangle

29. isosceles

31. 128.06 miles

33. distance ≈ 268 miles

35. midpoint $= (2003, 330)$

37. The values are misplaced. The correct distance would be $d = \sqrt{58}$.

39. The values were not used in the correct position. The correct midpoint would be: $\left(2, \dfrac{13}{2}\right)$.

41. true

43. true

45. The distance is: $d = \sqrt{(b-a)^2 + (a-b)^2} = \sqrt{2(a-b)^2}$
$= \sqrt{2}\,|a-b|$. The midpoint is: $m = \left(\dfrac{a+b}{2}, \dfrac{b+a}{2}\right)$.

47. $\sqrt{\left(x_1 - \dfrac{x_1+x_2}{2}\right)^2 + \left(y_1 - \dfrac{y_1+y_2}{2}\right)^2}$

$\sqrt{\left(\dfrac{2x_1 - x_1 - x_2}{2}\right)^2 + \left(\dfrac{2y_1 - y_1 - y_2}{2}\right)^2}$

$\sqrt{\left(\dfrac{x_1 - x_2}{2}\right)^2 + \left(\dfrac{y_1 - y_2}{2}\right)^2}$

$\dfrac{1}{2}\sqrt{(x_1 - x_2)^2 + (y_1 - y_2)^2}$

Using (x_2, y_2) with the midpoint yields the same result.

49. The distance is $d \approx 6.357$.

51. The distance is $d \approx 3.111$.

Section A.2

1. a. yes **b.** no

3. a. yes **b.** no

5.

x	y	(x, y)
-2	0	$(-2, 0)$
0	2	$(0, 2)$
1	3	$(1, 3)$

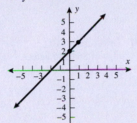

7.

x	y	(x, y)
-1	2	$(-1, 2)$
0	0	$(0, 0)$
$\dfrac{1}{2}$	$-\dfrac{1}{4}$	$\left(\dfrac{1}{2}, -\dfrac{1}{4}\right)$
1	0	$(1, 0)$
2	2	$(2, 2)$

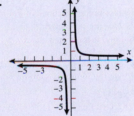

9.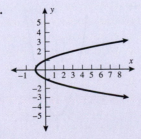

11.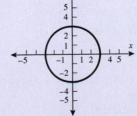

13.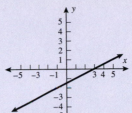

15. d

17. a

19. b

21. $(-1, -3)$

23. $(-7, 10)$

25. $(-3, -2), (3, 2), (-3, 2)$

27. x-axis

29. origin

31. x-axis

33. symmetric to x-axis, y-axis, and origin

35. symmetric to the y-axis

37. symmetric to y-axis

39. symmetric to origin

41.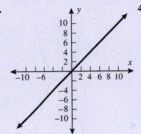

43.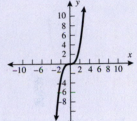

45.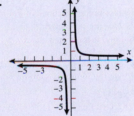

47.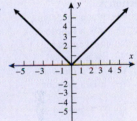

49.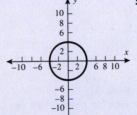

51.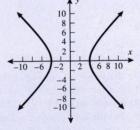

53.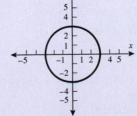

55.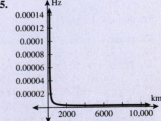

577

57.

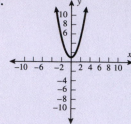

59. You are checking to see if the function is symmetric to the y-axis. Thus the substitution shown is incorrect. The correct substitution would be plugging in $-x$ for x into the function, not $-y$ for y.

61. false

63. true

65. Symmetric with respect to y-axis.

67. Symmetric with respect to x-axis, y-axis, and origin.

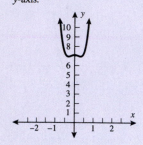

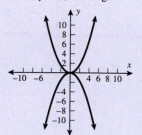

Section A.3

1. Yes, it is a function.

3. No, not a function.

5. Yes, it is a function.

7. No, not a function.

9. No, not a function.

11. Yes, it is a function.

13. No, not a function.

15. No, not a function.

17. Yes, it is a function.

19. No, not a function.

21. Yes, it is a function.

23. No, not a function.

25. $f(-2) = -7$

27. $g(1) = 6$

29. $f(-2) + g(1) = -1$

31. $2f(-1) - 2g(-3) = -33$

33. $\dfrac{f(-2)}{g(1)} = -\dfrac{7}{6}$

35. $\dfrac{f(0) - f(-2)}{g(1)} = \dfrac{2}{3}$

37. $f(x + 1) - f(x - 1) = 4$

39. $g(x + a) - f(x + a) = 8 - x - a$

41. $\dfrac{f(x + h) - f(x)}{h} = 2$

43. $\dfrac{g(t + h) - g(t)}{h} = 1$

45. $\dfrac{f(-2 + h) - f(-2)}{h} = 2$

47. $\dfrac{g(1 + h) - g(1)}{h} = 1$

49. domain: all $\mathbb{R}$, interval notation: $(-\infty, \infty)$

51. domain: all $\mathbb{R}$, interval notation: $(-\infty, \infty)$

53. domain: all $\mathbb{R}$, except 5, interval notation: $(-\infty, 5) \cup (5, \infty)$

55. domain: all $\mathbb{R}$, except -2 and 2, interval notation: $(-\infty, -2) \cup (-2, 2) \cup (2, \infty)$

57. domain: all $\mathbb{R}$, interval notation: $(-\infty, \infty)$

59. domain: $(-\infty, 7]$

61. domain: $\left[-\dfrac{5}{2}, \infty\right)$

63. domain: $(-\infty, -2] \cup [2, \infty)$

65. domain: $(3, \infty)$

67. $y = 45x$; domain: $(75, \infty)$

69. $T(12) = 90\ °F \qquad T(6) = 64.8\ °F$

71. $P(x) = 10 + \sqrt{400,000 - 100x}$
when $x = 10$, $P(10) = \$641.66$
when $x = 100$, $P(100) = \$634.50$

73. $V = x(10 - 2x)^2 \qquad$ domain: $0 < x < 5$

75. False, you must apply the vertical line test instead of the horizontal line test. Applying the vertical line test would show that the graph given is actually a function.

77. $f(x + 1) \neq f(x) + f(1)$
Given: $\qquad f(x) = x^2 + x$
$\qquad f(x + 1) = (x + 1)^2 - (x + 1)$
$\qquad\qquad\qquad = x^2 + 2x + 1 - x - 1$
$\qquad\qquad\qquad = x^2 + x$

79. $G(-1 + h) \neq G(-1) + G(h)$
Correct answer is $h - 2$.

81. false

83. false

85. $A = 2$

87. $C = -5$ and $D = -2$

89. Warmest at noon: 90 °F. Outside the interval [6, 18] the temperatures are too low.

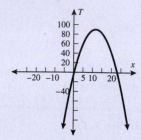

91. lowest price \$10, highest \$642.46 agrees.

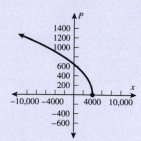

Section A.4

1. neither

3. even

5. odd

7. neither

9. odd

11. even

13. even

15. neither

17. neither

19. neither

21. neither

23. neither

25. 1

27. a

29. b

31. i

33. c

35. g

37. $y = |x| + 3$

39. $y = |-x|$

41. $y = 3|x|$

43. $y = x^3 - 4$

45. $y = (x + 1)^3 + 3$

47. $y = (-x)^3$

49. $y = -(x - 1)^2 + 2$

51. $y = \sqrt{-x - 1} - 2$

53. $y = \dfrac{-1}{x - 2} + 5$

55. $y = -x^5 - x + 2$

57.

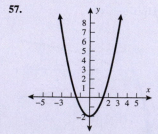

59.

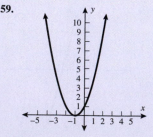

61.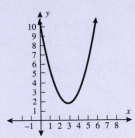

63.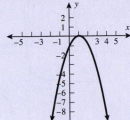

81.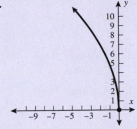

83. $S(x) = 10x$ and $S(x) = 10x + 50$

85. $T(x) = 0.33(1.25x - 6500)$

87. The following would be correct if it is shifted to the right 3.

89. $|3 - x| = |x - 3|$ Therefore, shift to the *right* 3 units.

91. true

93. true

95. $(a + 3, b + 2)$

97. Any part of the graph of $f(x)$ that is below the *x*-axis is reflected above it for $|f(x)|$.

99. If $0 < a < 1$, you have a vertical shrink. If $a > 1$, the graph is a vertical expansion.

65.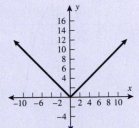

67.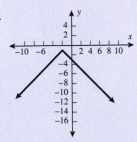

Section A.5

1. $f + g = x + 2$
$f - g = 3x$ $\Big\}$ domain: All real numbers
$f \cdot g = -2x^2 + x + 1$

$\dfrac{f}{g} = \dfrac{2x + 1}{1 - x}$ domain: $(-\infty, 1) \cup (1, \infty)$

3. $f + g = 3x^2 - x - 4$
$f - g = x^2 - x + 4$ $\Big\}$ domain: All real numbers
$f \cdot g = 2x^4 - x^3 - 8x^2 + 4x$

$\dfrac{f}{g} = \dfrac{2x^2 - x}{x^2 - 4}$ domain: $(-\infty, -2) \cup (-2, 2) \cup (2, \infty)$

5. $f + g = \dfrac{1 + x^2}{x}$

$f - g = \dfrac{1 - x^2}{x}$ $\Bigg\}$ domain: $(-\infty, 0) \cup (0, \infty)$

$f \cdot g = 1$
$\dfrac{f}{g} = \dfrac{\frac{1}{x}}{x} = \dfrac{1}{x^2}$

7. $f + g = 3\sqrt{x}$
$f - g = -\sqrt{x}$ $\Big\}$ domain: $[0, \infty)$
$f \cdot g = 2x$

$\dfrac{f}{g} = \dfrac{1}{2}$ domain: $(0, \infty)$

9. $f + g = \sqrt{4 - x} + \sqrt{x + 3}$
$f - g = \sqrt{4 - x} - \sqrt{x + 3}$ $\Big\}$ domain: $[-3, 4]$
$f \cdot g = \sqrt{4 - x}\sqrt{x + 3}$

$\dfrac{f}{g} = \dfrac{\sqrt{4 - x}}{\sqrt{x + 3}} = \dfrac{\sqrt{4 - x}\sqrt{x + 3}}{x + 3}$ domain: $(-3, 4]$

69.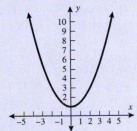

71.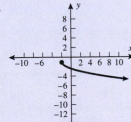

11. $f \circ g = 2x^2 - 5$ domain: All real numbers
$g \circ f = 4x^2 + 4x - 2$ domain: All real numbers

13. $f \circ g = \dfrac{1}{x + 1}$ domain: $(-\infty, -1) \cup (-1, \infty)$

$g \circ f = \dfrac{1}{x - 1} + 2$ domain: $(-\infty, 1) \cup (1, \infty)$

15. $f \circ g = \dfrac{1}{|x - 1|}$ domain: $(-\infty, 1) \cup (1, \infty)$

$g \circ f = \dfrac{1}{|x| - 1}$ domain: $(-\infty, -1) \cup (-1, 1) \cup (1, \infty)$

73.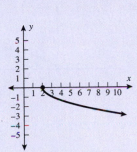

75.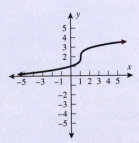

77.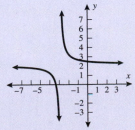

79.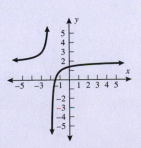

17. $f \circ g = \sqrt{x+4}$ domain: $[-4, \infty)$
$g \circ f = \sqrt{x-1} + 5$ domain: $[1, \infty)$

19. $f \circ g = x$ domain: $(-\infty, \infty)$
$g \circ f = x$ domain: $(-\infty, \infty)$

21. $(f+g)(2) = 15$

23. $(f \cdot g)(4) = 26\sqrt{3}$

25. $f(g(2)) = f(1) = 11$

27. not possible

29. $f(g(1)) = \dfrac{1}{3}$ $g(f(2)) = 2$

31. not possible

33. $f(g(1)) = \dfrac{1}{3}$ $g(f(2)) = 4$

41. $f(x) = 2x^2 + 5x$ $g(x) = 3x - 1$

43. $f(x) = \dfrac{2}{|x|}$ $g(x) = x - 3$

45. $f(x) = \dfrac{3}{\sqrt{x} - 2}$ $g(x) = x + 1$

47. $F = \dfrac{9}{5}(K - 273.15) + 32$

49. $A(x) = \left(\dfrac{x}{4}\right)^2$ x is the number of linear feet of fence.
$A(100) = 625$ square feet, $A(200) = 2500$ square feet

51. a. $C(p) = 62{,}000 - 20p$
b. $R(p) = 600{,}000 - 200p$
c. $P(p) = R(p) - C(p) = 538{,}000 - 180p$

53. area $= 150^2 \pi t$

55. $d(h) = \sqrt{h^2 + 4}$

57. domain: $x \neq -2$

59. The operation is composition, *not* multiplication.

61. false

63. true

65. $g \circ f = \dfrac{1}{x}$ $x \neq 0, a$

67. $g \circ f = x$ $x \geq -a$

29.

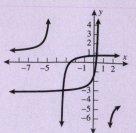

31.

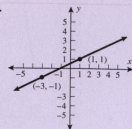

33.

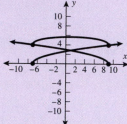

35.

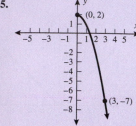

37.

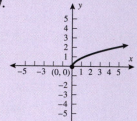

39. $f^{-1}(x) = x + 1$

41. $f^{-1}(x) = \dfrac{(x-2)}{-3}$

43. $f^{-1}(x) = (x-1)^{1/3}$

45. $f^{-1}(x) = x^2 + 3$

47. $f^{-1}(x) = \sqrt{x+1}$

49. $f^{-1}(x) = \sqrt{x+3} - 2$

51. $f^{-1}(x) = \dfrac{2}{x}$

53. $f^{-1}(x) = 3 - \dfrac{2}{x}$

55. $f^{-1}(x) = \dfrac{5x-1}{x+7}$

57. $f^{-1}(x) = \dfrac{5}{9}(x - 32)$ This now represents degrees Fahrenheit being turned into degrees Celsius.

59. $E(x) = 5.25x$ $E^{-1}(x) = \dfrac{x}{5.25}$ tells you how many hours you will have to work to bring home x dollars.

61. No, it's not a function because it fails the vertical line test.

63. The domain of the inverse function must be given.

65. false

67. false

69. $(b, 0)$

71. $m \neq 0$

73. no

75. no

69.

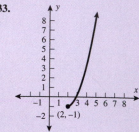

Section A.6

1. Not one-to-one.

3. Yes, one-to-one.

5. Yes, one-to-one.

7. Not a function, so it can't be one-to-one.

9. Not one-to-one.

11. Not one-to-one.

13. Yes, one-to-one.

15. Not one-to-one.

17. Not one-to-one.

19. Yes, one-to-one.

21.

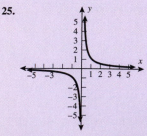

23.

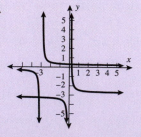

25.

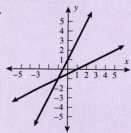

27.

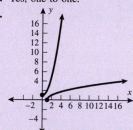

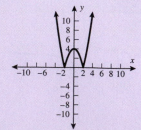

77.

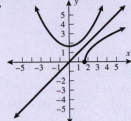

No, the functions are not inverses of each other. Had we restricted the domain of the parabola to $x > 0$, then they would be inverses.

Review Exercises

1. quadrant II
3. quadrant III

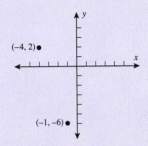

$(-4, 2)\bullet$

$(-1, -6)\bullet$

5. $d = \sqrt{45}$
7. $d = \sqrt{205}$
9. $\left(\dfrac{5}{2}, 6\right)$
11. $(3.85, 5.3)$
13. $d \approx 52.20$
15. y-axis
17. origin

19.

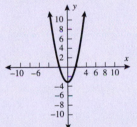

21.

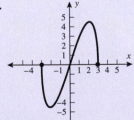

23.

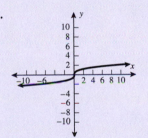

25.

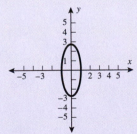

27. yes
29. yes
31. no
33. yes
35. no
37. 5
39. -665
41. -2
43. 4
45. domain: $(-\infty, \infty)$
47. $(-\infty, -4)\cup(-4, \infty)$
49. $[4, \infty)$
51. $D = 18$
53. neither
55. odd
57. neither
59. odd

61.

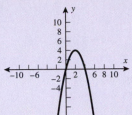

63.

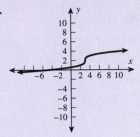

65.

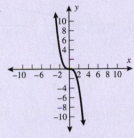

67. $y = \sqrt{x + 3}$
69. $y = \sqrt{x - 2} + 3$
71. $y = 5\sqrt{x} - 6$

73. $y = (x + 2)^2 - 12$

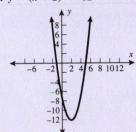

75. $g + h = -2x - 7$ domain: $(-\infty, \infty)$
$g - h = -4x - 1$ domain: $(-\infty, \infty)$
$g \cdot h = -3x^2 + 5x + 12$ domain: $(-\infty, \infty)$
$\dfrac{g}{h} = \dfrac{-3x - 4}{x - 3}$ domain: $(-\infty, 3)\cup(3, \infty)$

77. $g + h = \dfrac{1}{x^2} + \sqrt{x}$ domain: $(0, \infty)$

$g - h = \dfrac{1}{x^2} - \sqrt{x}$ domain: $(0, \infty)$

$g \cdot h = \dfrac{1}{x^{3/2}}$ domain: $(0, \infty)$

$\dfrac{g}{h} = \dfrac{1}{x^{5/2}}$ domain: $(0, \infty)$

79. $\left.\begin{array}{l} g + h = \sqrt{x - 4} + \sqrt{2x + 1} \\ g - h = \sqrt{x - 4} - \sqrt{2x + 1} \\ g \cdot h = \sqrt{x - 4}\sqrt{2x + 1} \\ \dfrac{g}{h} = \dfrac{\sqrt{x - 4}}{\sqrt{2x + 1}} \end{array}\right\}$ Domain: $[4, \infty)$ for all

81. $f \circ g = 6x - 1$ Domain: $(-\infty, \infty)$
$g \circ f = 6x - 7$ Domain: $(-\infty, \infty)$

83. $f \circ g = \dfrac{8 - 2x}{13 - 3x}$ Domain: $(-\infty, 4)\cup\left(4, \dfrac{13}{3}\right)\cup\left(\dfrac{13}{3}, \infty\right)$

$g \circ f = \dfrac{x + 3}{4x + 10}$ Domain: $(-\infty, -3)\cup\left(-3, -\dfrac{5}{2}\right)\cup\left(-\dfrac{5}{2}, \infty\right)$

85. $f(g(3)) = 857$ $g(f(-1)) = 51$

87. $f(g(3)) = \dfrac{17}{31}$ $g(f(-1)) = 1$

89. $f(x) = x - 2$ and $g(x) = 3x^2 + 4x + 7$

91. $f(x) = \dfrac{1}{\sqrt{x}}$ and $g(x) = x^2 + 7$

93. $A = 625\pi(t + 2)$ **95.** yes

97. yes   **99.** yes

101.   **103.**

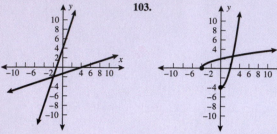

105. $f^{-1}(x) = \dfrac{x - 1}{2}$

domain f: $(-\infty, \infty)$ domain f^{-1}: $(-\infty, \infty)$
range f: $(-\infty, \infty)$ range f^{-1}: $(-\infty, \infty)$

107. $f^{-1}(x) = x^2 - 4$

domain f: $[-4, \infty)$ domain f^{-1}: $[0, \infty)$
range f: $[0, \infty)$ range: f^{-1}: $[-4, \infty)$

109. $f^{-1}(x) = \dfrac{6 - 3x}{x - 1}$

domain f: $(-\infty, -3)$ domain f^{-1}: $(-\infty, 1)\cup(1, \infty)$
$\cup(-3, \infty)$
range f: $(-\infty, 1)\cup(1, \infty)$ range f^{-1}: $(-\infty, -3)\cup(-3, \infty)$

111. $S(x) = 22{,}000 + 0.08x$ $S^{-1}(x) = \dfrac{(x - 22{,}000)}{0.08}$.

Sales required to earn desired income.

Practice Test

1. $d = \sqrt{82}$ **3.** $d = \sqrt{29}$ midpoint $= \left(\dfrac{1}{2}, 5\right)$

5. $(-3, -4)$ **7.** b

9. c   **11.** $\left(\dfrac{f}{g}\right)(x) = \dfrac{\sqrt{x - 2}}{x^2 + 11}$ domain: $[2, \infty)$

13. $g(f(x)) = x + 9$ domain: $[2, \infty)$

15. neither

17. domain: $[3, \infty)$
range: $(-\infty, 2]$

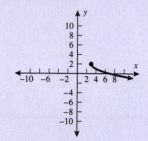

19. $f^{-1}(x) = x^2 + 5$
The domain and range of f is domain: $[5, \infty)$, range: $[0, \infty)$.
The domain and range of f^{-1} is domain: $[0, \infty)$, range: $[5, \infty)$.

21. $x \geq 0$

23. $c(x) = (0.70)(0.60)x$, where x is the original price of the suit.

APPLICATIONS INDEX

SUBJECT INDEX

GRAPHS OF THE TRIGONOMETRIC FUNCTIONS

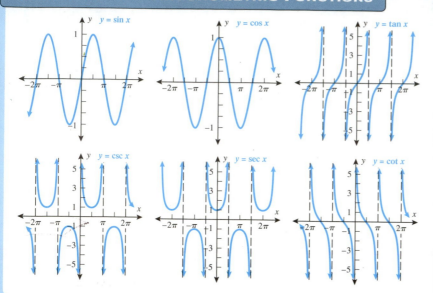

HERON'S FORMULA FOR AREA

If the semiperimeter s is

$$s = \frac{a + b + c}{2}$$

then

$$A = \sqrt{s(s - a)(s - b)(s - c)}$$

AMPLITUDE, PERIOD, AND PHASE SHIFT

	$A \sin(Bx + C)$	$A \cos(Bx + C)$	$A \tan(Bx + C)$
Amplitude	$|A|$	$|A|$	N/A
Period	$\dfrac{2\pi}{B}$	$\dfrac{2\pi}{B}$	$\dfrac{\pi}{B}$
Phase shift	$\dfrac{C}{B}$ units to the *left* if $\dfrac{C}{B} > 0$		
	$\dfrac{C}{B}$ units to the *right* if $\dfrac{C}{B} < 0$		

OBLIQUE TRIANGLE

Law of Sines

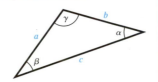

In any triangle, $\dfrac{\sin\alpha}{a} = \dfrac{\sin\beta}{b} = \dfrac{\sin\gamma}{c}$.

Law of Cosines

$$a^2 = b^2 + c^2 - 2bc\cos\alpha$$
$$b^2 = a^2 + c^2 - 2ac\cos\beta$$
$$c^2 = a^2 + b^2 - 2ab\cos\gamma$$

INVERSE TRIGONOMETRIC FUNCTIONS

$y = \sin^{-1}x$	$x = \sin y$	$-\dfrac{\pi}{2} \leq y \leq \dfrac{\pi}{2}$	$-1 \leq x \leq 1$
$y = \cos^{-1}x$	$x = \cos y$	$0 \leq y \leq \pi$	$-1 \leq x \leq 1$
$y = \tan^{-1}x$	$x = \tan y$	$-\dfrac{\pi}{2} < y < \dfrac{\pi}{2}$	x is any real number
$y = \cot^{-1}x$	$x = \cot y$	$0 < y < \pi$	x is any real number
$y = \sec^{-1}x$	$x = \sec y$	$0 \leq y \leq \pi, y \neq \dfrac{\pi}{2}$	$x \leq -1$ or $x \geq 1$
$y = \csc^{-1}x$	$x = \csc y$	$-\dfrac{\pi}{2} \leq y \leq \dfrac{\pi}{2}, y \neq 0$	$x \leq -1$ or $x \geq 1$